Mineral Processing Technology

AN INTRODUCTION TO THE PRACTICAL ASPECTS OF ORE TREATMENT AND MINERAL RECOVERY

FOURTH EDITION

International Series on
MATERIALS SCIENCE AND TECHNOLOGY

Volume 41—Editor: D. W. HOPKINS, M.Sc.✝

NOTICE TO READERS

Mineral Processing Technology

FOURTH EDITION

AN INTRODUCTION TO THE PRACTICAL ASPECTS OF
ORE TREATMENT AND MINERAL RECOVERY

by

B. A. WILLS, BSc, PhD, CEng, MIMM

Principal Lecturer
Camborne School of Mines, Cornwall, England

PERGAMON PRESS

OXFORD · NEW YORK · BEIJING · FRANKFURT
SÃO PAULO · SYDNEY · TOKYO · TORONTO

U.K.	Pergamon Press, Headington Hill Hall, Oxford OX3 0BW, England
U.S.A.	Pergamon Press, Maxwell House, Fairview Park, Elmsford, New York 10523, U.S.A.
PEOPLE'S REPUBLIC OF CHINA	Pergamon Press, Room 4037, Qianmen Hotel, Beijing, People's Republic of China
FEDERAL REPBLIC OF GERMANY	Pergamon Press, Hammerweg 6, D-6242 Kronberg, Federal Republic of Germany
BRAZIL	Pergamon Editora, Rua Eça de Queiros, 346, CEP 04011, Paraiso, Sao Paulo, Brazil
AUSTRALIA	Pergamon Press Australia, P.O. Box 544, Potts Point, N.S.W. 2011, Australia
JAPAN	Pergamon Press, 8th Floor, Matsuoka Central Building, 1-7-1 Nishishinjuku, Shinjuku-ku, Tokyo 160, Japan
CANADA	Pergamon Press Canada, Suite No. 271, 253 College Street, Toronto, Ontario, Canada M5T 1R5

First edition, 1979
Second edition, 1981
Reprinted 1982
Third edition, 1985
Fourth edition, 1988

Library of Congress Cataloging in Publication Data
Wills, B. A. (Barry Alan)
Mineral processing technology.
(International series on materials science and technology; v. 41)
Includes index.
1. Ore-dressing. I. Title. II. Series.
TN500.W54   1988   622'.7   87-14244

British Library Cataloging in Publication Data
Wills, B. A.
Mineral processing technology: an introduction to the practical aspects of ore treatment and mineral recovery.—4th ed.—(International series on materials and technology, v. 41).
1. Ore-dressing
I. Title
622.'7.   TN500

ISBN 0–08–034937–4 Hardcover
ISBN 0–08–034936–6 Flexicover

Printed in Great Britain by A. Wheaton & Co. Ltd., Exeter

This book is dedicated to my wife
Barbara for her tolerance and patience

PREFACE TO THE FOURTH EDITION

Since the publication of the third edition of this book, the fortunes of the world's minerals industry have not improved significantly, particularly in the base metals sector, where falling or stagnant metal prices have forced many mines to shut down or cut back on production.

In view of the low profit margins of many mining operations, and the increasing mineralogical complexity of mined ores, it has been necessary to seek improvements in existing processes, and to control them more efficiently. For example, much recent research effort has been directed towards improving the fundamental knowledge of the chemistry of froth flotation, the dominant method of mineral concentration, with a view to controlling the chemical environment more efficiently. Similarly, continuing efforts have been made to control flotation and other processes by computer, with good results in many cases.

This edition has been revised to incorporate the latest developments in mineral process engineering. Chapter 1 has been completely rewritten and includes examples of mill economics, as well as worked examples. The increasing use of computers is reflected in Chapter 3, which has been expanded to include worked examples and computer methods. Useful microcomputer programs are listed in an Appendix. Chapter 12 has also been thoroughly revised to reflect the latest developments in flotation.

Discussions with many minerals engineers have aided preparation of this edition, but I would particularly like to thank the following, who have provided valuable advice and assistance:

Dr. Alan Bromley, Dr. Julie Holl, and Prof. Ken Hosking of Camborne School of Mines; Prof. John Ralston of The South Australian Institute of Technology; Palabora Mining Co., South Africa; Mount Isa Mining Co., Australia.

Denver, Colorado   B. A. WILLS
26 February 1987

PREFACE TO THE THIRD EDITION

DURING the recent recession, which has had a great impact on the minerals industry, the priority of many mining companies has been on reducing costs, particularly energy costs, rather than on innovative new techniques. Of major impact during this period, however, has been the very rapid decline in cost, and the increase in reliability of microprocessors, which has led to the increasing use of automation in some milling processes.

This edition has been expanded to include an introduction to automatic control and descriptions of some of the control strategies which have been adopted on various concentrators. The section on mass balancing has also been extended to include an introduction to statistical methods, which are suited to evaluation by microcomputer.

The bibliography has been updated to include the latest references, and the section on froth flotation has been revised and expanded.

I would like to thank all those people who have provided valuable information and assistance in the preparation of this edition. Particularly, I would like to thank the following:

Gosta Diehl and Teppo Meriluoto of Outokumpu Oy, for arranging visits to various concentrators in Finland, and to the metallurgical staff of Pyhasalmi, Vihanti and Vuonos mines for invaluable discussions.

Prof. Peter Linkson, and Dennis Nobbs of the University of Sydney, Australia.

The metallurgical staff of Mount Isa Mines Ltd, Queensland, Australia.

Peter Lean and the metallurgical staff of the Broken Hill mines, New South Wales, Australia.

Dr. Don McKee of the Julius Kruttschnitt Mineral Research Centre, Brisbane, Australia.

Derek Ottley, Vice-President of Mineral Systems Inc., Stamford, USA.

Dr. Bob Barley of Camborne School of Mines.

Sydney, Australia   B. A. WILLS
26 September 1983

PREFACE TO THE SECOND EDITION

ALTHOUGH this edition is again basically intended to satisfy the needs of students in the minerals industry, it is also intended to provide practising engineers with a valuable source of reference. In this respect the bibliography has been considerably expanded and updated to include the latest published works. The text has been extended considerably to cover important subject areas more fully, and, where appropriate, worked examples have been added. Two appendices have been added for reference purposes.

I would like to thank the many people who have provided useful criticism regarding the first edition, which has been valuable in the preparation of this edition. Particularly, thanks are due to Mr. R. H. Parker, Vice-Principal of the Camborne School of Mines, Mr. F. B. Mitchell, formerly Vice-Principal and Head of Mineral Processing at the CSM, and Mr. R. S. Shoemaker, Metallurgical Manager of Bechtel Inc., San Francisco. I would also like to thank Prof. R. P. King, of the University of the Witwatersrand, Johannesburg, for many useful discussions.

Camborne, Cornwall   B. A. WILLS
18 September 1980

PREFACE TO THE FIRST EDITION

THE main objective of this book is to provide students of mineral processing, metallurgy, and mining with a review of the common physical ore-processing techniques utilised in today's mining industry. The need for such a book was highlighted during the preparation of the degree course in Mineral Processing Technology at the Camborne School of Mines. Although there are many excellent texts covering the unit processes, most of these are now out of date, and many of the methods described have been superseded by more modern techniques.

Descriptions of the treatments of specific ores have been deliberately restricted, except where to illustrate the applications of a particular process. Methods of processing differ in detail from mine to mine, even with very similar ores, and students should be strongly encouraged to peruse the mining periodicals, which regularly detail the treatment routes used on particular plants.

As in any book which deals with a very wide field, certain subject areas will undoubtedly be open to debate, and I would greatly appreciate comments and suggestions for improvement, which could be incorporated into later editions.

I am indebted to many organisations and individuals who have proved invaluable during the preparation of this book. The organisations who have provided illustrations for the text have been acknowledged separately. I would like to express sincere thanks to my friend and ex-colleague Dr. D. G. Osborne, without whose initial encouragement the book would not have been written. Thanks are also due to Mr. J. F. Turner and Mr. R. H. Parker, the Head of Mineral Processing Technology, and the Vice-Principal, respectively, of the Camborne School of Mines, for advice and encouragement. Acknowledgement and thanks are especially due to the Series Editor, Mr. D. W. Hopkins, of Swansea University, for invaluable discussions and criticism during the preparation of the text. Finally I would like to

xi

thank Mr. W. J. Watton for providing photographic services for the preparation of the illustrations, and Mrs. H. A. Taylor and Miss K. E. Bennett for patiently typing the draft manuscript.

Falmouth, Cornwall   BARRY ALAN WILLS
9 March 1978

ACKNOWLEDGEMENTS

THE author acknowledges with thanks the assistance given by the following companies and publishers in permitting the reproduction of illustrations from their publications:

Palabora Mining Co., for Figs. 1.2g, 12.60, 12.6.

Mount Isa Mines Ltd, for Fig. 1.2f.

AIMME, New York, for Figs. 1.8, 2.8, 2.9, from *Mineral Processing Plant Design* (ed. Mular and Bhappu).

Outokumpu Oy, for Figs. 2.2, 3.3, 3.7, 3.18.

Universal Engineering Corp., for Fig. 2.11.

Newell Dunford Engineering Ltd., for Fig. 2.12.

The Galigher Co., for Fig. 3.1.

Wiley Publishing Co., New York, for Figs. 3.2, 9.8, 10.24, from *Handbook of Mineral Dressing* by A. F. Taggart.

Chapman & Hall Ltd., London, for Figs. 3.4, 4.13, from *Particle Size Measurement* by T. Allen.

Gunson's Sortex Ltd., for Figs. 3.8, 14.1, 14.2.

George Kent Ltd., for Fig. 3.9.

Kay-Ray Inc., for Fig. 3.10.

Warman International Ltd., for Figs. 4.11, 4.12.

Armco Autometrics, for Fig. 4.14.

Pegson Ltd., for Figs. 6.6, 7.16, 7.17, 7.18.

Fuller Co., for Fig. 6.7(b).

Brown Lenox & Co. Ltd., for Figs. 8.13, 8.14.

Rexnord-Nordberg Machinery, for Figs. 6.9, 6.10, 6.12, 6.13, 6.15, 6.16, 8.3.

Pergamon Press Ltd., for Figs. 6.17, 15.5, from *Chemical Engineering*, vol. 2, by J. M. Coulson and J. F. Richardson.

Humboldt-Wedag Ltd., for Figs. 6.21, 7.25, 13.13, 13.14, 13.15, 13.16, 13.20.

GEC Mechanical Handling Ltd., for Fig. 6.22.

Joy Manufacturing Co. (Denver Equipment Div.), for Figs. 7.4, 7.7, 7.9, 7.12, 10.11, 12.16.

Head Wrightson Ltd., for Figs. 7.5, 7.11, 7.13.

NEI International Combustion for Figs. 7.22, 7.27, 7.28.

Koppers Co. Inc., for Figs. 7.6, 7.24.

Morgardshammer, for Fig. 7.8.

Allis-Chalmers Ltd., for Fig. 7.15.

Mineral Processing Systems Inc., for Fig. 7.26.

Boliden, for Fig. 7.35.

Dorr-Oliver, for Figs. 8.4, 9.10, 15.20, 15.24.

Bartles (Carn Brea) Ltd., for Fig. 8.5.

Locker Industries Ltd., for Fig. 8.8.

Sweco, for Fig. 8.10.

Tyler Inc., for Figs. 8.11, 8.12.
Mogensen Sizers Ltd., for Fig. 8.15.
NCB, for Figs. 8.16, 11.8.
N. Greening Ltd., for Fig. 8.19.
R. O. Stokes & Co. Ltd., for Figs. 9.6, 9.7.
Wemco, G. B., for Figs. 9.9, 9.11, 11.2, 11.3, 11.4, 12.40, 12.41.
IHC (Holland) for Figs. 10.12, 10.13, 10.14.
Applied Science Publishers Ltd. (London), for Figs. 10.17, 10.29, 12.14 from *Mineral Processing* by E. J. Pryor.
Mineral Deposits Ltd., for Figs. 10.20, 13.24, 13.25.
Humphreys Engineering Co., for Figs. 10.21, 10.22.
Wilfley Mining Machinery Co., for Fig. 10.28.
South African Coal Processing Society, Johannesburg, for Fig. 11.5, from *Coal Preparation Course*.
Minerals Separation Corps., for Figs. 11.10, 11.12.
Boxmag-Rapid Ltd., for Figs. 13.6, 13.10.
Eriez Magnetics, for Figs. 13.7, 13.9, 13.19.
Sala, for Figs. 13.8, 15.14, 15.15.
Dings Magnetic Separator Co., for Fig. 13.11.
RTZ Ore Sorters, for Figs. 14.3, 14.5.
Envirotech Corp., for Figs. 15.6, 15.7, 15.13.
Clarke Chapman Ltd. (International Combustion Div.), for Figs. 15.16, 15.17.
Rauma-Repola Oy, for Figs. 15.18, 15.19.
Filtres Vernay, for Fig. 15.22.
Krauss-Maffei Ltd., for Fig. 15.25.
Krebs Engineers, for Fig. 16.3.

Acknowledgements are also due to the following for illustrations taken from technical papers:

H. Heywood, for Fig. 4.6, from *Symposium on Particle Size Analysis* (Inst. of Chem. Eng., 1947).
A. C. Partridge, for Figs. 5.1 and 5.2, from *Mine & Quarry*, July/Aug. 1978.
W. Meintrup and F. Kleiner, for Fig. 7.10, from *Mining Engineering*, Sept. 1982.
L. R. Plitt, for Fig. 9.15, from *Canadian Mining and Metallurgical Bulletin*, Dec. 1976.
V. G. Renner and H. E. Cohen, for Fig. 9.16, from *Trans. IMM*, Sect. C, June 1978.
R. E. Zimmermann, for Figs. 10.15, 10.16, from *Mining Congress Journal*, May 1974.
R. L. Terry, for Fig. 10.25, from *Minerals Processing*, July/Aug. 1974.
T. Cienski and V. Coffin, for Fig. 12.34, from *Canadian Mining Journal*, March 1981.
P. Young, for Figs. 12.35, 12.38, 12.43 and 12.44, from *Mining Magazine*, January 1982.
F. Kitzinger et al., for Fig. 12.53, from *Mining Engineering*, April 1979.
J. H. Fewings et al., for Fig. 12.55, from *Proc. 13th Int. Min. Proc. Cong.*, Warsaw, 1979.
J. E. Lawyer and D. M. Hopstock, for Figs. 13.12, 13.18, from *Minerals Science Engineering*, vol. 6, 1974.
J. R. Goode, for Fig. 14.4, from *Canadian Mining Journal*, June 1975.
R. W. Shilling and E. R. May, for Fig. 16.7, from *Mining Congress Journal*, vol. 63, 1977.
C. G. Down and J. Stocks, for Fig. 16.8, from *Mining Magazine*, July 1977.

Figures 4.1, 4.3, and 4.10 from BS410: 1976 and BS3406 Part 2: 1963, are reproduced by permission of BSI, 2 Park Street, London W1A 2BS, from whom complete copies of the publications can be obtained.

CONTENTS

CHAPTER 1

INTRODUCTION

Minerals and Ores

Minerals

The forms in which metals are found in the crust of the earth and as sea-bed deposits depend on their reactivity with their environment, particularly with oxygen, sulphur, and carbon dioxide. Gold and the platinum metals are found principally in the *native* or metallic form. Silver, copper, and mercury are found native, as well as in the form of sulphides, carbonates, and chlorides. The more reactive metals are always in compound form, such as the oxides and sulphides of iron and the oxides and silicates of aluminium and beryllium. The naturally occurring compounds are known as *minerals*, most of which have been given names according to their composition (e.g. galena—lead sulphide, PbS; sphalerite—zinc sulphide, ZnS; cassiterite—tin oxide, SnO_2).

Minerals by definition are natural inorganic substances possessing definite chemical compositions and atomic structures. Some flexibility, however, is allowed in this definition. Many minerals exhibit *isomorphism*, where substitution of atoms within the crystal stucture by similar atoms takes place without affecting the atomic structure. The mineral olivine, for example, has the chemical composition $(Mg, Fe)_2SiO_4$, but the ratio of Mg atoms to Fe atoms varies in different olivines. The total number of Mg and Fe atoms in all olivines, however, has the same ratio to that of the Si and O atoms. Minerals can also exhibit *polymorphism*, different minerals having the same chemical composition, but markedly different physical properties due

1

to a difference in crystal structure. Thus the two minerals graphite and diamond have exactly the same composition, being composed entirely of carbon atoms, but have widely different properties due to the arrangement of the carbon atoms within the crystal lattice. The term "mineral" is often used in a much more extended sense to include anything of economic value which is extracted from the earth. Thus coal, chalk, clay, and granite do not come within the definition of a mineral, although details of their production are usually included in national figures for mineral production. Such materials are, in fact, *rocks*, which are not homogeneous in chemical and physical composition, as are minerals, but generally consist of a variety of minerals and form large parts of the earth's crust. Granite, for instance, which is the most abundant *igneous* rock, i.e. a rock formed by cooling of molten material, or *magma* within the earth's crust, is composed of three main mineral constituents, feldspar, quartz, and mica. These three homogeneous mineral components occur in varying proportions in different parts of the same granite mass.

Coals are not minerals in the geological sense, but a group of bedded rocks formed by the accumulation of vegetable matter. Most coal-seams were formed over 300 million years ago by the decomposition of vegetable matter from the dense tropical forests which covered certain areas of the earth. During the early formation of the coal-seams, the rotting vegetation formed thick beds of *peat*, an unconsolidated product of the decomposition of vegetation, found in marshes and bogs. This later became overlain with shales, sandstones, mud, and silt, and under the action of the increasing pressure and temperature and time, the peat-beds became altered, or *metamorphosed*, to produce the sedimentary rock known as coal. The degree of alteration is known as the *rank* of the coal, the lowest ranks (lignite or brown coal) showing little alteration, while the highest rank (anthracite) is almost pure graphite (carbon).

Metallic Ore Processing

The enormous growth of industrialisation from the eighteenth century onward, led to dramatic increases in the annual outputs of many metals, particularly the *base* metals (copper, lead, zinc and tin[1]) (Fig. 1.1). Copper output has grown by a factor of 250 over the past century and a half, while aluminium has shown an even greater production expansion; in terms of bauxite, output since the beginning of the twentieth century has increased by a factor approaching 900.[2]

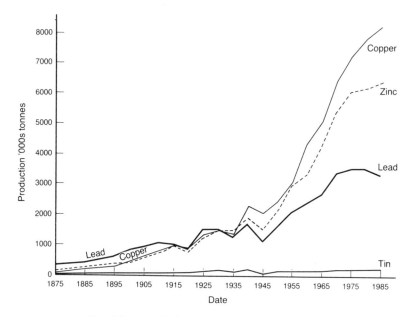

FIG. 1.1. Annual mined production of base metals.

All these metals suffered to a greater or lesser extent when the Organisation of Petroleum Exporting Countries (OPEC) quadrupled the price of oil in 1973–4, ending the great postwar industrial boom. The situation worsened in 1979–81, when the Iranian revolution forced the price of oil up from $13 to $34 a barrel, plunging the world into another and deeper recession, while early in 1986 a glut in the world's oil supply cut the price from $26 a barrel in December 1985 to below $15 in 1986.

These fluctuating, and often huge, increases in oil prices have had a great impact on metal ore mining, due to the large amounts of energy used to convert such natural resources into refined metal. For example, it has been estimated that the energy cost in copper production is about 35% of the selling price of the metal.[3]

Apart from energy, other major costs have increased dramatically. Table 1.1 shows how some of the important cost areas have increased in the USA.[4]

TABLE 1.1 COST INDICES 1970–83

	% increase
Labour—Construction	227
Mining & processing	293
Machinery	300
Rail transport	360
Energy—Coal	354
Electricity	453
Diesel fuel	720
Natural gas	1337

The price of most metals is governed by supply and demand, and the prices of many common metals, particularly copper, have not kept pace with inflation. This has had a drastic effect on the economics of many mines and the economy of whole nations, such as Zambia and Chile, which are heavily dependent on their mineral industries.

Estimates of the crustal abundancies of metals are given in Table 1.2,[5] together with the actual amounts of some of the most useful metals, to a depth of 3.5 km.[6]

The abundance of metals in the oceans is related to some extent to the crustal abundances, since they have come from the weathering of the crustal rocks, but superimposed upon this are the effects of acid rain-waters on mineral leaching processes; thus the metal availability from sea-water shown in Table 1.3[6] does not follow precisely that of the crystal abundance. The sea-bed is likely to become a heavily

TABLE 1.2 ABUNDANCE OF METALS IN THE EARTH'S CRUST

Element	Abundance (%)	Amount in 3.5 km of crust (tonnes)	Element	Abundance (%)	Amount in 3.5 km of crust (tonnes)
(Oxygen)	46.4		Vanadium	0.014	10^{14}–10^{15}
Silicon	28.2		Chromium	0.010	
Aluminium	8.2	10^{16}–10^{18}	Nickel	0.0075	
Iron	5.6		Zinc	0.0070	
Calcium	4.1		Copper	0.0055	10^{13}–10^{14}
Sodium	2.4		Cobalt	0.0025	
Magnesium	2.3	10^{16}–10^{18}	Lead	0.0013	
Potassium	2.1		Uranium	0.00027	
Titanium	0.57		Tin	0.00020	
Manganese	0.095	10^{15}–10^{16}	Tungsten	0.00015	10^{11}–10^{13}
Barium	0.043		Mercury	8×10^{-6}	
Strontium	0.038		Silver	7×10^{-6}	
Rare earths	0.023		Gold	$<5 \times 10^{-6}$	
Zirconium	0.017	10^{14}–10^{16}	Platinum	$<5 \times 10^{-6}$	$<10^{11}$
			metals		

exploited source of metals in the future for the so-called "manganese nodules" are rich in a variety of metals in addition to manganese.[7,8,9] These nodules are continuously being formed in oxygen-rich waters particularly in the Pacific Ocean.

TABLE 1.3 ABUNDANCE OF METAL IN THE OCEANS

Element	Abundance in sea-water (tonnes)	Element	Abundance in sea-water (tonnes)
Magnesium	10^{15}–10^{16}	Vanadium	10^{9}–10^{10}
Silicon	10^{12}–10^{13}	Titanium	
Aluminium		Cobalt	
Iron		Silver	10^{8}–10^{9}
Molybdenum	10^{10}–10^{11}	Tungsten	
Zinc		Chromium	
Tin		Gold	
Uranium		Zirconium	$<10^{8}$
Copper	10^{9}–10^{10}	Platinum	
Nickel			

It can be seen from Table 1.2 that eight elements account for over 99% of the earth's crust; 74.6% is silicon and oxygen and only three of the industrially important metals (aluminium, iron, and magnesium) are present in amounts above 2%. All the other useful metals occur in amounts below 0.1%; copper, for example, which is the most important non-ferrous metal, occurring only to the extent of 0.0055%. It is interesting to note that the so-called common metals, zinc and lead, are less plentiful than the rare-earth metals (cerium, thorium, etc.).

It is immediately apparent that if the minerals containing the important metals were uniformly distributed throughout the earth, they would be so thinly dispersed that their economic extraction would be impossible.

However, the occurrence of minerals in nature is regulated by the geological conditions throughout the life of the mineral. A particular mineral may be found mainly in association with one rock-type, e.g. cassiterite mainly associates with granite rocks, or may be found associated with both igneous and *sedimentary* rocks, i.e. those produced by the deposition of material arising from the mechanical and chemical weathering of earlier rocks by water, ice, and chemical decay. Thus, when granite is weathered, cassiterite may be transported and re-deposited as an *alluvial* deposit. Due to the action of these many natural agencies, mineral deposits are frequently found in sufficient concentrations to enable the metals to be profitably recovered. It is these concentrating agencies and the development of demand as a result of research and discovery which convert a mineral deposit into an *ore*. Most ores are mixtures of extractable minerals and extraneous rocky material described as *gangue*. They are frequently classed according to the nature of the valuable mineral. Thus in *native* ores the metal is present in the elementary form; *sulphide* ores contain the metal as sulphides, and in *oxidised* ores the valuable mineral may be present as oxide, sulphate, silicate, carbonate, or some hydrated form of these. *Complex* ores are those containing profitable amounts of more than one valuable mineral. Metallic minerals are often found in certain associations within which they may occur as mixtures of a wide range of particle sizes or as single-phase solid solutions or compounds. Galena and sphalerite, for example, commonly associate, as do copper sulphide minerals and sphalerite to a lesser extent. Pyrite (FeS_2) is very often associated with these minerals.

Ores are also classified by the nature of their gangues, such as *calcareous* or *basic* (lime rich), *siliceous* or *acidic* (silica rich). An ore can be described as an accumulation of mineral in sufficient quantity as to be capable of economic extraction. The minimum metal content (grade) required for a deposit to qualify as an ore varies from metal to metal. Many non-ferrous ores contain, as mined, as little as 1% metal, and often much less. Gold may be recovered profitably in ores containing only 5 parts per million (ppm) of the metal, whereas iron ores containing less than about 15% metal are regarded as low grade. Every tonne of material in the deposit has a certain *contained value* which is dependent on the metal content and current price of the contained metal. For instance, at a copper price of £980/t and a molybdenum price of £15/kg, a deposit containing 1% copper and 0.015% molybdenum has a contained value of £12/t. The deposit will be economic to work, and can be classified as an ore deposit if:

Contained value/t > (total processing costs + losses + other costs)/t.

A major cost is mining, and this can vary enormously, from only a few pence per tonne of ore to well over £50/t. High-tonnage operations are cheaper in terms of operating costs but have higher initial capital costs. These capital costs are paid off over a number of years, so that high-tonnage operations can only be justified for the treatment of deposits large enough to allow this. Small ore bodies are worked on a smaller scale, to reduce capital costs, but operating costs are correspondingly higher.

Alluvial mining is the cheapest method and, if on a large scale, can be used to mine ores of very low contained value due to low grade, low metal price, or both. For instance, in Malaysia, tin ores containing as little as 0.01% Sn are mined by alluvial methods. These ores have a contained value of less than £1/t, but very low processing costs allow them to be economically worked.

High-tonnage open-pit and underground block-caving methods are also used to treat ores of low-contained value, such as low-grade copper ores. Where the ore must be mined selectively, however, as is the case with underground vein-type deposits, mining methods become very expensive, and can only be justified on ores of high contained value. An underground selective mining cost of £30/t would obviously be hopelessly uneconomic on a tin ore of Malaysian grade, but may be economic on a hard-rock ore containing 1% tin, with a contained value of around £50/t.

In order to produce metals, the ore minerals must be broken down by the action of heat (pyrometallurgy), solvents (hydrometallurgy) or electricity (electrometallurgy), either alone or in combination, the most common method being the pyrometallurgical process of smelting. These chemical methods consume vast quantities of energy. Treatment of 1 t of copper ore, for instance, consumes in the region of 1500–2000 kWh of electrical energy which at a cost of, say, 3.5p per kWh is around £60/t, well above the contained value of all current copper ores.

Smelters are often remote from the mine site, being centred in areas where energy is relatively cheap, and where access to roads, rail or sea-links is available for shipment of fuel and supplies to, and products from, the smelter. The cost of transportation of mined ore to remote smelters could in many cases be greater than the contained value of the ore.

Mineral processing is usually carried out at the mine site, the plant being referred to as a *mill* or *concentrator*. The essential purpose is to reduce the bulk of the ore which must be transported to and processed by the smelter, by using relatively cheap, low-energy physical methods to separate the valuable minerals from the waste (gangue) minerals. This enrichment process considerably increases the contained value of the ore to allow economic transportation and smelting.

Compared with chemical methods, the physical methods used in mineral processing consume relatively small amounts of energy. For instance, to upgrade a copper ore from 1% to 25% metal would use in the region of 20–50 kWh t^{-1}. The corresponding reduction in weight of around 25:1 proportionally lowers transport costs and reduces smelter energy consumption to around 60–80 kWh in relation to the weight of mined ore. It is important to realise that, although the physical methods are relatively low energy users, the reduction in bulk lowers smelter energy consumption to the order of that used in mineral processing, and it is significant that as ore grades decline, the energy used in mineral processing becomes an important factor in deciding whether the deposit is viable to work or not.

Mineral processing not only reduces smelter energy costs but also smelter metal losses, due to the production of less metal-bearing slag. Although technically possible, the smelting of extremely low-grade ores, apart from being economically unjustifiable, would be very difficult due to the need to produce high-grade metal products free from deleterious impurities. These impurities are found in the gangue minerals and it is the purpose of mineral processing to reject them into the discard (tailings), as smelters often impose penalties according to their level. For instance, it is necessary to remove arsenopyrite from tin concentrates, as the contained arsenic is difficult to remove in smelting and produces a low-quality tin metal.

Against the economic advantages of mineral processing must be charged the losses occurred during milling, and the cost of the milling operations. The latter can vary over a wide range, depending on the method of treatment used, and particularly on the scale of the operation. As with mining, large-scale operations have higher capital, but lower operating costs (particularly labour and energy), than small-scale operations. As labour costs per tonne are most affected by the size of operation, so, as capacity increases, the energy costs per tonne become proportionally more significant, and these can be more than 25% of the total milling costs in a 10,000 t d^{-1} concentrator.

Losses to tailings are one of the most important factors in deciding whether a deposit is viable or not. Losses will depend very much on the ore mineralogy and dissemination, and on the technology available to achieve efficient concentration. Thus the development of froth flotation allowed the exploitation of vast low-grade copper deposits which were previously uneconomic to treat. Similarly, the introduction of solvent extraction enabled Nchanga Consolidated Copper Mines in Zambia to treat 9 million tonnes per year of flotation tailings, to produce 80,000 tonnes of finished copper from what was previously regarded as waste.[10]

In many cases not only is it necessary to separate valuable from gangue minerals, but it is also required to separate valuable minerals from each other. For instance, porphyry copper ores are an important source of molybdenum and the minerals of these metals must be separated for separate smelting. Similarly, complex sulphide ores containing economic amounts of copper, lead and zinc usually require separate concentrates of the minerals of each of these metals. The provision of clean concentrates, with little or no contamination with associated metals, is not always economically feasible, and this leads to another source of loss other than direct tailings loss. A metal which reports to the "wrong" concentrate may be difficult, or economically impossible, to recover, and never achieves its potential valuation. Lead, for example, is essentially irrecoverable in copper concentrates and is often penalised as an impurity by the copper smelter. The treatment of such polymetallic base metal ores, therefore, presents one of the great challenges to the mineral processor.

Mineral processing operations are often a compromise between improvements in metallurgical efficiency and milling costs. This is particularly true with ores of low contained value, where low milling costs are essential and cheap unit processes are necessary, particularly in the early stages, where the volume of material treated is relatively high. With such low-value ores, improvements in metallurgical efficiency by the use of more expensive methods or reagents cannot always be justified. Conversely, high metallurgical efficiency is usually of most importance with ores of high contained value, and expensive high-efficiency processes can often be justified on these ores.

Apart from processing costs and losses, other costs which must be taken into account are indirect costs such as ancillary services—power supply, water, roads, tailings disposal—which will depend much on the size and location of the deposit, as well as taxes, royalty payments, investment requirements, research and development, medical and safety costs, etc.

Non-metallic Ores

Ores of economic value can be classed as metallic or non-metallic, according to the use of the mineral. Certain minerals may be mined and processed for more than one purpose. In one category the mineral may be a metal ore, i.e. when it is used to prepare the metal, as when bauxite (hydrated aluminium oxide) is used to make aluminium. The alternative is for the compound to be classified as a non-metallic ore, i.e. when bauxite or natural aluminium oxide is used to make material for refractory bricks or abrasives.

Many non-metallic ore minerals associate with metallic ore minerals (Appendix II) and are mined and processed together, e.g. galena, the main source of lead, often associates with fluorite (CaF_2) and barytes $(BaSO_4)$, both important non-metallic minerals.

Diamond ores are the lowest grade of all ores presently mined, the diamond content usually being between 0.03 and 0.15 ppm. The ores are mined chiefly for industrial diamonds, which constitute about 80% of the total production, but also for gems. It is the mining of the very expensive gem diamonds, which are used for jewellery, which subsidises the costs of mining the relatively cheap industrial diamonds from such low-grade ore deposits.

Tailings Retreatment

There are many plants where minerals are recovered in secondary circuits, treating tailings, where the feed grades are much lower than would be economic on a mined ore. Typical ore grades for tungsten ores are in the range 0.5–1.5% WO_3, but the Climax Molybdenum plant in the United States treats 45 000 tonnes per day (t d^{-1}) of tailings, containing less than 0.1% WO_3, and is one of the two major producers of tungsten concentrate in the United States.[11]

Tin has been recovered from about 10 000 t d^{-1} each of tailings from the lead–zinc, and copper–lead–zinc plants of the Sullivan concentrator in British Columbia and from the Kidd Creek plant of Texasgulf, near Timmins, respectively. The plant feeds are of about 0.06 and 0.15% Sn respectively.[12]

The working costs for treating old tailings dumps are much lower than conventional mining costs, and operations have been set up in many old mining areas where large tonnages of mill tailings are amenable to processing.[13] The East Rand Gold and Uranium Company (ERGO) retreats old mill tailings from mines of South Africa's East Rand goldfields to produce gold, uranium, and sulphuric acid. The slimes dams are treated by *monitoring*, i.e. diverting high-pressure jets of water at the working face. The resultant slurry gravitates to transfer pump stations adjacent to each dam, and is pumped to a central processing plant. The plant treats 1.5 million tonnes per month of slimes, grading 0.53 ppm gold, 40 ppm U_3O_8, and 1.04% S.[14]

The world is now becoming aware of the finite nature of its resources at a price, and of the ever-increasing development costs of large new mines. Reprocessing of old tailings on a large scale must be worth examining very seriously by those with access to sufficient material of this type.

Mineral Processing Methods

"As-mined" or "run-of-mine" ore consists of valuable metallic minerals and gangue. Mineral processing, sometimes called *ore dressing, mineral dressing*, or *milling*, follows mining and prepares the ore for extraction of the valuable metal in the case of metallic ores, but produces a commercial end product of non-metallic minerals and coal. Apart from regulating the size of the ore, it is a process of physically separating the grains of valuable minerals from the gangue minerals, to produce an enriched portion, or *concentrate*, containing most of the valuable minerals, and a discard, or *tailing*, containing predominantly the gangue minerals.

It has been predicted that the importance of mineral processing of metallic ores may decline as the physical processes utilised are replaced by the hydro and pyrometallurgical routes used by the extractive metallurgist, because higher recoveries are obtained by some chemical methods. This may certainly apply when the useful mineral is very finely disseminated in the ore and adequate liberation from the gangue is not possible, in which case a combination of chemical and mineral processing techniques may be advantageous, as is the case with some highly complex ores containing economic amounts of copper, lead, zinc and precious metals.[15,16] However, in the majority of cases the energy consumed in direct smelting or leaching of low-grade ores would be so enormous as to make the cost prohibitive. Compared with these processes, mineral processing methods are inexpensive, and their use is readily justified on economic grounds.

If the ore contains worthwhile amounts of more than one valuable mineral, it is usually the object of mineral processing to separate them; similarly if undesirable minerals, which may interfere with subsequent refining processes, are present, it may be necessary to remove these minerals at the separation stage. Ore dressing is a necessary prelude with most nonferrous ores. However, there are so many high-grade iron ore deposits that, until relatively recently, there has been little incentive to upgrade them, with the possible exception of Swedish magnetite (Fe_3O_4) containing apatite. Milling of iron ores is, however, continually gaining importance with the working of lower-grade deposits, and most of the iron now produced in the United States is from low-grade taconite ores (finely disseminated iron minerals in silica).

There are two fundamental operations in mineral processing, namely the release, or *liberation*, of the valuable minerals from their waste gangue minerals, and separation of these values from the gangue, this latter process being known as *concentration*.

Liberation of the valuable minerals from the gangue is accomplished by *comminution*, which involves crushing, and, if necessary, grinding, to such a particle size that the product is a mixture of relatively clean particles of mineral and gangue. Grinding is often the greatest energy consumer, accounting for up to 50% of a concentrator's energy consumption. As it is this process which achieves liberation of values from gangue, it is also the process which is essential to efficient separation of the minerals, and is often said to be the key to good mineral processing. In order to produce clean concentrates with little contamination with gangue minerals, it is necessary to grind the ore fine enough to liberate the associated minerals. Fine grinding, however, increases energy costs, and can lead to the production of very fine untreatable "slime" particles which may be lost into the tailings. Grinding therefore becomes a compromise between clean (high-grade concentrates), operating costs and losses of fine minerals. If the ore is low grade, and the minerals have very small grain size and are disseminated through the rock, then grinding energy costs and fines losses can be high, unless the nature of the minerals is such that a pronounced difference in some property between the minerals and the gangue is available.

An intimate knowledge of the mineralogical asssembly of the ore is essential if efficient processing is to be carried out. Not only is a knowledge of the nature of the valuable and gangue minerals required, but also of the ore "texture". The texture refers to the aggregation (size), dissemination (distribution) and shape of the minerals within the ore. Every concentrator should be equipped with a microscope for examination of thin and polished ore sections, which can be used to predict grinding and concentration requirements, feasible concentrate grades, difficulties which may be encountered in the separation of valuable minerals and gangue due to their association and locking with other minerals, etc. Microscopic analysis of tailings particles and other mill products can also yield much valuable information regarding the response of the ore to treatment[17] (see Figs. 1.2a–i for examples).

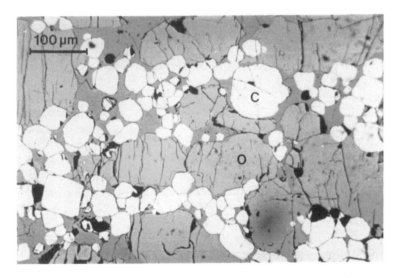

Fig. 1.2a. South African chromite ore. Relatively coarse grain size, and compact morphology of chromite (C) grains makes liberation from olivine (O) gangue fairly straightforward.

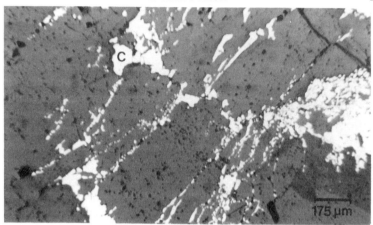

Fig. 1.2b. North American porphyry copper ore. Chalcopyrite (C) precipitated along fractures in quartz. Liberation of chalcopyrite fairly difficult due to "chain-like" distribution. Fracture is, however, likely to occur preferentially along the sealed fractures, producing particles with a surface coating of chalcopyrite, which can be effectively recovered into a low-grade concentrate by froth flotation.

Fig. 1.2c. Mixed sulphide ore, Wheal Jane, Cornwall. Chalcopyrite (C) and sphalerite (S), much of which is extremely finely disseminated in tourmaline (T), making a high degree of liberation impracticable.

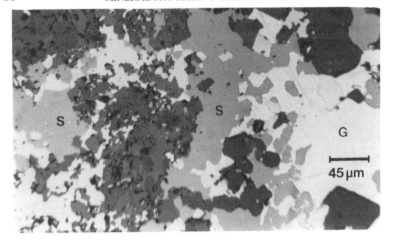

FIG. 1.2d. Hilton lead–zinc orebody, Australia. Gallena (G) and sphalerite (S) intergrown. Separate "clean" concentrates of lead and zinc will be difficult to produce, and contamination of concentrates with other metal is likely.

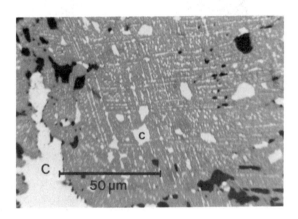

FIG. 1.2e. Copper–zinc ore. Grain of sphalerite with many minute inclusions of chalcopyrite (C) along cleavage planes. Fracturing during comminution takes place preferentially along the low coherence cleavage planes, producing a veneer of chalcopyrite on the sphalerite surface, making depression of the latter difficult in flotation.

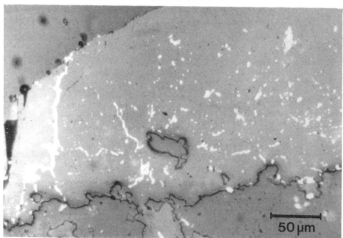

F<small>IG</small>. 1.2f. Lead–zinc ore. Fine grained native silver in vein networks and inclusions in carbonate host rock. Rejection of this material by heavy medium separation could lead to high silver loss.

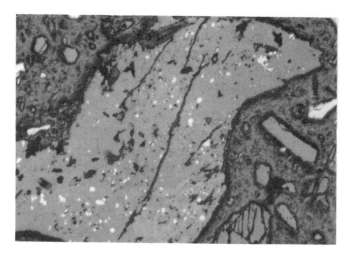

F<small>IG</small>. 1.2g. Flotation tailings, Palabora Copper Mine, South Africa. Finely disseminated grains of chalcopyrite enclosed in a grain of gangue, and irrecoverable by flotation. Maximum grain size of chalcopyrite is about 20 microns, such that attempts to liberate by further grinding would be impracticable.

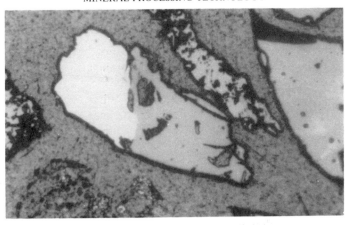

Fig. 1.2h. Gravity circuit tailings, tin concentrator. Cassiterite (light grey) locked with gangue (darker grey), mainly quartz. The composite particle is very fine (less than 20 μm), and has reported to tailings, rather than middlings, due to the inefficiency of gravity separation at this size. Loss of such particles to tailings is a major cause of poor recovery in gravity concentration. In this case, the composite tailings particles could be recovered by froth flotation into a low-grade concentrate.

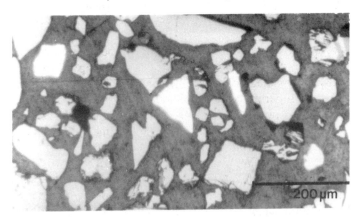

Fig. 1.2i. Tin concentrate, assaying about 60% tin. Although there is some limited locking of the cassiterite (light grey) with gangue (darker grey), the main contaminant is arsenopyrite (white), which, being a heavy mineral (S.G. 6), has partitioned with the cassiterite (S.G. 7) into the gravity concentrate. The arsenopyrite particles are essentially liberated, and can easily be removed by froth flotation, thereby increasing the tin grade of the concentrate and avoiding smelter penalties due to high arsenic levels.

The most important physical methods which are used to concentate ores are:

1. Separation dependent on optical and radioactive properties, etc. This is often called *sorting*, which commonly included hand selection of high-grade ores until relatively recently.

2. Separation dependent on specific gravity differences. This utilises the differential movement of minerals due to mass effects, usually in hydraulic currents. Although the method declined in importance with the development of the froth flotation process, it is now being increasingly used due to improved techniques and its relative simplicity compared with other methods. It also has the advantage of producing less environmental pollution.

3. Separation utilising the different surface properties of the minerals. *Froth flotation*, which is undoubtedly the most important method of concentration, is effected by the degree of affinity of the minerals for rising air-bubbles within the agitated pulp. By adjusting the "climate" of the pulp by various reagents, it is possible to make the valuable minerals air-avid (*aerophilic*) and the gangue minerals water-avid (*aerophobic*). This results in separation by transfer of the valuable minerals to the air-bubbles which form the froth floating on the surface of the pulp.

4. Separation dependent on magnetic properties. Low intensity magnetic separators can be used to concentrate ferromagnetic minerals such as magnetite (Fe_3O_4), while high-intensity separators are used to separate paramagnetic minerals from their gangue. Magnetic separation is an important process in the beneficiation of iron ores, but also finds application in the treatment of paramagnetic non-ferrous minerals. It is widely used to remove paramagnetic wolframite (($Fe,Mn)WO_4$) and hematite (Fe_2O_3) from tin ores, and has found considerable application in the processing of non-metallic minerals, such as those found in beach sand deposits.

5. Separation dependent on electrical conductivity properties. High-tension separation can be used to separate conducting from non-conducting minerals. This method is interesting, since theoretically it represents the "universal" concentrating method; almost all minerals show some difference in conductivity and it should be possible to separate almost any two by this process. However, the

method has fairly limited application, and its greatest use is in separating some of the minerals found in heavy sands from beach or stream placers. Minerals must be completely dry and the humidity of the surrounding air must be regulated, since most of the electron movement in dielectrics takes place on the surface and a film of moisture can change the behaviour completely. The biggest disadvantage of the method is that the capacity of economically sized units is very low.

Heat treatment is sometimes used to make the ore more suitable for subsequent processing. Roasting can be used to effect major chemical changes, such as the conversion of non-magnetic iron minerals to a ferromagnetic form. Calcination can be used to destroy the colloidal bond of clay minerals and to decompose hydrates and carbonates, making the ore easier to handle and treat.

Comminution and concentration are the two primary operations in mineral processing, but many other important steps are involved, among which are sizing of the ore at different stages in treatment, by the use of screens and classifiers, and dewatering of the mineral pulps, using thickeners, filters, and driers.

The Flowsheet

The flowsheet shows diagrammatically the sequence of operations in the plant. In its simplest form it can be presented as a block diagram in which all operations of one character are grouped (Fig. 1.3). In this case comminution deals with all crushing, grinding, and initial rejection. The next block, "separation", groups the various treatments incident to production of concentrate and tailing. The third, "product handling", covers the disposal of the products.

The simple line flowsheet (Fig. 1.4) is for most purposes sufficient, and can include details of machines, settings, rates, etc.

Milling Costs

It has been shown that the balance between milling costs and metal

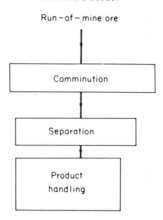

FIG. 1.3. Simple block flowsheet.

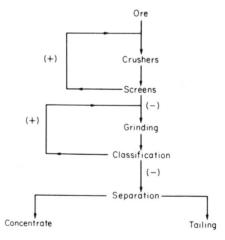

FIG. 1.4. Line flowsheet. (+) indicates oversized material returned for further treatment and (−) undersized material, which is allowed to proceed to the next stage.

losses is crucial, particularly with low-grade ores, and because of this, most mills keep detailed accounts of operating and maintenance costs, broken down into various sub-divisions, such as labour, supplies, energy, etc., for the various areas of the plant. This type of analysis is very useful in identifying high-cost areas where improvements in performance would be most beneficial. It is impossible to give typical operating costs for milling operations, as these vary enormously from mine to mine, and particularly from country to country, depending on local costs of energy, labour, water, supplies, etc., but Table 1.4 is an example of such a breakdown of costs on a 500 t d^{-1} concentrator.

Efficiency of Mineral Processing Operations

Liberation

One of the major objects of comminution is the liberation, or release, of the valuable minerals from the associated gangue minerals at the coarsest possible particle size. If such an aim is achieved, then not only is energy saved by the reduction of the amount of fines produced, but any subsequent separation stages become easier and cheaper to operate. If high-grade solid products are required, then good liberation is essential; however, for subsequent hydrometallurgical processes, such as leaching, it may also be necessary to *expose* the required mineral.

In practice complete liberation is seldom achieved, even if the ore is ground down to the grain size of the desired mineral particles. This is illustrated by Fig. 1.5, which shows a lump of ore which has been reduced to a number of cubes of identical volume and of a size below that of the grains of mineral observed in the original ore sample. It can be seen that each particle produced containing mineral also contains a portion of gangue; complete liberation has not been attained; the bulk of the major mineral—the gangue—has, however, been liberated from the minor mineral—the value.

The particles of "locked" mineral and gangue are known as *middlings*, and further liberation from this fraction can only be achieved by further comminution.

TABLE 1.4   TYPICAL BREAKDOWN OF MONTHLY MILLING COSTS

	Labour	Supplies	Power	Others	Total	Cost/t
Operating costs (£)						
Mill management	3300	—	—	400	3700	0.25
Crushing	3500	—	1320	—	4820	0.32
Grinding/classn	6020	10,200	7800	—	24,020	1.60
Concentration/pumps	8050	4800	4200	—	17,050	1.14
Dewatering	2250	80	900	—	3230	0.22
Tailings	1900	150	310	—	2360	0.16
Metallurgy	2200	120	120	150	2590	0.17
Assaying	1800	110	90	50	2050	0.14
Others	—	—	600	—	600	0.04
Total operating cost	29,020	15,460	15,340	600	60,420	
Cost/tonne milled (£)	1.94	1.03	1.02	0.04		4.03
Maintenance costs (£)						
Building	400	80	—	—	480	0.03
Crushers	1200	3500	—	—	4700	0.31
Bins/feeders/etc	550	500	—	—	1050	0.07
Motors/pumps	600	1200	—	—	1800	0.12
Grind/classn	700	7000	—	—	7700	0.51
Concentration	710	3500	—	—	4210	0.28
Dewatering	200	330	—	—	530	0.04
Tailings	250	80	—	—	330	0.02
Others	—	400	—	—	400	0.03
Total maintenance cost	4610	16,590	—	—	21,200	
Cost/tonne milled (£)	0.31	1.11	—	—		1.41
Total milling cost	33,630	32,050	15,340	600	81,620	
Milling cost/tonne	2.24	2.14	1.02	0.04		£5.44

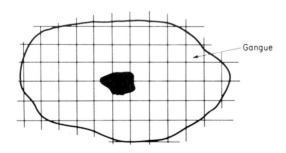

FIG. 1.5. "Locking" of mineral and gangue.

The "degree of liberation" refers to the percentage of the mineral occurring as free particles in the ore in relation to the total content. This can be high if there are weak boundaries between mineral and gangue particles, which is often the case with ores composed mainly of rock-forming minerals, particularly sedimentary materials. Usually, however, the adhesion between mineral and gangue is strong and, during comminution, the various constituents are cleft across. This produces much middlings and a low degree of liberation. New approaches to increasing the degree of liberation involve directing the breaking stresses at the mineral crystal boundaries, so that the rock can be broken without breaking the mineral grains.[18]

Many researchers have tried to quantify degree of liberation with a view to predicting the behaviour of particles in a separation process. The first attempt at the development of a model for the calculation of liberation was made by Gaudin;[19] King[20] developed an exact expression for the fraction of particles of a certain size that contain less than a prescribed fraction of any particular mineral. These models, however, suffer from many unrealistic assumptions that must be made with respect to the grain structure of the minerals in the ore, and as a result have not found much practical application. Attempts at quantifying liberation by means of automated optical image analysis have also been relatively unsuccessful due to the inherent inadequacies of the instrument in working with ore assemblies, although recent developments in scanning electron microscopy have made automated mineralogical assessment a real possibility for the near future.[21]

It should also be noted that a high degree of liberation is not necessary in certain processes, and, indeed, may be undesirable. For instance, it is possible to achieve a high recovery of values by gravity and magnetic separation even though the valuable minerals are completely enclosed by gangue, and hence the degree of liberation of the values is zero. As long as a pronounced density or magnetic susceptibility difference is apparent between the locked particles and the free gangue particles, the separation is possible. A high degree of liberation may only be possible by intensive fine grinding, which may reduce the particles to such a fine size that separation becomes very inefficient. On the other hand, froth flotation requires as much of the valuable mineral *surface* as possible to be exposed, whereas in a chemical leaching process, a portion of the surface must be exposed to provide a channel to the bulk of the mineral.

In practice, ores are ground to an *optimum mesh of grind*, determined by laboratory and pilot scale testwork, to produce an economic degree of liberation. The concentration process is then designed to produce a concentrate consisting predominantly of valuable mineral, with an accepted degree of locking with the gangue minerals, and a middlings fraction, which may require further grinding to promote optimum release of the minerals. The tailings should be mainly composed of gangue minerals.

Figure 1.6 is a cross-section through a typical ore particle, and illustrates effectively the liberation dilemma often facing the mineral processor.[22] Regions A represent valuable mineral, and region AA is rich in valuable mineral, but is highly *intergrown* with the gangue mineral. Comminution produces a range of fragments, ranging from fully liberated mineral and gangue particles, to those illustrated. Particles of type 1 are rich in mineral, and are classed as concentrate as they have an acceptable degree of locking with the gangue which limits the concentrate grade. Particles of type 4 would likewise be classed as tailings, the small amount of mineral present reducing the recovery of mineral into the concentrate. Particles of types 2 and 3, however, would probably be classed as middlings, although the degree of regrinding needed to promote economic liberation of mineral from particle 3 would be greater than that in particle 2.

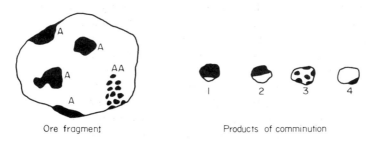

Ore fragment   Products of comminution

Fig. 1.6. Cross-sections of ore particles.

During the grinding of a low-grade ore the bulk of the gangue minerals is often liberated at a relatively coarse size (Fig. 1.5). In certain circumstances it may be economic to grind to a size much coarser than the optimum in order to produce in the subsequent concentration process a large middlings fraction and tailings which can be discarded at a coarse grain size. The middlings fraction can then be reground to produce a feed to the final concentration process (Fig. 1.7).

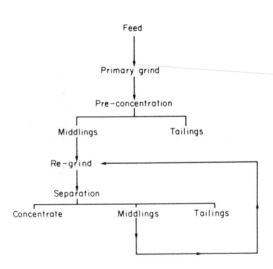

Fig. 1.7. Flowsheet for process utilising two-stage separation.

This method discards most of the coarse gangue early in the process, thus considerably reducing grinding costs, as needless comminution of liberated gangue is avoided. It is often used on minerals which can easily be separated from the free gangue, even though they are themselves locked to some extent with gangue. It is the basis of the heavy medium process of pre-concentration (Chapter 11).

Concentration

The object of mineral processing, regardless of the methods used, is always the same, i.e. to separate the minerals into two or more products with the values in the concentrates, the gangue in the tailings, and the "locked" particles in the middlings. Such separations are, of course, never perfect, so that much of the middlings produced are, in fact, *misplaced* particles, i.e. those particles which ideally should have reported to the concentrate or the tailings. This is often particularly serious when treating ultra-fine particles, where the efficiency of separation is usually low. In such cases, fine liberated valuable mineral particles often report in the middlings and tailings. The technology for treating fine-sized minerals is, as yet, poorly developed, and, in some cases, very large amounts of fines are discarded. For instance, it is common practice to remove material less than 10 μm in size from tin concentrator feeds and direct this material to the tailings, and, in the early 1970s, 50% of the tin mined in Bolivia, 30% of the phosphate mined in Florida, and 20% of the world's tungsten were lost as fines. Significant amounts of copper, uranium, fluorspar, bauxite, zinc, and iron were also similarly lost.[23]

Figure 1.8 shows the size range applicability of unit concentration processes.[24] It is evident that most mineral processing techniques fail in the ultra-fine size range. Gravity concentration techniques, especially, become unacceptably inefficient, and even flotation, the most important concentrating technique used today, usually fails when used alone in the ultra-fine size range.

It should be pointed out that the process is also limited by the mineralogical nature of the ore. For example, in an ore containing native copper it is theoretically possible to produce a concentrate containing 100% Cu, but, if the ore mineral was chalcopyrite ($CuFeS_2$), the best concentrate would contain only 34.5% Cu.

The *recovery*, in the case of the concentration of a metallic ore, is the percentage of the total metal contained in the ore that is recovered in

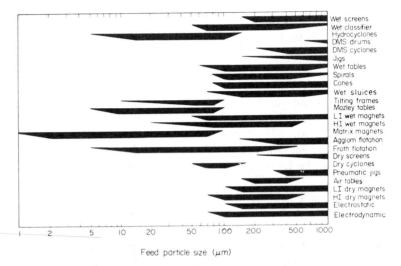

Feed particle size (μm)

FIG. 1.8. Effective range of application of conventional mineral processing techniques.

the concentrate; a recovery of 90% means that 90% of the metal in the ore is recovered in the concentrate and 10% is lost in the tailings. It is possible to speak of the recovery of metal in tailings, middlings, and other products, but normally the term refers to the amount of metal recovered in the concentrates—the valuable products. The recovery, when dealing with non-metallic ores, refers to the percentage of the total mineral contained in the ore that is recovered into the concentrate, i.e. recovery is usually expressed in terms of the valuable *end* product.

The *ratio of concentration* is the ratio of the weight of the feed (or *heads*) to the weight of the concentrates. It is a measure of the efficiency of the concentration process, and it is closely related to the *grade* or *assay* of the concentrate; the value of the ratio of concentration will generally increase with the grade of concentrate.

The grade, or assay, usually refers to the content of the marketable end product in the material. Thus in metallic ores, the per cent metal is often quoted, although in the case of very low-grade ores, such as gold, metal content may be expressed as parts per million (ppm), or its equivalent, grams per tonne $(g\,t^{-1})$. Some metals are sold in oxide form, and hence the grade may be quoted in terms of the marketable oxide content, e.g. $\%WO_3$, $\%U_3O_8$, etc. In non-metallic operations, grade usually refers to the mineral content, e.g. $\%CaF_2$ in fluorite ores; diamond ores are usually graded in *carats* per 100 tonnes (t), where 1 carat is 0.2 g. Coal is graded according to its *ash* content, i.e. the amount of incombustible mineral present within the coal; most of the coal produced in Britain is consumed in power stations, which require a feed with an ash content of between 15 and 20%.

The *enrichment ratio* is the ratio of the grade of the concentrate to the grade of the feed, and again is related to the efficiency of the process.

Ratio of concentration and recovery are essentially independent of each other, and in order to evaluate a given operation it is necessary to know both. For example, it is possible to obtain a very high grade of concentrate and ratio of concentration by simply picking a few lumps of pure galena from a lead ore, but the recovery would be very low. On the other hand, a concentrating process might show a recovery of 99% of the metal, but it might also put 60% of the gangue minerals in the concentrate. It is, of course, possible to obtain 100% recovery by not concentrating the ore at all.

There is an approximately inverse relationship between recovery and grade of concentrate in all concentrating processes. If an attempt is made to attain a very high-grade concentrate, the tailings assays are higher and the recovery is low. If a high recovery of metal is aimed for, there will be more gangue in the concentrate and the grade of concentrate and ratio of concentration will both decrease. It is impossible to give figures for representative values of recoveries and ratios of concentration. A concentration ratio of 2 to 1 might be satisfactory for certain high-grade non-metallic ores, but a ratio of 50 to 1 might be considered too low for a low-grade copper ore; ratios of concentration of several million to one are common with diamond ores. The aim of milling operations is to maintain the values of ratio of concentration and recovery as high as possible, all factors being considered.

Since concentrate grade and recovery are metallurgical factors, the *metallurgical efficiency* of any concentration operation can be expressed by a curve showing the recovery attainable for any value of concentrate grade. Figure 1.9 is a typical *recovery-grade* curve showing the characteristic inverse relationship between recovery and concentrate grade.

Concentrate grade and recovery, used simultaneously, are the most widely accepted measures of assessing metallurgical (not economic) performance. However, there is a problem in quantitatively assessing the technical performance of a concentration process whenever the results of two similar test runs are compared. If both the grade and recovery are greater for one case than the other, then the choice of process is simple, but if the results of one test show a higher grade but a lower recovery than the other, then the choice is no longer obvious. There have been many attempts to combine recovery and concentrate grade into a single index defining the metallurgical efficiency of the separation. These have been reviewed by Schulz,[25] who proposes the following definition:

$$\text{Separation efficiency (S.E.)} = Rm - Rg, \qquad (1.1)$$

where Rm = % recovery of the valuable mineral,
     Rg = % recovery of the gangue into the concentrate.

Suppose the feed material, assaying f% metal, separates into a concentrate assaying c% metal, and a tailing assaying t% metal, and

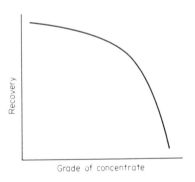

FIG. 1.9. Typical recovery-grade curve.

that C is the fraction of the total feed weight that reports to the concentrate, then:

$$Rm = 100 \; Cc/f \qquad (1.2)$$

i.e. recovery of valuable mineral to the concentrate is equal to metal recovery, assuming that all the valuable metal is contained in the same mineral.

The gangue content of the concentrate $= 100 - 100 \; c/m\%$, where m is the percentage metal content of the valuable mineral,

i.e. gangue content $= 100(m - c)/m$.

Therefore, $Rg = C \times$ gangue content of concentrate/gangue content of feed

$$= 100 \; C(m - c)/(m - f).$$

Therefore, $Rm - Rg = 100 \; Cc/f - 100 \; C(m - c)/(m - f)$

$$= \frac{100 \; Cm(c - f)}{(m-f)f}. \qquad (1.3)$$

Example 1.1

A tin concentrator treats a feed containing 1% tin, and three possible combinations of concentrate grade and recovery are:

High grade	63% tin at 62% recovery
Medium grade	42% tin at 72% recovery
Low grade	21% tin at 78% recovery

Determine which of these combinations of grade and recovery produce the highest separation efficiency.

Solution

Assuming that the tin is totally contained in the mineral cassiterite (SnO_2), which, when pure, contains 78.6% tin, then since mineral recovery (equation 1.2) is $100 \times C \times$ concentrate grade/feed grade, so for the high-grade concentrate:

$$62 = C \times 63 \times 100/1, \text{ and so } C \times 9.841 \times 10^{-3}.$$

Therefore, S.E. (equation 1.3) $= \dfrac{0.984 \times 78.6 \times (63 - 1)}{(78.6 - 1) \times 1} = 61.8\%.$

Similarly, for the medium-grade concentrate, from equation 1.2:

$$72 = 100 \times C \times 42/1.$$

Therefore, $C = 1.714 \times 10^{-2}$,
and S.E. (equation 1.3) = 71.2%.

For the low-grade concentrate, from equation 1.2:

$$78 = 100 \times C \times 21/1.$$

Therefore, $C = 3.714 \times 10^{-2}$,
and S.E. (equation 1.3) = 75.2%.

Therefore, the highest separation efficiency is achieved by the production of a low-grade (21% tin) concentrate at high (78%) recovery.

Although the value of separation efficiency can be useful in comparing the performance of different operating conditions on selectivity, it takes no account of economic factors, and, as will become apparent, a high value of separation efficiency does not necessarily lead to the most economic return.

Since the purpose of mineral processing is to increase the economic value of the ore, the importance of the recovery-grade relationship is in determining the most *economic* combination of recovery and grade which will produce the greatest financial return per tonne of ore treated in the plant. This will depend primarily on the current price of the valuable product, transportation costs to the smelter, refinery, or other further treatment plant, and the cost of such further treatment, the latter being very dependent on the grade of concentrate supplied. A high grade concentrate will incur lower smelting costs, but the lower recovery means lower returns in terms of final product. A low grade concentrate may achieve greater recovery of the values, but incur greater smelting and transportation costs due to the included gangue minerals. Also of importance are impurities in the concentrate which may be penalised by the smelter, although precious metals may produce a bonus.

The net return from the smelter (NSR) can be calculated for any recovery-grade combination from:

$$NSR = \text{Payment for contained metal} - (\text{Smelter charges} + \text{Transport costs}).$$

This is summarised in Fig. 1.10, which shows that the highest value of net smelter return is produced at an optimum concentrate grade. It is essential that the mill achieves a concentrate grade which is as close as possible to this target grade. Although the effect of moving slightly away from the optimum may only be of the order of a few pence per tonne treated, this can amount to very large financial losses, particularly on high-capacity plants treating thousands of tonnes per day. Changes in metal price, smelter terms, etc., obviously affect the net smelter return–concentrate grade curve, and the value of the optimum concentrate grade. For instance, if the metal price increases, then the optimum grade will be lower, allowing higher recoveries to be attained (Fig. 1.11).

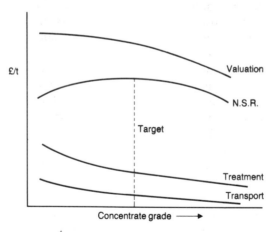

FIG. 1.10. Variation of payments and charges with concentrate grade.

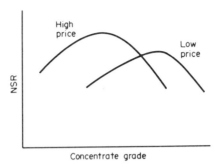

FIG 1.11. Effect of metal price on NSR–grade relationship.

It is evident that the terms agreed between the concentrator and smelter are of paramount importance in the economics of mining and milling operations. Such *smelter contracts* are usually fairly complex. Concentrates are sold under contract to "custom smelters" at prices based on quotations on metal markets such as the London Metal Exchange. The smelter, having processed the concentrates, disposes of the finished metal to the consumers. The proportion of the "free market" price of the metal received by the mine is determined by the terms of the contract negotiated between mine and smelter, and these terms can vary considerably. Table 1.5 summarises a typical low-grade smelter contract for the purchase of tin concentrates. As is usual in many contracts, one assay unit is deducted from the concentrate assay in assessing the value of the concentrates, and arsenic present in the concentrate is penalised. The concentrate assay is of prime importance in determining the valuation, and the value of the assay is usually agreed on the result of independent sampling and assaying performed by the mine and smelter. The assays are compared, and if the difference is no more than an agreed value, the mean of the two results may be taken as the agreed assay. In the case of a greater difference, an "umpire" sample is assayed at an independent laboratory. This umpire assay may be used as the agreed assay, or the mean of this assay and that of the party which is nearer to the umpire value may be chosen.

The use of smelter contracts, and the importance of by-products and changing metal prices, can be seen by briefly examining the economics

TABLE 1.5. SIMPLIFIED TIN SMELTER CONTRACT

Material
Tin concentrates, assaying no less than 15% Sn, to be free from deleterious impurities not stated, and to contain sufficient moisture as to evolve no dust when unloaded at our works.

Quantity
Total production of concentrates·

Valuation
Tin, less 1 unit per dry tonne of concentrates, at the lowest of the official London Metal Exchange prices.

Pricing
On the 7th market day after completion of arrival of each sampling lot into our works.

Treatment charge
£385 per dry tonne of concentrates.

Moisture
£24 per tonne of moisture.

Penalties
Arsenic £40 per unit per tonne.

Lot charge
£175 per lot sampled of less than 17 tonnes.

Delivery
Free to our works in regular quantities, loose on a tipping lorry, or in any other manner acceptable to both parties.

---

of processing two base metals—tin and copper—whose fortunes have fluctuated over the years for markedly different reasons.

Economics of Tin Processing

Almost a half of the world's supply of tin in the mid-nineteenth century was mined in south-west England, but by the end of the 1870s Britain's premium position was lost, with the emergence of Malaysia as the leading producer, and the discovery of rich deposits in Australia.

By the end of the century only nine mines of any consequence remained in Britain, where 300 had flourished 30 years earlier.

Unlike copper, zinc and lead, production has not risen dramatically over the years and has never exceeded 200,000 tonnes per year by a great margin (Fig. 1.1).

Until 1985 the real price of tin remained fairly stable over a number of years (Fig. 1.12).[2] This was because tin is unique among metals in that the greater part of its production and pricing has been regulated since 1956 by a series of international agreements between producers and consumers. The International Tin Council strived to ensure stability in the tin markets by buying and selling from the Council's own huge stockpiles—buying tin when the price fell below a certain level and selling from the buffer stock when the price rose to a certain agreed level. From the mid-70s, however, the price of tin was driven artificially higher and higher at a time of world recession, expanding production and falling consumption, the latter due mainly to the increasing use of aluminium, rather than tin-plated steel, cans. Although the ITC imposed restrictions on the amount of tin that could be produced by its member countries, the reason for the inflating tin price was that the price of tin was fixed by the Malaysian dollar, while the buffer stock manager's dealings on the London Metal Exchange were financed in sterling. The Malaysian dollar was tied to the American dollar, which strengthened markedly between 1982 and 1984, having the effect of increasing the price of tin in London simply because of the exchange rate. However, the American dollar began to weaken in early 1985, taking the Malaysian dollar with it, and effectively reducing the London Metal Exchange tin price from its historic peak. In October 1985 the buffer stock manager announced that the ITC could no longer finance the purchase of tin to prop up the price, as it had run out of funds, owing millions of pounds to the LME traders. This announcement caused near panic, the tin price fell to £8140 per tonne and the LME halted all further dealings. In 1986 many of the world's tin mines were forced to close down due to the depressed tin price, which had fallen to below £4000/t, although by 1987 the price on the open market had recovered to about £4500/t. The following discussion therefore relates to tin processing prior to the collapse.

It is fairly easy to produce concentrates containing over 70% tin (i.e.

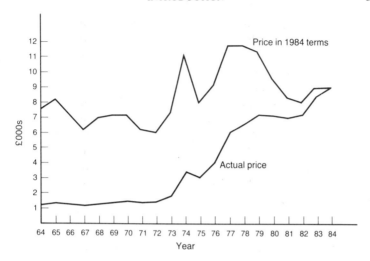

FIG. 1.12. Annual tin prices.

over 90% cassiterite) from alluvial ores, such as those worked in South-East Asia. Such concentrates present little problem in smelting and hence treatment charges are relatively low. Production of high-grade concentrates also incurs relatively low freight charges, which is important if the smelter is remote.

For these reasons it has been traditional in the past for lode tin concentrators to produce high-grade concentrates, but high tin prices, and the development of profitable low-grade smelting processes, changed the policy of many mines towards the production of lower-grade concentrates. The advantage of this is that the recovery of tin into the concentrate is increased, thus increasing smelter payments. However, the treatment of low-grade concentrates produces much greater problems for the smelter, and hence the treatment charges at "low-grade smelters" are normally much higher than those at the high-grade smelters. Freight charges are also correspondingly higher.

Suppose that a tin concentrator treats a feed containing 1% tin, and that three possible combinations of concentrate grade and recovery are (as in Example 1.1):

High grade   63% tin at 62% recovery
Medium grade   42% tin at 72% recovery
Low grade   21% tin at 78% recovery

Assuming that the concentrates are free of arsenic, and that the cost of transportation to the smelter is £20 per tonne of dry concentrate, then the return on each tonne of ore treated can be simply calculated, using the low-grade smelter terms set out in the contract in Table 1.5.

For instance, at a grade of 42% tin and 72% recovery, the weight of concentrate produced from 1 tonne of ore is 17.14 kg (Example 1.1).

The smelter payment for tin in this concentrate is

$$£P \times 17.14 \times (42 - 1)/100\,000$$

where P = L.M.E. tin price in £/tonne.

Assuming a tin price of £8500 per tonne, then the net smelter payment is £59.73.

The smelter treatment charge is £385 × concentrate weight = £6.59, and the transportation cost is £0.34.

The net smelter return for the processing of 1 tonne of concentrator feed is thus £59.73 − (6.59 + 0.34) = £52.80. Therefore, although the ore contains, at free market price, £85 worth of tin per tonne, the mine realises only 62% of the ore value in payments received.

Production of a lower-grade concentrate incurs higher smelter and freight charges, but increases the payment for contained metal, due to the higher recovery. Similar calculations show that at a grade of 21% tin and 78% recovery, the payment for tin is increased to £63.14, but the total deductions also increase to £15.04, producing a net smelter return of £48.10 per tonne of ore treated.

Clearly lowering the concentrate grade to 21% tin, in order to increase recovery, has increased the separating efficiency (Example 1.1), but has adversely affected the economic return from the smelter, the increased charges being more important than the increase in revenue from the metal.

Increasing the grade to 63% tin can obviously reduce charges even further, particularly if the concentrate can be sent to a high-grade smelter with lower treatment charges.

Assuming a treatment charge of £50 per tonne of concentrate, and identical payments and freight charges, the payment for metal in such a concentrate would be only £51.86, but the charges are reduced to £0.69 per tonne. The net smelter return per tonne of ore treated is thus £51.57. In this case, therefore, the return is highest from the low-grade smelter treating a medium-grade concentrate. This situation may change, however, if the metal price changes markedly. If the tin price falls and the terms of the smelter contracts remain the same, then the mine profits will suffer due to the reduction in payments. Rarely does a smelter share the risks of changing metal price, as it performs a service role, changes in smelter terms being made more on the basis of changing smelter costs rather than metal price. The mine does, however, reap the benefits of increasing metal price.

At a tin price of £6500/t, the net smelter return per tonne of ore from the low-grade smelter treating the 42% tin concentrate is £38.75, while the return from the high-grade smelter, treating a 63% Sn concentrate, is £38.96. Although this is a difference of only 21p per tonne of ore, to a $500 \, t \, d^{-1}$ tin concentrator this change in policy from relatively low- to high-grade concentrate, together with the subsequent change in concentrate market, would expect to increase the revenue by $£0.21 \times 500 \times 365 = £38,325$ per annum. The concentrator management must always be prepared to change its policies, both metallurgical and marketing, in this way if maximum returns are to be made, although production of a reliable grade-recovery relationship is often difficult due to the complexity of operation of lode tin concentrators and variations in feed characteristics.

It is, of course, necessary to deduct the costs of mining and processing from the net smelter return in order to deduce the profit achieved by the mine. Some of these costs will be indirect, such as salaries, administration, research and development, medical and safety, as well as direct costs, such as operating and maintenance, supplies and energy. The breakdown of milling costs varies enormously from mine to mine, depending very much on the size and complexity of the operation. Mines with very large ore reserves tend to have very high throughputs, and so although the capital outlay is higher, the operating and labour costs tend to be much lower than those on smaller plants, such as those treating lode tin ores. Mining costs also vary enormously, and are very much higher for underground than for open-pit operations.

If mining and milling costs of £40 and £8 respectively per tonne of ore are typical of underground tin operations, then it can be seen that at a tin price of £8500 the mine, producing a concentrate of 42% tin, which is sold to a low-grade smelter, makes a profit of £52.80 − 48 = £4.80 per tonne of ore, which at a throughput of 500 t d^{-1} corresponds to a gross annual profit of £867,000. It is also clear that if the tin price falls to £6500/t, the mine loses £48 − 38.96 = £9.04 for every tonne of ore treated.

The breakdown of revenue and costs at a tin price of £8500/t is summarised in Fig. 1.13. The mine profit per tonne of ore treated can be summarised as:

Contained value of ore−(costs+losses)

$$= £(85 − (40 + 8 + 23.80 + 8.40)) = £4.80/t.$$

Since 1 t of ore produces 0.0072 t of tin in concentrates, and the free market value of this contained metal is £61.20, the total effective cost of producing 1 t of tin in concentrates is £(61.20 − 4.80)/0.0072 = £7833.

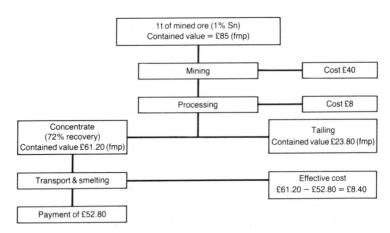

FIG. 1.13. Breakdown of costs and revenues for treatment of lode tin ore (fmp = free market price).

The importance of metal losses in tailings is shown clearly in Fig. 1.13. With ore of relatively high contained value, the recovery is often more important than the cost of promoting that recovery. Hence relatively high-cost unit processes can be justified if significant improvements in recovery are possible, and efforts to improve recoveries should always be made. For instance, suppose the concentrator, maintaining a concentrate grade of 42% tin, improves the recovery by 1% to 73% with no change in actual operating costs. The net smelter return will be £53.53/t, and after deducting mining and milling costs, the profit realised by the mine is £5.53 per tonne of ore. Since 1 t of ore now produces 0.0073 t of tin, having a contained value of £62.05, the cost of producing 1 t of tin in concentrates is thereby reduced to $£(62.05 - 5.53)/0.0073 = £7742$.

Due to the high processing costs and losses, lode tin mines, such as those in Cornwall and Bolivia, have the highest production costs, being above £7500/t in 1985. Alluvial operations, such as those in Malaysia, Thailand and Indonesia, have lower production costs (around £6000/t in 1985). Although these ores have much lower contained values (only about £1–2/t), mining and processing costs, particularly on the large dredging operations, are extremely low, as are smelting costs and losses, due to the high concentrate grades and recoveries produced. In 1985 the alluvial mines in Brazil produced the world's cheapest tin, having production costs of only about £2200/t.[26]

Economics of Copper Processing

In 1835 the United Kingdon was the world's largest copper producer, mining around 15,000 tonnes per year, just below half the world production. This leading position was held until the mid-1860s when the copper mines of Devon and Cornwall became exhausted and the great flood of American copper began to make itself felt. The United States produced about 10,000 tonnes in 1867, but by 1900 was producing over 250,000 tonnes per year. This output had increased to 1,000,000 tonnes per year in the mid-1950s, by which time Chile and Zambia had also become major producers.

Figure 1.14 shows that the price of copper has remained fairly

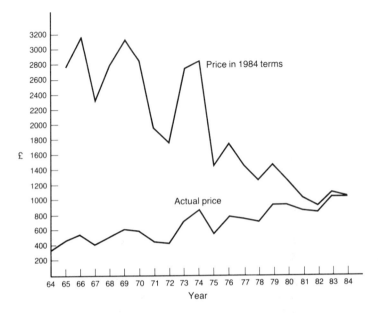

FIG. 1.14. Annual copper prices.

stagnant over the past 20 years, in real terms (index-linked to 1984 prices), its value having more than halved over this period.[2] It is against this economic background, and due also to decreasing ore grades, that in recent times American copper output has suffered, the leading producer now being Chile, having higher ore grades and lower costs.

Copper mining is now a very marginal operation, and many major operations have been forced to shut down. In many cases it is the by-products, such as gold, silver and molybdenum, which have made deposits economically workable, although the price of molybdenum suffered in the early eighties due to supply exceeding demand.

A typical smelter contract for copper concentrates is summarised in Table 1.6.

TABLE 1.6. SIMPLIFIED COPPER SMELTER CONTRACT

Payments	Copper:	Deduct from the agreed copper assay 1 unit, and pay for the remainder at the London Metal Exchange price for higher-grade copper.
	Silver:	If over 30 g t^{-1} pay for the agreed silver content at 90% of the LME silver price.
	Gold:	If over 1 g t^{-1} pay for agreed gold content at 95% of LME gold price.

Deductions
  Treatment charge: £30 per dry tonne of concentrates
  Refining charge: £115 per tonne of payable copper

Consider a porphyry copper mine treating an ore containing 0.6% Cu to produce a concentrate containing 25% Cu, at 85% recovery. This is a concentrate production of 20.4 kg t^{-1} of ore treated. Therefore, at a copper price of £980 per tonne:

Payment for copper = £20.4 × 0.24 × 980/1000 = £4.80
Treatment charge   = 30 × 20.4/1000 = £0.61
Refining charge   = £115 × 20.4 × 0.24/1000 = £0.56

Assuming a freight cost of £20 per tonne of concentrates, the total deductions are £(0.61 + 0.56 + 0.41) = £1.58, and the net smelter return per tonne of ore treated is thus £(4.80 − 1.58) = £3.22.

As mining, milling and other costs must be deducted from this figure, it is apparent that only those mines with very low operating costs can have any hope of profiting from such low-grade operations. Assuming a typical large open-pit mining cost of £1.25 per tonne of ore, a milling cost of £2/t and indirect costs of £2/t, the mine will lose £2.03 for every tonne of ore treated. The breakdown of costs and revenues is summarised in Fig. 1.15.

As each tonne of ore produces 0.0051 t of copper in concentrates, with a free market value of £5.00, the total effective production costs are £(5.00 + 2.03)/0.0051 = £1378 per tonne of copper in concentrates.

However, if the ore contains appreciable by-products, the effective production costs are reduced. Assuming the concentrate contains 25 g t^{-1} of gold and 70 g t^{-1} of silver, as is the case of Bougainville in

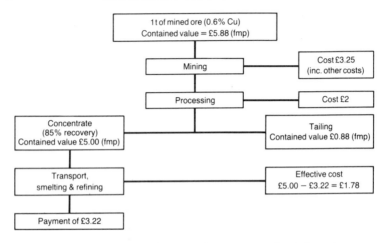

FIG. 1.15. Breakdown of costs and revenues for treatment of typical porphyry copper ore (fmp=free market price).

Papua New Guinea, the world's largest copper mine (see Chapter 12), then the payment for gold, at a LME price of £230 per troy ounce (1 troy oz = 31.1035 g),

$$= \frac{20.4}{1000} \times \frac{25}{31.1035} \times 0.95 \times 230 = £3.58.$$

and the payment for silver at a LME price of £4.5 per troy oz

$$= \frac{20.4}{1000} \times \frac{70}{31.1035} \times 0.9 \times 4.5 = £0.19.$$

The net smelter return is thus increased to £6.99/t of ore, and the mine makes a profit of £1.74 per tonne of ore treated. The effective cost of producing 1 tonne of copper is thus reduced to £(5.00 − 1.74)/0.0051 = £639.22.

By-products are thus extremely important in the economics of copper production, particularly for very low-grade operations. In this example, 42% of the mine's revenue is from gold, copper contributing 56%. This compares with the contributions to revenue realised at Bougainville Copper Ltd.[27]

Table 1.7 lists estimated effective costs per tonne of copper processed at some of the world's major copper mines, at a copper price of £980/t. [28]

TABLE 1.7. EFFECTIVE COSTS AT WORLD'S LEADING COPPER MINES IN 1985

Mine	Country	Effective cost £/t of processed copper
Chuquicamata	Chile	589
El Teniente	Chile	622
Bougainville	Papua New Guinea	664
Palabora	South Africa	725
Andina	Chile	755
Cuajone	Peru	876
El Salvador	Chile	906
Toquepala	Peru	1012
Inspiration	USA	1148
San Manuel	USA	1163
Morenci	USA	1193
Twin Buttes	USA	1208
Utah/Bingham	USA	1329
Nchanga	Zambia	1374
Gecamines	Zaire	1374

It is evident that, apart from Bougainville, which has a high gold content, and Palabora, a large open-pit operation with numerous heavy mineral by-products, the only economic copper mines in 1985 were the large SouthAmerican porphyries. The mines profited due to relatively low actual operating costs, by-product molybdenum production, and higher average grades (1.2% Cu) than the North America porphyries, which averaged only 0.6% Cu. Relatively high-grade deposits such as that at Nchanga failed to profit due partly to high operating costs, but mainly due to the lack of by-products. It is evident that if a large copper mine is to be brought into production in the present economic climate, then initial capitalisation on high-grade secondary ore and by-products must be made, as at Ok Tedi in Papua New Guinea, which commenced production in 1984, initially mining and processing the high-grade gold ore in the leached capping ore.

Since the profit margin involved in the processing of modern copper ores is usually only small, continual efforts must be made to try to reduce milling costs and metal losses. Even relatively small increases in return per tonne can have a significant economic effect, due to the very large tonnages that are often treated. There is, therefore, a constant search for improved flowsheets and flotation reagents.

Figure 1.15 shows that in the example quoted, the contained value in the flotation tailings is £0.88 per tonne of ore treated. The concentrate contains copper to the value of £5.00, but the smelter payment is £3.22. Therefore, the mine realises only 64.4% of the free market value of copper in the concentrate. On this basis, the actual metal loss into the tailings is only about £0.57 per tonne of ore. This is relatively small compared with milling costs, and an increase in recovery of 0.5% would raise the net smelter return by only £0.01. Nevertheless, this can be significant; to a mine treating 50,000 t d^{-1}, this is an increase in revenue of £500 per day, which is extra profit, providing that it is not offset by any increased milling cost. For example, improved recovery may be possible by the use of a more effective reagent, or by increasing the dosage of an existing reagent, but if the increased reagent cost is greater than the increase in smelter return, then the action is not justified.

This balance between milling cost and metallurgical efficiency is very critical on a concentrator treating an ore of low contained value, where it is crucial that milling costs be as low as possible. Reagent costs are typically around 10% of the milling costs on a large copper mine, but energy costs may contribute well over 25% of these costs. Grinding is by far the greatest energy consumer and this process undoubtedly has the greatest influence on metallurgical efficiency. Grinding is essential for the liberation of the minerals in the assembly, but it should not be carried out any finer than is justified economically. Not only is fine grinding energy intensive, but it also leads to increased media costs. Grinding steel often contributes as much as, if not more than, the total mill energy cost, and the quality of grinding medium used often warrants special study. Figure 1.16 shows the effect of fineness of grind on net smelter return and grinding costs for a typical low-grade copper ore. Although flotation recovery, and hence net smelter return, increases with fineness of grind, it is evident that there is no economic benefit in grinding finer than 105 microns. Even this

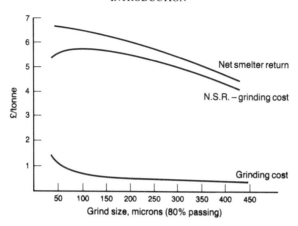

FIG. 1.16. Effect of fineness of grind on net smelter return and grinding costs.

fineness will probably be beyond the economic limit because of the additional capital cost of the grinding equipment required to achieve it.

Economic Efficiency

It is evident from the foregoing that the metallurgical significance of grade and recovery is of less importance than the economic consideration. It is apparent that a certain combination of grade and recovery produces the highest economic return under certain conditions of metal price, smelter terms, etc. However, this metallurgical efficiency combination may not promote the highest return if those conditions change. *Economic efficiency* compares the *actual* net smelter return per tonne of ore milled with the *theoretical* return, thus taking into account all the financial implications. The theoretical return is the maximum possible return that could be achieved, assuming "perfect milling", i.e. complete separation of the valuable mineral into the concentrate, with all the gangue reporting to tailings. Using economic efficiency, plant efficiencies can be compared even during periods of fluctuating market conditions.

MPT—C

Example 1.2

Calculate the economic efficiency of a tin concentrator, treating an ore grading 1% tin and producing a concentrate grading 42% tin at 72% recovery, under the conditions of the smelter contract shown in Table 1.5. The cost of transportation to the smelter is £20 per tonne of concentrate. Assume a tin price of £8500 per tonne.

Solution

It was shown in the section on tin processing that this concentrate would realise a net smelter return of £52.80.

Assuming perfect milling, 100% recovery of the tin would be achieved, into a concentrate grading 78.6% tin (i.e. pure cassiterite).

The weight of concentrate produced from 1 tonne of feed = 12.72 kg.

Therefore, transport cost = £12.71 × 20/1000 = £0.25.

Treatment charge = £385 × 12.72/1000 = £4.90.

Valuation = £12.72 × (78.6 − 1) × 8500/100000 = £83.90.

Therefore, net smelter return = £(83.90 − 4.90 − 0.25) = £78.75, and Economic efficiency = 100 × 52.80/78.75 = 67.0%.

In recent years attempts have been made to optimise the performance of some concentrators by controlling plant conditions to achieve maximum economic efficiency (see Chapters 3 and 12). A dilemma often facing the metallurgist on a complex flotation circuit producing more than one concentrate is: how much contamination of one concentrate by the mineral that should report to the other concentrate can be tolerated? For instance, on a plant producing separate copper and zinc concentrates, copper is always present in the zinc concentrate, as is zinc in the copper concentrate. Metals misplaced into the wrong concentrate are rarely paid for by the specialist smelter, and are sometimes penalised. There is, therefore, an optimum "degree of contamination" that can be tolerated. The most important reagent controlling this factor is often the depressant, which inhibits flotation of the zinc minerals. Increase in the addition of this reagent produces a cleaner copper concentrate, but also reduces copper recovery into this concentrate, as it also has a lesser depressing effect on the copper minerals. The depressed copper minerals are likely to report to the zinc concentrate, so the addition rate of depressant needs to be carefully monitored and controlled to produce an optimum compromise. This should occur when the economic efficiency is maximised.

Example 1.3

The following assay data was collected from a copper–zinc concentrator:

Feed   0.7% copper, 1.94% zinc
Cu concentrate   24.6% copper, 3.40% zinc
Zn concentrate   0.4% copper, 49.7% zinc

Calculate the overall economic efficiency under the following simplified smelter terms:

> *Copper:*   Copper price: £1000/tonne
> Smelter payment: 90% of Cu content
> Smelter treatment charge; £30/tonne of concentrate
> Transport cost: £20/tonne of concentrate

> *Zinc:*   Zinc price; £400/tonne
> Smelter payment: 85% of zinc content
> Smelter treatment charge: £100/tonne of concentrate
> Transport cost: £20/tonne of concentrate

Solution

1. Assuming *Perfect milling*

(a) *Copper*
Assuming that all the copper is contained in the mineral chalcopyrite, then maximum copper grade is 34.6% Cu (pure chalcopyrite).

If C is weight of copper concentrate per 1000 kg of feed, then for 100% recovery of copper into this concentrate:

$$100 = \frac{34.6 \times C \times 100}{0.7 \times 1000} \quad \text{and } C = 20.2 \text{ kg.}$$

Transport costs = £20 × 20.2/1000 = £0.40.
Treatment cost = £30 × 20.2/1000 = £0.61.
Revenue   = £20.2 × 0.346 × 1000 × 0.9/1000 = £6.29.

Therefore, net smelter return for copper concentrate = £5.28/t of ore.

(b) *Zinc*

Assuming that all the zinc is contained in the mineral sphalerite, then maximum zinc grade is 67.1% (pure sphalerite).

If Z is weight of zinc concentrate per 1000 kg of feed, then for 100% recovery of zinc into this concentrate:

$$100 = \frac{67.1 \times Z \times 100}{1000 \times 1.94} \quad \text{and } Z = 28.9 \text{ kg.}$$

Transport cost = £20 × 28.9/1000 = £0.58.
Treatment cost = £100 × 28.9/1000 = £2.89.
Revenue   = £28.9 × 0.671 × 0.85 × 400/1000 = £6.59.

Therefore, net smelter return for zinc concentrate = £3.12/t of ore.

Total net smelter return for perfect milling = £(5.28 + 3.12) = *£8.40/t.*

2. *Actual milling*

Similar calculations give:

Net copper smelter return = £4.46/t of ore.
Net zinc smelter return = £1.71/t of ore.
Total net smelter return = £6.17/t.

Therefore, overall economic efficiency = 100 × 6.17/8.40 = *73.5%.*

References

1. Evans, A. M., *An Introduction to Ore Geology*, Blackwell Scientific Publications, Oxford, 1980.
2. Anon., Metals output and prices—a historical perspective, *Mining Annual Review*, 22 (1985).
3. Dahlstrom, D. A., Impact of changing energy economics on mineral processing, *Mining Engng.*, **38**, 45 (Jan. 1986).
4. McNulty, T. P., Changing energy economics in extractive metallurgy, *Annual Meeting, SME-AIME*, Salt Lake City, Oct. 1983.
5. Taylor, S. R., Abundance of chemical elements in the continental crust, *Geochim. Cosmochim. Acta* **28**, 1280 (1964).
6. Lee Tan Yaa Chi-Lung, Abundance of the chemical elements in the Earth's crust, *Int. Geology Rev.* **12**, 778 (1970).
7. Boin, U., Limits and possibilities of deep sea mining for the extraction of mineral raw materials—the case of manganese nodules, *Min. Mag.* 43 (Jan. 1980).
8. Agarwal, J. C., *et al.*, Kennecott process for recovery of copper, nickel, cobalt, and molybdenum from ocean nodules, *Mining Engng.* **31**, 1704 (Dec. 1979).
9. Anon., *Analysis of Processing Technology for Manganese Nodules*, Graham & Trotman Ltd., London, 1986.
10. Anon., Nchanga Consolidated Copper Mines, *Engng. Min. J.* **180**, 150 (Nov. 1979).
11. Bazzanella, F. L., and Weyler, P. A., Examples of gravity concentration flowsheets, in *Mineral Processing Plant Design* (ed. Mular and Bhappu), AIMME, New York, 1978.
12. Ottley, D. J., Technical, economic and other factors in the gravity concentration of tin, tungsten, columbium, and tantalum ores, *Minerals Sci. Engng.* **11**, 99 (Apr. 1979).
13. Down, C. G., and Stocks, J., Positive uses of mill tailings, *Min. Mag.* 213 (Sept. 1977).
14. Chadwick, J. R., Ergo-gold tailings reclamation and retreatment, *World Mining*, **33**, 48 (June 1980).
15. Barbery, G., Complex sulphide ores—processing options, in *Mineral Processing at a Crossroads—problems and prospects*, eds. B. A. Wills and R. W. Barley, p. 157, Martinus Nijhoff, Dordrecht, 1986.
16. Gray, P. M. J., Metallurgy of the complex sulphide ores, *Mining Mag.*, 315 (Oct. 1984).
17. Petruk, W., Applied mineralogy in ore dressing in, *Mineral Processing Design*, eds. B. Yarar and Z. M. Dogan, p. 2, Martinus Nijhoff, Dordrecht, 1987.
18. Wills, B. A., Developments and research needs in mineral processing, *Proc. 1st Int. Min. Proc. Symp.*, p. 1, Izmir, Turkey, 1986.
19. Gaudin, A. M., *Principles of Mineral Dressing*, McGraw-Hill, London, 1939.
20. King, R. P., The prediction of mineral liberation from mineralogical textures, XIVth International Mineral Process Congress, paper VII-1, CIM, Toronto, Canada (Oct. 1982).
21. Binns, D. G., Parker, R. H., and Wills, B. A., Limitations of optical image analysis of mineralogical materials, IMM Meeting: *Update on Mineral Identification Techniques*, London (Oct. 13, 1983).

22. Cutting, G. W., The characterisation of mineral release during comminution processes. *The Chem. Engr.* 845 (Dec. 1979).
23. Somasundaran, P., An overview of the ultrafine problem, in *Mineral Processing at a Crossroads—problems and prospects*, eds. B. A. Wills and R. W. Barley, p. 1, Martinus Nijhoff Publishers, Dordrecht, 1986.
24. Mills, C., Process design, scale-up and plant design for gravity concentration, in *Mineral Processing Plant Design* (ed. A. L. Mular and R. B. Bhappu), AIMME, New York, 1978.
25. Schulz, N. F., Separation efficiency, *Trans. SME-AIME*, **247,** 81 (March 1970).
26. Anon., Tin—paying the price, *Mining Journal*, 477 (Dec. 27, 1985).
27. Sassos, M. P., Bougainville Copper, *Engng. & Mining J.*, **184,** 56 (Oct. 1983).
28. Anon., Room for improvement at Chuquicamata, *Mining Journal*, 249 (Sept. 27, 1985).

CHAPTER 2

ORE HANDLING

Introduction

Ore handling, which may account for 30–60% of the total delivered price of raw materials, covers the processes of transportation, storage, feeding, and washing of the ore *en route* to, or during, its various stages of treatment in the mill.

Since the physical state of ores *in situ* may range from friable, or even sandy material, to monolithic deposits with the hardness of granite, the methods of mining and provisions for the handling of the freshly excavated material will vary extremely widely. Ore that has been well broken can be transported by trucks, belts, or even by sluicing, but large lumps of hard ore may need individual blasting. Modern developments in microsecond delay fuses and plastic explosive have resulted in more controllable primary breakage and easier demolition of occasional very large lumps. At the same time, crushers have become larger and lumps up to 2 m in size can now be fed into some primary units.

Open-pit ore tends to be very heterogeneous, the largest lumps often being about 1.5 m across. The broken ore from the pit, after blasting, is loaded directly into trucks, holding up to 200 t of ore in some cases, and is transported directly to the primary crushers. Storage of such ore is not always practicable, due to its "long-ranged" particle size which causes segregation during storage, the fines working their way down through the voids between the larger particles; extremely coarse ore is sometimes difficult to start moving once it has been stopped. Sophisticated storage and feed mechanisms are therefore often dispensed with, the trucks depositing their loads directly into the mouth of the primary crusher.

53

The operating cycle on an underground mine is complex. Drilling and blasting are often performed on one shift, the ore broken in this time being hoisted to the surface during the other two shifts of the working day. The ore is transported through the passes via chutes and tramways and is loaded into skips, holding as much as 30 t of ore, to be hoisted to the surface. Large rocks are often crushed underground by primary breakers in order to facilitate loading and handling at this stage. The ore, on arrival at the surface, having undergone some initial crushing, is easier to handle than that from an open pit mine and storage and feeding is usually easier, and indeed essential, due to the intermittent arrival of skips at the surface.

The Removal of Harmful Materials

Ore entering the mill from the mine (run-of-mine ore) normally contains a small proportion of material which is potentially harmful to the mill equipment and processes. For instance, large pieces of iron and steel broken off from mine machinery can jam in the crushers. Wood is a major problem in many mills as this is ground into a fine pulp and causes choking or blocking of screens, etc. It can also choke flotation cell ports, consume flotation reagents by absorption, and decompose to give depressants, which render valuable minerals unfloatable.

Clays and slimes adhering to the ore are also harmful as they hinder screening, filtration, and thickening, and again consume valuable flotation reagents.

All these must be removed as far as possible at an early stage in treatment.

Hand sorting from conveyor belts has declined in importance with the development of mechanised methods of dealing with large tonnages, but it is still used when plentiful cheap labour is available.

Crushers can be protected from large pieces of "tramp" iron and steel by electromagnets suspended over conveyor belts (Fig. 2.1). These powerful electromagnets can pick up large pieces of iron and steel travelling over the belt and, at intervals, can be swung away from the belt and unloaded. Guard magnets, however, cannot be used to

FIG. 2.1. Conveyor guard magnet.

remove tramp iron from magnetic ores, such as those containing magnetite, nor will they remove non-ferrous metals or non-magnetic steels from the ore. Metal detectors which measure the electrical conductivity of the material being conveyed can be fitted over or around conveyor belts (Fig. 2.2). The electrical conductivity of ores is much lower than that of metals and fluctuations in electrical conductivity in the conveyed material can be detected by measuring the change that tramp metal causes in a given electromagnetic field.

FIG. 2.2. Metal detector over conveyor belt.

When a metal object causes an alarm, the belt automatically stops and the object can be removed. It is advantageous with non-magnetic ores to precede the metal detector with a heavy guard magnet which will remove the ferro-magnetic tramp metals and thus minimise belt stoppages.

Large pieces of wood which have been "flattened out" by passage through a primary crusher can be removed by passing the ore feed over a vibrating scalping screen. Here the apertures of the screen are slightly larger than the maximum size of particle in the crusher discharge, allowing the ore to fall through the apertures and the flattened wood particles to ride over the screen and be collected separately.

Wood can be further removed from the pulp discharge from the grinding mills by passing the pulp through a fine screen. Again, the ore particles pass through the apertures and the wood collects on top of the screen and can be periodically removed.

Washing of run-of-mine ore can be carried out to facilitate hand-sorting by removing obscuring dirt from the surfaces of the ore particles. However, washing to remove very fine material, or *slimes*, of little or no value, is more important.

Washing is normally performed after primary crushing as the ore is then of a suitable size to be passed over washing screens. It should always precede secondary crushing as slimes severely interfere with this stage.

The ore is passed through high-pressure jets of water on mechanically vibrated screens. The screen apertures are usually of similar size to the particles in the feed to the grinding mills, the reason for which will become apparent.

In the circuit shown in Fig. 2.3 material passing over the screen, i.e. washed ore, is transported to the secondary crushers. Material passing through the screens is classified into coarse and fine fractions by a mechanical classifier, or hydrocyclone (Chapter 9) or both. It may be beneficial to classify initially in a mechanical classifier as this is more able to smooth out fluctuations in flow than is the hydrocyclone and it is better suited to handling coarse material.

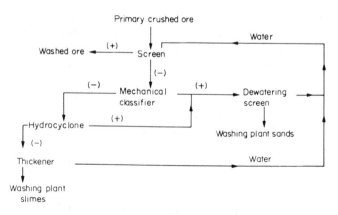

FIG. 2.3. Typical washing plant flowsheet.

The coarse product from the classifier, designated "washing plant sands", is either routed direct to the grinding mills or is dewatered over vibrating screens before being sent to mill storage. A considerable load, therefore, is taken off the dry crushing section.

The fine product from classification, i.e. the "slimes", may be partially dewatered in shallow large diameter settling tanks known as thickeners (Chapter 15) and the thickened pulp either pumped to tailings disposal or, if containing values, direct to the concentration process, thus removing load from the grinding section. In the circuit shown, the thickener overflows are used to feed the high-pressure washing sprays. Water conservation in this manner is practised in most mills.

Wood pulp may again be a problem in the above circuit, as it will tend to float in the thickener, and will choke the water spray nozzles unless it is removed by retention on a fine screen.

Ore Transportation

In a mineral processing plant operating at the rate of 40 000 t d^{-1}, this is equivalent to about 28 t of solid per minute, requiring up to 75 m^3 min^{-1} of water. It is therefore important to operate with the minimum upward or horizontal movement and with the maximum practicable pulp density in all of those stages subsequent to the addition of water to the system. The basic philosophy requires maximum use of gravity and continuous movement over the shortest possible distances between processing units.[1]

Dry ore can be moved through chutes, provided they are of sufficient slope to allow easy sliding, and sharp turns are avoided. Clean solids slide easily on a 15–25° steel-faced slope, but for most ores, a 45–55° working slope is used. The ore may be difficult to control if the slope is too steep.

The belt conveyor is the most widely used method of handling loose bulk materials.[2] Belts are now in use with capacities up to 20 000 t h^{-1} and single flight lengths exceeding 5000 m, with feasible speeds of up to 10 m s^{-1}.[3]

The standard rubber conveyor belt has a foundation of sufficient strength to withstand the driving tension and loading strains. This foundation, which may be of cotton, nylon, or steel cord, is bound together with a rubber matrix and completely covered with a layer of vulcanised rubber.

The carrying capacity of the belt is increased by passing it over troughing idlers. These are support rollers set normal to the travel of the belt and inclined upward from the centre so as to raise the edges and give it a troughlike profile. There may be three or five in a set and they will be rubber-coated under a loading point, so as to reduce the wear and damage from impact. Spacing along the belt is at the maximum interval which avoids excessive sag. The return belt is supported by horizontal straight idlers which overlap the belt by a few inches at each side.

To induce motion without slipping requires good contact between the belt-and-drive pulley. This may not be possible with a single 180° turn over a pulley and some form of "snubbed pulley" drive or "tandem" drive arrangement may be more effective (Fig. 2.4).

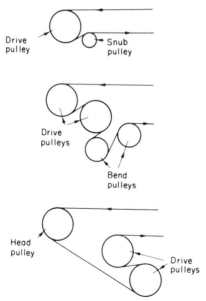

FIG. 2.4. Conveyor-belt drive arrangements.

The belt system must incorporate some form of tensioning device to adjust the belt for stretch and shrinkage and thus prevent undue sag between idlers and slip at the drive pulley. In most mills, gravity-operated arrangements are used which adjust the tension continuously (Fig. 2.5). Hydraulics have also been used extensively, and when more refined belt-tension control is required, especially in starting and stopping long conveyors, load-cell controlled electrical tensioning devices are used.

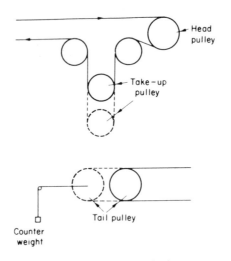

FIG. 2.5. Conveyor-belt tensioning systems.

The reliability of belt systems has been enhanced by advances in control technology, making possible a high degree of fail-safe automation. A series of belts should incorporate an interlock system, such that failure of any particular belt will automatically stop preceding belts. Interlock with devices being fed by the belt is important for the same reasons. It should not be possible to shut down any machine in the system without arresting the feed to the machine at the same time and, similarly, motor failure should lead to the automatic tripping of all preceding belts and machines. Sophisticated electrical, pneumatic and hydraulic circuits have been widely employed to replace all but a few manual operations.

Several methods can be used to minimise loading shock on the belt. A typical arrangement is shown in Fig. 2.6 where the fines are screened on to the belt first and provide a cushion for the larger pieces of rock.

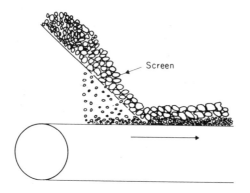

Screen

FIG. 2.6. Belt-loading system.

Feed chutes must be designed to deliver the bulk of the material to the centre of the belt and at a velocity close to that of the belt. Ideally it should be the same, but in practice this condition is very seldom obtained, particularly with wet sand or sticky materials. Where conditions will allow, the angle of the chute should be as great as possible, thereby allowing it to be gradually placed at lesser angles to the belt until the correct speed of flow is attained. The material, particularly if it is heavy, or lumpy, should never be allowed to strike the belt vertically. Baffles in transfer chutes, to guide material flow, are now often remotely controlled by hydraulic cylinders.

The conveyor may discharge at the head pulley, or the load may be removed before the head pulley is reached. The most satisfactory device for achieving this is a tripper. This is an arrangement of pulleys by which the belt is raised and doubled back so as to give a localised discharge point. It is usually mounted on wheels, running on tracks, so that the load can be delivered at several points, as over a long bin or into several bins. The discharge chute on the tripper can deliver to one or both sides of the belt. The tripper may be moved by hand, by head and tail ropes from a reversible hoisting drum, or by a motor. It may be automatic, moving backwards and forwards under power from the belt drive.

Shuttle belts are reversible self-contained conveyor units mounted on carriages, which permit them to be moved lengthwise to discharge to either side of the feed point. The range of distribution is approximately twice the length of the conveyor. They are often preferred to trippers for permanent storage systems because they require less head room and, being without reverse bends, are much easier on the belts.

Where space limitation does not permit the installation of a belt conveyor, gravity bucket elevators can be used (Fig. 2.7). These provide only low handling rates with both horizontal conveying and elevating of the material. The elevator consists of a continuous line of buckets attached by pins to two endless roller chains running on tracks and driven by sprockets. The buckets are pivoted so that they always remain in an upright position and are dumped by means of a ramp placed to engage a shoe on the bucket, thus turning it into the dumping position.

Hydraulic transport of the ore stream normally takes over from dry transportation at the grinding stage in most modern mills. Pulp may be made to flow through open launders by gravity in some cases. Launders are gently sloping troughs of rectangular, triangular, or semicircular section, in which the solid is carried in suspension, or by sliding or rolling. The slope must increase with particle size, with the solid content of the suspension, and with specific gravity of the solid. The effect of depth of water is complex; if the particles are carried in suspension, a deep launder is advantageous because the rate of solid transport is increased. If the particles are carried by rolling, a deep flow may be disadvantageous.

In plants of any size, the pulp is moved through piping via centrifugal pumps. Pipelines should be as straight as possible to prevent abrasion at bends. The use of oversize pipe is dangerous whenever slow motion might allow the solids to settle and hence choke the pipe. The factors involved in pipeline design and installation are complex and include the solid–liquid ratio, the average pulp density, the density of the solid constituents, the size analysis and particle shape, and the fluid viscosity.[4]

FIG. 2.7. Gravity bucket elevator.

Centrifugal pumps are cheap in capital cost and maintenance and occupy little space.[5,6] Single-stage pumps are normally used, lifting up to 30 m, and 100 m in extreme cases. Their main disadvantage is the high velocity produced within the impeller chamber, which may result in serious wear of the impeller and chamber itself, especially when a coarse sand is being pumped.

Ore Storage

The necessity for storage arises from the fact that different parts of the operation of mining and milling are performed at different rates, some being intermittent and some continuous, some being subject to frequent interruption for repair, and others being essentially batch processes. Thus, unless reservoirs for material are provided between the succeeding different steps, the whole operation is rendered spasmodic and, consequently, uneconomical.

The amount of storage necessary depends on the equipment of the plant as a whole, its method of operation, and the frequency and duration of regular and unexpected shut-downs of individual units.

For various reasons, at most mines, ore is hoisted for only a part of each day. On the other hand, grinding and concentration circuits are most efficient when running continuously. Mine operations are more subject to unexpected interruption than mill operations, and coarse-crushing machines are more subject to clogging and breakage than fine crushers, grinding mills, and concentration equipment. Consequently, both the mine and the coarse-ore plant should have a greater hourly capacity than the fine crushing and grinding plants, and storage reservoirs should be provided between them. Ordinary mine shut-downs, expected or unexpected, will not generally exceed a 24-h duration, and ordinary coarse-crushing plant repairs can be made within an equal period if a good supply of spare parts is kept on hand. Therefore, if a 24-h supply of ore that has passed the coarse crushing plant is kept in reserve ahead of the mill proper, the mill can be kept running independent of shut-downs of less than a 24-h duration in mine and coarse-crushing plant. It is wise to provide for a similar mill shut-down and, in order to do this, the reservoir between coarse-crushing plant and mill must contain at all times unfilled space capable of holding a day's tonnage from the mine. This is not economically possible, however, with many of the modern very large mills; there is a trend now to design such mills with smaller storage reservoirs, often supplying less than a two-shift supply of ore, the philosophy being that storage does not *do* anything to the ore, and can, in some cases, have an adverse effect by allowing the ore to oxidise. Unstable sulphides must be treated with minimum delay, and wet ore cannot be left exposed to extreme cold as it will freeze and be difficult to move.

Storage has the advantage of allowing blending of different ores so as to provide a consistent feed to the mill. Both tripper and shuttle conveyors can be used to blend the material into the storage reservoir. If the units shuttle back and forth along the pile, the materials are layered and mix when reclaimed. If the units form separate piles for each quality of ore, a blend can be achieved by combining the flow from selected feeders onto a reclaim conveyor.

Depending on the nature of the material treated, storage is accomplished in stock piles, bins, or tanks.

Stock piles are often used to store coarse ore of low value outdoors. In designing stock piles, it is merely necessary to know the angle of repose of the ore, the volume occupied by the broken ore, and the tonnage.

Although material can be reclaimed from stock piles by front-end loaders or by bucket-wheel reclaimers, the most economical method is by the reclaim tunnel system, since it requires a minimum of manpower to operate.[7] It is especially suited for blending by feeding from any combination of openings. Conical stock piles can be reclaimed by a tunnel running through the centre, with one, or more, feed openings discharging via gates, or feeders, onto the reclaim belt. The amount of reclaimable material, or the *live storage*, is about 20–25% of the total (Fig. 2.8). Elongated stock piles are reclaimed in a similar manner, the live storage being 30–35% of the total (Fig. 2.9).

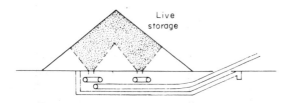

FIG. 2.8. Reclamation from conical stock pile.

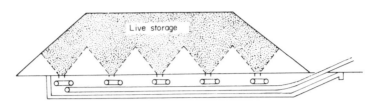

FIG. 2.9. Reclamation from elongated stock pile.

For continuous feeding of crushed ore to the grinding section, feed bins are used for transfer of the coarse material from belts and rail and road trucks. They are made of wood, concrete, or steel. They must be easy to fill and must allow steady fall of the ore through to the discharge gates with no "hanging up" of material or opportunity for it to segregate into coarse and fine fractions. The discharge must be adequate and drawn from several alternative points if the bin is large. Flat-bottomed bins cannot be emptied completely and must retain a substantial tonnage of dead rock. This, however, provides a cushion to protect the bottom from wear, and such bins are easy to construct. This type of bin, however, should not be used with easily oxidised ore which might age dangerously and mix with the fresh ore supply. Bins with sloping bottoms are better in such cases.

Pulp storage on a large scale is not as easy as dry ore storage. Conditioning tanks are used for storing suspensions of fine particles to provide time for chemical reactions to proceed. These tanks must be agitated continuously, not only to provide mixing, but also to prevent settlement and choking up. Surge tanks are placed in the pulp flow-line when it is necessary to smooth out small operating variations of feed rate. Their content can be agitated by stirring, by blowing in air, or by circulation through a pump.

Feeding

Feeders are necessary whenever it is desired to deliver a uniform stream of dry or moist ore, since such ore will not flow evenly from a storage reservoir of any kind through a gate, except when regulated by some type of mechanism.

Feeding is essentially a conveying operation in which the distance travelled is short and in which close regulation of the rate of passage is required. Where succeeding operations are at the same rate, it is unnecessary to interpose feeders. Where, however, principal operations are interrupted by a storage step, it is necessary to provide a feeder.

A typical feeder consists of a small bin, which may be an integral part of a large bin, with a gate and a suitable conveyor. Feeders of many types have been designed, notably apron, belt, chain, roller, rotary, revolving disc, and vibrating feeders.

In the primary crushing stage, the ore is normally crushed as soon as possible after its arrival at the surface. Skips, lorries, trucks, and other handling vehicles are intermittent in arrival whereas the crushing section, once started, calls for steady feed. Surge bins provide a convenient holding arrangement able to receive all the intermittent loads and to feed them steadily through gates at controllable rates. The chain-feeder (Fig. 2.10) is sometimes used for smooth control of bin discharge.

This consists of a curtain of heavy loops of chain, lying on the ore at the outfall of the bin at approximately the angle of repose. The rate of feed is controlled automatically, or manually, by the chain sprocket drive such that when the loops of chain move, the ore on which they rest begins to slide.

Primary crushers depend for normal operation on the fact that broken rock contains a certain amount of voidage. If all the feed goes to a jaw crusher without a preliminary removal of fines, there can be danger when there has been segregation of coarse and fine material in the bin. Such fines could pass through the upper zones of the crusher and drop into the final sizing zone so as to fill the voids. Should the bulk arriving at any level exceed that departing, it is as though an attempt is being made to compress solid rock. This so-called "packing of the crushing chamber" is just as serious as tramp iron in the crusher and can cause major damage. It is common practice, therefore, to "scalp" the feed to the crusher, heavy-duty screens known as *grizzlies* normally preceding the crushers and removing fines and undersize.

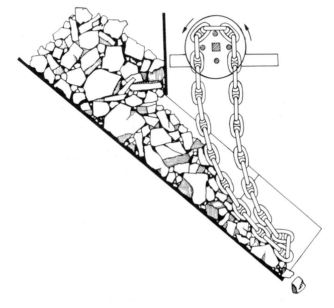

FIG. 2.10. Chain-feeder.

Primary crusher feeders, which scalp and feed in one operation, have been developed, such as the vibrating grizzly feeder. The elliptical bar feeder (Fig. 2.11) consists of elliptical bars of steel which form the bottom of a receiving hopper and are set with the long axes of the ellipses in alternate vertical and horizontal positions. Material is dumped directly onto the bars which rotate in the same direction, all at the same time, so that the spacing remains constant. As one turns down, the succeeding one turns up, imparting a rocking, tumbling motion to the load. This works loose the fines, which sift through the load directly on to a conveyor belt, while the oversize is moved forward to deliver to the crusher. This type of feeder is probably better suited to handling high clay or wet materials such as laterite, rather than hard, abrasive ores.

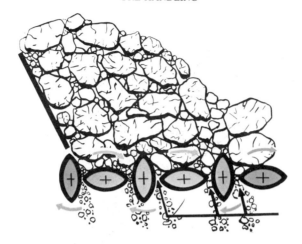

Fig. 2.11. Cross-section of elliptical bar feeder.

The apron feeder (Fig. 2.12) is one of the most widely used feeders for handling coarse ore, especially jaw crusher feed. It is ruggedly constructed consisting of a series of high carbon or manganese steel pans, bolted to strands of heavy-duty chain which run on steel sprockets. The rate of discharge is controlled by varying the speed or by varying the height of the ribbon of ore by means of an adjustable gate.

Apron feeders are often preferred to reciprocating plate feeders which push forward the ore lying at the bottom of the bin with strokes controllable for rate and amplitude, as they require less driving power and provide a steadier, more uniform feed.

Belt feeders are essentially short-belt conveyors, used to control the discharge of material from inclined chutes. They frequently replace apron feeders for fine ore and are increasingly being used to handle coarse, primary crushed ore. Such feeders have been installed to handle crushed ore at Similkameen in British Columbia, and Palabora in South Africa. These feeders are 2.45 m wide, require less installation height, cost substantially less, and can be operated at substantially higher speeds than apron feeders.[8]

FIG. 2.12. Apron feeder

References

1. Gould, W. D., Basic design problems for a large concentrator, *AIME Annual Meeting New York City* (Feb. 1975).
2. Cabrera, V., Conveyor belts make sense for long distance haulage, *World Mining* **35**, C3 (July 1982).
3. White, L., Processing responding to new demands, *Engng Min. J.* 219 (June 1976).
4. Loretto, J. C., and Laker, E. T., Process piping and slurry transportation, in *Mineral Processing Plant Design* (ed. Mular and Bhappu), AIMME, New York, 1978.
5. Wilson, G., Selecting centrifugal slurry pumps to resist abrasive wear, in *Mining Engng* **33**, 1323 (Sept. 1981).
6. Pearse, G., Pumps for the minerals industry, *Min. Mag.* 299 (April 1985).
7. Dietiker, F. D., Belt conveyor selection and stockpiling and reclaiming applications, in *Mineral Processing Plant Design* (ed. Mular and Bhappu), AIMME, New York, 1978.
8. Shoemaker, R. S., Aspects of mineral processing plant design, *J. Camborne Sch. Mines* **78**, 29 (1978).

CHAPTER 3

METALLURGICAL ACCOUNTING AND CONTROL

Introduction

Metallurgical accounting is an essential feature of all efficient metallurgical operations. Not only is it used to determine the distribution of the various products of a mill, and the values contained in them, but it is also used to control the operations, since values of recovery and grade obtained from the accounting procedure are indications of process efficiency.

The essential requirements of a good accounting and control system are efficient and representative sampling of the process streams, upon which accurate analyses of the valuable components can be undertaken, and reliable and accurate measurement of the mass flowrate of important flowstreams.

Computer control of mineral processing plants requires continuous measurement of such parameters, and the development of real-time on-line sensors, such as magnetic flowmeters, nuclear density gauges and chemical and particle size analysers has made important contributions to the rapid developments in this field since the early 1970s, as has the increasing availability and reliability of cheap microprocessors.

Sampling and Weighing the Ore

Ideally, weighing and sampling should be carried out before the material is subject to losses in the mill. For this to be absolutely the

71

case, these operations must be carried out on run-of-mine ore entering the primary crusher stage. Weighing can be carried out satisfactorily, but accurate sampling is not possible on account of the wide range of particle size and heterogeneity of the material being handled. This difficulty applies particularly to the preparation of a moisture sample, an essential requirement, since all calculations are carried out on the basis of dry weights of material. Run-of-mine, and coarsely crushed ore, tends to segregate, and it is very likely that the fines are of a different grade and moisture content from the coarse material. It will probably be necessary to take at least 5% of the total weight of ore as a primary sample if the required degree of accuracy is to be obtained. This must be reduced in size by successive stages, the bulk being reduced by a sample division or "cut" between each stage.

For this reason, and also because of the high capital cost of a sample plant designed to operate on coarse feed, accurate sampling and weighing of coarse rock is confined usually to those cases where two ores must be accounted for separately, and it is not possible to operate parallel crushing sections. Weighing and sampling are, therefore, wherever possible undertaken when the ore is in its most finely divided state.

Moisture Sampling

While all metallurgical accounting requires accurate knowledge of the dry weights of solids handled, actual materials may contain moisture to varying degrees and samples must be taken for this constituent to be measured accurately. Ideally, the moisture sample and the assay sample should be prepared from the same quantity of material, both being taken from a point near to the weighing equipment. With proper handling the errors due to subsequent wetting or drying can be reduced to very low levels. In practice, it is common to find that some form of *grab sampling* is used for moisture determination. This is the least accurate of the common sampling methods, but the cheapest and most rapid. By this method, small quantities of material are chosen at random from different spots in the large bulk, and these are mixed together to form the base for the final sample. Grab sampling ensures that the sample can be quickly collected and placed in sealed containers, the assumption being that error due to crudeness of method is less than the error introduced by longer exposure of material during more elaborate sampling.

Grab samples for moisture determination are frequently taken from the end of a conveyor belt after material has passed over the weighing device. The samples are immediately weighed wet, dried at a suitable temperature until all hygroscopic water is driven off, and then weighed again. The difference in weight represents moisture and is expressed as:

$$\% \text{ moisture} = \frac{\text{wet weight} - \text{dry weight}}{\text{wet weight}} \times 100. \qquad (3.1)$$

The drying temperature should not be so high that breakdown of the minerals, either physically or chemically, occurs. Sulphide minerals are particularly prone to lose sulphur dioxide if overheated; samples should not be dried at temperatures above 105°C.

Assay Sampling

Sampling is the means whereby a small amount of material is taken from the main bulk in such a manner that it is representative of that larger amount. Great responsibility rests on a very small sample, so it is essential that samples are truly representative of the bulk.[1]

Wherever possible samples should be taken of the material when it has been reduced to the smallest particle size consistent with the process. For instance, the ground ore pulp will be easier to sample, and will give more accurate results than the feed to the primary crusher.

In practice, the most satisfactorily method of minimising variables in the feed stream, such as particle size variation in belt loading, settling out of particles in the pulp due to velocity change, surges, etc., is to sample the material while it is in motion at a point of free fall discharge, making a cut at right angles to the stream. Since there may be segregation or changed composition within the stream, good practice demands a sample of all the stream. When a sample cutter moves continuously across the stream at a uniform speed, the sample taken represents a small portion of the entire stream. If the cutter moves through the stream at regular intervals it produces incremental samples that are considered representative of the stream at the time the sample was taken.

Sampling is dependent on probability, and the more frequently the incremental sample is taken the more accurate the final sample will be. The sampling method devised by Gy[2] is often used to calculate the size of sample necessary to give the required degree of accuracy. The method takes into account the particle size of the material, the content and degree of liberation of the minerals, and the particle shape.

Gy's basic sampling equation can be written as:

$$\frac{ML}{L - M} = \frac{Cd^3}{s^2}$$ (3.2)

where M is the minimum weight of sample required (g), L is the gross weight of material to be sampled (g), C is the sampling constant for the material to be sampled (g cm^{-3}), d is the dimension of the largest pieces in the material to be sampled (cm), and s is the measure of the statistical error committed by sampling. In most cases, M is small in relation to L, and equation 3.2 approximates to:

$$M = \frac{Cd^3}{s^2}.$$ (3.3)

The term s is used to obtain a measure of confidence in the results of the sampling procedure. The *relative* standard deviation of a normal distribution curve representing the random assay-frequency data for a large number of samples taken from the ore is s and the relative variance is s^2.

Assuming normal distribution, 67 out of 100 assays of samples would lie within ± s of the true assay; 95 out of 100 assays would be within ± $2s$ of the true assay, and 99 out of 100 would be within ± $3s$ of the true assay. As sampling is a statistical problem, there can never be complete confidence in the result of a sampling exercise, and for most practical purposes, a 95-times-in-a-100 chance of being within prescribed limits is an acceptable probability level. Table 3.1 shows the results of a computer simulation of a random unbiased sampling exercise on an ore containing exactly 50% valuable material. It is apparent that it can never be guaranteed that the assay result will lie within the prescribed limits, but that the more sample is taken, the greater is the confidence. The effect of undersampling is, however, clearly illustrated.

TABLE 3.1. RESULTS OF SAMPLING A HYPOTHETICAL ORE CONTAINING 50% VALUE. The ore was sampled 100 times at each sample weight, and the value content of each sample assessed. The number of "assays" within 5% of the true value content is shown, as is the maximum assay error encountered, and the mean assay of the 100 samples taken

Sample weight (g)	Mean assay (%)	Number of assays within 5%	Maximum error (%)
10	46.70	14	88.55
100	49.70	24	45.60
500	50.35	37	18.38
1000	50.08	74	14.80
2500	50.18	86	9.94
3500	49.82	93	7.09
5000	50.12	98	5.10
10000	49.97	99	5.01

The actual variance determined by Gy's equation may differ from that obtained in practice because it is usually necessary to carry out a number of sampling steps in order to obtain the assay sample, and there are also errors in assaying. The practical variance (or total variance) would therefore be the sum of all other variances, i.e.

$$S_t^2 = s^2 + S_s^2 + S_a^2.$$

The values of S_s (sampling) and S_a (assay) would normally be small, but could be determined by assaying a large number of portions of the same sample (at least 50) to give S_a^2 and by cutting a similar number of samples in an identical manner and assaying each one to give $(S_a^2 + S_s^2)$. However, for routine plant sampling, s^2 can be assumed to equal S_t^2, and wherever possible, or practical, two to three times the minimum weight of sample should be taken to allow for the many unknowns, although, of course, over-sampling has to be avoided to preclude problems in handling and preparation.

The sampling constant C is specific to the material being sampled, taking into account the mineral content, and its degree of liberation,

$$C = fglm,$$

where f is a shape factor, which is taken as 0.5, except for gold ores, where it is 0.2; g is a factor which is dependent on the particle size range. If approximately 95% of the sample weight contains particles of size less than d cm, and 95% of size greater than d' cm, then if:

$$d/d' > 4, \qquad g = 0.25$$
$$d/d' \text{ is } 2\text{--}4, \qquad g = 0.5$$
$$d/d' < 2, \qquad g = 0.75$$
$$d/d' = 1, \qquad g = 1$$

l is a liberation factor, which has values between 0 for completely homogeneous material and 1.0 for completely heterogeneous material. Gy devised a table (shown below) based on d, the dimension of the largest pieces in the ore to be sampled, which can be taken as the screen aperture which passes 90–95% of the material, and L, the size in centimetres at which, for practical purposes, the mineral is essentially liberated. This can be estimated microscopically. Values of l can be estimated from the table, from corresponding values of d/L, or can be calculated from the expression:

$$l = (L/d)^{1/2}$$

d/L	< 1	1–4	4–10	10–40	40–100	100–400	> 400
l	1	0.8	0.4	0.2	0.1	0.05	0.02

m is a mineralogical composition factor which can be calculated from the expression

$$m = \frac{1 - a}{a}[(1 - a)r + at],$$

where r and t are the mean densities of the valuable mineral and gangue minerals respectively, and a is the fractional average mineral content of the material being sampled. This value could be determined by assaying a number of samples of the material.

Gy's equation assumes that samples are taken at random, and without bias, and is most applicable to streams of ore transported on conveyors or in pulp streams rather than heap deposits which are partly inaccessible to the sampler.

The equation gives the minimum theoretical weight of sample which must be taken, but does not state how the sample is to be taken. The size of each increment taken, in the case of stream sampling, and the increment between successive cuts must be such that sufficient weight is recovered to be representative.

Gy's equation can be used to illustrate the benefits of sampling material when it is in its most finely divided state.

Consider, for instance, a lead ore, assaying about 5% Pb, which must be routinely sampled for assay to a confidence level of $\pm$ 0.1% Pb, 95 times out of 100. The galena is essentially liberated from the quartz gangue at a particle size of 150 μm.

If sampling is undertaken during crushing, when the top size of the ore is 25 mm, then

$$d = 2.5 \text{ cm,}$$

$$2s = \frac{0.1}{5} = 0.02.$$

Therefore $s = 0.01$.

$$l = (0.015/2.5)^{1/2} = 0.077.$$

Assuming the galena is stoichiometrically PbS, then the ore is composed of 5.8% PbS.

Therefore $\quad a = 0.058, \quad r = 7.5, \quad t = 2.65.$

Therefore $m = 117.8$ g cm^{-3}

$$C = fglm = 0.5 \times 0.25 \times 0.077 \times 117.8$$
$$= 1.13 \text{ g cm}^{-3}$$
$$M = Cd^3/s^2 = 176.6 \text{ kg.}$$

In practice, therefore, about 350 kg of ore would have to be sampled in order to give the required degree of confidence, and to allow for assay and mechanical sampling errors. No account has been taken here of further sample divisions required prior to assay.

If, however, the sampling takes place from the pulp stream after grinding to the liberation size of the ore, then $d = 0.015$ cm, and assuming that classification has given fairly close sizing,

$$C = 0.5 \times 0.5 \times 1 \times 117.8$$
$$= 29.46 \text{ g cm}^{-3}.$$

Therefore $M = 1$ g.

Such a small weight of sample could, however, not be cut for assay from a pulp stream, as it makes no provision for segregation within the stream, variations in assay, and particle size, etc., with time. It may, however, be used as a guide for the increment to be cut at each passage of the cutter, the interval between cuts being decided from the fluctuations in the quality of the pulp stream.

Sampling Systems

Most automatic samplers operate by moving a collecting device through the material as it falls from a conveyor or a pipe. It is important that:

(1) The face of the collecting device or cutter is presented at right angles to the stream.
(2) The cutter covers the whole stream.
(3) The cutter moves at constant speed.
(4) The cutter is large enough to pass the sample.

The width of the cutter w will be chosen to give an acceptable weight of sample, but must not be made so small that the biggest particles have difficulty in entering. Particles that strike the edges of the receiver are likely to bounce out and not be collected, so that the effective width is $(w - d)$, where d is the diameter of the particles. The effective width is therefore greater for small particles than for large ones. To reduce this error to a reasonable level, the ratio of cutter width to the diameter of the largest particle should be made as large as possible, with a minimum value of 20:1.

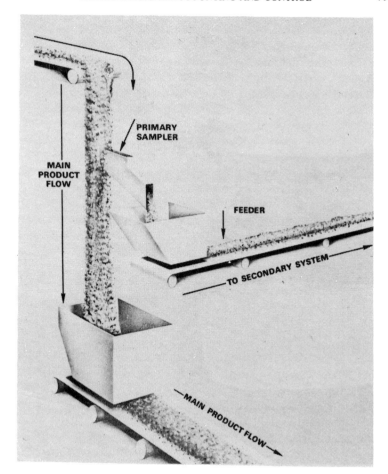

FIG. 3.1. Typical sampling system.

All sampling systems require a primary sampling device or cutter, and a system to convey the collected material to a convenient location for crushing and further sample division (Fig. 3.1).

There are many different types of sample cutter; the Vezin type sampler (Fig. 3.2) is widely used to sample a falling ore stream.

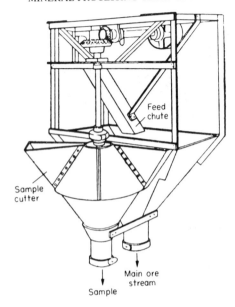

FIG. 3.2. Vezin sampler.

This consists of a revolving cutter in the shape of a circular sector of such dimensions as to cut the whole stream of ore, and divert the sample into a separate sample chute.

Figure 3.3 shows four types of Outokumpu slurry sampler which are commonly used to deliver samples to on-line analysis systems.

Automatic samplers known as *poppet valves* are used in some plants.[3] They consist essentially of a pneumatically operated piston immersed directly into the pipeline, usually a rising main, carrying the pulp stream, the piston in the "open" position allowing the transfer of a sample from the pulp stream, and in the "closed" position preventing the passage of pulp to the sample line. The cycle of opening and closing is controlled by an automatic timer, sample level controller, or other means, depending on the circumstances, the volume of sample taken at each cut being determined by the time elapsed with the valve raised from its seat.

The bulk sample requires thorough drying and mixing before further

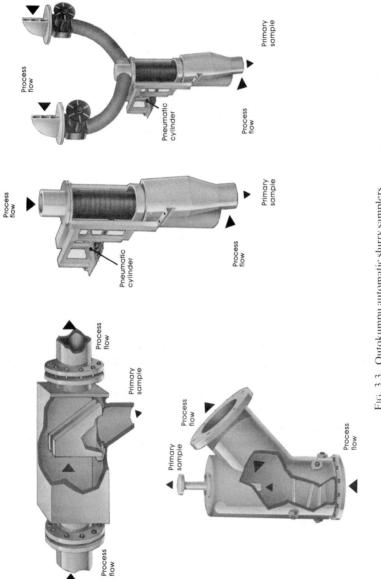

FIG. 3.3. Outokumpu automatic slurry samplers.

division to produce a reasonable size for assay. The principles involved in reducing the material down to the assay sample are the same as those discussed when considering the collection of the gross sample. To obtain the best results, the bulk sample should be made as homogeneous as possible. If complete homogeneity of the material is achieved, then every increment obtained by the sampling method will be representative of the materials. Ores and concentrates containing coarse particles are less homogeneous than those containing fine particles, and it is always necessary to take a larger sample of coarse material in order for it to be representative. Wherever possible a sampling step is preceded by a reduction in particle size, the number of steps being dependent on the size of the original sample and the equipment available for crushing. The weight of sample required at each stage in sample division can be determined using Gy's formula. For example, consider the sampling of the lead ore discussed earlier, from the crushing circuit, at a top size of 25 mm.

Each incremental sample taken, representative of the ore at the time of sampling, is conveyed to the secondary sample system for further division and sampling, either automatically or manually.

Assuming that the sample is crushed in three stages, to 5 mm, 1 mm, and, finally, for assay, to 40 μm, and is sampled after each stage, then there are in total four sampling stages, and the square of the total error produced in sampling is the sum of the squares of the errors incurred at each stage, i.e.

$$S_t^2 = S_1^2 + S_2^2 + S_3^2 + S_4^2.$$

If an equal error is assumed at each stage, then

$$S_t^2 = 4S_1^2.$$

Therefore

$$S_1^2 = S_2^2 = S_3^2 = S_4^2 = \frac{S_t^2}{4}.$$

Since a confidence level in assaying of 5% $\pm$ 0.1% Pb 95 times of 100 is to be achieved,

$$S_t = 0.01.$$

Therefore

$$S_1^2 = S_2^2 = S_3^2 = S_4^2 = 0.25 \times 10^{-4}.$$

For the primary sampling stage at 25 mm top size,

$$M = \frac{1.13 \times (2.5)^3}{0.25 \times 10^{-4}} = 706.3 \text{ kg}.$$

For the second sampling stage at 5 mm,

$$M = 12.8 \text{ kg}.$$

For the third sampling stage at 1 mm,

$$M = 228.2 \text{ g}.$$

For the fourth stage, at 40 μm,

$$M = 0.04 \text{ g}.$$

The sampling system could be designed from this information. For instance, the following weights might be taken to allow for assay and other errors: 1.5 tonne of ore per shift is taken from the crushed ore stream, is crushed to 5 mm, and a 25 kg sample taken. This sample is further crushed to 1 mm and a 500-g sample taken, which is finely ground to 40 microns, from which a sample of about 0.5 g is cut for assay.

The sample weights calculated above assume equal statistical errors at each sampling stage. The primary sample weight required can, however, be reduced by taking more than the calculated requirement of the finer sizes. For instance, in the above example, 0.5 g of sample is taken for final assay, which is well above the sample requirement calculated from Gy's formula. This means that the statistical error at this stage is relatively low, allowing larger errors (smaller sample weights) in the early sample stages, since $S_t^2 = S_1^2 + S_2^2 + S_3^2 + S_4^2$. A computer program (GY) which allows manipulation of any number of sample stage weights, and calculates the remaining sample stage requirements to accommodate the acceptable statistical error, is listed in Appendix 3. The table below (Table 3.2) shows the effect of increasing the weight of the finer sized samples on the amount of primary sample required (safety factor of 2 applied).

TABLE 3.2

	Stage 1	Stage 2	Stage 3	Stage 4
Equal sampling error at each stage	1412.6 kg	25.6 kg	456.4 g	0.08 g
Fixed weight at stages 2, 3 and 4	570.1 kg	50.0 kg	500.0 g	1.0 g

Sample Division Methods

Some of the common methods of sample division are:

Coning and quartering. This is an old Cornish method which is often used in dividing samples of material. It consists of pouring the material into a conical heap and relying on its radial symmetry to give four identical samples when the heap is flattened and divided by a cross-shaped metal cutter. Two opposite corners are taken as the sample, the other two corners being discarded. The portion chosen as the sample may again be coned and quartered, and the process continued until a sample of the required size is produced. Although accuracy is increased by crushing the sample between each division, the method is very dependent on the skill of the operator and should not be used for accurate sampling.

Table sampling. In a sampling table (Fig. 3.4) the material is fed to the top of an inclined plane in which there are a series of holes. Prisms placed in the path of the stream break it into fractions, some material falling through the hole to be discarded, while the material remaining on the plane passes onto the next row of prisms and holes, and more is recovered, and so on. The material reaching the bottom of the plane is the sample. Table samplers are often used to divide samples of 5 kg or over.

The Jones riffle. This splitter (Fig 3.5) is an open V-shaped box in which a series of chutes is mounted at right angles to the long axis to

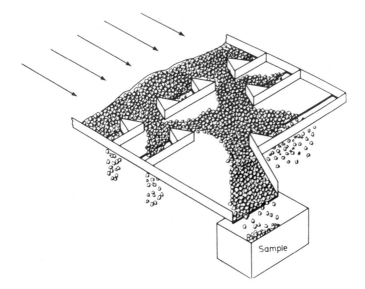

FIG. 3.4. Table sampler.

give a series of rectangular slots of equal area alternately feeding two trays placed on either side of the trough. The laboratory sample is poured into the chute and split into equal portions by the slots, until after repeated cycles a sample of the desired size is obtained.

The ore or concentrate sample must now be analysed, or assayed, so that the exact chemical composition of the material is obtained. Assays are of great importance, as they are used to control operations, calculate throughput and reserves, and to calculate profitability. Modern methods of assaying are very sophisticated and accurate, and are beyond the scope of this book. They include chemical methods, X-ray fluorescence, and atomic absorption spectrometry.[4]

On-line Analysis

The benefits of continuous analysis of process streams in mineral processing plants led to the development in the early 1960s of devices

FIG. 3.5. Jones riffle sampler.

for X-ray fluorescence (XRF) analysis of flowing slurry streams. On-line analysis enables a change of quality to be detected and corrected rapidly and continuously, obviating the delays involved in off-line laboratory testing. This method also frees skilled staff for more productive work than testing of routine samples. The whole field of on-line chemical analysis applied to concentrator automation has been comprehensively reviewed elsewhere. [5, 6, 7]

The principle of on-line chemical analysis is shown in Fig. 3.6. Basically it consists of a source of radiation which is absorbed by the sample and causes it to give off fluorescent response radiation characteristic of each element. This enters a detector which generates a quantitative output signal as a result of measuring the characteristic radiation of one element from the sample. The detector output signal is generally used to obtain an assay value which can be used for process control.

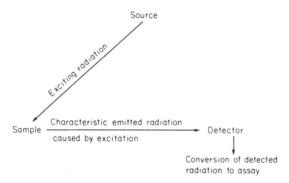

FIG. 3.6. Principle of on-stream X-ray analysis.

The two practical methods of on-line X-ray fluorescence analysis are centralised X-ray (on-stream) and in-stream probe systems. Centralised, on-stream analysis employs a single high-energy excitation source for analysis of several slurry samples delivered to a central location where the complete equipment is installed. In-stream analysis employs sensors installed in, or near, the slurry stream, and sample excitation is carried out with convenient low-energy sources, usually radioactive isotopes.[8, 9] This overcomes the problem of transporting representative samples of slurry to the analyser. The excitation sources are packaged with a detector in a compact device called a probe.

Centralised analysis is usually installed in large plants, requiring continuous monitoring of many different pulp streams, whereas smaller plants with fewer streams may incorporate probes on the basis of lower capital investment.

One of the major problems in on-stream X-ray analysis is ensuring that the samples presented to the radiation are representative of the bulk, and that the response radiation is obtained from a representative fraction of this sample. The exciting radiation interacts with the slurry by first passing through a thin plastic film window and then penetrating the sample which is in contact with this window. Response radiation takes the opposite path back to the detector. Most of the radiation is absorbed in a few millimetres depth of the sample, so that the layer of slurry in immediate contact with the window has the greatest influence on the assays produced. Accuracy and reliability depend on this very thin layer being representative of the bulk material. Segregation in the slurry can take place at the window due to flow patterns resulting from the presence of the window surface. This can be eliminated by the use of high turbulence flow cells in the centralised X-ray analyser. Operating costs and design complexities of sampling and pumping can be largely avoided with the probe-measuring devices positioned near the bulk stream to be assayed, noting the requirement for turbulent flow of representative slurry sample at the measurement interface of the probe.

There are many different types of sampling systems available for on-stream analysis. A typical one has been developed by Outokumpu for use with the Courier 300 analysis system,[10,11] the oldest and perhaps most widely known process analyser. In this system a continuous sample flow is taken from each process slurry to be analysed. The final sample flow is obtained by abstractions from two or three parts depending on the volume of the process flow. Each slurry flows through a separate cell in the analyser, where fluorescence intensities are measured through thin windows in the cells, the measurement time for each slurry being 20 s. The system comprises up to fourteen sampling circuits, although by duplexing two slurry streams into each sample cell, a sequence of up to 28 samples can be set up. The XRF measuring head containing the X-ray tube and the crystal spectrometer channels is mounted on a trolley, which travels along the bank of 14 sample cells, analysing each slurry sample in sequence (Fig. 3.7). The unit analyses up to 7 elements plus a reading of the slurry density, and completion of one full cycle takes 7 minutes, meaning that the analysis of each sample is updated every 7 or 14 minutes, depending on the system version (14 or 28 lines).

A primary sample of about 200 l min^{-1} is taken from the process flow, different types of samplers having been developed for different sampling situations (Fig. 3.3). The flow through the sample cells of the analyser is about 20 l min^{-1}, and the final sample is cut from the primary sample flow in the secondary sampling system, which is a compact unit consisting of up to fourteen sample circuits, depending on the number of slurries. Each circuit has a continuous sample

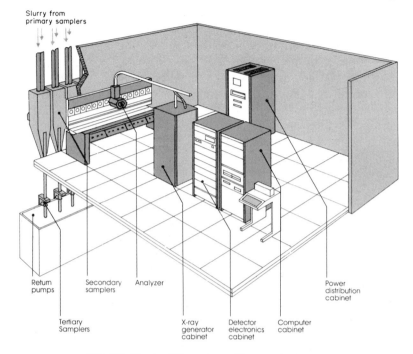

Slurry from
primary samplers

Return
pumps

Secondary
samplers

Analyzer

Power
distribution
cabinet

Tertiary
Samplers

X-ray
generator
cabinet

Detector
electronics
cabinet

Computer
cabinet

FIG. 3.7. Courier 300 on-stream X-ray analysis system.

stream. The circuit comprises a secondary sampler, a cleaner screen to remove wood and other refuse, a feeding tank, and a level-control assembly.

In the cutter-type system, the ends of the primary sample pipelines are rubber tubes that swing synchronously to and fro over the cutters of the secondary samplers. The jaws of the cutters are controlled by the slurry level in the feeding tank. The slurry is fed from the feeding tank into the sample cell of the analyser by syphon action. The sample cell return flows and rejects from the secondary samplers are pumped separately, or together back into the process pump sumps.

On-stream Ash Analysis

On-stream monitoring of the ash content of coals is being increasingly used in coal preparation plants to automatically control the constituents which make up a constant ash blend.[12] The operating principle of the monitor is based on the concept that when a material is subjected to irradiation by X-rays, a portion of this radiation is absorbed, with the remainder being reflected. The radiation absorbed by elements of low atomic number (carbon and hydrogen) is lower than that absorbed by elements of high atomic number (silicon, aluminium, iron), which form the ash in coal, so the variation of absorption coefficient with atomic number can be directly applied to the ash determination.

A number of analysers have been designed, and a typical one is shown in Fig. 3.8. A representative sample of coal is collected and

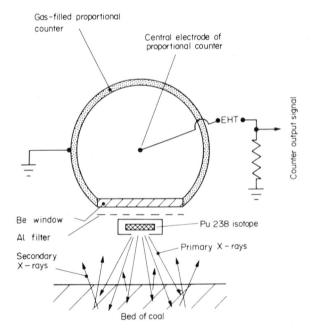

FIG. 3.8. Sensor system of ash monitor.

crushed, and fed as a continuous stream into the presentation unit of the monitor, where it is compressed into a compact uniform bed of coal, with a smooth surface and uniform density. The surface is irradiated with X-rays from a plutonium 238 isotope and the radiation is absorbed or back-scattered in proportion to the elemental composition of the sample, the back-scattered radiation being measured by a proportional counter. At the low-energy level of the plutonium 238 isotope (15–17 keV), the iron is excited to produce fluorescent X-rays that can be filtered, counted, and compensated. The proportional counter simultaneously detects the back-scattered and fluorescent X-rays after they have passed through an aluminium filter. This absorbs the fluorescent X-rays preferentially, its thickness preselected to suit the iron content and its variation. The proportional counter converts the radiation to electrical pulses that are then amplified and counted in the electronic unit. The count-rate of these pulses is converted to a voltage which is displayed and is also available for process control.

A key sensor required for the development of an effective method of controlling coal flotation is one which can measure the ash content of coal slurries. Production units have recently been manufactured and installed in coal preparation plants,[13,14] and due to the development of on-line ash monitors, coupled with an improved knowledge of process behaviour, control strategies for coal flotation are beginning to emerge and be adopted.[15,16,17]

Weighing the Ore

Many schemes are used for determination of the tonnage of ore delivered to or passing through different sections of a mill. The general trend is towards weighing materials on the move. The predominant advantage of continuous weighing over batch weighing is its ability to handle large tonnages without interrupting the material flow. The accuracy and reliability of continuous-weighing equipment have improved greatly over recent years. However, static weighing equipment is still used for many applications because of its greater accuracy.

Belt scales, or *weightometers*, are the most common type of continuous-weighing devices and consist of one or more conveyor idlers mounted on a weighbridge. The belt load is transmitted from the weighbridge either direct or via a lever system to a load-sensing device, which can be either electrically, mechanically, hydraulically, or pneumatically activated. The signal from the load-sensing device is usually combined with another signal representing belt speed. The combined output from the load and belt-speed sensors provides the flow rate of the material passing over the scale. A totaliser can integrate the flow-rate signal with time, and the total tonnage carried over the belt scale can be registered on a digital read-out. Accuracy is normally 1–2%.

Periodic testing of the weightometer can be made either by passing known weights over it or by causing a length of heavy roller chain to trail from an anchorage over the suspended section while the empty belt is running.

Most simple concentrators use one master weigher only, and in the case of a weightometer this will probably be located at some convenient point between the crushing and grinding sections. The conveyor feeding the fine-ore bins is often selected, as this normally contains the total ore feed of the plant.

Weighing of concentrates is usually carried out after dewatering, before the material leaves the plant. Weighbridges can be used for material in wagons, trucks, or ore cars. They may require the services of an operator who balances the load on the scale beam and notes the weight on a suitable form. After tipping the load, the tare (empty) weight of the truck must be determined. This method gives results within 0.5% error, assuming that the operator has balanced the load carefully and noted the result accurately. With recording scales, the operator merely balances the load, then turns a screw which automatically punches a card and records the weight. Modern scales weigh a train of ore automatically as it passes over the platform, which removes the chance of human error entirely except for occasional standardisation. Sampling, of course, must be carried out at the same time for moisture determination. Assay samples should be taken, whenever possible, from the moving stream of material, as described earlier, before loading the material into the truck.

Sampling is very unsatisfactory from a wagon or container because of the severe segregation that occurs during filling and in motion. Sampling should be performed to a pre-set pattern by *augering* the loaded truck. This is achieved by pushing in a sample probe which extracts the sample in the form of a cylinder extending the full depth of the load. This avoids selective removal of particles which slide down the surface and avoids the surface layer in which extreme segregation will probably have occurred due to vibration.

Tailings weights are rarely, if ever, measured. They are calculated from the difference in feed and concentrate weights. Accurate sampling of tailings is essential, and is easy to carry out accurately, automatic sample cutters often being used.

Slurry Streams

From the grinding stage onwards, most mineral processing operations are carried out on slurry streams, the water and solids mixture being transported through the circuit via pumps and pipelines.

As far as the mineral processor is concerned, the water is acting as a transportation medium, such that the *weight* of slurry flowing through the plant is of little consequence. What is of importance is the *volume* of slurry flowing, as this will affect residence times in unit processes. For the purposes of metallurgical accounting, the weight of dry solids contained within the slurry is important.

If the volumetric flowrate is not excessive, it can be measured by diverting the stream of pulp into a suitable container for a measured period of time. The ratio of volume collected to time gives the flowrate of pulp. This method is ideal for most laboratory and pilot scale operations, but is impractical for large-scale operations, where it is usually necessary to measure the flowrate by on-line instrumentation.

Volumetric flowrate is important in calculating retention times in processes. For instance, if 20 m^3/h of slurry is fed to a flotation conditioning tank of volume 120 m^3, then *on average*, the retention time of particles in the tank will be:

Flowrate/Tank volume = 20 × 60/120 = 10 min.

Slurry, or pulp, density is most easily measured in terms of weight of pulp per unit volume (kg/m³). As before, on flowstreams of significant size, this is usually measured continuously by on-line instrumentation.

Small flowstreams can be diverted into a container of known volume, which is then weighed to give slurry density directly. This is probably the most common method used for routine assessment of plant performance, and is facilitated by using a density can of known volume which, when filled, is weighed on a specially graduated balance giving a direct reading of pulp density.

The composition of a slurry is often quoted as the % solids by weight (100−% moisture), and can be determined by sampling the slurry, weighing, drying and reweighing, and comparing wet and dry weights (equation 3.1). This is time-consuming, however, and most routine methods for computation of % solids require knowledge of the density of the solids in the slurry. There are a number of methods used to measure this, each method having its relative merits and disadvantages. For most purposes the use of a standard density bottle has been found to be a cheap and, if used with care, accurate method. A 25-ml or 50-ml bottle can be used, and the following procedure adopted:

(1) Wash the density bottle with acetone to remove traces of grease.
(2) Dry at about 40°C.
(3) After cooling, weigh the bottle and stopper on a precision analytical balance, and record the weight, $M1$.
(4) Thoroughly dry the sample to remove all moisture.
(5) Add approximately 5–10 g of sample to the bottle and reweigh. Record the weight, $M2$.
(6) Add double distilled water to the bottle until half-full. If appreciable "slimes" (minus 45 micron particles) are present in the sample, there may be a problem in wetting the mineral surfaces. This may also occur with certain hydrophobic mineral species, and can lead to falsely low density readings. The effect may be reduced by adding one drop of wetting agent, which is insufficient to significantly affect the density of water. For solids with extreme wettability problems, an organic liquid such as toluene can be substituted for water.

(7) Place the density bottle in a desiccator to remove air entrained within the sample. This stage is essential to prevent a low reading. Evacuate the vessel for at least 2 min.

(8) Remove the density bottle from the desiccator, and top up with double distilled water (do not insert stopper at this stage).

(9) When close to the balance, insert the stopper and allow it to fall into the neck of the bottle under its own weight. Check that water has been displaced through the stopper, and wipe off excess water from the bottle. Record the weight, $M3$.

(10) Wash the sample out of the bottle.

(11) Refill the bottle with double distilled water, and repeat procedure 9. Record the weight, $M4$.

(12) Record the temperature of the water used, as temperature correction is essential for accurate results.

The density of the solids (s) is given by:

$$s = \frac{M2 - M1}{(M4 - M1) - (M3 - M2)} \times Df \text{ kg/m}^3 \qquad (3.4)$$

where Df = density of fluid used.

Knowing the densities of the pulp and dry solids, the % solids by weight can be calculated. Since the total pulp volume is equal to the volume of the solids plus the volume of water, then for 1 m^3 of pulp:

$$1 = xD/100s + (100 - x)D/100w \qquad (3.5)$$

where x = % solids by weight,
  D = pulp density (kg/m^3),
  s = density of solids (kg/m^3),
  w = density of water.

Assigning a value of 1000 kg/m^3 to the density of water, which is sufficiently accurate for most purposes, equation 3.5 gives:

$$x = \frac{100s (D - 1000)}{D(s - 1000)}. \qquad (3.6)$$

Having measured the volumetric flowrate (F m^3/h), the pulp density (D kg/m^3), and the density of solids (s kg/m^3), the weight of slurry can now be calculated ($FD/100$ kg/h), and, of more importance, the mass flowrate of dry solids in the slurry, M kg/h:

$$M = FDx/100 \qquad (3.7)$$

or combining equations 3.6 and 3.7:

$$M = \frac{Fs\,(D - 1000)}{(s - 1000)} \text{ kg/h} \qquad (3.8)$$

Example 3.1

A slurry stream containing quartz is diverted into a 1-litre density can. The time taken to fill the can is measured as 7 sec. The pulp density is measured by means of a calibrated balance, and is found to be 1400 kg/m^3. Calculate the % solids by weight, and the mass flowrate of quartz within the slurry.

Solution

The density of quartz is 2650 kg/m^3. Therefore, from equation 3.6,

% solids by weight, $x = \dfrac{100 \times 2650 \times (1400 - 1000)}{1400 \times (2650 - 1000)}$

$$= 45.9\%.$$

The volumetric flowrate, $F = 1/7$ litres/s

$$= 3600/7000 \text{ m}^3/\text{h}$$

$$= 0.51 \text{ m}^3/\text{h}.$$

Therefore, mass flowrate $M = \dfrac{0.51 \times 1400 \times 45.9}{100}$

$$= 330.5 \text{ kg/h}.$$

Example 3.2

A pump is fed by two slurry streams. One stream has a flowrate of 5.0 m^3/h and contains 40% solids by weight. The other stream has a flowrate of 3.4 m³/h and contains 55% solids. Calculate the tonnage of dry solids pumped per hour. (Density of solids is 3000 kg/m³.)

Solution

Slurry stream 1 has a flowrate of 5.0 m³/h and contains 40% solids. Therefore, from equation 3.6:

$$D = \frac{1000 \times 100s}{s(100 - x) + 1000x} \qquad (3.9)$$

or

$$D = \frac{1000 \times 100 \times 3000}{(3000 \times 60) + (1000 \times 40)}$$

$$= 1364 \text{ kg/m}^3.$$

Therefore, from equation 3.8, the mass flowrate of solids in slurry stream 1

$$= \frac{5.0 \times 3000 \times (1364 - 1000)}{(3000 - 1000)} \text{ kg/h}$$

$$= 2.73 \text{ } t/h.$$

Slurry stream 2 has a flowrate of 3.4 m³/h and contains 55% solids. From equation 3.9, the pulp density of the stream = 1579 kg/m³. Therefore, from equation 3.8, the mass flowrate of solids in slurry stream 2 = 1.82 t/h. The tonnage of dry solids pumped is thus:

$$1.82 + 2.73 = 4.55 \text{ } t/h.$$

In some cases it is necessary to know the % solids by volume. This is a parameter sometimes used in mathematical models of unit processes.

$$\% \text{ solids by volume} = xD/s. \qquad (3.10)$$

Also of use in milling calculations is the ratio of the weight of water to the weight of solids in the slurry, or the *dilution ratio*. This is defined as:

$$\text{Dilution ratio} = (100 - x)/x. \qquad (3.11)$$

This is particularly important as the product of dilution ratio and weight of solids in the pulp is equal to the weight of water in the pulp.

Example 3.3

A flotation plant treats 500 tonnes of solids per hours. The feed pulp, containing 40% solids by weight, is conditioned for 5 min with reagents before being pumped to flotation. Calculate the volume of conditioning tank required. (Density of solids is 2700 kg/l.)

Solution

The volumetric flowrate of *solids* in the slurry stream

$$= \text{Mass flowrate/Density}$$
$$= 500 \times 100/2700 = 185.2 \text{ m}^3/\text{h.}$$

The mass flowrate of water in the slurry stream

$$= \text{Mass flowrate of solids} \times \text{Dilution ratio}$$
$$= 500 \times (100 - 40)/40$$
$$= 750 \text{ t/h.}$$

Therefore, volumetric flowrate of water $= 750 \text{ m}^3/\text{h.}$
Volumetric flowrate of slurry $= 750 + 185.2 = 935.2 \text{ m}^3/\text{h.}$
Therefore, for a nominal retention time of 5 min, the conditioning tank should have a volume of $935.2 \times 5/60$,

$$= 77.9 \text{ m}^3.$$

Example 3.4

Calculate the % solids content of the slurry pumped from the sump in Example 3.2.

Solution

The mass flowrate of solids in slurry stream 1 is 2.73 t/h. The slurry contains 40% solids, hence the mass flowrate of water

$$= 2.73 \times 60/40 = 4.10 \text{ t/h}.$$

Similarly, the mass flowrate of water in slurry stream 2

$$= 1.82 \times 45/55 = 1.49 \text{ t/h}.$$

Total slurry weight pumped

$$= 2.73 + 4.10 + 1.82 + 1.49 = 10.14 \text{ t/h}.$$

Therefore, % solids by weight

$$= 4.55 \times 100/10.14 = 44.9\%.$$

On-line Instrumentation for Mass Flow Measurement

Many modern plants now use *mass-flow integration* to obtain a continuous recording of dry tonnage of material from pulp streams.

The mass-flow unit consists essentially of an electromagnetic flowmeter and a radioactive source density gauge fitted to the vertical pipeline carrying the upward-flowing ore stream.

The fundamental operating principle of the magnetic flowmeter (Fig. 3.9) is based on Faraday's law of electromagnetic induction, which states that the voltage induced in any conductor as it moves across a magnetic field is proportional to the velocity of the conductor. Thus, providing the pulp completely fills the pipeline, its velocity will be directly proportional to the flow rate. Generally, most aqueous solutions are adequately conductive for the unit and, as the liquid flows through the metering tube and cuts through the magnetic field, an emf is induced in the liquid and is detected by two small measuring electrodes fitted virtually flush with the bore of the tube, the flow rate then being recorded on a chart or continuously on an integrator. The coil windings are excited by a single-phase AC mains supply and are arranged around the tube to provide a uniform magnetic field across

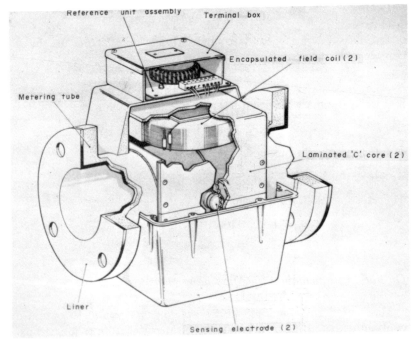

FIG. 3.9. Magnetic flowmeter.

the bore. The unit has many advantages over conventional flow-measuring devices, notable ones being that there is no obstruction to flow; pulps, and aggressive liquids can be handled; and it is immune to variations in density, viscosity, pH, pressure, or temperature.

The density of the slurry is measured automatically and continuously in the nucleonic density gauge (Fig. 3.10) by using a radioactive source. The gamma-rays produced by this source pass through the pipe walls and the slurry at an intensity that is inversely proportional to the pulp density. The rays are detected by a high-efficiency ionisation chamber and the electrical signal output is recorded directly as pulp density. The instrument must be calibrated initially "on stream" using

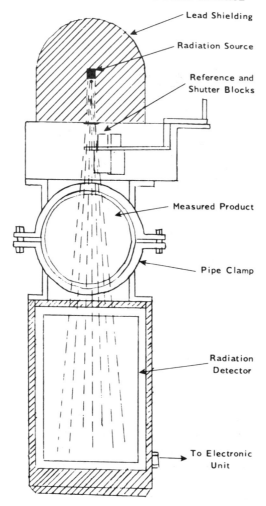

FIG. 3.10. Nucleonic density gauge.

conventional laboratory methods of density analysis from samples withdrawn from the line.

The mass-flow unit integrates the rate of flow provided by the magnetic flowmeter and the pulp density to provide a continuous record of tonnage of dry solids passing through the pipeline, providing that the specific gravity of the solids comprising the ore stream is known. The method offers a reliable, accurate means of weighing the ore stream and entirely removes any chance of operator error and errors due to moisture sampling. Another advantage is that accurate sampling points, such as poppet valves, can be incorporated at the same location as the mass-flow unit. Mass-flow integrators are less practicable, however, with concentrate pulps, especially after flotation, as the pulp contains many air-bubbles, which lead to erroneous values of flow rate and density.

Automatic Control in Mineral Processing

Important advances have been made since the early 1970s in the field of automatic control of mineral processing operations, particularly in grinding and flotation. The main reasons for this rapid development are:

(i) The development of reliable instrumentation for process control systems. On-line sensors such as flowmeters, density gauges, and chemical composition analysers are of greatest importance, and on-line particle size analysers (Chapter 4) have been successfully utilised in grinding circuit control.

Other important sensors are pH meters, and level and pressure transducers, all of which provide a signal relating to the measurement of the particular process variable. This allows the final control elements, such as servo valves, variable speed motors and pumps, to manipulate the process variable based on a signal from the controllers. These sensors and final control elements are used in many industries beside the minerals industry, and are described elsewhere.[18,19]

(ii) The availability of sophisticated digital computers at very low cost. During the 1970s the real cost of computing virtually halved each year, and the development of the microprocessor allowed very powerful computer hardware to be housed in increasingly smaller units. The development of high-level languages allowed relatively easy access to software, providing a more flexible approach to changes in control strategy within a particular circuit.

(iii) A more thorough knowledge of process behaviour, which has led to more reliable mathematical models of various important unit processes being developed.[20] Many of the mathematical models which have been developed theoretically, or "off-line", have had limited value in automatic control, the most successful models having been developed "on-line" by empirical means. Often the improved knowledge of the process gained during the development of the model has led to improved techniques for the control of the system.

(iv) The increasing use of very large grinding mills and flotation cells has facilitated control, and reduced the amount of instrumentation required.

Financial models have been developed for the calculation of costs and benefits of the installation of automatic control systems,[21] and benefits reported include significant energy savings, increased metallurgical efficiency and throughput, and decreased consumption of reagents, as well as increased process stability.[22]

The concepts, terminology, and practice of process control in mineral processing have been comprehensively reviewed by Ulsoy and Sastry;[23] in relation to crushing and grinding by Lynch,[24] and in relation to flotation by Lynch et al.[25] and Wills[26] and will be briefly reviewed here, and in the relevant later chapters.

The basic function of a control system is to stabilise the process performance at a desired level, by preventing or compensating for disturbances to the system. Ultimately the objective is not only stabilisation but also optimisation of the process performance based on economic considerations.[27]

In order to achieve these objectives, there are various "levels" of control used to make up a "distributed hierarchial computer system". At the lowest level—regulatory control—the process variables are controlled at defined "set-points" by the various individual control loops.

A simple feed-back control loop, which can be used to control the level of slurry in a pump, is shown in Figure 3.11. Feedback, or "closed-loop", control is by far the most widely used in practice. The signal from the level transmitter is fed to the controller where the measured value is compared with the desired value or set-point. The controller then transmits a signal which adjusts the opening of the feed

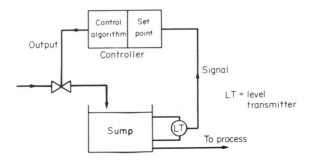

FIG. 3.11. Simple feed-back control loop.

valve according to the difference between the measured and desired value. The output signal is controlled by a standard control algorithm which is most commonly a proportional plus integral plus derivative (PID) algorithm. The PID, or "three-term" controller, carries out signal processing according to the following equation:

$$m = K_c[e + 1/T_i \int e\, dt + T_d\, de/dt]$$

where m = controller output,
     K_c = proportional sensitivity (gain),
     e = deviation of measured variable from set-point,
     T_i = integral time constant,
     T_d = derivative time constant,
     t = time.

The three-term controller generates an output that is intended to drive the process so that the deviation from set-point (e) decreases to zero. The proportional action ($K_c e$) provides an action which is directly proportional to the deviation, with the constant of proportionality, or "gain", equal to K_c. It is a feature of proportional controllers that they can produce an exact correction only for one load condition. With all other load conditions there must be some deviation of corrected output from set-point, and this is called "offset".

Offset can be eliminated by the second term in the equation, $K_c/T_i \int e\, dt$, which is *integral*, or "reset" control. This changes the value of the manipulated variable (*m*) at a *rate* which is proportional to the deviation of the measured variable from the set-point. Thus, if the deviation is double its previous value, the final control element (i.e. the valve, feeder or pump) responds twice as fast. When the measured value is at its set-point, i.e. $e = 0$, only then does the final control element remain stationary. The integral time, T_i, is the time of change of manipulated variable caused by unit change of deviation, and is usually expressed as "reset-rate" (inverse of integral time in minutes per repeat), defined as the number of times per minute that the proportional part of the response is duplicated.

Problems can occur with integral control if large, sustained load changes occur, where the measured variable deviates from the set-point for long periods. This can occur, for example, under automatic start-up of the process, where integral action results in large overshooting above the set-point by the control variable. At start-up, the measured variable is zero, so the controller output is at a maximum as there is maximum deviation from measured value and set-point. The control element, therefore, does not close until the variable crosses the set-point. *Derivative* action reduces the controller output proportional to the rate of change of the deviation with time, and helps avoid overshooting the set-point. However, integral and derivative modes interact when operating successively on a signal and most commercial controllers provide only proportional and integral terms, and are known as "PI controllers".

The constants K_c and T_i (and T_d) can be adjusted on controllers to suit process conditions, and such adjustment is known as "tuning", the objective being to determine values for the parameters such that the deviation between set-point and the actual value of the process output is most efficiently corrected. For instance, if the controller has a low value of gain (K_c), then the control action will be small and the time for stabilisation of the process will be relatively long . However, above a certain value for the gain, the system becomes unstable. A high value of gain, which should cause the output to track the set-point better, causes the controller to adjust the control action by a greater amount than is actually necessary. When the output responds, it overshoots the set-point and goes so far that the deviation is not only reversed, but is even larger than it was originally. The control action is thus reversed, again by too much, and after a delay due to the transient nature of the system, the output reverses its magnitude by an amount even greater than before. The optimum gain is usually at the point where oscillations start. In most systems the integral time (T_i) is of the same order, or slightly less than the period of the oscillation. Empirical rules for establishing the optimum values of the controller parameters have evolved from the original work of Ziegler and Nichols.[28]

Although the various processes may be controlled by separate analogue controllers, it is now more common to control the variables at their set-points by a process computer. The requirement of such *direct digital control* is that the analogue signals from the measuring devices are converted into digital pulses by an analogue-digital converter, and that the digital signals from the computer are converted to analogue form and transmitted to the transducers which convert the analogue signals to mechanical form in order to operate the control devices.

Although set-points may be adjusted manually, it is preferable to control them automatically by other computers in the hierarchy. Such *supervisory*, or *cascade*, control may use feed-back loops, such as in the regulation of reagent feed to a flotation process in response to the metal content of the flotation tailing (Fig. 3.12).

A change in the metal content of the tailing is acted upon by the supervisory computer which, by means of a process algorithm, modifies the set-point used by the process computer in the regulatory control loop. The greatest problem found, which has never been fully

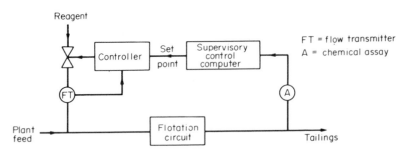

FIG. 3.12. Feed-back loop with supervisory control.

overcome in flotation control, is in developing algorithms which accommodate changes in ore type and which can define flexible limits to the maximum and minimum amounts of reagent added.

The disadvantage of feed-back control loops is that they compensate for disturbances in the system only after the disturbances have occurred, and the effect of the control system on the manipulated variables is not observed until after a time delay roughly equivalent to the residence time of the process flow between the control device and the measuring device.

Feed-forward control loops do not suffer from this disadvantage, and they are sometimes used in flotation processes to control the addition of a specific reagent to the process feed. Such a control loop is shown in Figure 3.13.

The flowrate of slurry and slurry density are continuously measured by means of a magnetic flowmeter and density gauge respectively, the two signals being computed to mass flowrate of dry solids. The chemical analysis signal is incorporated to obtain the mass flowrate of valuable metal to the process, the resultant value being used to calculate the required reagent flowrate in order to maintain constant reagent addition per unit weight of metal. The new set-point is then compared with the measured value of reagent flowrate and adjustments made as before. The success of such control loops

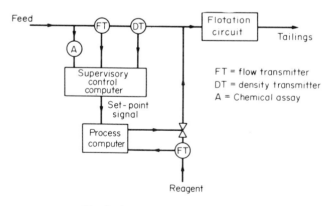

FIG. 3.13. Feed-forward control loop.

depends upon a consistent and predictable relationship between the controlled variable, i.e. the reagent flow, and the measured variable (the metal flowrate to the plant), and they often fail when significant changes occur in the nature of the plant feed. Since it is really *mineral* content which controls reagent requirements, and this can only be inferred from the metal content, changes in mineralogy or concentration of recycled reagents can cause unmeasured disturbances, which cannot be compensated for in such pure feed-forward loops, and hence some form of feed-back trim is sometimes incorporated,[29] often only with limited success.

Figure 3.14 shows a simple feed-forward/feed-back system which has been used to control the reagent addition to maintain an optimum tailings assay.

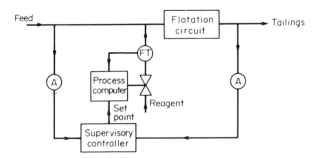

FIG. 3.14. Feed-forward/feed-back system.

The optimum tailings assay is calculated using the assay of the feed to the process and a feed-forward algorithm. The reagent addition is controlled at the set-point, which is adjusted according to the difference between the calculated and actual tailings assays. The disadvantage of this system, as with the simple feed-back loop, is the time lag between changes in head grade and tailings grade response.

Although single-variable control, with PI or PID-type controllers, is the approach used for the majority of process control applications, it does have certain limitations. One such limitation is in the overall control of processes that show severe interaction, or "coupling behaviour", between loops. Normal decoupling techniques, which may involve "detuning", i.e. slowing the response, of one controller, only partially eliminates the interactions, and a solution may be some form of multi-variable control,[30,31] where the interactions are specifically accommodated in the design of the control system.

PI control alone also has considerable limitations because of the long time delays of many mineral processes; unless the controller is detuned, the controller output will be such as to cause process oscillations. A detuned PI controller provides very slow response, which can be unsuitable for many processes where input disturbances result in process upsets, and so dynamic compensation is added to the PI controller in cases of long time delays.

The problem of PI controller tuning is compounded by the non-linear nature of most mineral processing operations, and variability of process response is frequently encountered in mineral control systems, this often being caused by changes in the nature of the ore. Some of these disturbances, such as ore hardness, slurry viscosity and liberation characteristics, are difficult or impossible to measure. The necessary controller gain settings are not fixed in such situations, and imposition of fixed gains to suit the average dynamic behaviour of the process frequently results in poor controller performance. In order to overcome these deficiencies, a range of "modern" model-based control techniques is being developed, which can not only tune the controller automatically but can also optimise and control its set-point.

The development of reliable on-line models has led to "adaptive" control methods, which overcome some of the weaknesses of classical PI control techniques. It is now possible for a microprocessor-based controller to review the performance of the control loop, and to modify the controller parameters to suit the current dynamic process responses. This type of "self-tuning" control adapts not only to load-dependent dynamic changes, but also to time-related and/or random dynamic characteristic changes.[32,33]

The ultimate aim of control is not only to stabilise the process, but also to optimise the process performance and hence increase the economic efficiency. This higher level of control has been attempted in a few concentrators with somewhat limited success, as optimisation can only be achieved when reliable supervisory stabilisation of the plant has been fully effective. The *evolutionary optimisation* (EVOP) approach is based on control action to either continue or reverse the direction of movement of a process efficiency by manipulating the set-point of the controller in order to achieve some higher level of performance. The efficiency of the process is determined by a suitable economic criterion which in the case of a concentrator can be the economic efficiency attained (Chapter 1). The controlled variables are altered according to a predetermined strategy, and the effect on the process efficiency is computed on-line. If the efficiency is increased as a result of the change, then the next change is made in the same direction, otherwise the direction is reversed; eventually the efficiency should converge to the optimum.

A logic flow diagram for the control of reagent to a flotation circuit is shown in Fig. 3.15.[34] The system incorporates an alarm to alert the

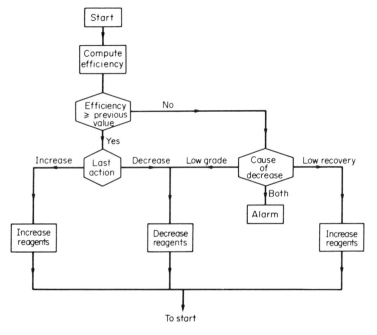

FIG. 3.15. Optimisation control logic.

process operator when the computer's efforts fail to halt a falling efficiency. Action can then be initiated to adjust process variables not influenced by the control strategy. In searching for peaks in efficiency, the controller is beginning to act in an intelligent manner. It is seeking a performance peak in a similar way as do operators, by reacting according to changes in process conditions.

The EVOP approach is fairly simple, requiring no process mathematical model. However, it is realised that some form of process model which is capable of providing more accurate predictions of process behaviour is really essential for effective optimisation, and it is for this reason that most optimising control systems are now model-based strategies.[35,36]

The application of model-based control systems is summarised in Fig. 3.16.[37] The function of the estimator is to combine both model predictions and process measurements to define the state of the system. The optimiser selects a set of control actions which are

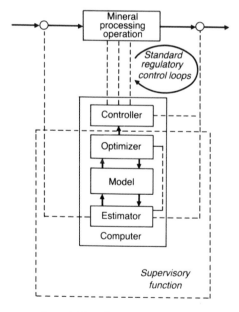

FIG. 3.16. Components of a model-based control system (after Herbst and Rajamani[37]).

calculated to maximise or minimise an objective function. Set-points for standard **PI** controllers are calculated in the optimisation procedure. In the estimator a model of the process is run in parallel with the actual process, and the inputs to the process are also fed to the model. The measurements coming from the actual process are compared to the measurements predicted by the model, and the difference is used as the basis for the correction to the model. This correction alters the parameters and states in the model so as to make its predictions match those of the actual process better. The basis of parameter estimation is "recursive estimation", or Kalman filtering,[38] where the estimator updates its estimates continuously with time as each input arrives, rather than collecting all the information together and processing in a single batch (Fig. 3.17). A simple recursive least-squares algorithm is most commonly used to update the model parameters:

$$y(t + 1) = a_1 y(t) + \ldots + a_n y(t - n + 1) + b_o u(t) + \ldots$$
$$+ b_n u(t - n + 1) + z(t) \qquad (3.12)$$

where
$$y(t) = \quad \text{process output at time } t,$$
$$u(t) = \quad \text{process input at time } t,$$
$$z(t) = \quad \text{output noise at time } t,$$
$$a_1 - a_n, b_1 - b_n \text{ are model parameters.}$$

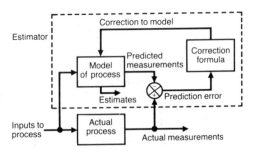

FIG. 3.17. Schematic diagram of the operation of recursive estimation (after Barker[31]).

Since the model parameters $a_1 \ldots a_n$ and $b_1 \ldots b_n$ are being updated continuously, then equation 3.12 should give a reasonably accurate prediction of the dynamic behaviour of the process. The model parameters can then be used to calculate controller parameters. With this technique a control law is used which is always suitably tuned for the current process characteristics.[35]

In addition to the process and supervisory control computers, a general purpose computer may be incorporated in the hierarchy. This computer is generally housed in the central control room (Fig. 3.18) and coordinates the activities of the supervisory computers, as well as performing such tasks as logging and evaluation of plant data, preparation and printing of shift, daily and monthly reports, and supervision of shut-down and start-up. The computer can allow the operator to input information such as changes in metal prices, smelter terms, reagent costs, etc., which can aid optimisation of the set-points of the supervisory controllers.

FIG. 3.18. Concentrator central control room.

Mass Balancing Methods

In order to assess plant performance, and to control the operation using the evaluated results, it is necessary to account for the products in terms of material and contained component weights. Mass balancing is particularly important in accounting for valuable mineral or metal distributions, and the *two-product formula* is of great use in this respect.

If the weights of the feed, concentrate and tailings are F, C, and T respectively, and their corresponding assays f, c, and t, then

$$F = C + T \qquad (3.13)$$

i.e. material input = material output

and
$$Ff = Cc + Tt \qquad (3.14)$$

i.e. the valuable metal (or mineral) is balanced.

Therefore, $Ff = Cc + (F - C)t$

which gives $F/C = (c - t)/(f - t) \qquad (3.15)$

where F/C represents the *ratio of concentration*.

The plant *recovery* is $(Cc/Ff) \times 100\%$

or $\text{Recovery} = 100c(f - t)/f(c - t)\% . \qquad (3.16)$

As values of recovery, ratio of concentration and *enrichment ratio* (c/f) can be determined from the assay results alone, the two-product formula method is often used to provide information for plant control, although this will be retrospective, dependent on the time taken to receive and process the assay results. Direct control can be affected using on-stream analysis systems, where values of c, f and t can be continuously computed to provide up-to-date values of metallurgical performance.

Example 3.5

The feed to a flotation plant assays 0.8% copper. The concentrate produced assays 25% Cu, and the tailings 0.15% Cu. Calculate the recovery of copper to the concentrate, the ratio of concentration, and the enrichment ratio.

Solution

The concentrator recovery (equation 3.16) is:

$$\frac{100 \times 25(0.8 - 0.15)}{0.8(25 - 0.15)}\%$$

$$= 81.7\%.$$

The ratio of concentration (equation 3.15) is:

$$\frac{25 - 0.15}{0.8 - 0.15}$$

$$= 24.8.$$

The enrichment ratio (c/f) is:

$$25/0.8$$

$$= 31.3.$$

Metallurgical Accounting

There are many methods used to account for a plant's production. Most concentrators produce a metallurgical balance showing the performance of each shift, the shift results being cumulated over a longer period (daily, monthly, annually) to show the overall performance.

Although concentrates can be weighed accurately prior to loading into rail cars or lorries, it is improbable that concentrate weighed in this way during a particular shift will correspond with the amount actually produced, as there is often a variable inventory of material between the concentrator and the final disposal area. This inventory may consist of stockpiled material, and concentrates in thickeners, filters, agitators, etc. For shift accounting, the feed tonnage to the concentrator is usually accurately measured, this providing the basis for the calculation of product stream weights. Samples are also taken periodically of the feed, concentrates and tailings streams, the composite samples being collected and assayed at the end of the shift. Equation 3.15 can be used to calculate the concentrate weight produced, allowing a metallurgical balance to be prepared. Suppose, for instance, that a plant treats 210.0 tonnes of material during a shift, assaying 2.5% metal, to produce a concentrate of 40% metal, and a tailing of 0.20% metal.

From equation 3.15:

$$F/C = (40 - 0.20)/(2.5 - 0.20).$$

Hence $C = 12.1$ t.
The tailing weight is thus $210.0 - 12.1 = 197.9$ t.
The metallurgical balance for the shift is tabulated in Table 3.3.

TABLE 3.3. SHIFT 1 PERFORMANCE

Item	Weight t	Assay %	Weight metal t	Distribution metal %
Feed	210.0	2.5	5.25	100.0
Concentrate	12.1	40.0	4.84	92.2
Tails	197.9	0.20	0.40	7.8

The distribution of the metal into the concentrate (i.e. recovery) is $4.84 \times 100/5.25 = 92.2\%$, this value corresponding to that obtained from equation 3.16. Suppose that on the next shift 305.0 tonnes of material are treated, assay 2.1% metal, and that a concentrate of 35.0% metal is produced, leaving a tailing of 0.15% metal. The metallurgical balance for the shift is shown in Table 3.4.

TABLE 3.4. SHIFT 2 PERFORMANCE

Item	Weight t	Assay %	Weight metal t	Distribution metal %
Feed	305.0	2.1	6.41	100.0
Concentrate	17.1	35.0	5.99	93.45
Tails	287.9	0.15	0.42	6.55

The composite balance for the two shifts can be produced by adding all material and metal weights, and then weighting the assays and distributions accordingly (Table 3.5).

TABLE 3.5. COMBINED PERFORMANCE

Item	Weight t	Assay %	Weight metal t	Distribution metal %
Feed	515.0	2.3	11.66	100.0
Concentrate	29.2	37.1	10.83	92.9
Tails	485.8	0.17	0.83	7.1

Similarly, the shift balances can be cumulated over a weekly, monthly, or annual accounting period, in order to obtain the true feed weight and the weighted average assays of the streams. It is apparent that the calculated weights of concentrate and tailings are those which "fit" the available assay values, and a "perfect" balance is always produced by this method, as the two-product equation is consistent with the available data,

i.e. $$Ff - Cc - Tt = 0.$$

A more realistic assessment can be made by accurately weighing one more stream, and comparing this with the calculated value (*check in–check out* method). For instance, if, in the previous example, the concentrate produced in the two shifts is accurately weighted, and is 28.8 tonnes, then it is possible to obtain values for theoretical and actual plant recovery. The balance is shown in Table 3.6.

TABLE 3.6.

Item	Weight t	Assay %	Weight metal t	Distribution metal %
Feed	515.0	2.3	11.66	100.0
Concentrate	28.8	37.1	10.68	91.6
Unaccounted loss	—	—	0.15	1.3
Tails	486.2	0.17	0.83	7.1

The *actual* recovery (91.6%) is declared, and any discrepancy in metal weight is regarded as an *unaccounted loss* (1.3%). The weights of material are accepted and it is assumed that there are no *closure errors* in the material balance, i.e. $F - C - T = 0$. Physical losses will, of course, occur on any plant and it should always be endeavoured to keep these as low as possible. If the third weight is also accurately known (which it rarely is), then any closure error can also be reported as an unaccounted loss (or gain) of material.

As was mentioned earlier, it is not easy, particularly in a large plant, to obtain accurate concentrate production weights over a short period because of the inventory between the concentrator and the concentrates disposal area or smelter, where weighing is undertaken.

TABLE 3.7.

	Weight t	% metal	Weight metal t
(a) *Beginning of month*			
Thickeners	210	44.1	92.6
Agitators & filters	15	43.9	6.6
In transit to smelter	207	46.9	97.1
Total inventory	432	45.4	196.3
(b) *End of month*			
Thickeners	199	39.6	78.8
Agitators & filters	25	40.8	10.2
In transit to smelter	262	39.3	103.0
Total inventory	486	39.5	192.0

If over a monthly accounting period, for example, the change in inventory can be quantified by assessing at the beginning of each month the amount of concentrate retained in the plant in thickeners, filters, etc., and in trucks or other containers in transit between mill and smelter, then this inventory change can be used to adjust the smelter receipts to calculate the production figure.

For example, suppose that a smelter's monthly receipts from a mill are 3102 t of concentrates assaying 41.5% metal. The inventory of concentrates in the products disposal section is determined at the beginning and the end of the month and is as shown in Table 3.7. The monthly concentrate production can now be calculated thus:

	Weight t	Assay (%)	Weight metal t
Concentrate received	3102	41.5	1287.3
Inventory change	+54	—	−4.3
Production	3156	40.7	1283.0

Note that in this case there has been an increase in the inventory of material, which obviously increases the production, but a decrease in the inventory of metal, which lowers the metal production.

Although the metal content of the inventory can be assessed to some degree by sampling and assaying stockpiles and truck loads, it may be more accurate to rely on the weighted concentrate assay produced from the shift balances, rather than to adjust the value as above.

Example 3.6

From cumulative shift balances, the total monthly feed to a copper flotation plant was 28,760 dry tonnes at a grade of 1.1% copper. The weighted concentrate and tailings assays were 24.9% and 0.12% Cu respectively. The weight of concentrate received at the smelter during the month was 1090.7 tonnes, assaying 24.7% Cu.

At the beginning of the month, 257 t of concentrate were in transit to the smelter, and 210 t were in transit at the month end. Tabulate a metallurgical balance for the month's production, showing unaccounted losses of copper and compare actual and theoretical recoveries.

Solution

Inventory change during month = 210 − 257 = −47 t. Therefore, concentrate production is:

$$1090.7 - 47 = 1043.7 \text{ t.}$$

Metallurgical balance:

	Weight t	Assay % Cu	Weight Cu t	Distribution Cu %
Feed	28,760.0	1.1	316.4	100.0
Concentrate	1043.7	24.9	259.9	82.1
Unaccounted loss	—	—	23.2	7.3
Tails	27,716.3	0.12	33.3	10.5

The actual recovery is 82.1%, and the theoretical recovery is 82.1% + 7.3 = 89.4%. Clearly there is a large discrepancy, which could indicate poor sampling, assaying or weighing of flowstreams.

The Use of Size Analyses in Mass Balancing

Many unit process machines, such as hydrocyclones and certain gravity separators, produce a good degree of particle size separation, and size analysis data can often be effectively used in the two-product formula.

Example 3.7

In the circuit shown in Fig. 3.19, the rod mill is fed at the rate of 20 t/h of dry solids (density 2900 kg/m³). The cyclone feed contains 35% solids by weight, and size analyses on the rod mill discharge, ball mill discharge and cyclone feed gave:

Rod mill discharge 26.9% + 250 μm
Ball mill discharge 4.9% + 250 μm
Cyclone feed   13.8% + 250 μm

Calculate the volumetric flowrate of feed to the cyclone.

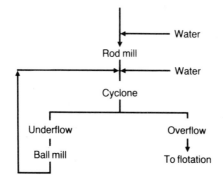

FIG. 3.19. Rod mill-ball mill-cyclone circuit.

Solution

A material balance on the cyclone feed junction gives:

$$F = 20 + B$$

where F = cyclone feed, and B = ball mill discharge.
   Therefore $F = 20 + (F - 20)$, and a balance of +250 μm material gives:

$$13.8F = (26.9 \times 20) + (F - 20) \times 4.9$$

from which $F = 49.4$ t/h.

Volumetric flowrate of solids = 49.4 × 1000/2900 = 17.0 m^3/h
Volumetric flowrate of water = 49.4 × 65/35 = 91.7 m^3/h
Therefore, flowrate of feed to the cyclone = 17.0 + 91.7

$$=108.7 \ m^3/h.$$

The Use of Dilution Ratios in Mass Balancing

Water plays a very important role in mineral processing operations. Not only is it used as a transportation medium for the solids in the circuit, but it is also the medium in which most of the mineral separations take place. Individual processes require different optimum water contents. Ball mills, for instance, rarely operate below about 65% solids by weight, and the discharge may need diluting before being fed to hydrocyclones. Most flotation operations are performed at between 25–40% solids by weight, and some gravity concentration devices such as Reichert cones operate most efficiently on slurries containing 55–70% solids. A mineral processing plant is a large consumer of water. In a plant treating 10,000 tonnes per day of ore, about 20 m^3/min of water is required, which is expensive if some form of conservation is not practised. If the slurry must be dewatered before feeding to a unit process, then the water should be used to dilute the feed as required elsewhere in the circuit. For optimum performance, therefore, there is a water requirement which produces optimum slurry composition in all parts of the circuit. The two-product formula is of great use in assessing water balances.

Consider a hydrocyclone fed with a slurry containing f % solids by weight, and producing two products—an underflow containing u% solids, and an overflow containing v% solids. If the weight of solids per unit time in the feed, underflow and overflow are F, U and V respectively, then providing the cyclone is operating under equilibrium conditions:

$$F = U + V. \tag{3.17}$$

The dilution ratio of the feed slurry = $(100 - f)/f = f'$. Similarly, the dilution ratio of the underflow = $(100 - u)/u = u'$ and dilution ratio of overflow = $(100 - v)/v = v'$.

Since the weight of water entering the cyclone must equal the weight leaving in the two products, the water balance is:

$$Ff' = Uu' + Vv'. \tag{3.18}$$

Combining equations 3.17 and 3.18:

$$U/F = (f' - v')/(u' - v'). \tag{3.19}$$

Example 3.8

A cyclone is fed at the rate of 20 t/h of dry solids. The cyclone feed contains 30% solids, the underflow 50% solids, and the overflow 15% solids by weight. Calculate the tonnage of solids per hour in the underflow.

Solution

Dilution ratio of feed slurry = 70/30 = 2.33
Dilution ratio of underflow = 50/50 = 1.00
Dilution ratio of overflow   = 85/15 = 5.67

A material balance on the cyclone gives:

$$20 = U + V$$

where U = tonnes of dry solids per hour in underflow,
V = tonnes of dry solids per hour in overflow.

Since the weight of water entering the cyclone equals the weight of water leaving:

$$20 \times 2.33 = 1.00U + 5.67V$$

or

$$46.6 = U + 5.67(20 - U)$$

which gives

$$U = 14.3 \ t/h.$$

Example 3.9

A laboratory hydrocyclone is fed with a slurry of quartz (density 2650 kg/m^3) at a pulp density of 1130 kg/m^3. The underflow has a pulp density of 1280 kg/m^3 and the overflow 1040 kg/m^3.

A 2-litre sample of underflow was taken in 3.1 sec. Calculate the mass flowrate of feed to the cyclone.

Solution

% solids content of feed (equation 3.6)

$$= \frac{100 \times 2650 \times 130}{1130 \times 1650} = 18.5\%.$$

Similarly:

% solids content of underflow = 35.1%
% solids content of overflow = 6.2%

Therefore, dilution ratios of feed, underflow and overflow are 4.4, 1.8, and 15.1 respectively.

Volumetric flowrate of underflow = 2/3.1 litres/sec

$$= \frac{2 \times 3600}{3.1 \times 1000} \text{m}^3/\text{h} = 2.32 \text{ m}^3/\text{h}.$$

Therefore, mass flowrate of underflow (equation 3.7)

$$= 2.32 \times 1280 \times 35.1/100 = 1.04 \text{ t/h}.$$

Therefore, from a water balance on the cyclone:

$$4.4F = 1.04 \times 1.8 + (F - 1.04) \times 15.1$$

which gives mass flow rate of feed (F)

$$= 1.29 \text{ t/h}.$$

In the example shown above, a two-product balance can be performed using pulp densities alone, obviating the need to convert to % solids and dilution ratios. The density of solids, therefore, need not be measured (assuming that this is the same in all three streams).

Since a material balance on the cyclone gives:

$$F = U + V$$

a balance of *slurry* weights gives:

$$\frac{F}{\% \text{ solids in feed}} = \frac{U}{\% \text{ solids in underflow}} + \frac{V}{\% \text{ solids in overflow}} .$$

If f, u and v are the pulp densities of feed, underflow and overflow respectively, then from equation 3.6:

$$\frac{Ff(s - 1000)}{100s(f - 1000)} = \frac{Uu(s - 1000)}{100s(u - 1000)} + \frac{Vv(s - 1000)}{100s(v - 1000)}$$

or

$$\frac{Ff}{f - 1000} = \frac{Uu}{u - 1000} + \frac{(F - U)v}{v - 1000}$$

which gives

$$\frac{U}{F} = \frac{(f - v)(u - 1000)}{(u - v)(f - 1000)} . \qquad (3.20)$$

Therefore, in example 3.9,

$$\frac{U}{F} = \frac{(1130 - 1040)(1280 - 1000)}{(1280 - 1040)(1130 - 1000)}$$

$$= 0.81.$$

Water balances can be used to calculate circuit water requirements, and to determine the value of circulating loads.

Example 3.10

The flowsheet shown in Fig. 3.20 illustrates a conventional closed circuit grinding operation.

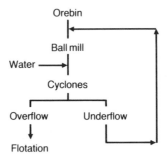

FIG. 3.20. Closed circuit grinding flowsheet.

The cyclone overflow line is instrumented with a magnetic flowmeter and nuclear density gauge, and the mass of dry ore fed to flotation is 25 t/h.

The feed from the fine ore bins is sampled, and is found to contain 5% moisture.

The cyclone feed contains 33% solids, the cyclone underflow 65% solids, and the overflow 15% solids.

Calculate the circulating load on the circuit and the amount of water required to dilute the ball mill discharge.

Solution

A water balance on the cyclone gives:

$$\frac{67F}{33} = \frac{85}{15} \times 25 + \frac{35U}{65}$$

where F = cyclone feed (dry t/h),
     U = cyclone underflow (dry t/h).

The mass flowrate of feed from the ore bin = 25 t/h (since input to circuit = output).

Therefore, $F = 25 + U$

and

$$(25 + U)\frac{67}{33} = 25 \times \frac{85}{15} + \frac{35U}{65}$$

from which $U = 61.0$ dry t/h.
The circulating load is therefore *61.0 t/h*, and the circulating load ratio
is $61.0/25 = 2.44$.

The ball mill feed = ore from bin + circulating load
Water in ball mill feed = $25 \times 5/95 + 61.0 \times 35/75$
                        = 34.2 m³/h.
Water in cyclone feed = $(25 + 61.0)67/33$
                      = 174.6 m³/h.
Therefore, water requirement at cyclone feed
                      = $174.6 - 34.2$
                      = *140.4 m³/h.*

Example 3.11

Calculate the circulating load in the grinding circuit shown in Fig.
3.19 and the amounts of water added to the rod mill and cyclone feed.

Feed to rod mill   = 55 tonnes of dry ore per hour
Rod mill discharge = 62% solids
Cyclone feed   = 48% solids
Cyclone overflow   = 31% solids
Cyclone underflow = 74% solids

Solution

Since input to circuit = output, the cyclone overflow contains 55 t/h
of solids.
A water balance on the cyclone gives:

$$(U + 55)\frac{52}{48} = \frac{26U}{74} + \frac{69}{31} \times 55$$

which gives $U = 85.8$ t/h.

The circulating load ratio is thus $85.8/55 = 1.56$.

Water in rod mill discharge $= 55 \times 38/62 = 33.7$ t/h.

Therefore, water addition to rod mill is 33.7 m^3/h.

Water in ball mill discharge $= 85.8 \times 26/74 = 30.1$ t/h
Water in cyclone feed $= (55 + 85.8) \times 52/48 = 152.5$ t/h.

Therefore, water requirement to cyclone feed

$= 152.5 - (33.7 + 30.1) = 88.7$ t/h.

$= 88.7$ m^3/h.

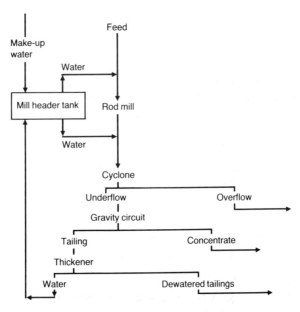

FIG. 3.21. Tin concentrator circuit.

Example 3.13

The flowsheet shown in Fig. 3.21 is that of a tin concentrator treating 30 dry tonnes per hour of ore. The ore, containing 10% moisture, is fed into a rod mill which discharges a pulp containing 65% solids by weight. The rod mill discharge is diluted to 30% solids before being pumped to cyclones. The cyclone overflows, at 15% solids, are pumped to the slimes treatment plant.

The cyclone underflows, at 40% solids, and containing 0.9% tin, are fed to a gravity concentration circuit, which produces a tin concentrate containing 45% tin, and a tailing containing 0.2% tin.

The tailing slurry, containing 30% solids by weight, is dewatered to 65% solids in a thickener, the overflow being routed to the mill header tank, which supplies water to the rod mill feed and rod mill discharge.

Calculate the flow rate of make-up water required for the header tank, and the water addition needed to the rod mill feed and discharge.

Solution

Water content of plant feed $= 30 \times 10/90 = 3.33$ t/h
Water content of rod mill feed $= 30 \times 35/65 = 16.2$ t/h

Therefore, water addition to rod mill feed

$$= 16.2 - 3.33 = 12.9 \ m^3/h.$$

Water content of cyclone feed $= 30 \times 70/30 = 70$ t/h.
Therefore, water addition to cyclone feed

$$= 70 - 16.2 = 53.8 \ m^3/h.$$

A water balance on the cyclone gives:

$$30 \times 70/30 = U \times 60/40 + (30 - U)85/15$$

which gives $U = 24.0$ t/h.

The feed to the gravity concentrator is thus 24.0 t/h, containing 0.9% tin.

A mass balance on the gravity concentrator gives:

$$24.0 = C + T$$

where C = concentrate weight (t/h),
T = tailings weight (t/h).

Since the weight of tin entering the plant equals the weight leaving:

$$24.0 \times 0.9/100 = (24.0 - T) \times 45.0/100 + T \times 0.2/100,$$

which gives $T = 23.6$ t/h.

Water in thickener feed = $23.6 \times 70/30 = 55.1$ m³/h. Assuming no solids are lost in the thickener overflow, water in thickener underflow = $23.6 \times 35/65 = 12.7$ m³/h. Therefore, water in thickener overflow

$$= 55.1 - 12.7 = 42.4 \text{ m}^3/\text{h}.$$

Therefore, make-up water required to header tank

$$= 53.8 + 12.9 - 42.4$$
$$= 24.3 \ m^3/h.$$

Limitations of the Two-product Formula

Although of great use, the two-product formula does have limitations in plant accounting and control. The equations assume steady-state conditions, the fundamental assumption being that input is equal to output. While this may be true over a fairly long period, and so may be acceptable for daily, or shift, accounting, such dynamic equilibrium may not exist over a shorter period, such as the interval between successive on-stream analyses of products.

Sensitivity of the Recovery Equation

Equation 3.16, defining recovery of a unit operation, is very sensitive to the value of t, as the equation represents the ratio of the two expressions c/f and $(c - t)/(f - t)$, which differ only by the presence of t in the latter. Equation 3.16 can be partially differentiated with respect to f, c and t to give:

$$\frac{\partial R}{\partial f} = \frac{100ct}{f^2(c - t)} \qquad \frac{\partial R}{\partial c} = \frac{-100t(f - t)}{f(c - t)^2} \qquad \frac{\partial R}{\partial t} = \frac{-100c(c - f)}{f(c - t)^2} .$$

Since the variance of a function can be found from its derivatives:

$$V_{F_{(x)}} = \sum_i \left(\frac{\partial F}{\partial x_i}\right)^2 V_{x_i} \tag{3.21}$$

$$V_R = \left(\frac{\partial R}{\partial f}\right)^2 V_f + \left(\frac{\partial R}{\partial c}\right)^2 V_c + \left(\frac{\partial R}{\partial t}\right)^2 V_t$$

where V_R, V_f, V_c and V_t are the variances in R, f c and t respectively. Therefore:

$$V_R = \frac{100^2}{f^2(c - t)^2}\left[\frac{c^2t^2}{f^2}V_f + \frac{(f - t)^2t^2}{(c - t)^2}V_c + \frac{c^2(c - f)^2}{(c - t)^2}V_t\right] . \tag{3.22}$$

Equation 3.22 is useful in assessing the error that can be expected in the calculated value of recovery due to errors in the measurement of f, c and t.

For instance, in a concentrator which treats a feed containing 2.0% metal to produce a concentrate grading 40% metal and a tailing of 0.3% metal, the calculated value of recovery (equation 3.16) is 85.6%, and:

$$V_R = 57.1 \, V_f + 0.0003 \, V_c + 2325.2 \, V_t. \tag{3.23}$$

It is immediately apparent that the calculated value of recovery is most sensitive to the variance of the tailings assay, and is extremely insensitive to the variance of the concentrate assay.

If it is assumed that all the streams can be assayed to a relative standard deviation of 5%, then the standard deviations of feed, concentrate and tailings streams are 0.1%, 2% and 0.015% respectively, and, from equation 3.23, $V_R = 1.1$, or the standard deviation of R is 1.05. This means that, to within 95% confidence limits, the recovery is 85.6 ± 2.1%.

It is interesting to compare the expected error in the calculation of recovery from data in which little separation of the component values takes place. Suppose the feed, of 2.0% metal, separates into a concentrate of 2.2% metal and a tailing of 1.3% metal. The calculated recovery is, as before, 85.6%, and:

$$V_R = 6311 \ V_f + 3155 \ V_c + 738 \ V_t,$$

the value of recovery, in this case, being more dependent on the accuracy of the feed and concentrate assays than on the tailings assay. However, a relative standard deviation of 5% on each of the assay values produces a standard deviation in the recovery of 10.2%. The degree of accuracy obtained using the two-product formula is therefore dependent on the *extent* of the separation process and there must always be a significant difference between the component values (in this case assays) if reliable results are to be obtained.

Example 3.14

A copper concentrator has installed an on-stream analysis system on its process streams. The accuracy of the system is estimated to be:

% Cu	Relative standard deviation (%)
0.05–2.0	6–12
2.0–10.0	4–10
>10	2–5

The feed to a rougher bank is measured as 3.5% Cu, the concentrate as 18% Cu, and the tailing as 1% Cu. Calculate the recovery, and the uncertainty in its value.

Solution

Assuming relative standard deviations on feed, concentrate and tailings assays of 4%, 2% and 8% respectively, then the standard deviations are:

Feed $\quad\quad$ 4 × 3.5/100 = 0.14%
Concentrate 18 × 2/100 = 0.36%
Tailings $\quad\quad$ 1 × 8/100 = 0.08%

The calculated value of recovery (equation 3.16) is 75.6%, and the variance in this value (equation 3.22) is 5.7. Therefore, standard deviation on recovery is 2.4%, and, to a 95% confidence level, the uncertainty in the calculation of recovery is ±2 × 2.4

$$= \pm 4.8\%.$$

(A computer program—RECVAR—for such calculations is listed in Appendix 3.)

Sensitivity of the Mass Equation

Equation 3.15 can be used to calculate the concentrate weight as a fraction or percentage (C) of the feed weight:

$$C = 100(f - t)/(c - t). \quad\quad\quad (3.24)$$

Although expression 3.24 is very useful in material balancing, it is, like the recovery equation, prone to considerable error if the component values are not well separated. For example, a hydrocyclone is a separator which produces good separation in terms of contained water content, and of certain size fractions, but not necessarily in terms of contained metal values. When all such data is available, the problem is often deciding which component will produce the most accurate material balance.

If equation 3.24 is partially differentiated with respect to f, c and t respectively, then:

$$\frac{\partial C}{\partial f} = \frac{100}{c - t} \qquad \frac{\partial C}{\partial c} = \frac{-100(f - t)}{(c - t)^2} \qquad \frac{\partial C}{\partial t} = \frac{-100(c - f)}{(c - t)^2} \ .$$

From equation 3.21, the variance in C, V_c can be determined from:

$$V_c = \left(\frac{\partial C}{\partial f}\right)^2 V_f \qquad + \left(\frac{\partial C}{\partial f}\right)^2 V_c \qquad + \left(\frac{\partial C}{\partial t}\right)^2 V_t$$

$$= \left(\frac{100}{(c - t)}\right)^2 V_f + 100^2 \left[\frac{(f - t)}{(c - t)^2}\right]^2 V_c + 100^2 \left[\frac{(c - f)}{(c - t)^2}\right]^2 V_t. \quad (3.25)$$

Example 3.15

A unit spiral concentrator in a grinding circuit was sampled, and tin assays on feed and products were:

Feed   0.92% ± 0.02% Sn
Concentrate 0.99% ± 0.02% Sn
Tailings   0.69% ± 0.02% Sn

Pulp densities were also measured, and water–solid ratios were:

Feed   4.87 ± 0.05
Concentrate 1.77 ±0.05
Tailings   15.73 ± 0.05

By means of sensitivity analysis, calculate the percentage of feed material reporting to the concentrate, and the uncertainty in this value. Which component should be chosen for subsequent routine evaluations?

Solution

Assuming 95% confidence limits, the standard deviation in the tin assays is 0.01, and the variance is thus 1×10^{-4}. The value of C

determined from tin assays (equation 3.24) is 76.7%, and, from equation 3.25, $V_c = 18.2$. The standard deviation, s, is thus 4.3, and the relative standard deviation in the mass calculation (s/C) is 0.06.

The standard deviation on measurement of water–solids ratio is 0.025, and the variance is 6.25×10^{-4}. The value of C calculated from water–solids ratio is 77.8%, and V_c is 0.05; therefore, s is 0.23. Relative standard deviation (s/C) is thus 0.003, this being lower than that obtained using tin assays. Water–solids ratio is therefore chosen for subsequent evaluations, being the less sensitive component.

Using water–solids ratio,

$$C = 77.8\% \pm 0.46\% \text{ to 95\% confidence limits.}$$

(A computer program—MASSVAR—for such calculations is listed in Appendix 3.)

Maximising the Accuracy of Two-product Recovery Computations

It has been shown that the recovery equation (3.16) is very sensitive to the accuracy of the component values, and to the degree of separation that has taken place. Equation 3.16 can also be written as:

$$R = \frac{Cc}{f} \tag{3.26}$$

where

$$C = \frac{100(f - t)}{c - t}. \tag{3.24}$$

C represents the percentage of the total feed weight which reports to the concentrate. This value can often be calculated by using components other than the component whose recovery is being determined.[39]

For instance, in a flotation concentrator treating a copper–gold ore, the recovery of gold into the concentrate may have to be assessed. If the gold assays are low (particularly the tailings assay), and not well separated, then the recovery from equation 3.16 is prone to much uncertainty. However, if the copper assays are used to determine the value of C, then only the assays of gold in the concentrate and feed are required, the recovery being evaluated from equation 3.26. The choice of "mass-fraction" component can be determined by sensitivity analysis. Equation 3.24 can be written as:

$$M = \frac{100(a - d)}{b - d} \qquad (3.27)$$

where a, b and d are the mass-fraction components in feed, concentrate and tailings respectively, these components being independent of f, c and t, and M is the value of C calculated from these components. Hence:

$$R = Mc/f. \qquad (3.28)$$

From equation 3.25:

$$V_M = \frac{100^2}{(b - d)^2} \left[V_a + \left(\frac{a - d}{b - d} \right)^2 V_b + \left(\frac{b - a^2}{b - d} \right) V_d \right] \qquad (3.29)$$

Providing that estimates of component variance are known, then V_M can be calculated. If a number of components (e.g. a complete size analysis) are available, equation 3.29 can be used to select the least sensitive component as the mass-fraction component. The component will be that which produces the lowest value of relative standard deviation (RSD) in the mass calculations:

$$\text{RSD}(M) = V_M^{1/2}/M. \qquad (3.30)$$

Having chosen the mass-fraction component, the value of required component recovery can be calculated from:

$$R = \frac{100c(a - d)}{f(b - d)}. \qquad (3.31)$$

The variance in calculation of recovery can be found from equation 3.28, i.e.:

$$V_R = \left(\frac{\partial R}{\partial M} \right)^2 V_M + \left(\frac{\partial R}{\partial c} \right)^2 V_c + \left(\frac{\partial R}{\partial f} \right)^2 V_f.$$

Therefore:

$$V_R = \left(\frac{c}{f}\right)^2 V_M + \left(\frac{M}{f}\right)^2 V_c + \left(\frac{Mc}{f^2}\right)^2 V_f \qquad (3.32)$$

providing that c and f are independent of b and a.
  Combining equations 3.29 and 3.32:

$$V_R = \frac{100^2 c^2}{(b-d)^2 f^2}\left[V_a + \left(\frac{a-d}{b-d}\right)^2 V_b + \left(\frac{b-a}{b-d}\right)^2 V_d \right.$$

$$\left. = \left(\frac{a-d}{c}\right)^2 V_c + \left(\frac{a-d}{f}\right)^2\right] V_f. \qquad (3.33)$$

  Should the mass-fraction component correspond to the recovery component, then equation 3.22 must be used to express recovery variance.

Example 3.16

  Calculate the recovery of tin into the concentrate of the spiral described in Example 3.15. Show how the accuracy of the recovery calculation is improved by using water–solids ratio as the mass-fraction component.

Solution

  Using tin assays, the recovery of tin into the concentrate (equation 3.16) is 82.5%, and, from equation 3.22, V_R is 11.6. The standard deviation is thus 3.16, and:

  Recovery = 82.5 ± 6.8% to 95% confidence limits.

Since it was shown in Example 3.15 that the relative standard deviation in the mass calculation is lower when using water–solids ratio rather than tin assays, water–solids ratio is chosen as the mass-fraction component, and, from equation 3.28, $R = 83.7\%$, and, from equation

3.33, $V_R = 1.60$, and standard deviation is 1.27. Therefore, recovery = $83.7 \pm 2.5\%$ to 95% confidence limits.

This method of maximising two-product recovery calculations is useful for assessment of unit operations, as once a preliminary survey has been made on the separator to determine the most suitable mass-fraction component, then this component can be routinely assessed for further evaluations with a high degree of confidence. A computer program—MAXREC—for performing such computations is listed in Appendix 3.

Introduction to Mass Balances on Complex Circuits

A concentrator, no matter how complex, can be broken down into a series of unit operations, each of which can be assessed by a two-product balance. For instance, in a complex flotation process, such a unit operation might be the composite feed to the roughing circuit, which splits into a rougher concentrate (cleaner feed) and rougher tailings (scavenger feed).

Example 3.17

25 t/h of ore containing 5% lead is fed to a bank of flotation cells.

A high-grade concentrate is produced, assaying 45% lead. The high-grade tailings assay 0.7% lead, and feed the low-grade cells, which produce a concentrate grading 7% lead. The low-grade tailings contain 0.2% lead. Calculate the weight of high- and low-grade concentrates produced per hour, and the recovery of lead produced in the bank of cells.

Solution

(a) *Balance on high-grade circuit*

From equation 3.15:

$$\frac{C}{25} = \frac{5 - 0.7}{45 - 0.7},$$

from which, $C = 2.43$ t/h.

Mass flowrate of high-grade tailings

$$= 25 - 2.43 = 22.57 \text{ t/h.}$$

(b) *Balance on low-grade circuit*

From equation 3.15:

$$\frac{C}{22.57} = \frac{0.7 - 0.2}{7 - 0.2}.$$

from which, $C = 1.66$ *t/h.*

Weight of lead in concentrates

$$= \frac{2.43 \times 45}{100} + \frac{1.66 \times 7}{100}$$

$$= 1.21 \text{ t/h.}$$

Therefore, recovery of lead to concentrates

$$= \frac{1.21 \times 100 \times 100}{25 \times 5} = 96.8\%.$$

In this simple example it is not difficult to assess the number of streams which must be sampled in order to produce data for a unique set of equations for the system. However, in order to calculate a steady-state mass balance for an entire complex circuit, a more analytical method of generating n linear equations for n unknowns is required. Any plant flowsheet can be reduced to a series of *nodes*, where process streams either join or separate. *Simple nodes* have either one input and two outputs (a *separator*) or two inputs and one output (a *junction*) (Fig. 3.22). It has been shown[40] that, providing the mass flow of a reference stream (usually the feed) is known, the minimum number of streams which must be sampled to ensure production of a complete circuit mass balance is:

$$N = 2(F + S) - 1 \tag{3.34}$$

(a)   (b)

FIG. 3.22. Simples nodes: (a) separator, (b) junction.

where F = number of feed streams,
     S = number of simple separators.

The flowsheet described in Example 3.17 can be reduced to node form (Fig. 3.23).

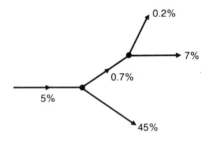

FIG. 3.23. Simple node flowsheet.

The circuit contains two simple separator nodes, and hence the minimum number of streams that must be sampled is:

$$2 (1 + 2) - 1 = 5,$$

i.e. all streams must be sampled in order to produce a balance.

Separators which produce more than two products, or junctions which are fed by more than two streams, can be cascaded into simple nodes by connecting them with streams which have no physical existence. For instance, the flotation bank shown in Fig. 3.24a can be reduced to node form (Fig. 3.24b) and cascaded into simple nodes (Fig. 3.24c).

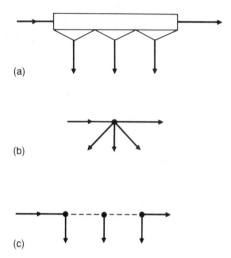

FIG. 3.24. Flotation bank: (a) flowsheet, (b) in node form, (c) in simple nodes.

The minimum number of streams that must be sampled is thus:

$$2 (1 + 3) - 1 = 7$$

and since only five streams can be sampled, two more weights are required to supplement the reference weight.

It can be seen from Figs. 3.24b and 3.24c that a node producing two products can be cascaded to three simple separator nodes, and, in general, if a separator produces n products, then this can be cascaded to $n - 1$ simple nodes. This is useful, as reducing a very complex plant to simple nodes, cascaded with non-existent streams, can lead to confusion and error. A procedure has been developed by Frew[41] which allows easy automation and provides a check on the count-up of nodes from the flow diagram.

The method involves the use of the *connection-matrix C*,[42] where each element in the matrix is

$$C_{ij} = \begin{cases} + \text{ for stream } j \text{ flowing into the } i\text{th node} \\ - 1 \text{ for stream } j \text{ flowing out of the } i\text{th node} \\ 0 \text{ for stream } j \text{ not appearing at the } i\text{th node} \end{cases}$$

A computer program (CONMAT) for preparing and analysing the connection matrix is listed in Appendix 3, and the following examples illustrate the use of the method:

Consider the flowsheet shown in Fig. 3.25a. This can be reduced to the node flowsheet shown in Fig. 3.25(b).

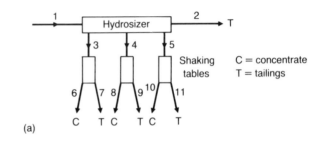

(a)

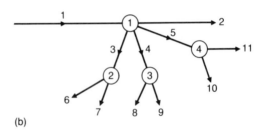

(b)

FIG. 3.25. (a) circuit flowsheet, (b) flowsheet in node form.

There are eleven flowstreams and four nodes. The connection-matrix thus has eleven rows and four columns as shown below:

$$C = \begin{array}{ccccccccccc} 1 & -1 & -1 & -1 & -1 & 0 & 0 & 0 & 0 & 0 & 0 \\ 0 & 0 & 1 & 0 & 0 & -1 & -1 & 0 & 0 & 0 & 0 \\ 0 & 0 & 0 & 1 & 0 & 0 & 0 & -1 & -1 & 0 & 0 \\ 0 & 0 & 0 & 0 & 1 & 0 & 0 & 0 & 0 & -1 & -1 \end{array}$$

The contents of each column represent the individual streams, and when summed must equal $+1$, -1 or 0, any other result indicating an error in the input of data, i.e. Column sum = $\begin{cases} +1 & - \text{ stream is a feed} \\ -1 & - \text{ stream is a product} \\ 0 & - \text{ stream is internal stream} \end{cases}$

Therefore, summation of columns shows that stream 1 is a feed, streams 2, 6, 7, 8, 9, 10, 11 are products, and streams 3, 4 and 5 are internal streams.

The elements of each row represent the individual nodes, and if the number of "+1" entries (n_p) and the number of "−1" entries (n_n) are counted, then n_p and n_n can be used to assess the number of simple nodes:

Number of simple junctions $(J) = n_p - 1$
Number of simple separators $(S) = n_n - 1$.

The nodes can now be classified as below:

Node	n_p	n_n	J	S
1	1	4	0	3
2	1	2	0	1
3	1	2	0	1
4	1	2	0	1
			0	6

There are thus six simple separators, and no junctions, and the minimum number of streams that must be sampled is:

$$2(1 + 6) - 1 = 13.$$

Since there are only eleven available streams, two additional mass flows are required to supplement the reference stream in order to produce a balance. It is important that when additional mass flow measurements are required, no subset of flow measurements includes all streams at a node or group of nodes. Mass measurements on flowstreams 6 and 7, for example, will provide complete mass data for node 2 and a unique balance will not be produced.

Consider the circuit shown in Fig. 3.26a. The circuit has been reduced to node form in Fig. 3.26b. Note that the ball mill has been left out of this circuit, as it is an *abnormal node* where no separation takes place, and so there is no change in the overall assay or flowrate at steady state. Note that there will be a change in size distribution at this abnormal node, so size analysis data should not be used as the balancing components between nodes linked by this stream. Only components that are conserved at nodes can be used.

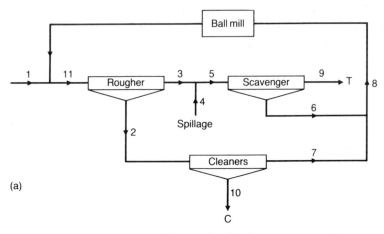

(a)

FIG. 3.26. (a) Flotation flowsheet.

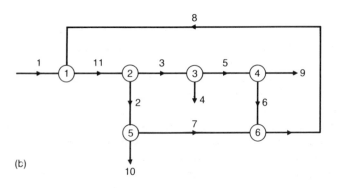

(b)

FIG. 3.26. (b) Flotation circuit in node form.

There are eleven flowstreams and six nodes, which can be represented by the connection-matrix:

$$C = \begin{bmatrix} 1 & 0 & 0 & 0 & 0 & 0 & 0 & 1 & 0 & 0 & -1 \\ 0 & -1 & -1 & 0 & 0 & 0 & 0 & 0 & 0 & 0 & 1 \\ 0 & 0 & 1 & 1 & -1 & 0 & 0 & 0 & 0 & 0 & 0 \\ 0 & 0 & 0 & 0 & 1 & -1 & 0 & 0 & -1 & 0 & 0 \\ 0 & 1 & 0 & 0 & 0 & 0 & -1 & 0 & 0 & -1 & 0 \\ 0 & 0 & 0 & 0 & 0 & 1 & 1 & -1 & 0 & 0 & 0 \end{bmatrix}$$

The column identifies streams 1 and 4 as feeds, streams 9 and 10 as products, and the other streams as internal flows.

The node classification is:

Node	n_p	n_n	J	S
1	2	1	1	0
2	1	2	0	1
3	2	1	1	0
4	1	2	0	1
5	1	2	0	1
6	2	1	1	0
			3	3

The system thus consists of three simple junctions and three simple separators, and $N = 2\,(2 + 3) - 1 = 9$. Therefore, although there are eleven available streams, only nine need be sampled to produce a balance. In choosing the nine streams, all feeds and products should be part of the set and the connection matrix can be used to determine the choice of remaining streams. If, in this example, stream 1 is the reference stream, then streams 2 to 11 are of unknown weight. Ten independent linear equations are therefore required to determine the mass flows of each stream relative to stream 1. A material balance can be performed on each node, giving six equations, and a component balance on the plant feeds and products provides an extra equation. Three component balances are thus required on the circuit nodes. It is apparent that if streams 3 and 7 are not sampled, then component balances are not possible on nodes which include either of these streams, i.e. nodes 2, 3, 5 and 6. Only two nodes are available for component balances, and insufficient independent equations are available. If, however, streams 3 and 5 are not sampled, then component balances are possible on nodes 1, 5 and 6, and a consistent set of equations is produced. It is also apparent that if sampling of only stream 3 is omitted, then ten linear equations can be produced from the six separate nodes, and the feed and product component balance becomes redundant. If the experimental data were entirely free of error, then the choice, if it exists, of the set of nine streams required would be of no consequence, as each complete set would yield an identical balance. Since experimental error does exist, the choice of flowstreams required to produce the balance is important, as certain streams may increase the sensitivity to error. For instance, a balance at a junction where little component separation takes place is prone to error. Smith and Frew[40] have developed a sensitivity analysis technique which indicates which equations should be used in a minimum variance mass balance to obtain least sensitivity to data error. The procedures used also show that, where possible, measurement of mass flow should be performed, as this reduces sensitivity to experimental error. Each additional mass flow measurement reduces N by one, providing that, as stated earlier, the location for mass flow measurement is not chosen such that all the mass flows at any node are known, i.e. mass flows should not produce data which can be calculated from the available component measurements. In this respect, a concentrator can be reduced to a single separator node, such that if the feed mass flowrate is known, measurement of the concentrate mass flowrate enables the tailings mass flowrate to be directly calculated, so that, although this information may be of use, it cannot be used in the overall balance.

It has been indicated that the connection matrix can be used to provide the set of linear equations that must be solved in order to produce the stream mass flowrates.

A material matrix, M can be defined, where each element in the matrix is

$$M_{ij} = C_{ij}B_j$$

where B_j represents the mass flowrate of solids in stream j.

Using the flotation circuit (Fig. 3.26) connection matrix as an example, each row in the matrix generates a linear equation representing a material balance. For instance, row 2 is:

$$C_{2j} = 0 -1 -1\ 0\ 0\ 0\ 0\ 0\ 0\ 0\ 1$$

and the material matrix M_{2j} at node 2 is thus:

$$-B_2 - B_3 - B_{11} = 0.$$

A component matrix, A, can also be defined, where each matrix element is

$$A_{ij} = C_{ij}B_j a_j = M_{ij}a_j$$

a_j representing the component value (assay, % in size fraction, dilution ratio, etc.) in stream j, which gives at node 2:

$$-B_2 a_2 - B_3 a_3 - B_{11}a_{11} = 0.$$

At any particular node, it is important that the same component is used to assess each stream, and the component should be chosen so as to produce an equation with least sensitivity to error. The component can be selected by sensitivity analysis, and providing that the same component is used at any particular node, other components can be used to balance other nodes in the circuit. This means that in a complex circuit balance, components such as metal content, dilution ratios, and size analyses may be utilised in various parts of the circuit.[43]

Combining M_{ij} and A_{ij} into one matrix produces:

$$M_{11}\ M_{12}............................M_{1s}$$
$$M_{21}\ M_{22}............................M_{2s}$$
$$.$$
$$.$$
$$.$$
$$M_{n1}\ M_{n2}............................M_{ns}$$
$$A_{11}\ A_{12}............................A_{1s}$$
$$.$$
$$.$$
$$A_{n1}\ A_{n2}............................A_{ns}$$

where s = number of streams, and n = number of nodes.
If stream s is the reference stream (preferably a feed), and $B_s = 1$,

then B_j represents the fraction of the reference stream reporting to stream j. Since $B_s = 1$, $M_{1s} = C_{1s}$, and $A_{1s} = C_{1s} a_s$.

Hence, in matrix form, the set of linear equations that must be solved is:

$$
\begin{vmatrix}
C_{11} \ldots \ldots C_{1(s-1)} \\
C_{21} \ldots \ldots C_{2(s-1)} \\
\cdot \\
\cdot \\
\cdot \\
C_{n1} \ldots \ldots C_{n(s-1)} \\
C_{11}a_1 \ldots \ldots C_{1(s-1)}a_{(s-1)} \\
C_{21}a_1 \ldots \ldots C_{2(s-1)}a_{(s-1)} \\
\cdot \\
\cdot \\
\cdot \\
C_{n1}a_1 \ldots \ldots C_{n(s-1)}a_{(s-1)}
\end{vmatrix}
\begin{vmatrix}
B_1 \\
B_2 \\
\cdot \\
\cdot \\
\cdot \\
\cdot \\
\cdot \\
\cdot \\
\cdot \\
\cdot \\
B_{(s-1)}
\end{vmatrix}
=
\begin{vmatrix}
-C_{1s} \\
-C_{2s} \\
\cdot \\
\cdot \\
\cdot \\
-C_{ns} \\
-C_{1s}a_s \\
-C_{2s}a_s \\
\cdot \\
\cdot \\
-C_{ns}a_s
\end{vmatrix}
$$

A further equation can be included in the set. The plant can be represented as a single node, such that the weight of contained component in the feed is equal to the component weight in the products. This equation should be used if possible, as there is usually very good component separation at this node. The material balance on this node cannot, however, be included in the set, as it is not independent of the set of material balance equations on the internal nodes.

Example 3.18

The circuit shown in Fig. 3.27 is sampled, and the following results obtained:

Stream	Assay % metal
1	Not sampled
2	0.51
3	0.12
4	16.1
5	4.2
6	25.0
7	Not sampled
8	2.1
9	1.5

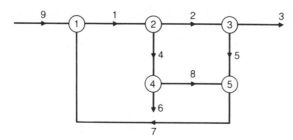

FIG. 3.27. Flotation circuit in node form.

Confirm, by the use of the connection matrix, that sufficient data has been obtained to calculate all the mass flowrates, and use the connection matrix to calculate the flows.

Solution

The connection matrix is:

$$\begin{matrix} -1 & 0 & 0 & 0 & 0 & 0 & 1 & 0 & 1 \\ 1 & -1 & 0 & -1 & 0 & 0 & 0 & 0 & 0 \\ 0 & 1 & -1 & 0 & -1 & 0 & 0 & 0 & 0 \\ 0 & 0 & 0 & 1 & 0 & -1 & 0 & -1 & 0 \\ 0 & 0 & 0 & 0 & 1 & 0 & -1 & 1 & 0 \end{matrix}$$

The matrix confirms stream 9 as a feed, streams 3 and 6 as products, and the remaining streams as internal streams. The number of simple separators is equal to three, and so the minimum number of streams to be sampled (equation 3.34) is

$$2(3 + 1) - 1 = 7.$$

Assuming stream $B_9 = 1$, then the material matrix (from the connection matrix) is:

$$\begin{matrix} -B_1 & 0 & 0 & 0 & 0 & 0 & B_7 & 0 & 1 \\ B_1 & -B_2 & 0 & -B_4 & 0 & 0 & 0 & 0 & 0 \\ 0 & B_2 & -B_3 & 0 & -B_5 & 0 & 0 & 0 & 0 \\ 0 & 0 & 0 & B_4 & 0 & -B_6 & 0 & -B_8 & 0 \\ 0 & 0 & 0 & 0 & B_5 & 0 & -B_7 & B_8 & 0 \end{matrix}$$

Since streams 1 and 7 were not sampled, component balances on nodes containing these streams cannot be performed. The component matrix (from nodes 3 and 4) is thus:

$$\begin{matrix} 0 & 0.51B_2 & -0.12B_3 & 0 & -4.2B_5 & 0 & 0 & 0 & 0 \\ 0 & 0 & 0 & 16.1B_4 & 0 & -25.0B_6 & 0 & -2.1B_8 & 0 \end{matrix}$$

In order to produce a square matrix, one further equation is required. If the whole circuit is considered as a simple node, then a component balance gives:

$$0.12B_3 + 25.0B_6 - 1.5 = 0.$$

Thus the streams sampled provide sufficient data, and the matrix which must now be solved is:

$$\begin{pmatrix} -1 & 0 & 0 & 0 & 0 & 0 & 1 & 0 \\ 1 & -1 & 0 & -1 & 0 & 0 & 0 & 0 \\ 0 & 1 & -1 & 0 & -1 & 0 & 0 & 0 \\ 0 & 0 & 0 & 1 & 0 & -1 & 0 & -1 \\ 0 & 0 & 0 & 0 & 1 & 0 & -1 & 1 \\ 0 & 0.51 & -0.12 & 0 & -4.2 & 0 & 0 & 0 \\ 0 & 0 & 0 & 16.1 & 0 & -25.0 & 0 & -2.1 \\ 0 & 0 & -0.12 & 0 & 0 & -25.0 & 0 & 0 \end{pmatrix} \begin{pmatrix} B_1 \\ B_2 \\ B_3 \\ B_4 \\ B_5 \\ B_6 \\ B_7 \\ B_8 \end{pmatrix} = \begin{pmatrix} -1 \\ 0 \\ 0 \\ 0 \\ 0 \\ 0 \\ 0 \\ -1.5 \end{pmatrix}$$

This matrix can be solved by Gaussian elimination, and back-substitution (see GAUSSEL in Appendix 3) to give:

$$B_1 = 1.14$$
$$B_2 = 1.04$$
$$B_3 = 0.94$$
$$B_4 = 0.09$$
$$B_5 = 0.10$$
$$B_6 = 0.06$$
$$B_7 = 0.14$$
$$B_8 = 0.04$$

The above example illustrates clearly the advantage of using the connection matrix to produce the necessary set of linear equations to evaluate the circuit. The seven streams sampled produced sufficient data for the evaluation. If, however, streams 2 and 8 had not been sampled, then component balances on nodes 2, 3, 4 and 5 would not have been possible, and insufficient linear equations would have been available.

The connection matrix is the basis for generalised computer packages for mass balancing which have been produced in recent years.[44]

Reconciliation of Excess Data

It has been shown that it is common practice in mass balancing computations to reduce the circuit to simple nodes, and to calculate relative mass flowrates by means of measured components. In many cases an excess of data is available at each node, such as multi-component size analyses, dilution ratios, metal assays, etc., so that it is possible to calculate C (equation 3.24) by a variety of routes, each route being independent of the others, and of apparently equal validity. The problem thus arising is which of these components should be used in order to produce a component balance, and hence which of the components becomes redundant. The approach which has become increasingly adopted is to use all the available data to compute a best estimate of C, and to adjust the data to make the component values consistent with this estimate. These methods are complex when applied to circuits of arbitrary configuration, and require powerful computational facilities.[44-47] For simplicity, therefore, the techniques will be described in relation to simple nodes, as the computer programs required can readily be accommodated by microcomputer.

Two basic methods have commonly been adopted, both of which use a least-squares approach, and they can be broadly classified as:

(a) Minimisation of the sum of squares of the residuals in the component closure equations.
(b) Minimisation of the sum of squares of the component adjustments.

Minimisation of the Sum of the Squares of the Closure Residuals

In this method, the best-fit values of mass flowrates are calculated from the experimental data, after which the data is adjusted to accommodate these estimates.[48–51] If the simple separator streams are each sampled, and assayed for n components, then:

$$f_k - Cc_k - (1 - C)t_k = r_k \tag{3.35}$$

for $k = 1$ to n, and where:

f_k represents the value of component k in the feed stream,
c_k represents the value of component k in the concentrate stream,
t_k represents the value of component k in the tailings stream,
and r_k is the residual in the closure equation generated by experimental errors in the measurements of component k.

Equation 3.35 can be written as:

$$(f_k - t_k) - C(c_k - t_k) = r_k. \tag{3.36}$$

The objective of this method is to choose a value of C which minimises the sum of the squares of the closure errors; i.e. to minimise S, where:

$$S = \sum_{k=1}^{n} (r_k)^2 \tag{3.37}$$

and by substitution from equation 3.36:

$$S = \sum_{k=1}^{n} (f_k - t_k)^2 + C^2 \sum_{k=1}^{n} (c_k - t_k) - 2C \sum_{k=1}^{n} (f_k - t_k)(c_k - t_k). \tag{3.38}$$

The value of S cannot be zero at any value of C unless the data is consistent. However, it has a minimum value when $dS/dC = 0$ (Fig. 3.28), i.e. when:

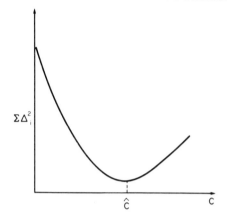

FIG. 3.28. Plot of the sum of the squares of the component errors versus values of C.

$$2\hat{C} \sum_{k=1}^{n} (c_k - t_k) - 2 \sum_{k=1}^{n} (f_k - t_k)(c_k - t_k) = 0$$

where $\hat{C}$ = best-fit estimate of C.

Therefore:

$$\hat{C} = \frac{\sum_{k=1}^{n} (f_k - t_k)(c_k - t_k)}{\sum_{k=1}^{n} (c_k - t_k)^2} \qquad (3.39)$$

this value being most influenced by the values of the components which are most altered by the separation process.

Having determined $\hat{C}$, the next stage is to adjust the component values to make them consistent with the calculated flowrates. The closure errors (equation 3.35) must be distributed between the component values such that:

$$\hat{f}_k - \hat{C}\hat{c}_k - (1 - \hat{C})\hat{t}_k = 0 \qquad (3.40)$$

where $\hat{f}_k$, $\hat{c}_k$, and $\hat{t}_k$ are the best-fit values of component k in the three streams, i.e.

$$(f_k - f_{ka}) - \hat{C}(c_k - c_{ka}) - (1 - \hat{C})(t_k - t_{ka}) = 0 \qquad (3.41)$$

where f_{ka}, c_{ka}, t_{ka} are the adjustments to the kth component values in the three streams.

The closure equation (3.35) can be written as:

$$f_k - \hat{C}c_k - (1 - \hat{C})t_k = r_k \qquad (3.42)$$

and subtracting equation 3.41 from this gives:

$$r_k = f_{ka} - \hat{C}c_{ka} - (1 - \hat{C})t_{ka}. \qquad (3.43)$$

Using a least-squares approach, the sum of squares to be minimised is S_a, where:

$$S_a = \sum_{k=1}^{n} (f_{ka}^2 + c_{ka}^2 + t_{ka}^2) \qquad (3.44)$$

subject to the constraint of equation 3.43.

This minimisation problem can be most conveniently solved by the method of Lagrangian multipliers. In this method, the constraints are expressed in such a way that they equal zero; i.e. equation 3.43 becomes:

$$r_k - f_{ka} + \hat{C}c_{ka} + (1 - \hat{C})t_{ka} = 0. \qquad (3.45)$$

The minimisation problem requires all the adjustments to be as small as possible, and the Lagrangian technique involves minimisation of the "Lagrangian L" defined as:

$$L = \sum_{k=1}^{n} (f_{ka}^2 + c_{ka}^2) + 2 \sum_{k=1}^{n} \lambda_k (\text{constraint } k) \qquad (3.46)$$

where $2\lambda_k$ is the Lagrangian multiplier for the constraint equation k. Thus:

$$L = S_a + 2 \sum_{k=1}^{n} \lambda_k (r_k - f_{ka} + \hat{C}c_{ka} + (1 - \hat{C})t_{ka}). \qquad (3.47)$$

L is differentiated with respect to each of the unknowns (adjustments and multipliers), and the derivatives are set to zero.

Hence, $\partial L / \partial f_{ka} = 2f_{ka} - 2\lambda_k = 0$

i.e. $f_{ka} = \lambda_k$ $\qquad\qquad\qquad\qquad\qquad\qquad\qquad$ (3.48)

$$\partial L/\partial c_{ka} = 2c_{ka} + 2\lambda_k \hat{C} = 0$$

i.e. $c_{ka} = -\lambda_k \hat{C}$ \hfill (3.49)

$$\partial L/\partial t_{ka} = 2t_{ka} + 2\lambda_k(1 - \hat{C}) = 0$$

i.e. $t_{ka} = -\lambda_k(1 - \hat{C})$ \hfill (3.50)

$$\partial L/\partial \lambda_k = 2(r_k - f_{ka} + \hat{C}c_{ka} + (1 - \hat{C})t_{ka} = 0$$

and substituting for f_{ka}, c_{ka}, t_{ka}:

$$r_k = \lambda_k(1 + \hat{C}^2 + (1 - \hat{C})^2)$$

$$= h\lambda_k,$$

where $h = 1 + \hat{C}^2 + (1 - \hat{C})^2$. \hfill (3.51)

Hence:

$$f_{ka} = r_k/h \qquad (3.52)$$

$$c_{ka} = -\hat{C}r_k/h \qquad (3.53)$$

$$t_{ka} = -(1 - \hat{C})r_k/h. \qquad (3.54)$$

Therefore, once $\hat{C}$ has been determined, h can be calculated from equation 3.51, and r_k from equation 3.42. The appropriate component value adjustments can then be calculated from equations 3.52–54. The results of a six-component separation on a single node are presented in Table 3.8, together with adjusted data values, and a best-fit estimate of C (see Appendix 3 for listing of computer program—LAGRAN).

TABLE. 3.8. SEPARATOR EVALUATION USING MINIMISATION OF CLOSURE ERRORS

Component	Actual assay %			C	Adjusted assay %		
	Feed	Conc.	Tails	(%)	Feed	Conc.	Tails
Tin	21.90	43.00	6.77	41.76	20.78	43.41	7.47
Iron	3.46	5.50	1.76	45.45	3.25	5.58	1.89
Silica	58.00	25.10	75.30	34.46	57.16	25.41	75.83
Sulphur	0.11	0.12	0.09	66.67	0.10	0.12	0.09
Arsenic	0.36	0.38	0.34	50.00	0.36	0.38	0.34
TiO$_2$	4.91	9.24	2.07	39.61	4.79	9.28	2.15

Best-fit value of $C(\hat{C}) = 37.03\%$.

Experimental and adjusted size analysis data on the streams from a hydrocyclone are tabulated in Table 3.9, U representing the proportion of the feed mass reporting to the cyclone underflow. The best estimate of U from the available data is 84.6%.

It is interesting to compare the results that are obtained if the data are presented in cumulative form rather than as a fractional distribution. Table 3.10 expresses the size analyses as cumulative oversize. It can be seen that the adjusted data differ slightly from the non-cumulative adjusted data, and the best-fit estimate of U is 85.2%. Identical results to those shown in Table 3.10 are produced by using cumulative undersize data. This clearly illustrates that the mass balance results obtained depend not only on the method which is used to deal with the data, but also on the manner in which the information is presented. This is due to the difference in error structure in non-cumulative and cumulative data. On a cumulative basis, the errors are added and biases are introduced into the data, and because of this it is preferable to use non-cumulative data.[48]

TABLE 3.9. HYDROCYCLONE STREAM ANALYSES—WEIGHT % IN EACH FRACTION

Size range microns	Experimental data				Adjusted data		
	Feed	U/flow	O/flow	U (%)	Feed	U/flow	O/flow
+425	3.6	2.4	13.3	88.99	3.88	2.17	13.26
355–425	3.2	2.0	13.4	89.47	3.52	1.73	13.35
300–355	3.9	2.1	11.1	80.00	3.66	2.30	11.14
250–300	3.5	2.1	10.4	83.13	3.43	2.16	10.41
212–250	5.5	3.4	18.7	86.27	5.65	3.28	18.68
180–212	5.3	3.3	12.9	79.17	5.00	3.55	12.95
150–180	6.6	5.9	12.9	90.00	6.82	5.72	12.87
125–150	7.9	8.0	7.1	88.89	7.88	8.02	7.10
106–125	10.2	11.0	0.1	92.66	9.69	11.43	0.18
−106	50.3	59.8	0.1	84.09	50.48	59.65	0.07

Best estimate of $U(\%) = 84.60$.

TABLE 3.10. HYDROCYCLONE STREAM ANALYSES—CUMULATIVE OVERSIZE %

Size microns	Experimental data				Adjusted data		
	Feed	U/flow	O/flow	U(%)	Feed	U/flow	O/flow
425	3.6	2.4	13.3	88.99	3.84	2.20	13.27
355	6.8	4.4	26.7	89.24	7.32	3.96	26.62
300	10.7	6.5	37.8	86.58	10.95	6.29	37.76
250	14.2	8.6	48.2	85.86	14.35	8.47	48.18
212	19.7	12.0	66.9	85.97	19.94	11.79	66.86
180	25.0	15.3	79.8	84.96	24.91	15.38	79.81
150	31.6	21.2	92.7	85.45	31.70	21.11	92.68
125	39.5	29.2	99.8	85.41	39.59	29.13	99.79
106	49.7	40.2	99.9	84.09	49.32	40.52	99.96
−106	50.3	59.8	0.1	84.09	50.68	59.48	0.04

Best estimate of $U(\%) = 85.20$.

Minimisation of the Sum of Squares of the Component Adjustments

In this method, a residual is defined in each measured component, so that the calculated component values are consistent with estimated values of C.[52] The residuals are squared and summed, and the best-fit value of C is that which minimises this sum of squares. The objective, therefore, is to minimise S, where:

$$S = \sum_{i=1}^{3} \sum_{k=1}^{n} (x_{ik} - \hat{x}_{ik})^2 \qquad (3.55)$$

where n = number of component assays on each stream,

x_{ik} = measured value of component k in stream i,

$\hat{x}_{ik}$ = adjusted value of component k in stream i.

C is chosen iteratively such that equation 3.40 is satisfied. The data adjustments which must be made in order to satisfy this condition can be calculated from equations 3.42 and 3.51–54, and are used to calculate S. The method searches for a value of C which minimises the value of S.

Some workers[53,54] solve this minimisation problem directly, by differentiating the Lagrangian equation (equation 3.47) not only with respect to f_{ka}, c_{ka}, t_{ka} and λ_{ka}, but also with respect to $\hat{C}$, which is not separately calculated from the raw data. The results obtained are identical to those achieved by iteration, but the direct method has computational advantages over iteration, particularly when assessing complex circuits. Combining equations 3.42 and 3.47:

$$L = S_a + 2 \sum_{k=1}^{n} \lambda_k [f_k - t_k - f_{ka} + t_{ka} - \hat{C}(c_k - t_k - c_{ka} + t_{ka})].$$

(3.56)

Differentiating with respect to $\hat{C}$, and substituting for c_{ka} and t_{ka} (from equations 3.53 and 3.54):

$$\partial L / \partial \hat{C} = -2 \sum_{k=1}^{n} \lambda_k [c_k - t_k + \hat{C} r_k / h - (1 - \hat{C}) r_k / h].$$

Setting this derivative to zero

$$\sum_{k=1}^{n} \lambda_k (c_k - t_k) + \left[\hat{C} \sum_{k=1}^{n} \lambda_k r_k / h \right] - \left[(1 - \hat{C}) \sum_{k=1}^{n} \lambda_k r_k / h \right] = 0$$

and since $\lambda_k = r_k / h$:

$$\sum_{k=1}^{n} r_k (c_k - t_k) / h + \hat{C} \sum_{k=1}^{n} (r_k / h)^2 - (1 - \hat{C}) \sum_{k=1}^{n} (r_k / h)^2 = 0 \quad (3.57)$$

where $r_k = (f_k - t_k) - \hat{C}(c_k - t_k)$ (equation 3.36)

and $h = 1 + \hat{C}^2 + (1 - \hat{C})^2$.   (equation 3.51)

Equation 3.57 can be expanded and reduced to a quadratic equation, which can be written as:

$$\hat{C}^2 (X - 2Z) + 2\hat{C}(Y - Z) + (2Z - Y) = 0 \quad (3.58)$$

where

$$X = \sum_{k=1}^{n} (c_k - t_k)^2 \quad (3.59)$$

$$Y = \sum_{k=1}^{n} (f_k - t_k)^2 \tag{3.60}$$

$$Z = \sum_{k=1}^{n} (c_k - t_k)(f_k - t_k). \tag{3.61}$$

This can be solved by applying the general solution to a quadratic:

$$C = \frac{-2(Y - Z) \pm \{[2(Y - Z)]^2 - 4(X - 2Z)(2Z - Y)\}^{1/2}}{2(X - 2Z)}. \tag{3.62}$$

Equation 3.62 has two roots, the +ve fraction being the true result. Once $\hat{C}$ is found, the data adjustment proceeds as before, according to equations 3.52–54.

The two methods described have been compared by Wills and Manser,[55] who conclude that for most practical purposes the results achieved by both methods are essentially the same.

Weighting the Adjustments

It is evident from the results of the six-component separation (Table 3.8) that $\hat{C}$ is biased towards the silica assay data, i.e. towards the assay values which are relatively high, but not necessarily most effectively separated. This is due to the assumption that the total absolute error in the experimental data is distributed equally to each assay value, or in other words that each asssay value contains the same absolute error, which is highly unlikely in practice. It is more likely that the absolute error in each value is proportional to the assay itself (i.e. the *relative* error is constant), and where multi-component assays are used, as in the example discussed, each component may have a different relative error, which may also be dependent on the assay value. It is therefore preferable to weight the component adjustments such that good data is adjusted relatively less than poor data. The weighting factor which is most used is the inverse of the estimated component variance, such that equation 3.55 becomes:

$$S = \sum_{i=1}^{3} \sum_{k=1}^{n} (x_{ik} - \hat{x}_{ik})^2/V_{ik} \qquad (3.63)$$

where V_{ik} is the variance in the measurement of component k in stream i.

Similarly, the Lagrangian equation (equation 3.56) becomes:

$$L = \sum_{k=1}^{n} \left(\frac{f_{ka}^2}{V_{fk}} + \frac{c_{ka}^2}{V_{ck}} + \frac{t_{ka}^2}{V_{tk}} \right) + 2 \sum_{k=1}^{n} \lambda_k [f_k - t_k - f_{ka} + t_{ka}$$

$$- \hat{C}(c_k - t_k - c_{ka} + t_{k_a})] \qquad (3.64)$$

which can be solved by partial differentiation and reducing to a quadratic, as before.

If the variance is assumed proportional to the value of the assay, then equation 3.63 becomes:

$$S = \sum_{i=1}^{3} \sum_{k=1}^{n} \left(\frac{x_{ik} - \hat{x}_{ik}}{e_k x_{ik}} \right)^2 \qquad (3.65)$$

where e_k = relative error in measurement of component k.

In the minimisation of closure residuals method, equation 3.37 is weighted with the variances in the closure errors (V_{rk}):

$$S = \sum_{k=1}^{n} (r_k)^2/V_{rk} \qquad (3.66)$$

and since the variance of a function can be found from its derivatives (equation 3.21), then for random changes (i.e. measurement errors) in f_k, c_k and t_k:

$$V_{rk} = \left(\frac{\partial r_k}{\partial f_k} \right)^2 V_{fk} + \left(\frac{\partial r_k}{\partial c_k} \right)^2 V_{ck} + \left(\frac{\partial r_k}{\partial t_k} \right)^2 V_{tk}$$

so from equation 3.42 as shown by Lynch[49]:

$$V_{rk} = V_{fk} + \hat{C}^2 V_{ck} + (1 - \hat{C})^2 V_{tk} \qquad (3.67)$$

where V_{fk}, V_{ck} and V_{tk} are the variances in measurement of f, c and t respectively for component k.

Equation 3.39 becomes:

$$\hat{C} = \frac{\sum_{k=1}^{n} (f_k - t_k)(c_k - t_k)/V_{rk}}{\sum_{k=1}^{n} (c_k - t_k)^2/V_{rk}} . \qquad (3.68)$$

The expression for V_{rk} (equation 3.67) contains $\hat{C}$, so the calculation must be performed iteratively. An estimated value of $\hat{C}$ is introduced into equation 3.68, and a new value of $\hat{C}$ calculated. This value is used in the calculation until the estimated and calculated values converge, which is usually after only a few iterations. Once $\hat{C}$ has been determined, the data is adjusted as before, according to the following equations:

$$f_{ka} = r_k V_{fk}/h_k \qquad (3.69)$$

$$c_{ka} = -\hat{C} r_k V_{ck}/h_k \qquad (3.70)$$

$$t_{ka} = -(1 - \hat{C})r_k V_{tk}/h_k \qquad (3.71)$$

where

$$h_k = V_{fk} + \hat{C}^2 V_{ck} + (1 - \hat{C})^2 V_{tk} \qquad (3.72)$$

equations 3.69–72 being the weighted equivalents of equations 3.51–54. Analysis of the six-component separation, assuming equal relative error on all assay data, produces very similar results using the methods of direct data adjustment and the closure residual minimisation method (Table 3.11), although this is not a general observation.[55] It is evident, however, that the bias towards the high value components (silica assays) has now been removed (see Appendix 3 for computer program—WEGHTRE—which reconciles excess data by a weighted least-squares method).

It has been proposed by Wills and Manser[55] that a more significant estimate of $\hat{C}$ can be obtained by weighting the mean of the component calculations of C by the standard deviation in the calculation of each value of C, according to the equation:

TABLE. 3.11. SEPARATION EVALUATION ASSUMING EQUAL RELATIVE ERROR ON ASSAY
VALUES

Component	Actual assay % Feed	Conc.	Tails	C (%)	Adjusted assay % Feed	Conc.	Tails
Tin	21.90	43.00	6.77	41.76	21.80	43.16	6.78
Iron	3.46	5.50	1.76	45.45	3.36	5.61	1.78
Silica	58.00	25.10	75.30	34.46	55.87	25.26	77.41
Sulphur	0.11	0.12	0.09	66.67	0.10	0.12	0.09
Arsenic	0.36	0.38	0.34	50.00	0.36	0.38	0.34
TiO$_2$	4.91	9.24	2.07	39.61	4.98	9.13	2.06

Best-fit value of $C(\hat{C}) = 41.30\%$.

$$\hat{C} = \sum_{k=1}^{n} \frac{(f_k - t_k)}{(c_k - t_k)(Vc_k)^{1/2}} \bigg/ \sum_{k=1}^{n} \left(\frac{1}{Vc_k}\right)^{1/2}$$

where

$$Vc_k = \frac{V_{fk}}{(c_k - t_k)^2} + \left[\frac{f_k - t_k}{(c_k - t_k)^2}\right]^2 Vc_k + \left[\frac{c_k - f_k}{(c_k - t_k)^2}\right]^2 Vt_k .$$

Once $\hat{C}$ has been determined, the data values are adjusted as before, using equations 3.69–72. A computer program (WILMAN) which reconciles excess data in this manner is listed in Appendix 3, together with an evalution of the six-component separation.

References

1. Lister, B., Sampling for quantitative analysis, *Min. Mag.* 221 (Sept. 1980).
2. Gy, P. M., *Sampling of Particulate Materials: Theory and Practice*, Elsevier Scientific Publishing Co., Amsterdam, 1979.
3. Carson, R., and Acornley, D., Sampling of pulp streams by means of a pneumatically operated poppet valve, *Trans. IMM* Sec. C **82**, 46 (Mar. 1973).
4. Strasham, A., and Steele, T. W., *Analytical Chemistry in the Exploration, Mining and Processing of Materials*, Pergamon Press, Oxford, 1978.
5. Cooper, H. R:, Recent developments in on-line composition analysis of process streams, in *Control '84 Minerals/Metallurgical Processing*, ed. J. A. Herbst, p. 29, AIMME, New York, 1984.

6. Kawatra, S. K., and Cooper, H. R., On-stream composition analysis of mineral slurries, in *Design and Installation of Concentration and Dewatering circuits*, eds. A. L. Mular and M. A. Anderson, p. 641, SME Inc., Littleton, 1986.

7. Lyman, G. J., On-stream analysis in mineral processing, in *Mineral and Coal Flotation Circuits*, (A. J. Lynch *et al.*), Elsevier Scientific Publishing Co., Amsterdam, 1981.

8. Bergeron, A., and Lee, D. J., A practical approach to on-stream analysis, *Min. Mag.*, 257 (Oct. 1983).

9. Toop, A., *et al.*, Advances in in-stream analysis, in *Mineral Processing and Extractive Metallurgy*, eds. M. J. Jones and P. Gill, p. 187, IMM, London, 1984.

10. Leskinen, T., *et al.*, Performance of on-stream analysers at Outokumpu concentrators, Finland, *CIM Bull.* 66, 37 (Feb. 1973).

11. Lundan, A., On-stream analyzers improve the control of the mineral concentrator process, *ACIT '82 Convention*, Sydney, Australia (Nov. 1982).

12. Bernatowicz, *et al.*, On-line coal analysis for control of coal preparation plants, in *Coal Prep. '84*, p. 347, Industrial Presentations Ltd, Houston, 1984.

13. Kawatra, S. K., The development and plant trials of an ash analyser for control of a coal flotation circuit, in *Proc. XVth Int. Min. Proc. Cong.*, Cannes, **3**, 176 (1985).

14. Jenkinson, D. E., Coal preparation into the 1990s, *Colliery Guardian*, **223**, 301 (July 1985).

15. Herbst, J. A., and Bascur, O. A., Alternative control strategies for coal flotation, *Minerals and Metallurgical Processing*, **2**, 1 (Feb. 1985).

16. Salama, A. I. A., *et al.*, Coal preparation process control, *CIM Bulletin*, **78**, 59 (Sept. 1985).

17. Clarkson, C. J., The potential for automation and process control in coal preparation, in *Automation for Mineral Resource Development—Proc. 1st IFAC Symposium*, p. 247, Pergamon, Oxford, 1986.

18. Considine, D. M. (ed.), *Process Instruments and Controls Handbook*, McGraw Hill Book Co., New York, 1974.

19. Instrumentation and process control, *Chemical Engng.* (Deskbook Issue), *86* (Oct. 15, 1979).

20. Herbst, J. A., and Mular, A. L., Modelling and simulation of mineral processing unit operations, in *Computer Methods for the 80s*, ed. A. Weiss, AIMME, New York, 1979.

21. Purvis, J. R., and Erickson, I., Financial models for justifying computer systems, *INTECH*, 45, Nov. 1982.

22. Chang, J. W., and Bruno, S. J., Process control systems in the mining industry, *World Mining* 37 (May 1983).

23. Ulsoy, A. G., and Sastry, K. V. S., Principal developments in the automatic control of mineral processing systems, *CIM Bulletin*, **74**, 43 (Dec. 1981).

24. Lynch, A. J., *Mineral Crushing and Grinding Circuits*, Elsevier Scientific Publishing Co., Amsterdam, 1977.

25. Lynch, A. J., Johnson, N. W., Manlapig, E. V., and Thorne, C. G., *Mineral and Coal Flotation Circuits*, Elsevier Scientific Publishing Co., Amsterdam, 1981.

26. Wills, B. A., Automatic control of flotation, *Engng. & Min. J.*, **185**, 62 (June 1984).

27. Mular, A. L., Process optimisation, in *Computer Methods for the 80s*, ed. A. Weiss, AIMME, New York, 1979.

28. Ziegler, J. G., and Nichols, N. B., Optimum settings for automatic controllers, *Trans. AIME.*, **64**, 759 (1942).

29. Gault, G. A., *et al*, Automatic control of flotation circuits, theory and operation, *World Mining* **54** (Dec. 1979).
30. Owens, D. H., *Multivariable and Optimal Systems*. Academic Press, London, 1981.
31. Barker, I. J., Advanced control techniques. *School on Measurement and Process Control in the Minerals Industry*, South Afr. IMM, Joburg (July 1984).
32. Hughes, F. M., Self-tuning and adaptive control—a review of some basic techniques. *Proc. Advances in Process Control*, Inst. of Chem. Engs., Bradford, UK (Sept. 1985).
33. Putman, R. E. J., Self-tuning regulators for control of comminution circuits, in *Design and Installation of Comminution Circuits*, eds A. L. Mular and G. V. Jergensen, Chap, 43, AIMME, New York, 1982.
34. King, R. P., Computer controlled flotation plants in Canada and Finland, *NIM Report No. 1517*, South Africa (1973).
35. McKee, D. J., and Thornton, A. J., Emerging automatic control approaches in mineral processing, in *Mineral Processing at a Crossroads—Problems and Prospects*, eds. B. A. Wills and R. W. Barley, p. 117, Martinus Nijhoff, Dordrecht (1986).
36. Herbst, J. A., and Bascur, O. A., Mineral processing control in the 1980s—realities and dreams, in *Control '84 Mineral/Metallurgical Processing*, eds. J. A. Herbst, *et al.*, p. 197, SME, New York, 1984.
37. Herbst, J. A., and Rajamani, K., The application of modern control theory to mineral processing operation, *Proc. 12th CMMI Cong.* p. 779, ed. H. W. Glenn, S. Afri., IMM, Joburg, 1982.
38. Bozic, S. M., *Digital and Kalman Filtering*. Edward Arnold, London, 1979.
39. Wills, B. A., Maximising accuracy of two-product recovery calculations, *Trans. Inst. Min. Metall.* **94**, C101 (June 1985).
40. Smith, H. W., and Frew, J. A., Design and analysis of sampling experiments—a sensitivity approach, *Int. J. Min. Proc.* **11**, 267 (1983).
41. Frew, J. A., Computer-aided design of sampling systems, *Int. J. Min. Proc.*, **11**, 255 (1983).
42. Cutting, G. W., Estimation of interlocking mass-balances on complex mineral beneficiation plants, *Int. J. Min. Proc.* **3**, 207 (1976).
43. Wills, B. A., Complex circuit mass balancing—a simple, practical, sensitivity analysis method, *Int. J. Min. Proc.* **16**, 245 (1986).
44. Reid, K. J., *et al.*, A survey of material balance computer packages in the mineral industry, in *Proc. 17th APCOM Symposium*, ed. T. B. Johnson and R. J. Barnes, p. 41, AIME, New York, 1982.
45. Smith, H. W. and Ichiyen H. W., Computer adjustment of metallurgical balances, *CIM Bulletin*, 97 (Sept. 1973).
46. Hodouin, D. and Everell, M. D., A hierarchical procedure for adjustment and material balancing of mineral processes data, *Int. J. Min. Proc.* **7**, 91 (1980).
47. Hodouin, D., Kasongo, T., Kouame, E. and Everell, M. D., BILMAT: An algorithm for material balancing mineral processing circuits, *CIM Bulletin*, **74**, 123 (Sept. 1981).
48. Klimpel, A., Estimation of weight ratios given component make-up analyses of streams, Paper 79–24, AIME Annual Meeting, New Orleans, 1979.
49. Lynch, A. J., *Mineral Crushing and Grinding Circuits: their simulation, optimization and control*, Elsevier, Amsterdam, 1977.
50. Tipman, R., Burnett, T. C., and Edwards, C. R., Mass balances in mill

metallurgical operations. *10th Annual Meeting of the Canadian Mineral Processors*, CANMET, Ottawa, 1978.

51. Finch, J., and Matwijenko, O., Individual mineral behaviour in a closed-grinding circuit, *AIME Annual Meeting*, Atlanta, Paper 77-B-62 (March 1977).
52. Mular, A. L., Data and adjustment procedures for mass balances, in *Computer Methods for the 80's in the Minerals Industry*, ed. A. Weiss, AIMME, New York, 1979.
53. Cutting, G. W., Material balances in metallurgical studies: current use at Warren Spring Laboratory, *AIME Annual Meeting*, New Orleans, Paper 79-3 (1979).
54. Wiegel, R. L. Advances in mineral processing material balances, *Can. Metall. Q.*, **11**(2), 413 (1972).
55. Wills, B. A., and Manser, R. J., Reconciliation of simple node excess data, *Trans. Inst. Min. Metall.* **94**, C209 (Dec. 1985).

CHAPTER 4

PARTICLE-SIZE ANALYSIS

Introduction

Size analysis of the various products of a mill constitutes a fundamental part of laboratory testing procedure. It is of great importance in determining the quality of grinding and in establishing the degree of liberation of the values from the gangue at various particle sizes. In the separation stage, size analysis of the products is used to determine the optimum size of the feed to the process for maximum efficiency and to determine the size range at which any losses are occurring in the plant, so that they may be reduced.

It is essential, therefore, that methods of size analysis must be accurate and reliable, as important changes in plant operation may be made on the results of the laboratory tests. Since only relatively small amounts of material are used in the sizing tests, it is essential that the sample is representative of the bulk material and the same care should be taken over sampling for size analysis as for assay (Chapter 3).

Particle Size and Shape

The primary function of precision particle analysis is to obtain quantitative data about the size and size distribution of particles in the material. However, the exact size of an irregular particle cannot be measured. The terms length, breadth, thickness, or diameter have little meaning because so many different values of these quantities can be determined. The size of a spherical particle is uniquely defined by its diameter. For a cube, the length along one edge is characteristic, and for other regular shapes there are equally appropriate dimensions.

For irregular particles, it is desirable to quote the size of a particle in terms of a single quantity, and the expression most often used is the "equivalent diameter". This refers to the diameter of a sphere that would behave in the same manner as the particle when submitted to some specified operation.

The assigned equivalent diameter usually depends on the method of measurement, hence the *particle-sizing technique should, wherever possible, duplicate the process one wishes to control.*

Several equivalent diameters are commonly encountered. For example, the Stokes' diameter is measured by sedimentation and elutriation techniques; the projected area diameter is measured microscopically and the sieve-aperture diameter is measured by means of sieving. The last refers to the diameter of a sphere equal to the width of the aperture through which the particle just passes. If the particles under test are not true spheres, and they rarely are in practice, this equivalent diameter refers only to their second largest dimension.

Recorded data from any size analysis should, where possible, be accompanied by some remarks which indicate the approximate shape of the particles. Descriptions such as "granular" or "acicular" are usually quite adequate to convey the approximate shape of the particle in question.

Some of these terms are given below:[1]

Acicular	needle-shaped.
Angular	sharp-edged or having roughly polyhedral shape.
Crystalline	freely developed in a fluid medium of geometric shape.
Dendritic	having a branched crystalline shape.
Fibrous	regular or irregularly thread-like.
Flaky	plate-like.
Granular	having approximately an equidimensional irregular shape.
Irregular	lacking any symmetry.
Modular	having rounded, irregular shape.
Spherical	global shape.

A short list of some of the more common methods of size analysis, together with their effective size ranges, is given in Table 4.1.[2]

Test sieving is the most widely used method for particle-size analysis. It covers a very wide range of particle size, this range being the one of most industrial importance. So common is test sieving as a method of size analysis that particles finer than about 75 μm are often referred to as being in the "sub-sieve" range, although modern sieving methods allow sizing to be carried out down to about 5 μm.

TABLE 4.1. SOME METHODS OF PARTICLE-SIZE ANALYSIS

Method	Approximate useful range (microns)*
Test sieving	100 000–10
Elutriation	40–5
Microscopy (optical)	50–0.25
Sedimentation (gravity)	40–1
Sedimentation (centrifugal)	5–0.05
Electron microscopy	1–0.005

*A micron (μm) is 10^{-6} m.

Sieve Analysis

Sieve analysis is one of the oldest methods of size analysis and is accomplished by passing a known weight of sample material successively through finer sieves and weighing the amount collected on each sieve to determine the percentage weight in each size fraction. Sieving is carried out with wet or dry materials and the sieves are usually agitated to expose all the particles to the openings.

Sieving, when applied to irregularly shaped particles, is complicated by the fact that a particle with a size near that of the nominal aperture of the test sieve may pass only when presented in a favourable position. As there is inevitably a variation in the size of sieve apertures, due to irregularity of weaving, prolonged sieving will cause the larger apertures to exert an unduly large effect on the sieve analysis. Given time, every particle small enough could find its way through a very few such holes. The procedure is also complicated in many cases by the presence of "near-size" particles which cause "blinding", or obstruction of the sieve apertures, and reduce the effective area of the sieving medium. Blinding is most serious with test sieves of very small aperture size.

The process of sieving may be divided into two stages; first the elimination of particles considerably smaller than the screen apertures, which should occur fairly rapidly and, secondly, the separation of the so-called "near-size" particles, which is a gradual process rarely reaching final completion. Both stages require the sieve to be manipulated in such a way that all particles have opportunities for passing the apertures, and so that any which blind an aperture may be removed from it. Ideally, each particle should be presented individually to an aperture, as is permitted for the largest aperture sizes, but for most sizes this is impracticable.

The effectiveness of a sieving test depends on the amount of material put on the sieve (the "charge") and the type of movement imparted to the sieve.

A comprehensive account of sampling techniques for sieving is given in BS 1017.[3] Basically, if the charge is too large, the bed of material will be too many particles deep to allow each one a chance to meet an aperture in the most favourable position for sieving in a reasonable time. The charge, therefore, is limited by a requirement for the maximum amount of material retained at the end of sieving appropriate to the aperture size. On the other hand, the sample must contain enough particles to be representative of the bulk, so a minimum size of sample is specified. In some cases, the sample will have to be subdivided into a number of charges if the requirements for preventing overloading of the sieves is to be satisfied.

Test Sieves

Test sieves are designated by the nominal aperture size, which is the nominal central separation of opposite sides of a square aperture or the nominal diameter of a round aperture. A variety of sieve aperture ranges are currently used, the most popular being the German Standard, DIN 4188; ASTM standard, E11; the American Tyler series; the French series, AFNOR; and the British Standard, BSS 410.

Until relatively recently woven-wire sieves were designated by a mesh number, which referred to the number of wires per inch, which is the same as the number of square apertures per inch. This method was widely used, and until 1962 was the basic designation in BS 410. It has the serious disadvantage that the same mesh number on the various standard ranges corresponds to different aperture sizes, depending on the thickness of wire used in the woven-wire cloth. Sieves are now designated by aperture size, which gives the user directly the information he needs.

Since many workers still refer to sieve sizes in terms of mesh number, Table 4.2 lists mesh numbers for the BSS 410 series against nominal aperture size.

TABLE 4.2. BSS 410 WIRE-MESH SIEVES

Mesh number	Nominal aperture size (μm)	Mesh number	Nominal aperture size (μm)
3	5600	36	425
3.5	4750	44	355
4	4000	52	300
5	3350	60	250
6	2800	72	212
7	2360	85	180
8	2000	100	150
10	1700	120	125
12	1400	150	106
14	1180	170	90
16	1000	200	75
18	850	240	63
22	710	300	53
25	600	350	45
30	500	400	38

Wire-cloth screens are woven to produce nominally uniform square apertures within required tolerances.[4] Wire cloth in sieves with a nominal aperture of 75 μm and greater are plain woven, while those in cloths with apertures below 63 μm may be twilled (Fig. 4.1).

Standard test sieves are not available with aperture sizes smaller than about 37 μm. Micromesh sieves are available in aperture sizes from 5 μm to 150 μm, and are made by electroforming nickel in

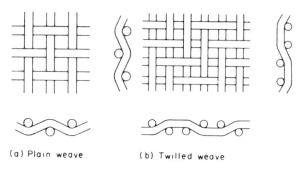

(a) Plain weave   (b) Twilled weave

FIG. 4.1. Weaves of wire cloth.

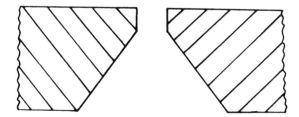

FIG. 4.2. Cross-section of a micro-plate aperture.

square and circular mesh. Another popular type is the "micro-plate sieve" which is fabricated by electroetching a nickel plate. The apertures are in the form of truncated cones with the small circle uppermost (Fig. 4.2). This reduces blinding but also reduces the percentage open area, i.e. the percentage of the total area of the sieving medium occupied by the apertures.

Micro-sieves are used for wet or dry sieving where a very high degree of accuracy is required in particle-size analysis down into the very fine size range.[5] The tolerances in these sieves are much better than those for woven-wire sieves, the aperture being guaranteed to within 2 μm of nominal size.

For aperture sizes above about 1 mm, perforated plate sieves are often used, with round or square holes (Fig. 4.3). Square holes are arranged in line with the centre points at the vertices of squares, while

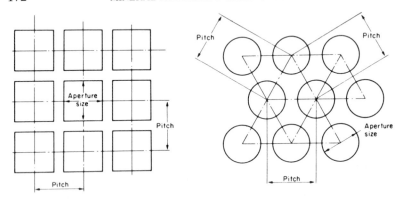

FIG. 4.3. Arrangement of square and round holes in perforated plate sieves.

round holes are arranged with the centres at the apices of equilateral triangles.

Choice of Sieve Sizes

In each of the standard series the apertures of consecutive sieves bear a constant relationship to each other.

It has long been realised that a useful sieve scale is one in which the ratio of the aperture widths of adjacent sieves is the square root of $2(\sqrt{2} = 1.414)$. The advantage of such a scale is that the aperture areas double at each sieve, facilitating graphical presentation of results.

Most modern sieve series are based on a fourth root of 2 ratio $(\sqrt[4]{} = 1.189)$ or, on the metric scale, a tenth root of 10 $(\sqrt[10]{10} = 1.259)$, which makes possible much closer sizing of particles.

For most size analyses it is usually impracticable and unnecessary to use all the sieves in a particular series. For most purposes, alternative sieves, i.e. a $\sqrt{2}$ series, are quite adequate, whereas over certain size ranges of particular interest, or for accurate work, consecutive sieves, i.e. a $\sqrt[4]{2}$ series, may be used. Intermediate sieves should never be chosen at random, as the data obtained will be difficult to interpret.

In general, the sieve range shold be chosen such that no more than about 5% of the sample is retained on the coarsest sieve, or passes the finest sieve. These limits, of course, may be lowered for more accurate work.

Testing Methods

The general procedures for test sieving are comprehensively covered in BS 1796 (1976).[6]

Machine sieving is almost universally used, as hand sieving is long and tedious, and its accuracy depends to a large extent on the operator.

The sieves chosen for the test are arranged in a stack, or *nest*, with the coarsest sieve on the top and the finest at the bottom. A tight-fitting pan or receiver is placed below the bottom sieve to receive the final undersize, and a lid is placed on top of the coarsest sieve to prevent escape of the sample.

The material to be tested is placed in the uppermost, coarsest sieve, and the nest is then placed in a sieve shaker which vibrates the material in a vertical plane (Fig. 4.4), and, on some models, a horizontal plane. The duration of screening can be controlled by an automatic timer. During the shaking, the undersize material falls through successive sieves until it is retained on a sieve having apertures which are slightly smaller than the diameter of the particles. In this way the sample is separated into size fractions.

After the required time, the nest is taken apart and the amount of material retained on each sieve weighed. Most of the near mesh particles, which block the openings, can be removed by inverting the sieve and tapping the frame gently. Failing this, the underside of the gauze may be brushed gently with a soft brass wire or nylon brush. Blinding becomes more of a problem the finer the aperture, and brushing, even with a soft hair brush, of sieves finer than about 150-μm aperture tends to distort the individual meshes.

Wet sieving can be used on material already in the form of a slurry, or it may be necessary for powders which form aggregates when dry-sieved. A full description of the techniques is given in BS 1796.[6]

Water is the liquid most frequently used in wet sieving, although for

FIG. 4.4. Vibrating sieve shaker.

materials which are water-repellent, such as coal or sulphide ores, a wetting agent may be necessary.

The test sample may be washed down through a nest of sieves, with the finest sieve at the bottom. At the completion of the test the sieves, together with the retained oversize material, are dried at a suitable low temperature and weighed.

Presentation of Results

There are several ways in which the results of a sieve test can be tabulated. The three most convenient methods are shown in Table 4.3.[6]

TABLE 4.3. RESULTS OF TYPICAL SIEVE TEST

1 Sieve size range (μm)	2 Sieve fractions wt (g)	3 %wt	4 Nominal aperture size (μm)	5 Cumulative % undersize	6 Cumulative % oversize
+250	0.02	0.1	250	99.9	0.1
−250 + 180	1.32	2.9	180	97.0	3.0
−180 + 125	4.23	9.5	125	87.5	12.5
−125 +90	9.44	21.2	90	66.3	33.7
−90 + 63	13.10	29.4	63	36.9	63.1
−63 + 45	11.56	26.0	45	10.9	89.1
−45	4.87	10.9			

Table 4.3 shows:

(1) The sieve size ranges used in the test.
(2) The weight of material in each size range, e.g. 1.32 g of material passed through the 250-μm sieve, but was retained on the 180-μm sieve: the material therefore is in the size range −250 + 180 μm.
(3) The weight of material in each size range expressed as a percentage of the total weight.
(4) The nominal aperture sizes of the sieves used in the test.
(5) The cumulative percentage of material passing through the sieves, e.g. 87.5% of the material is less than 125 μm in size.
(6) The cumulative percentage of material retained on the sieves.

The results of a sieving test should always be plotted graphically in order to assess their full significance.

There are many different ways of recording the results, the most common being that of plotting cumulative undersize (or oversize) against particle size. Although arithmetic graph paper can be used, it suffers from the disadvantage that points in the region of the finer aperture sizes tend to become congested. A semi-logarithmic plot avoids this, with a linear ordinate for percentage oversize or undersize and a logarithmic abscissa for particle size. Figure 4.5 shows graphically the results of the sieve test tabulated in Table 4.3.

It is not necessary to plot both cumulative oversize and undersize curves as they are mirror images of each other. A valuable quantity which can be determined from such curves is the "median size" of the sample. This refers to the mid-point in the size distribution—50% of the particles are smaller than this size and 50% are larger.

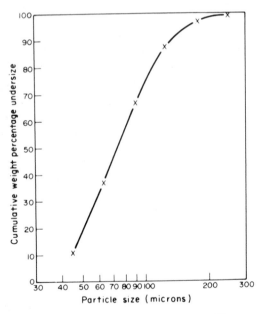

FIG. 4.5. Screen analysis graph (Table 4.3).

Size analysis is very important in assessing the performance of grinding circuits. The "mesh of grind" is usually quoted in terms of one point on the cumulative undersize curve, this often being the 80% passing size. Although this does not show the overall size *distribution* of the material, it does facilitate routine control of the grinding circuit. For instance, if the target mesh of grind is 80% minus 250 μm, then, for routine control, the operator need only screen a fraction of the mill product at this one size. If it is found that, say, 50% of the sample is minus 250 μm, then the product is too coarse, and control steps to remedy this can swiftly be made.

Many curves of cumulative oversize or undersize against particle size are S-shaped, leading to congested plots at the extremities of the graph. More than a dozen methods of plotting in order to proportion the ordinate are known. The two most common methods, which are often applied to comminution studies, where non-uniform size distributions are obtained, are the Gates–Gaudin–Schuhmann[7] and the Rosin–Rammler methods.[8] Both methods are derived from attempts to represent particle-size distribution curves by means of equations, which results in scales which, relative to a linear scale, are expanded in some regions and contracted in others.

In the Gates–Gaudin–Schuhmann method, cumulative undersize data are plotted against sieve aperture on log–log graph paper. As with most log–log plots, this very frequently leads to a straight line, over a wide size range, particularly over the finer sizes. Interpolation is much easier from a straight line than it is from a curve. Thus, if it is known that data obtained from the material usually yields a straight-line plot, the burden of routine analysis can be greatly relieved, as relatively few sieves will be needed to check essential features of the size distribution.

Plotting on a log–log scale considerably expands the region below 50% in the cumulative undersize curve, especially that below 25%. It does, however, severely contract the region above 50%, and especially above 75%, which is a major disadvantage of the method (Fig. 4.6).

The Rosin–Rammler method is often used for representing the results of sieve analyses performed on material which has been ground in ball mills. Such products have been found to obey the following relationship:

$$100 - P = 100 \exp{(bd^n)}, \qquad (4.1)$$

where P is the cumulative undersize in per cent, b is a constant, d is the particle size, and n is a constant.

This can be rewritten as

$$\log\left[\ln\frac{100}{100-P}\right] = \log b + n \log d. \qquad (4.2)$$

Thus a plot of $\ln{(100/100 - P)}$ versus d on log–log axes gives a line of slope n.

In comparison with the log–log method, the Rosin–Rammler plot expands the regions below 25% and above 75% cumulative undersize (Fig. 4.6) and it contracts the 30–60% region. It has been shown,

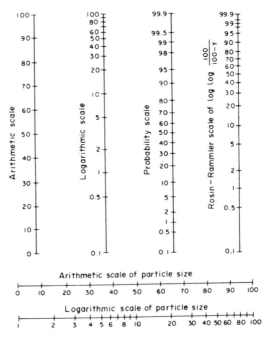

FIG. 4.6. Comparison of scales.

however,[9] that this contraction is insufficient to cause adverse effects. The method is tedious to plot unless charts having the axes divided proportionally to $\log [\ln(100/100 - P)]$ and $\log d$ are used.

The Gates–Gaudin–Schuhmann plot is often preferred to the Rosin–Rammler method in mineral processing applications, the latter being more often used in coal-preparation studies. The two methods have been assessed by Harris,[9] who suggests that the Rosin–Rammler is in fact the better method for mineral processing applications. It is useful for monitoring grinding operations for highly skewed distributions, but, as noted by Allen,[10] it should be used with caution, since the method of taking logs always reduces scatter, hence taking logs twice is not to be recommended.

Although cumulative size curves are used almost exclusively, the particle-size distribution curve is sometimes more informative. Ideally, this is derived by differentiating the cumulative undersize curve and plotting the gradient of the curve obtained against particle size. In practice, the size distribution curve is obtained by plotting the retained fraction on the sieves against size. The points on the frequency curve may be plotted in between two sieve sizes. For example, material which passes a 250-μm sieve, but is retained on a 180-μm sieve, may be regarded as having a mean particle size of 215 μm for the purpose of plotting. If the distribution is represented on a histogram, then the horizontals on the columns of the histogram join the various adjacent sieves used in the test. Unless each size increment is of equal width, however, the histogram has little value. Figure 4.7 shows the size distribution of the material in Table 4.3 represented on a frequency curve and a histogram.

Fractional curves or histograms are useful and rapid ways of visualising the relative frequency of occurrence of the various sizes present in the material. The only numerical parameter that can be obtained from these methods is the "mode" of the distribution, i.e. the most commonly occurring size.

For assessment of the metal losses in the tailings of a plant, or for preliminary evaluation of ores, assaying must be carried out on the various screen fractions. It is important, therefore, that the bulk sample satisfies the minimum sample weight requirement given by Gy's equation (3.3).

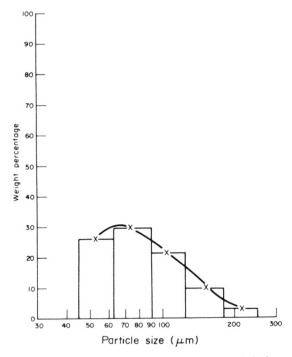

FIG. 4.7. Fractional representation of screen analysis data.

Table 4.4 shows the results of a screen analysis performed on an alluvial tin deposit for preliminary evaluation of its suitability for treatment by gravity concentration. Columns 1, 2, and 3 show the results of the sieve test and assays, which are evaluated in the other columns. It can be seen that the calculated overall assay for the material is 0.21% Sn, but that the bulk of the tin is present within the finer fractions. The results show that, for instance, if the material was initially screened at 210 μm and the coarse fraction discarded, then the bulk required for further processing would be reduced by 24.9%, with a loss of only 4.6% Sn. This may be acceptable if mineralogical analysis shows that the tin is finely disseminated in this coarse fraction, which would necessitate extensive grinding to give reasonable

TABLE 4.4.

(1) Size range (μm)	(2) Weight (%)	(3) Assay (% Sn)	Distribution (% Sn)	Size (microns)	Cumulative oversize (%)	Cumulative distribution (% Sn)
+422	9.7	0.02	0.9	422	9.7	0.9
−422 + 300	4.9	0.05	1.2	300	14.6	2.1
−300 + 210	10.3	0.05	2.5	210	24.9	4.6
−210 + 150	23.2	0.06	6.7	150	48.1	11.3
−150 + 124	16.4	0.12	9.5	124	64.5	20.8
−124 + 75	33.6	0.35	56.5	75	98.1	77.3
−75	1.9	2.50	22.7			
	100.0	0.21	100.00			

liberation. Heavy liquid analysis (Chapter 11) on the −210-μm fraction would determine the theoretical grades and recoveries possible, but the screen analysis results also show that much of the tin (22.7%) is present in the −75-μm fraction, which constitutes only 1.9% of the total bulk of the material. This indicates that there may be difficulty in processing this material, as gravity separation techniques are not very efficient at such fine sizes.

Sub-sieve Techniques

Sieving is rarely carried out on a routine basis below about 40 μm; below this size the operation is referred to as *sub-sieving*. The most widely used procedures are sedimentation, elutriation, and microscopy, although a whole new field of electronic techniques is now in use.

There are many concepts in use for designating particle size within the sub-sieve range, and it is important to be aware of them, particularly when combining size distributions determined by different methods. It is preferable to cover the range of a single distribution with a single method, but this is not always possible.

Approximate conversion factors, which should be used with caution, are given below:[8]

	Multiplying factor
Sieve size to Stokes' diameter (sedimentation, elutriation)	0.94
Sieve size to projected diameter (microscopic)	1.4

Stokes' Equivalent Diameter

In sedimentation techniques the material to be sized is dispersed in a fluid and allowed to settle under carefully controlled conditions; in elutriation techniques, samples are sized by allowing the dispersed material to settle against a rising fluid velocity. Both techniques separate the particles on the basis of resistance to motion in a fluid. This resistance to motion determines the terminal velocity which the particle attains as it is allowed to fall in a fluid under the influence of gravity.

For particles within the sub-sieve range, the terminal velocity is given by the equation derived by Stokes,[11] namely:

$$v = \frac{d^2 g (D_s - D_f)}{18\eta}, \qquad (4.3)$$

where v is the terminal velocity of the particle (m s^{-1}), d is the particle diameter (m), g is the acceleration due to gravity (m s^{-2}), D_s is the particle density (kg m^{-3}), D_f is the fluid density (kg m^{-3}), and η is the fluid viscosity (N s m^{-2}); $\eta = 0.001$ N s m^{-2} for water).

Stokes' law is derived for spherical particles; non-spherical particles will attain a terminal velocity, but this velocity will be influenced by the shape of the particles. Nevertheless, this velocity can be substituted in the Stokes' equation to give a value of d, which can be used to characterise the particle. This value of d is referred to as the "Stokes' equivalent spherical diameter". It is also known as the "Stokes' diameter" or the "sedimentation diameter".

Stokes' law is only valid in the region of viscous flow (Chapter 9), which sets an upper size limit to the particles which can be tested by sedimentation and elutriation methods in a given liquid. The limit is determined by the Reynolds' number, a dimensionless quantity defined by

$$R = \frac{vd\,D_f}{\eta}\,.$$ (4.4)

The Reynolds' number should not exceed 0.2 if the error in using Stokes' law is not to exceed 5%.[12] In general, Stokes' law will hold for all particles below 40 μm dispersed in water; particles above this size should be removed by sieving beforehand. The lower limit may be taken as 1 μm, below which the settling times are too long, and also the effects of unintentional disturbances, such as may be caused by convection currents, are far more likely to produce serious errors.

Sedimentation Methods

Sedimentation methods are based on the measurement of the rate of settling of the powder particles uniformly dispersed in a fluid and the principle is well illustrated by the common laboratory method of "beaker decantation".

The material under test is uniformly dispersed in water contained in a beaker or similar parallel-sided vessel. A wetting agent may need to be added to ensure complete dispersion of the particles. A syphon tube is immersed into the water to a depth of h below the water-level, corresponding to about 90% of the liquid depth L.

The terminal velocity v is calculated from Stokes' law for the various sizes of particle in the material, say 35, 25, 15, and 10 μm. For an ore, it is usual to fix D_s for particles which are most abundant in the sample.

The time required for a 10-μm particle to settle from the water level to the bottom of the syphon tube, distance h, is calculated ($t = h/v$). The pulp is gently stirred to disperse the particles through the whole volume of water and it is then allowed to stand for the calculated time. The water above the end of the tube is syphoned off and all particles in this water are assumed to be smaller than 10-μm diameter (Fig. 4.8). However, a fraction of the -10-μm material, which commenced settling from various levels below the water-level, will be present in the

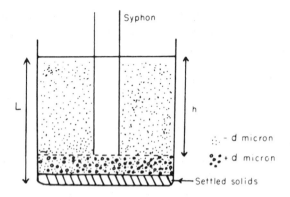

FIG. 4.8. Beaker decantation.

material below the syphon level. In order to recover these particles, the pulp remaining must be diluted with water to the original level, and the procedure repeated until the decant liquor is essentially clear. In theory, this requires an infinite number of decantations, but in practice at least five treatments are needed, depending on the accuracy required. The settled material can be treated in a similar manner at larger separating sizes, i.e. at shorter decanting times, until a sufficient number of fractions is obtained.

The method is simple and cheap, and has the advantage over many other sub-sieve techniques in that it produces a true fractional size analysis, i.e. reasonable quantities of material in specific size ranges are collected, which can be analysed chemically and mineralogically.

The method is, however, extremely tedious, as long settling times are required for very fine particles, and separate tests must be performed for each particle size. For instance, a 25-μm particle of quartz has a settling velocity of 0.056 cm s^{-1}, and therefore takes about 3½ min to settle 12 cm, a typical immersion depth for the syphon tube. Five separate tests to ensure a reasonably clear decant therefore require a total settling time of about 18 min. A 5-μm particle, however, has a settling velocity of 0.0022 cm s^{-1}, and therefore takes about 1½ h to settle 12 cm. The total time for evaluation of such material is thus about 8 h. A complete analysis may therefore take an operator several days.

Another problem is the large quantity of water which dilutes the undersize material due to repeated decantation. Theoretically an infinite number of decantations are required to produce a 100% efficient separation into oversize and undersize fractions, and the number of practical decantations must be chosen according to the accuracy required and the width of the size range required in each of the fractions.

In the system shown in Fig. 4.8, after time t, all particles larger than size d have fallen to a depth below the level h.

All particles of a size d_1, where $d_1 < d$, will have fallen below a level h_1 below the water-level, where $h_1 < h$.

The efficiency of removal of particles of size d_1 into the decant is thus:

$$\frac{h - h_1}{L}$$

since at time $t = 0$ the particles were uniformly distributed over the whole volume of liquid, corresponding to depth L, and the fraction removed into the decant is the volume above the syphon level, $h - h_1$.

Now, since $t = h/v$, and $v \propto d^2$,

$$\frac{h}{d^2} = \frac{h_1}{d_1^2}.$$

Therefore the efficiency of removal of particles of size d_1

$$= \frac{h - h(d_1/d)^2}{L} = \frac{[h(1 - (d_1/d^2)]}{L} = a\left[1 - \left(\frac{d_1}{d}\right)^2\right] = E,$$

where $a = h/L$.

If a second decantation step is performed, the amount of $-d_1$ material in the dispersed suspension is $1 - E$, and the efficiency of removal of $-d_1$ particles after two decantations is thus

$$E + (1 - E)E = 2E - E^2 = 1 - [1 - E]^2.$$

In general, for n decantation steps, the efficiency of removal of particles of size d_1, at a separation size of d,

$$= 1 - [1 - E]^n$$

$$\text{or efficiency} = 1 - \left[1 - a\left\{1 - \left(\frac{d_1}{d}\right)^2\right\}\right]^n. \tag{4.5}$$

Table 4.5 shows the number of decantation steps required for different efficiencies of removal of various sizes of particle expressed relative to d, the separating size, where the value of $a = 0.9$. It can be shown[13] that the value of a has relatively little effect, therefore there is nothing to be gained by attempting to remove the suspension adjacent to the settled particles, thus risking disturbance and re-entrainment.

Table 4.5 shows that a large number of decantations is necessary for effective removal of particles close to the separation size, but that relatively small particles are quickly eliminated. For most purposes, unless very narrow size ranges are required, no more than about twelve decantations are necessary for the test.

TABLE 4.5. NUMBER OF DECANTATIONS REQUIRED FOR REQUIRED
EFFICIENCY OF REMOVAL OF FINE PARTICLES

Relative particle size d_1/d	Efficiency of removal of particles of size d_1 into decant		
	90%	95%	99%
0.95	25	33	50
0.9	12	16	25
0.8	6	8	12
0.5	2	3	4
0.1	1	1	2

A much quicker, and less-tedious method of sedimentation analysis is the *Andreasen pipette technique*.[12]

The apparatus consists of a half-litre graduated cylindrical flask (Fig. 4.9) and a pipette connected to a 10-ml reservoir by means of a two-way stopcock. The tip of the pipette is in the plane of the zero mark when the ground glass stopper is properly seated.

A 3–5% suspension of the sample, dispersed in the sedimentation fluid, usually water, is added to the flask. The pipette is introduced and

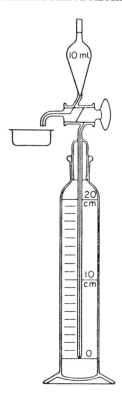

FIG. 4.9. Andreasen pipette.

the suspension agitated by inversion. The suspension is then allowed to settle, and at given intervals of time, samples are withdrawn by applying suction to the top of the reservoir, manipulating the two-way cock so that the sample is drawn up as far as the calibration mark on the tube above the 10-ml reservoir. The cock is then reversed, allowing the sample to drain into the collecting dish. After each sample is taken, the new liquid level is noted.

The samples are then dried and weighed, and the weights compared with the weight of material in the same volume of the original suspension.

There is a definite particle size, D, corresponding to each settling distance h and time t, and this represents the size of the largest particle that can still be present in the sample. These particle sizes are calculated from Stokes' law for the various sampling times. The weight of solids collected, g, compared with the corresponding original weight, g_0, i.e. g/g_0, then represents the fraction of the original material having a particle size smaller than D, which can be plotted on the size-analysis graph.

The method is much quicker than beaker decantation, as samples are taken off successively throughout the test for increasingly finer particle sizes. For example, although 5-μm particles of quartz will take about 2½ h to settle 20 cm, once this sample is collected, all the coarser particle-size samples will have been taken, and so the complete analysis, in terms of settling times, is only as long as the settling time for the finest particles.

The disadvantage of the method is that the samples taken are each representative of the particles smaller than a particular size, which is not as valuable, for mineralogical and chemical analysis, as samples of various size *ranges*, as are produced by beaker decantation.

Sedimentation techniques tend to be very tedious, due to the long settling times required for fine particles (up to 5 h for 3-μm particles) and the time required to dry and weigh the samples. The main difficulty, however, lies in completely dispersing the material within the suspending liquid, such that no agglomeration of particles occurs. Combinations of suitable suspending liquids and dispersing agents for various materials are given in BS 3406, Part 2.[12]

Although the Andreasen pipette is perhaps the most widely used method of sizing by sedimentation, various other techniques have been developed, which attempt to speed up testing. Examples of these methods, which are comprehensively reviewed by Allen,[10] are the photo-sedimentometer, which combines gravitational settling with photo-electric measurement, and the sedimentation balance, in which the weight of material settling out onto a balance pan is recorded against time, to produce a cumulative sedimentation size analysis.

Elutriation Techniques

Elutriation is a proces of sizing particles by means of an upward current of fluid, usually water or air. The process is the reverse of gravity sedimentation, and Stokes' law applies.

All elutriators consist of one or more "sorting columns" (Fig. 4.10) in which the fluid is rising at a constant velocity. Feed particles introduced into the sorting column will be separated into two fractions, according to their terminal velocities, calculated from Stokes' law.

Those particles having a terminal velocity less than that of the velocity of the fluid will report to the overflow, while those particles having a greater terminal velocity than the fluid velocity will sink to the underflow. Elutriation is carried out until there are no visible signs of further classification taking place or the rate of change in weights of the products is negligible.

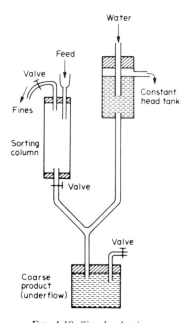

FIG. 4.10. Simple elutriator.

This involves the use of much water, involving dilution of the undersize fraction, but it can be shown that this is not as serious as in beaker decantation. Consider a sorting column of depth h, sorting material at a separating size of d. If the upward velocity of water flow is v, then by Stokes' law, $v \propto d^2$.

Particles smaller than the separating size d will move upward in the water flow at a velocity dependent on their size.

Thus particles of size d_1, where $d_1 < d$, will move upwards in the sorting column at a velocity v_1, where $v_1 \propto (d^2 - d_1^2)$.

The time required for a complete volume change in the sorting column is h/v, and the time required for particles of size d_1 to move from the bottom to the top of the sorting column is h/v_1.

Therefore the number of volume changes required to remove all particles of size d_1 from the sorting column

$$= h/v_1/h/v = \frac{d^2}{d^2 - d_1^2} = \frac{1}{1 - (d_1/d)^2} .$$

The number of volume changes required for various values of d_1/d are shown below:

d_1/d	Number of volume changes required
0.95	10.3
0.9	5.3
0.8	2.8
0.5	1.3
0.1	1.0

Comparing these figures with those in Table 4.1, it can be seen that the number of volume changes required is far less with elutriation than it is with decantation. It is also possible to achieve complete separation by elutriation, whereas this can only be achieved in beaker decantation by an infinite number of volume changes.

Elutriation thus appears more attractive than decantation, and has certain practical advantages in that the volume changes need no operator attention. It suffers from the disadvantage, however, that the fluid velocity is not constant across the sorting column, being a minimum at the walls of the column, and a maximum at the centre. The separation size is calculated from the mean volume flow, so that some coarse particles are misplaced into the overflow, and some fines are misplaced into the coarse underflow. The fractions thus have a considerable overlap in particle size and are not sharply separated. Although the decantation method never attains 100% efficiency of separation, the lack of sharpness of the division into fractions is much less than that due to velocity variation in elutriation.[13]

Elutriation is limited at the coarsest end by the validity of Stokes' law, but most materials in the sub-sieve range exhibit laminar flow. At the fine end of the scale, separations become impracticable below about 10 μm, as the material tends to agglomerate, or extremely long separating times are required. Separating times can be considerably decreased by utilisation of centrifugal forces, and one of the most widely used methods of sub-sieve sizing in modern mineral processing laboratories is the Warman cyclosizer,[5,14] which is extensively used for routine testing and plant control in the size range 8–50 μm for materials of specific gravity similar to quartz (sp. gr. 2.7), and down to 4 μm for particles of high specific gravity, such as galena (sp. gr. 7.5).

The cyclosizer unit consists of five cyclones (see Chapter 9 for a full description of the principle of the hydrocyclone), arranged in series such that the overflow of one unit is the feed to the next unit (Fig. 4.11).

The individual units are inverted in relation to conventional cyclone arrangements, and at the apex of each a chamber is situated so that the discharge is effectively closed (Fig. 4.12).

Water is pumped through the units at a controlled rate, and a weighed sample of solids is introduced ahead of the cyclones.

The tangential entry into the cyclones induces the liquid to spin, resulting in a portion of the liquid, together with the faster-settling particles, reporting to the apex opening, while the remainder of the liquid, together with the slower settling particles, is discharged through the vortex outlet and into the next cyclone in the series. There is a successive decrease in the inlet area and vortex outlet diameter of each cyclone in the direction of the flow, resulting in a corresponding increase in inlet velocity and an increase in the centrifugal forces within

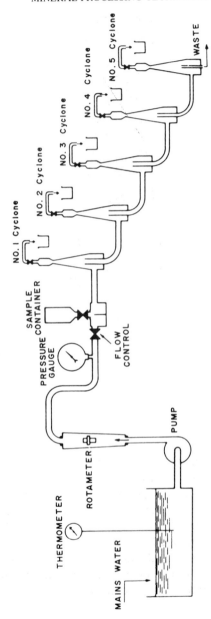

FIG. 4.11. Warman cyclosizer.

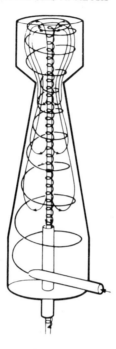

FIG. 4.12. Flow pattern inside a cyclosizer cyclone unit.

the cyclone, resulting in a successive decrease in the limiting particle-separation size of the cyclones.

The cyclosizer is manufactured to have definite limiting separation sizes at standard values of the operating variables, viz. water flow-rate, water temperature, particle density, and elutriation time. To correct for practical operation at other levels of these variables, a set of correction graphs is provided.

Complete elutriation normally takes place after about 20 min, after which the sized fractions are collected by discharging the contents of each apex chamber into separate beakers.

Microscopic Sizing

Microscopy can be used as an absolute method of particle-size analysis since it is the only method in which individual mineral particles are observed and measured.[10,15] The image of a particle seen in a microscope is two-dimensional and from this image an estimate of particle size must be made. Microscopic sizing involves comparing the projected area of a particle with the areas of reference circles, or graticules, of unknown sizes, and it is essential for meaningful results that the mean projected areas of the particles are representative of the particle size. This requires a random orientation in three dimensions of the particle on the microscope slide, which is unlikely in most cases.

The optical microscope method is applicable to particles in the size range 0.8–150 μm, and down to 0.001 μm using electron microscopy.

Basically, all microscopy methods are carried out on extremely small laboratory samples, which must be collected with great care in order that they may be truly representative.

The dispersed particles are viewed by transmission microscopy, and the areas of the magnified images are compared with the areas of circles of known sizes inscribed on a graticule.

The relative numbers of particles are determined in each of a series of size classes. These represent the size distribution by number, from which it is possible to calculate the distribution by volume and, if all the particles have the same density, the distribution by weight.

Manual analysis of microscope slides is tedious and error prone; semi-automatic and automatic systems have been developed which speed up analyses and reduce the tedium of manual methods.[10]

The development of quantitative image analysis has made possible the rapid sizing of fine particles using small laboratory samples. Image analysers accept samples in a variety of forms—photographs, electron micrographs, and direct viewing. The particles in the field are scanned by a television camera and displayed on a console, while electrical information from the sample is passed to a detector. At the detector, a specific image analysis function is performed by means of chosen modules, the features to be measured then being passed to an analog computer which counts the chosen data and stores or presents it in the desired manner.

Electrical Resistance Method

The electrical resistance method makes use of current changes in an electrical circuit as a result of the presence of a particle.[16] The measuring system of a *Coulter counter* is shown in Fig. 4.13. The particles, suspended in a known volume of electrically conductive liquid, flow through a small aperture having an immersed electrode on either side; the particle concentration is such that the particles traverse the aperture substantially one at a time.

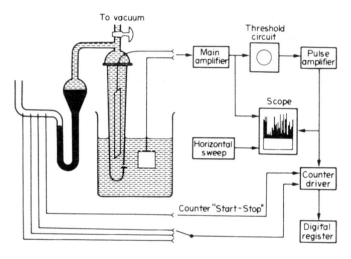

FIG. 4.13. Coulter counter.

Each particle passage displaces electrolyte within the aperture, momentarily changing the resistance between the electrodes and producing a voltage pulse of magnitude proportional to particle volume. The resultant series of pulses is electronically amplified, scaled, and counted.

The amplified pulses are fed to a threshold circuit having an adjustable screen-out voltage level, and those pulses which reach or exceed this level are counted, this count representing the number of particles larger than some determinable volume proportional to the appropriate threshold setting. By taking a series of counts at various amplification and threshold settings, data is directly obtained for determining number frequency against volume, which can be used to determine the size distribution.

The method has become popular in certain applications since size analyses can be carried out rapidly with good reproducibility using semi-skilled operators. It is applicable to the size range 0.5–400 μm.

Laser Beam Particle-size Analysis

Recently, the application of lasers has been introduced to the particle-size measurement of powdered products. The principle used is based on the diffraction of a coherent light beam by the grains of powder to be analysed, which results in the presence of light outside the geometrical limits of the beam. The powder to be examined is dispersed in a liquid which is then circulated through a glass cell. A parallel beam from a low-power laser lights up the cell, and the beam which leaves is focused by means of a convergent optical system. The values of illumination with and without sample are read by an electronic detector and fed into a programmed processor, which then displays the results as cumulative percentage undersize.

This method has many attractive advantages for routine analysis, such as simplicity of use, extremely quick results, and reproducibility. Its main disadvantage, at present, is its high capital cost.

On-line Particle-size Analysis

Continuous measurement of particle size has been available since 1971, the PSM system produced by Armco Autometrics having been installed in a number of mineral processing plants.[17]

The PSM system consists of three sections: the air eliminator, the sensor section, and the electronics section (Fig. 4.14). The air eliminator draws a sample from the process stream and removes entrained air-bubbles. The de-aerated pulp then passes between the sensors. Measurement depends on the varying absorption of ultrasonic waves in suspensions of different particle sizes. Since solids concentration also affects the absorption of ultrasonic radiation, two

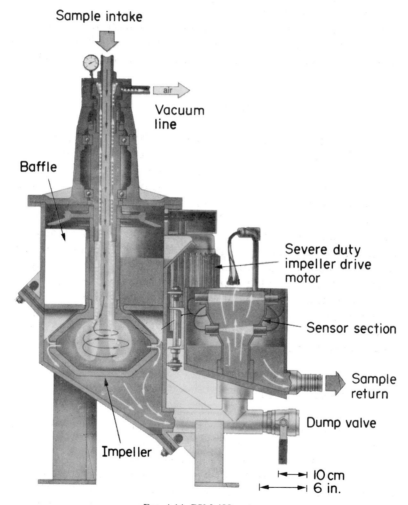

Fig. 4.14. PSM-400 system.

pairs of transmitters and receivers, operating at different frequencies, are employed to measure particle size and solids concentration of the pulp, the processing of this information being performed by the electronics section.

Attempts have been made to use small calibrated hydrocyclones[18,19] and elutriation columns[20] as continuous size analysers, a typical cyclonic device developed by Holland-Batt[21] having been rigorously investigated by Osborne.[22] The device consists of a helical tube of rectangular cross-section. Sample pulp densities and volume flow-rates to the device are held constant, thus keeping the size of separation at a fixed level. Size distribution parameters can be determined by beta-ray attenuation measurements of the densities, at different locations, of the size segregated slurry emerging from the classifying helix. The main limitation to the accuracy of all these classifying devices as size analysers is their sensitivity to changes in pulp density.[23]

References

1. British Standard 2955: Glossary of terms relating to powders.
2. *Test Sieve Data*, Endecott (Filters) Ltd., London.
3. British Standard 1017: 1960, The sampling of coal and coke.
4. British Standard 410: 1976, Test sieves.
5. Finch, J. A., and Leroux, M., Fine sizing by cyclosizer and micro-sieve, *CIM Bulletin*, **75**, 235 (March 1982).
6. British Standard 1796: 1976, Test sieving.
7. Schuhmann, R., Jr., Principles of comminution, 1—Size distribution and surface calculation, *Tech. Publs. AIME*, no. 1189, 11 (1940).
8. Rosin, P., and Rammler, E., The laws governing the fineness of powdered coal, *J. Inst. Fuel* **7**, 29 (1933–4).
9. Harris, C. C., Graphical presentation of size distribution data: an assessment of current practice, *Trans. IMM* Sect. C, **80**, 133 (Sept. 1971).
10. Allen, T., *Particle Size Measurement*, Chapman & Hall, London, 1974.
11. Stokes, Sir G. G., *Mathematical and Physical Paper III*, Cambridge University Press, 1891.
12. British Standard 3406: Part 2 (1963), Liquid sedimentation methods.
13. Heywood, H., Fundamental principles of sub-sieve particle size measurement, in *Recent Developments in Mineral Dressing*, IMM, London, 1953.
14. Particle size analysis in the sub-sieve range, Bulletin WCS/2, Warman International.
15. British Standard 3406: Part 4 (1963), Optical microscope method.

16. Coulter Industrial Bibliography, Coulter Electronics Ltd.
17. Hathaway, R. E., and Guthnals, D. L., The continuous measurement of particle size in fine slurry processes, *CIM Bull.* **766,** 64 (Feb. 1976).
18. Putman, R. E. J., Optimising grinding-mill loading by particle-size analysis, *Min. Cong. J.* 68 (Sept. 1973).
19. Plitt, L. R., and Kawatra, S. K., Estimating the cut (d_{50}) size of classifiers without product particle size measurement, *Int. J. Min. Proc.* **5,** 369 (1979).
20. Hinde, A. L., A review of real-time particle size analysers, *Jl. S. Afric. IMM* **73,** 258 (1973).
21. Holland-Batt, A. B., Further developments of the Royal School of Mines on-stream size analyzer, *Trans. IMM* Sect. C, **77,** 185 (1968).
22. Osborne, B. F., Practical wide-ranging continuous particle-size analyzer, *Trans. Soc. AIME* **256,** 277 (Dec. 1974).
23. Hinde, A. L., and Lloyd, P. J. D., Real-time particle size analysis in wet closed-circuit milling, *Powder Technol.* **12,** 37 (1975).

CHAPTER 5

COMMINUTION

Introduction

Because most minerals are finely disseminated and intimately associated with the gangue, they must be initially "unlocked" or "liberated" before separation can be undertaken. This is achieved by *comminution*, in which the particle size of the ore is progressively reduced until the clean particles of mineral can be separated by such methods as are available. Comminution in its earliest stages is carried out in order to make the freshly excavated material easier to handle by scrapers, conveyors, and ore carriers, and in the case of quarry products to produce material of controlled particle size.

Explosives are used in mining to remove ores from their natural beds, and blasting can be regarded as the first stage in comminution. Comminution in the mineral processing plant, or "mill", takes place as a sequence of crushing and grinding processes. Crushing reduces the particle size of run-of-mine ore to such a level that grinding can be carried out until the mineral and gangue are substantially produced as separate particles.

Crushing is accomplished by compression of the ore against rigid surfaces, or by impact against surfaces in a rigidly constrained motion path. This is contrasted with grinding which is accomplished by abrasion and impact of the ore by the free motion of unconnected media such as rods, balls, or pebbles.

Crushing is usually a dry process, and is performed in several stages, with small *reduction ratios* ranging from three to six in each stage. The reduction ratio of a crushing stage can be defined as the ratio of maximum particle size entering to maximum particle size leaving the crusher, although other definitions are sometimes used.

Tumbling mills with either steel rods or balls, or sized ore as the grinding media, are used in the last stages of comminution. Grinding is usually performed "wet" to provide a slurry feed to the concentration process, although dry grinding has limited applications. There is an overlapping size area where it is possible to crush or grind the ore. From a number of case studies, it appears that at the fine end of crushing operations equivalent reduction can be achieved for roughly half the energy and costs required by tumbling mills.[1]

Principles of Comminution

Most minerals are crystalline materials in which the atoms are regularly arranged in three-dimensional arrays. The configuration of atoms is determined by the size and types of physical and chemical bonds holding them together. In the crystalline lattice of minerals, these inter-atomic bonds are effective only over small distances, and can be broken if extended by a tensile stress. Such stresses may be generated by tensile or compressive loading (Fig. 5.1).

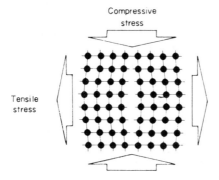

FIG. 5.1. Strain of a crystal lattice resulting from tensile or compressive stresses.

Even when rocks are uniformly loaded, the internal stresses are not evenly distributed, as the rock consists of a variety of minerals dispersed as grains of various sizes. The distribution of stress depends upon the mechanical properties of the individual minerals, but more

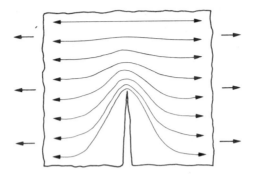

FIG. 5.2. Stress concentration at a crack tip.

importantly, upon the presence of cracks or flaws in the matrix, which act as sites for stress concentration (Fig. 5.2).

It has been shown[2] that the increase in stress at such a site is proportional to the square root of the crack length perpendicular to the stress direction. Therefore there is a critical value for the crack length at any particular level of stress at which the increased stress level at the crack tip is sufficient to break the atomic bond at that point. Such rupture of the bond will increase the crack length, thus increasing the stress concentration and causing a rapid propagation of the crack through the matrix, thus causing fracture.

Although the theories of comminution assume that the material is brittle, crystals can, in fact, store energy without breaking, and release this energy when the stress is removed. Such behaviour is known as *elastic*. When fracture does occur, some of the stored energy is transformed into free surface energy, which is the potential energy of atoms at the newly produced surfaces. Due to this increase in surface energy, newly formed surfaces are often more chemically active, and are more amenable to the action of flotation reagents, etc., as well as oxidising more readily.

Griffith[3] showed that materials fail by crack propagation when this is energetically feasible, i.e. when the energy released by relaxing the strain energy is greater than the energy of the new surface produced. Brittle materials relieve the strain energy mainly by crack propagation, whereas "tough" materials can relax strain energy without crack propagation, by the mechanism of *plastic flow*, where the atoms or molecules slide over each other and energy is consumed in distorting the shape of the material. Crack propagation can also be inhibited by encounters with other cracks or by meeting crystal boundaries. Fine-grained rocks, such as taconites, are therefore usually tougher than coarse-grained rocks.

The energy required for comminution is reduced in the presence of water, and can be further reduced by chemical additives which adsorb onto the solid.[4] This may be due to the lowering of the surface energy on adsorption providing that the surfactant can penetrate into a crack and reduce the bond strength at the crack tip before rupture.

Real particles are irregularly shaped, and loading is not uniform, but is achieved through points, or small areas, of contact. Breakage is achieved mainly by crushing, impact, and attrition, and all three modes of fracture (compressive, tensile, and shear) can be discerned depending on the rock mechanics and the type of loading.

When an irregular particle is broken by compression, or crushing, the products fall into two distinct size ranges—coarse particles resulting from the induced tensile failure, and fines from compressive failure near the points of loading, or by shear at projections (Fig. 5.3). The amount of fines produced can be reduced by minimising the area of loading and this is often done in compressive crushing machines by using corrugated crushing surfaces.[5]

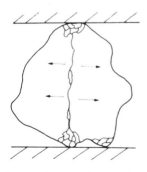

FIG. 5.3. Fracture by crushing.

In impact breaking, due to the rapid loading, a particle experiences a higher average stress while undergoing strain than under slow loading. As a result the particle absorbs more energy than is necessary to achieve simple fracture, and tends to break apart rapidly, mainly by tensile failure. The products are often very similar in size and shape.

Attrition (shear failure) produces much fine material, and is usually undesirable. Attrition occurs mainly in practice due to particle–particle interaction (inter-particle comminution), which may occur if a crusher is fed too fast, contacting particles thus increasing the degree of compressive stress and hence shear failure.

Comminution Theory

Comminution theory is concerned with the relationship between energy input and the product particle size made from a given feed size. Various theories have been expounded, none of which is entirely satisfactory.

The greatest problem lies in the fact that most of the energy input to a crushing or grinding machine is absorbed by the machine itself, and only a small fraction of the total energy is available for breaking the material. It is to be expected that there is a relationship between the energy required to break the material and the new surface produced in the process, but this relationship can only be made manifest if the energy consumed in creating new surface can be separately measured.

In a ball mill, for instance, it has been shown that less than 1% of the total energy input is available for actual size reduction, the bulk of the energy being utilised in the production of heat.

Another factor is that a material which is plastic will consume energy in changing shape, a shape which it will retain without creating significant new surface. All the theories of comminution assume that the material is brittle, so that no energy is absorbed in processes such as elongation or contraction which is not finally utilised in breakage.

The oldest theory is that of Rittinger (1867),[6] which states that the energy consumed in the size reduction is proportional to the area of new surface produced. The surface area of a known weight of particles of uniform diameter is inversely proportional to the diameter, hence Rittinger's law equates to

$$E = K \left(\frac{1}{D_2} - \frac{1}{D_1} \right), \qquad (5.1)$$

where E is the energy input, D_1 is the initial particle size, D_2 is the final particle size, and K is a constant.

The second theory (1885) is that of Kick.[7] He stated that the work required is proportional to the reduction in volume of the particles concerned. Where f is the diameter of the feed particles and p the diameter of the product particles, the reduction ratio R is f/p. According to Kick's law, the energy required for comminution is proportional to $\log R/\log 2$.

Bond[8] developed an equation which is based on the theory that the work input is proportional to the new crack tip length produced in particle breakage, and equals the work represented by the product minus that represented by the feed. In particles of similar shape, the surface area of unit volume of material is inversely proportional to the diameter. The crack length in unit volume is considered to be proportional to one side of that area and therefore inversely proportional to the square root of the diameter.

For practical calculations the size in microns which 80% passes is selected as the criterion of particle size. The diameter in microns which 80% of the product passes is designated as P, the size which 80% of the feed passes is designated as F, and the work input in kilowatt hours per short ton is W. Bond's third theory equation is

$$W = \frac{10W_i}{\sqrt{P}} - \frac{10W_i}{\sqrt{F}}, \qquad (5.2)$$

where W_i is the *work index*. The work index is the comminution parameter which expresses the resistance of the material to crushing and grinding; numerically it is the kilowatt hours per short ton required to reduce the material from theoretically infinite feed size to 80% passing 100 μm.

Various attempts have been made to show that the relationships of Rittinger, Kick, and Bond are interpretations of single general equations. Hukki[9] suggests that the relationship between energy and particle size is a composite form of the three laws. The probability of breakage in comminution is high for large particles, and rapidly diminishes for fine sizes. He shows that Kick's law is reasonably accurate in the crushing range above about 1 cm in diameter; Bond's theory applies reasonably in the range of conventional rod-mill and ball-mill grinding, and Rittinger's law applies fairly well in the fine grinding range of 10–1000 μm.

Grindability

Ore grindability refers to the ease with which materials can be comminuted and data from grindability tests are used to evaluate crushing and grinding efficiency.

Probably the most widely used parameter to measure ore grindability is the Bond work index W_i. If the breakage characteristics of a material remain constant over all size ranges, then the calculated work index would be expected to remain constant since it expresses the resistance of material to breakage. However, for most naturally occurring raw materials, differences exist in the breakage characteristics depending on particle size, which can result in variations in the work index. For instance, when a mineral breaks easily at the boundaries but individual grains are tough, then grindability increases with fineness of grind. Consequently work index values are generally obtained for some specified grind size which typifies the comminution operation being evaluated.[10]

Grindability is based upon performance in a carefully defined piece of equipment according to a strict procedure. Bond has devised several methods for predicting ball-mill and rod-mill energy requirements,[11] which provide an accurate measure of ore grindability.

Table 5.1. lists standard Bond work indices for a selection of materials.

The standard Bond test is time consuming, and a number of methods have been used to obtain the indices related to the Bond work index. Smith and Lee[12] used batch-type grindability tests to arrive at the work index, and compared their results with work indices from the

TABLE 5.1. SELECTION OF BOND WORK INDICES

Material	Work index	Material	Work index
Barite	4.73	Fluorspar	8.91
Bauxite	8.78	Granite	15.13
Coal	13.00	Graphite	43.56
Dolomite	11.27	Limestone	12.74
Emery	56.70	Quartzite	9.58
Ferro-silicon	10.01	Quartz	13.57

standard Bond tests, which require constant screening out of undersize material in order to simulate closed-circuit operation. The batch-type tests compared very favourably with the standard grindability test data, the advantage being that less time is required to determine the work index.

Berry and Bruce[13] developed a comparative method of determining the grindability of an ore. The method requires the use of a reference ore of known grindability. The reference ore is ground for a certain time and the power consumption recorded. An identical weight of the test ore is then ground for a length of time such that the power consumed is identical with that of the reference ore. Then if r is the reference ore and t the ore under test, from Bond's equation (5.2),

$$ W_r = W_t = W_{ir} \left[\frac{10}{\sqrt{P_r}} - \frac{10}{\sqrt{F_r}} \right] = W_{it} \left[\frac{10}{\sqrt{P_t}} - \frac{10}{\sqrt{F_t}} \right]. $$

Therefore

$$ W_{it} = \left(\frac{10}{\sqrt{P_r}} - \frac{10}{\sqrt{F_r}} \right) \bigg/ \left(\frac{10}{\sqrt{P_t}} - \frac{10}{\sqrt{F_t}} \right). \tag{5.3} $$

Reasonable values for the work indices are obtained by this method as long as the reference and test ores are ground to about the same product size distribution.

The low efficiency of grinding equipment in terms of the energy actually used to break the ore particles is a common feature of all types of mill, but there are substantial differences between various designs. Some machines are constructed in such a way that much energy is absorbed in the component parts and is not available for breaking.

Work indices have been obtained[14] from grindability tests on different sizes of several types of equipment, using identical feed materials. The values of work indices obtained are indications of the efficiencies of the machines. Thus the equipment having the highest work indices, and hence the largest power consumers, are found to be jaw and gyratory crushers and tumbling mills; intermediate consumers are impact crushers and vibration mills, and roll crushers are the smallest consumers. The smallest consumers of energy are those machines which apply a steady, continuous, compressive stress on the material.

Values of operating work indices obtained from specific units can be used to assess the effect of operating variables, such as mill speeds, size of grinding media, type of liner, etc. The higher the value of W_i, the lower is the grinding efficiency. It should be noted that the value of W is the power applied to the pinion shaft of the mill. Motor input power thus has to be converted to power at the mill pinion shaft unless the motor is coupled direct to the pinion shaft.

Simulation of Comminution Processes and Circuits

Perhaps the most important future use of the computer in mineral processing will be in the simulation of processes as an aid to circuit design and optimisation. Computer simulation is intimately associated with mathematical modelling and realistic simulation relies heavily on the availability of accurate and physically meaningful models. Simulation studies often highlight inadequacies in models and can be used to modify existing ones. Comminution has received the greatest attention, particularly grinding/classification, due to the fact that this is by far the most important unit operation both in terms of energy consumption and overall plant performance. Other aspects of mineral processing have not received the same intensive research accorded to grinding.

The Bond Work Index has little use in simulation, as it does not predict the complete product size distribution, only the 80% passing size, nor does it predict the effect of operating variables on mill circulating load, nor classification performance. The complete size distribution is required in order to simulate the behaviour of the product in ancillary equipment such as screens and classifiers, and for this reason population balance models are finding increased usage in the design, optimisation and control of grinding circuits. In the model formulation the particulate assembly that undergoes breakage in a mill is divided into several narrow size intervals (e.g. root 2 sieve intervals). The size reduction process is defined by the matrix equation:

$$p = K \cdot f \tag{5.4}$$

where p represents the product and f the feed elements. The element p_{ij} in the product array is given by:

$$p_{ij} = K_{ij} \cdot f_j$$

where K_{ij} represents the mass fraction of the particles in the jth size range which fall in the ith size range in the product. The product array for n size ranges can thus be written as:

$$
\begin{array}{lllll}
K_{11}f_1 & 0 & 0 & \dots\dots\dots\dots\dots\dots\dots\dots\dots\dots & 0 \\
K_{21}f_1 & K_{22}f_2 & 0 & \dots\dots\dots\dots\dots\dots\dots\dots\dots\dots & 0 \\
K_{31}f_1 & K_{32}f_2 & K_{33}f_3 & \dots\dots\dots\dots\dots\dots\dots\dots & 0 \\
\phantom{K_{11}f_1}\cdot & \phantom{K_{22}f_2}\cdot & & & \cdot \\
K_{n1}f_1 & K_{n2}f_2 & K_{n3}f_3 & \dots\dots\dots\dots\dots\dots\dots & K_{nn}f_n
\end{array}
$$

The product array is only useful if K is known. The behaviour of particles in each size interval is characterised by a size-discretised selection, or breakage rate, function, S, which is the probability of particles in that size range being selected for breakage, the remainder passing through the process unbroken, and a set of size-discretised breakage functions, B, which give the distribution of breakage fragments produced by the occurrence of a primary breakage event in that size interval. $S \cdot f$ represents the portion of particles which are broken, $(1 - S) \cdot f$ thus representing the unbroken fraction. K in equation 5.4 is thus replaced by B, and the equation for a primary breakage process becomes:

$$p = B \cdot S \cdot f + (1 - S) \cdot f.$$

The model can be combined with information on the distribution of residence times in the mill to provide a description of open-circuit grinding, which can be coupled with information concerning the classifier to produce closed-circuit grinding conditions.[15] These models can only realise their full potential, however, if accurate methods of estimating model parameters are available for a particular system. The complexity of the breakage environment in a tumbling mill precludes the calculation of these values from first principles, so that successful application depends on the development of efficient techniques for the estimation of model parameters from experimental data. The methods used for the determination of model parameters have been compared by Lynch.[16] This comparison shows that, while all the modern ball mill models use a similar method for describing the breakage rate and the breakage distribution functions, each model has its own way of representing the material transport mechanisms.

Parameter estimation techniques can be classified into three broad categories:

(a) Graphical methods which are based mainly on the grinding of narrow size distributions.

(b) Tracer methods, involving the introduction of a tracer into one of the size intervals of the feed, followed by analysis of the product for the tracer.

(c) Non-linear regression methods which allow all parameters to be computed from a minimum of experimental data.

Rajamani and Herbst[17] report the development of an algorithm for simultaneous estimation of selection and breakage functions from experimental data with the use of non-linear regression, and present the results of estimation for batch and continuous operations. The estimated parameter values show good agreement with parameters determined by direct experimental methods, and a computer program based on the algorithm has been developed. The program is said to be capable of simulating tumbling mill grinding behaviour for a specified set of model parameters, and of estimating the model parameters from experimental data.

Single particle breakage tests have been used by a number of researchers to investigate some salient features of the complex comminution process. A comparison of the results from single particle breakage tests with grindability and ball mill tests is given by

COMMINUTION   211

Narayanan.[18] Application of the results from single particle breakage tests to modelling industrial comminution processes is described, and the necessity for further research into single particle breakage tests to develop a simple but comprehensive technique for estimating the breakage characteristics of ores is discussed.

Although selection and breakage functions for homogeneous materials can be determined on a small scale and used to predict large-scale performance, it is more difficult to predict the behaviour of mixtures of two or more components. Furthermore, the relationship of mineral size reduction to subsequent processing is even more difficult to predict, due to the complexities of mineral release. However, recent work at the University of Utah[19] has focused on the development of a grinding model which includes mineral liberation in the size reduction description. It has been incorporated into a total plant simulator and has been used as a stand-alone model for simulating two-component grinding. The results of initial testwork indicate that the model might have wide applicability for cases where a knowledge of liberation during milling is important, such as where flotation follows grinding.

Despite the practical difficulties, computer simulation is beginning to be increasingly used. It offers clear advantages in terms of the accurate predictions of the metallurgical performances of alternative circuits, which can be used to optimise design, and the flowrates of process streams, which can be used to size pump and pipelines, largely eliminating the need for laborious and expensive plant trials. However, the dangers of computer simulation come from its great computational power and relative ease of use. It is always necessary to bear in mind the realistic operating range over which the models are used, as well as the realistic limits which must be placed on equipment operation, such as pumping capacity. Simulation studies are a powerful and useful tool, complementary to sound metallurgical judgement and familiarity with the circuit being simulated and its metallurgical objectives.

References

1. Flavel, M. D., Control of crushing circuits will reduce capital and operating costs. *Min. Mag.* 207 (March 1978).

2. Inglis, C. E., Stresses in a plate due to the presence of cracks and sharp corners, *Proc. Inst. Nav. Arch.* (1913).
3. Griffith, A. A., *Phil. Trans. R. Soc.* **221**, 163 (1921).
4. Hartley, J. N., Prisbrey, K. A., and Wick, O. J., Chemical additives for ore grinding: how effective are they? *Engng. Min. J.* 105 (Oct. 1978).
5. Partridge, A. C., Principles of comminution, *Mine & Quarry* **7**, 70 (July/Aug. 1978).
6. Von Rittinger, P. R., *Lehrbuch der Aufbereitungs Kunde*, Ernst and Korn, Berlin, 1867.
7. Kick, F., *Des Gesetz der Proportionalem widerstand und Seine Anwendung*, Felix, Leipzig, 1885.
8. Bond, F. C., The third theory of comminution, *Trans. AIME* **193**, 484 (1952).
9. Hukki, R. T., The principles of comminution; an analytical summary, *Eng. Min. J.* **176**, 106 (May 1975).
10. Horst, W. E., and Bassarear, J. H., Use of simplified ore grindability technique to evaluate plant performance, *Trans. SME/AIME* **260**, 348 (Dec. 1976).
11. Bond, F. C., Crushing and grinding calculations, Bulletin 07R9235B, Allis-Chalmers, 1961.
12. Smith, R. W., and Lee, K. H., A comparison of data from Bond type simulated closed-circuit and batch type grindability tests, *Trans. SME/AIME* **241**, 91 (1968).
13. Berry, T. F., and Bruce, R. W., A simple method of determining the grindability of ores, *Can. Min. J.* 63 (July 1966).
14. Lowrison, G. C., *Crushing and Grinding*, Butterworths, London, 1974.
15. Lynch, A. J., and Narayanan, S. S., Simulation—the design tool for the future, in *Mineral Processing at a Crossroads—problems and prospects*, eds. B. A. Wills and R. W. Barley, p. 89, Martinus Nijhoff Publishers, Dordrecht, 1986.
16. Lynch, A. J. *et al.*, Ball mill models: their evolution and present status, in *Advances in Mineral Processing*, ed. P. Somasundaran, chap. 3. p. 48, SME Inc., Littleton, 1986.
17. Rajamani, K. and Herbst, J. A., Simultaneous estimation of selection and breakage functions from batch and continuous grinding data, *Trans. Inst. Min. Metall.*, **93**, C74 (June 1984).
18. Narayanan, S. S., Single particle breakage tests: a review of principles and applications to comminution modelling, *Bull. Proc. Australas. Inst. Min. Metall.*, **291**, 49 (June 1986).
19. Peterson, R. D. and Herbst, J. A., Estimation of kinetic parameters of a grinding-liberation model, *Int. J. Miner. Proc.*, **14**, 111 (1985).

CHAPTER 6

CRUSHERS

Introduction

Crushing is the first mechanical stage in the process of comminution in which the main objective is the liberation of the valuable minerals from the gangue.

It is generally a dry operation and is usually performed in two or three stages. Lumps of run-of-mine ore can be as large as 1.5 m across and these are reduced in the primary crushing stage to 10–20 cm in heavy-duty machines.

In most operations, the primary crushing schedule is the same as the mining schedule. When primary crushing is performed underground, this operation is normally a responsibility of the mining department; when primary crushing is on the surface, it is customary for the mining department to deliver the ore to the crusher and for the mineral processing department to crush and handle the ore from this point through the successive ore-processing unit operations. Primary crushers are commonly designed to operate 75% of the available time, mainly because of interruptions caused by insufficient crusher feed and by mechanical delays in the crusher.[1]

Secondary crushing includes all operations for reclaiming the primary crusher product from ore storage to the disposal of the final crusher product, which is usually between 0.5 cm and 2 cm in diameter. The primary crusher product from most metalliferous ores can be crushed and screened satisfactorily, and secondary plant generally consists of one or two size-reduction stages with appropriate crushers and screens. If, however, the ore tends to be slippery and tough, the tertiary crushing stage may be substituted by coarse grinding in rod mills. On the other hand, more than two size-reduction

213

stages may be used in secondary crushing if the ore is extra-hard, or in special cases where it is important to minimise the production of fines.

A basic flowsheet for a crushing plant is shown in Fig. 6.1, incorporating two stages of secondary crushing. A washing stage is

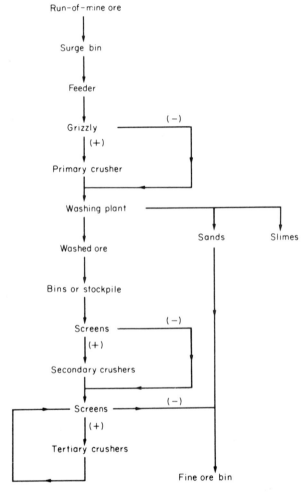

FIG. 6.1. Basic crushing plant flowsheet.

included, which is often necessary for sticky ores containing clay, which may lead to problems in crushing and screening (see Chapter 2).

Vibrating screens are sometimes placed ahead of the secondary crushers to remove undersize material, or *scalp* the feed, and thereby increase the capacity of the secondary crushing plant. Undersize material tends to pack the voids between the large particles in the crushing chamber, and can choke the crusher, causing damage, because the packed mass of rock is unable to swell in volume as it is broken.

Crushing may be in open or closed circuit depending on product size (Fig. 6.2). In open-circuit crushing, undersize material from the screen is combined with the crusher product and is then routed to the next operation. Open-circuit crushing is often used in intermediate crushing stages, or when the secondary crushing plant is producing a rod mill feed. If the crusher is producing ball-mill feed it is good practice to use closed-circuit crushing in which the undersize from the screen is the finished product. The crusher product is returned to the screen so that any over-size material will be recirculated. One of the main reasons for closing the circuit is the greater flexibility given to the crushing plant as a whole. The crusher can be operated at a wider setting if necessary, thus altering the size distribution of the product and by making a selective cut on the screen, the finished product can be adjusted to give the required specification. There is the added factor that if the material is wet or sticky (and climatic conditions can vary), then it is possible to open the setting of the crusher to prevent the

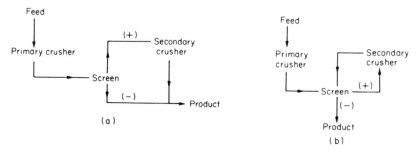

FIG. 6.2. (a) Open-circuit crushing, (b) closed-circuit crushing.

possibility of packing, and by this means the throughput of the machine is increased, which will compensate for the additional circulating load. Closed-circuit operation also allows compensation for wear which takes place on liners, and generally gives greater freedom to meet changes in requirements from the plant.

Surge bins precede the primary crusher to receive dumped loads from skips or lorries and should have enough storage capacity to maintain a steady feed to the crusher. In most mills the crushing plant does not run for 24 h a day, as hoisting and transport of ore is usually carried out on two shifts only, the other shift being used for drilling and blasting. The crushing section must therefore have a greater hourly capacity than the rest of the plant, which is run continuously. Ore is always stored after the crushers to ensure a continuous supply to the grinding section. The obvious question is, why not have similar storage capacity before the crushers and run this section continuously also? Apart from the fact that it is cheaper in terms of power consumption to crush at off-peak hours, large storage bins are expensive, so it is uneconomic to have bins at the crushing *and* grinding stage. It is not practicable to store large quantities of run-of-mine ore, as it is "long-ranged", i.e. it consists of a large range of particle sizes and the small ones move down in the pile and fill the voids. This packed mass is difficult to move after it has settled. Run-of-mine ore should therefore be kept moving as much as possible, and surge bins should have sufficient capacity only to even out the flow to the crusher.

Primary Crushers

Primary crushers are heavy-duty machines, used to reduce the run-of-mine ore down to a size suitable for transport and for feeding the secondary crushers. They are always operated in open circuit, with or without heavy-duty scalping screens (grizzlies). There are two main types of primary crusher in metalliferous operations—jaw and gyratory crushers—although the impact crusher has limited use as a primary crusher and will be considered separately.

Jaw Crushers

The distinctive feature of this class of crusher is the two plates which open and shut like animal jaws.[2] The jaws are set at an acute angle to each other, and one jaw is pivoted so that it swings relative to the other fixed jaw. Material fed into the jaws is alternately *nipped* and released to fall further into the crushing chamber. Eventually it falls from the discharge aperture.

Jaw crushers are classified by the method of pivoting the swing jaw (Fig. 6.3). In the *Blake crusher* the jaw is pivoted at the top and thus has a fixed receiving area and a variable discharge opening. In the *Dodge crusher* the jaw is pivoted at the bottom, giving it a variable feed area but fixed delivery area. The Dodge crusher is restricted to laboratory use, where close sizing is required, and is never used for heavy-duty crushing as it chokes very easily. The *Universal crusher* is pivoted in an intermediate position, and thus has a variable delivery and receiving area.

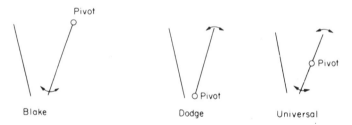

FIG. 6.3. Jaw-crusher types.

The Blake crusher was patented by W. E. Blake in 1858 and variations in detail on the basic form are found in most of the jaw crushers used today.

There are two forms of the Blake crusher—double toggle and single toggle.

Double-toggle Blake crushers. In this model (Fig. 6.4) the oscillating movement of the swinging jaw is effected by vertical

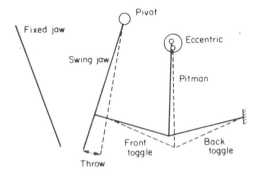

FIG. 6.4. Blake jaw crusher (functional diagram).

movement of the pitman. This moves up and down under the influence of the eccentric. The back toggle plate causes the pitman to move sideways as it is pushed upward. This motion is transferred to the front toggle plate and this in turn causes the swing jaw to close on the fixed jaw. Similarly, downward movement of the pitman allows the swing jaw to open.

The important features of the machine are:

(1) Since the jaw is pivoted from above, it moves a *minimum* distance at the entry point and a *maximum* distance at the delivery. This maximum distance is called the *throw* of the crusher.

(2) The horizontal displacement of the swing jaw is greatest at the bottom of the pitman cycle and diminishes steadily through the rising half of the cycle as the angle between the pitman and the back toggle plate becomes less acute.

(3) The crushing force is *least* at the start of the cycle, when the angle between the toggles is most acute, and is strongest at the top, when full power is delivered over a reduced travel of the jaw.

Figure 6.5 shows a cross-section through a double-toggle jaw crusher. All jaw crushers are rated according to their receiving areas, i.e. the *width* of the plates and the *gape*, which is the distance between

FIG. 6.5. Cross-section through double-toggle crusher.

the jaws at the feed opening. For example, an 1830 × 1220-mm crusher has a width of 1830 mm and a gape of 1220 mm.

Consider a large piece of rock falling into the mouth of the crusher. It is nipped by the jaws, which are moving relative to each other at a rate depending on the size of the machine and which usually varies inversely with the size. Basically, time must be given for the rock broken at each "bite" to fall to a new position before being nipped again. The ore falls until it is arrested. The swing jaw closes on it, quickly at first and then more slowly with increasing power towards the end of the stroke. The fragments now fall to a new arrest point as the jaws move apart and are then gripped and crushed again. During each "bite" of the jaws the rock swells in volume due to the creation of voids between the particles. Since the ore is also falling into a gradually reducing cross-sectional area of the crushing chamber, choking of the crusher would soon occur if it were not for the increasing amplitude of swing towards the discharge end of the crusher. This accelerates the material through the crusher, allowing it to discharge at a rate sufficient to leave space for material entering above. This is *arrested or free crushing* as opposed to *choked crushing*, which occurs when the volume of material arriving at a particular cross-section is greater than that leaving. In arrested crushing, crushing is by the jaws only, whereas in choked crushing, particles break each other. This *interparticle comminution* can lead to excessive production of fines and if choking is severe can damage the crusher.

The discharge size of material from the crusher is controlled by the *set*, which is the *maximum* opening of the jaws at the discharge end. This can be adjusted by using toggle plates of the required length. Wear on the jaws can be taken up by adjusting the back pillow into which the back toggle plate bears. A number of manufacturers offer jaw setting by hydraulic jacking, and some fit electro-mechanical systems which allow remote control[3]

A feature of all jaw crushers is the heavy flywheel attached to the drive, which is necessary to store energy on the idling half of the stroke and deliver it on the crushing half. Since the jaw crusher works on half-cycle only, it is limited in capacity for its weight and size. Due to its alternate loading and release of stress, it must be very rugged and needs strong foundations to accommodate the vibrations.

Single-toggle jaw crushers. In this type of crusher (Fig. 6.6) the swing jaw is suspended on the eccentric shaft, which allows a lighter, more compact design than with the double-toggle machine. The motion of the swing jaw also differs from that of the double-toggle design. Not only does the swing jaw move towards the fixed jaw, under the action of the toggle plate, but it also moves vertically as the eccentric rotates. This elliptical jaw motion assists in pushing rock through the crushing chamber. The single-toggle machine therefore has a somewhat higher capacity than the double-toggle machine of the same gape. The eccentric movement, however, increases the rate of wear on the jaw plates. Direct attachment of the swing jaw to the eccentric imposes a high degree of strain on the drive shaft, and so maintenance costs tend to be higher than with the double-toggle machine.

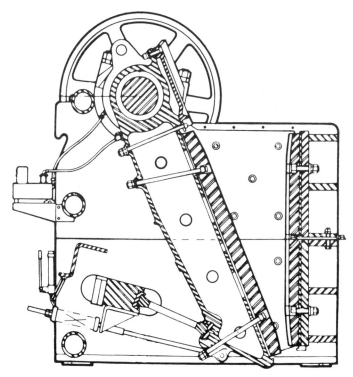

FIG. 6.6. Cross-section of single-toggle jaw crusher.

Double-toggle machines cost about 50% more than single-toggle machines of the same size, and are usually used on tough, hard, abrasive materials, although the single-toggle crusher is used in Europe, especially Sweden, for heavy-duty work on tough taconite ores, and is often choke fed, since the jaw movement tends to make it self-feeding.

Jaw-crusher construction. Jaw crushers are extremely heavy-duty machines and hence must be robustly constructed. The main frame is often made from cast iron or steel, connected with tie-bolts. It is often made in sections so that it can be transported underground for installation. Modern jaw crushers may have a main frame of mild steel plate welded together.

The jaws themselves are usually constructed from cast steel and are fitted with replaceable liners, made from manganese steel, or "Ni-hard", a Ni–Cr alloyed cast iron. Apart from reducing wear, hard liners are essential in that they minimise crushing energy consumption, reducing the deformation of the surface at each contact point. They are bolted in sections on to the jaws so that they can be removed easily and reversed periodically to equalise wear. Cheek plates are fitted to the sides of the crushing chamber to protect the main frame from wear. These are also made from hard alloy steel and have similar lives to the jaw plates. The jaw plates themselves may be smooth, but are often corrugated, the latter being preferred for hard, abrasive materials. Patterns on the working surface of the crushing members also influence capacity, especially at small settings. Laboratory tests have demonstrated that the capacity is reduced about 50 times when a corrugated profile is used rather than a smooth surface. Nevertheless, some type of pattern is desirable for the jaw plate surface in a jaw crusher, partly to reduce the risk of undesired large flakes easily slipping through the straight opening, and partly to reduce the contact surface when crushing flaky blocks. In several installations, a slight wave shape has proven successful. The angle between the jaws is usually less than 26°, as the use of a larger angle than this causes slipping, which reduces capacity and increases wear.

In order to overcome problems of choking near the discharge of the crusher, which is possible if fines are present in the feed, curved plates are sometimes used. The lower end of the swing jaw is concave, whereas the opposite lower half of the fixed jaw is convex. This allows a more gradual reduction in size as the material nears the exit, hence minimising the chances of packing. Less wear is also reported on the jaw plates, since the material is distributed over a larger area.

The speed of jaw crushers varies inversely with the size, and usually lies in the range of 100–350 rev min^{-1}. The main criterion in determining the optimum speed is that particles must be given sufficient time to move down the crusher throat into a new position before being nipped again.

The maximum amplitude of swing of the jaw, or "throw", is

determined by the type of material being crushed and is usually adjusted by changing the eccentric. It varies from 1 to 7 cm depending on the machine size, and is highest for tough, plastic material and lowest for hard, brittle ore. The greater the throw, the less danger is there of chokage, as material is removed more quickly. This is offset by the fact that a large throw tends to produce more fines, which inhibits arrested crushing. Large throws also impart higher working stresses to the machine.

In all crushers, provision must be made for avoiding the damage which could result from uncrushable material entering the chamber. Many jaw crushers are protected from such "tramp" material (usually metal objects) by a weak line of rivets on one of the toggle plates, although automatic trip-out devices are now becoming more common, and one manufacturer uses automatic overload protection based on hydraulic cylinders between the fixed jaw and the frame. In the event of excessive pressure caused by an overload, the jaw is allowed to open, normal gap conditions being reasserted after clearance of the blockage. This allows a full crusher to be started under load.[3]

Jaw crushers range in size up to 1680-mm gape by 2130-mm width. This size machine will handle ore with a maximum size of 1.22 m at a crushing rate of approximately 725 t h^{-1} with a 203-mm set. However, at crushing rates above 545 t h^{-1} the economic advantage of the jaw crusher over the gyratory diminishes; and above 725 t h^{-1} jaw crushers cannot compete with gyratory crushers.[1]

Gyratory Crushers

Gyratory crushers are principally used in surface-crushing plants, although a few currently operate underground. The gyratory crusher (Fig. 6.7) consists essentially of a long spindle, carrying a hard steel conical grinding element, the head, seated in an eccentric sleeve. The spindle is suspended from a "spider" and, as it rotates, normally between 85 and 150 rev min^{-1}, it sweeps out a conical path within the fixed crushing chamber, or shell, due to the gyratory action of the eccentric. As in the jaw crusher, maximum movement of the head occurs near the discharge. This tends to relieve the choking due to

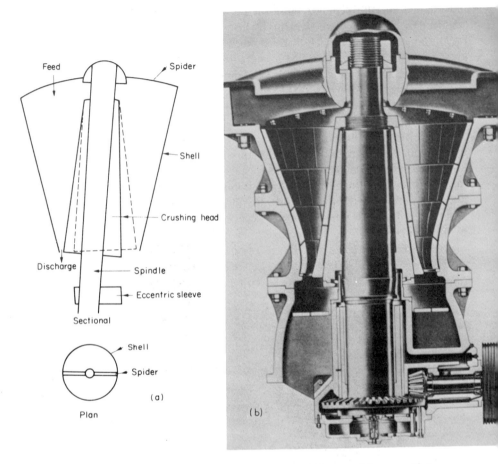

Fig. 6.7. Gyratory crusher: (a) functional diagram, (b) cross-section.

swelling, the machine thus being a good arrested crusher. The spindle is free to turn on its axis in the eccentric sleeve, so that during crushing the lumps are compressed between the rotating head and the top shell segments, and abrasive action in a horizontal direction is negligible.

At any cross-section there are in effect two sets of jaws opening and shutting like jaw crushers. In fact, the gyratory crusher can be regarded as an infinitely large number of jaw crushers each of infinitely small width. Since the gyratory, unlike the jaw crusher, crushes on full cycle, it has a much higher capacity than a jaw crusher of the same gape, and is usually favoured in plants handling very large throughputs. In mines with crushing rates above 900 t h^{-1}, gyratory crushers are always selected.

Crushers range in size up to gapes of 1830 mm and can crush ores with a top size of 1370 mm at a rate of up to 5000 t h^{-1} with a 200-mm set. Power consumption is as high as 750 kW on such crushers. Large gyratories often dispense with expensive feeding mechanisms and are often fed direct from trucks (Fig. 6.8). They can be operated satisfactorily with the head buried in feed. Although excessive fines may have to be "scalped" from the feed, the modern trend in large-capacity plants is to dispense with grizzlies if the ore allows. This

FIG. 6.8. Gyratory crusher fed direct from truck.

reduces capital cost of the installation and reduces the height from which the ore must fall into the crusher, thus minimising damage to the spider. Choked crushing is encouraged to some extent, but if this is not serious, the rock-to-rock crushing produced in the primaries reduces the rock-to-steel crushing required in the secondaries, thus reducing steel consumption.[4]

Gyratory-crusher construction. The outer shell of the crusher is constructed from heavy steel casting or welded steel plate, with at least one constructional joint, the bottom part taking the drive shaft for the head, the top and lower top shells providing the crushing chamber. If the spindle is carried on a suspended bearing, as in the bulk of primary gyratories, then the spider carrying the bearing forms a joint across the top of the shell. The crushing shell is protected by manganese steel or reinforced alloyed white cast-iron (Ni-hard) liners or *concaves*. In small crushers the concave is one continuous ring bolted to the shell. Large machines use sectionalised concaves, called *staves*, which are wedge-shaped, and either rest on a ring fitted between the upper and lower shell, or are bolted to the shell. The concaves are backed with some soft filler material, such as white metal, zinc, or plastic cement, which ensures even seating against the steel bowl.

The head is one of the steel forgings which make up the spindle (Fig. 6.9). The head is protected by a manganese steel *mantle*, which is fastened on to the head by means of nuts, on threads which are pitched so that they are self-tightening during operation. The mantle is backed with zinc, plastic cement, or, more recently, with an epoxy resin. The vertical profile is often bell-shaped to assist the crushing of material having a tendency to choke.

Some gyratory crushers have a hydraulic mounting and, when overloading occurs, a valve is tripped which releases the fluid, thus dropping the spindle and allowing the "tramp" material to pass out between the head and the bowl. This mounting is also used to adjust the set of the crusher at regular intervals so as to compensate for wear on the concaves and mantle. Many crushers use simple mechanical means to control the set, the most common method being by the use of a ring nut on the main shaft suspension.

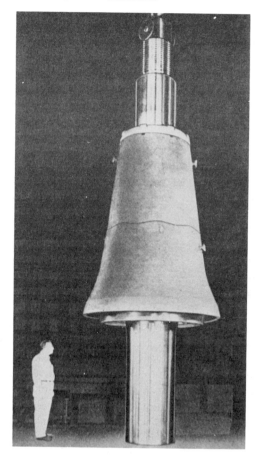

FIG. 6.9. Crusher head.

Comparison of Primary Crushers

In deciding whether a jaw or a gyratory crusher should be used in a particular plant, the main factor is the maximum size of ore which the crusher will be required to handle and the capacity required.

Gyratory crushers are, in general, used where high capacity is required. Since they crush on full cycle, they are more efficient than jaw crushers, provided that the chamber can be kept full, which is normally easy, since the crusher can work with the head buried in ore.

Jaw crushers tend to be used where the crusher gape is more important than the capacity. For instance, if it is required to crush material of a certain maximum diameter, then a gyratory having the required gape would have a capacity about three times that of a jaw crusher of the same gape. If high capacity is required, then a gyratory is the answer. If, however, a large gape is needed but not capacity, then the jaw crusher will probably be more economical as it is a smaller machine and the gyratory would be running idle most of the time. A useful relationship, which is often used in plant design, is that given by Taggart:[5]

If $t\ h^{-1} < 161.7$ (gape in metres)^2 use a jaw crusher.

Conversely, if the tonnage is greater than this value, use a gyratory crusher.

Because of the complex nature of jaw and gyratory crushers, exact formulae expressing their capacities have never been entirely satisfactory. Crushing capacity depends on many factors, such as the angle of nip (i.e. the angle between the crushing members), stroke, speed, and the liner material, as well as on the feed material, and its initial particle size. Capacity problems do not usually occur in the upper and middle sections of the crushing cavity, providing the angle of nip is not too great. It is normally the discharge zone, the narrowest section of the crushing chamber, which determines the crushing capacity. Broman[6] describes the development of simple models for optimising the performance of jaw and gyratory crushers. The volumetric capacity of a jaw crusher is expressed as:

$$Q = BS_s . \cot[a . k . 60n]\ m^3/h$$

where B = inner width of crusher (m),
$\quad S$ = open side setting (m),
$\quad s$ = throw (m),
$\quad a$ = angle of nip,
$\quad n$ = speed of crusher (rpm),
and k is a material constant, the size of which varies with the characteristics of the crushed material, the feeding method, liner type, etc., normally having values between 1.5 and 2.

For gyratory crushers, the corresponding formula is:

$$Q = (D - S)\pi Ss \cot[a \ . \ k \ . \ 60n] \ \text{m}^3/\text{h}$$

where D = diameter of the outer head mantle at the discharge point (m), and k the material constant normally varying between 2 and 3.

The capital and maintenance costs of a jaw crusher are slightly less than those of the gyratory, but they may be offset by the installation costs, which are lower with the gyratory, since it occupies about two-thirds the volume and has about two-thirds the weight of a jaw crusher of the same capacity. This is because the circular crushing chamber allows a more compact design with a larger proportion of the total volume being accounted for by the crushing chamber than in the jaw crusher. Jaw-crusher foundations need to be much more rugged than those of the gyratory, due to the alternating working stresses.

The better self-feeding capability of the gyratory compared with the jaw results in a capital cost saving in some cases, with the elimination of expensive feeding devices, such as the heavy-duty chain feeder. This is, however, often false economy as the capital cost saving is considered of less importance in many cases than the improved performance and the pre-crusher scalping which is available with separate feeding devices.

In some cases, the jaw crusher has found favour, due to the ease with which it can be sectionalised. Thus, because of the need for transportation to remote locations and for underground use, it may be advantageous to install jaw crushers.

The type of material being crushed may also determine the crusher used. Jaw crushers perform better than gyratories on clayey, plastic material, due to their greater throw. Gyratories have been found to be particularly suitable for hard, abrasive material, and they tend to give a more cubic product than jaw crushers if the feed is laminated or "slabby".

Secondary Crushers

Secondary crushers are much lighter than the heavy-duty, rugged primary machines. Since they take the primary crushed ore as feed, the maximum feed size will normally be less than 15 cm in diameter and, because most of the harmful constituents in the ore, such as tramp metal, wood, clays, and slimes have already been removed, it is much easier to handle. Similarly, the transportation and feeding arrangements serving the crushers do not need to be as rugged as in the primary stage. Secondary crushers also operate with dry feeds, and their purpose is to reduce the ore to a size suitable for grinding. In those cases where size reduction can be more effectively carried out by crushing, there may be a tertiary stage before the material is passed to the grinding mills.

Tertiary crushers are, to all intents and purposes, of the same design as secondaries, except that they have a closer set.

The bulk of secondary crushing of metalliferous ores is performed by cone crushers, although crushing rolls and hammer mills are used for some applications.

The Cone Crusher

The cone crusher is a modified gyratory. The essential difference is that the shorter spindle of the cone crusher is not suspended, as in the gyratory, but is supported in a curved, universal bearing below the gyratory head or *cone* (Fig. 6.10).

Power is transmitted from the source to the countershaft through a V-belt or direct drive. The countershaft has a bevel pinion pressed and keyed to it, and drives the gear on the eccentric assembly. The eccentric has a tapered, offset bore and provides the means whereby the head and main shaft follow an eccentric path during each cycle of rotation.

Since a large gape is not required, the crushing shell or "bowl" flares outwards which allows for the swell of broken ore by providing an increasing cross-sectional area towards the discharge end. The cone crusher is therefore an excellent arrested crusher. The flare of the bowl allows a much greater head angle than in the gyratory crusher, while retaining the same angle between the crushing members (Fig. 6.11). This gives the cone crusher a high capacity, since the capacity of gyratory crushers is roughly proportional to the diameter of the head.

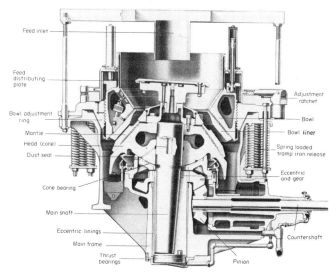

FIG. 6.10. Cross-section of heavy-duty Symons cone crusher.

The head is protected by a replaceable mantle, which is held in place by a large locking nut threaded onto a collar bolted to the top of the head. The mantle is backed with plastic cement, or zinc, or more recently with an epoxy resin.

Unlike a gyratory crusher, which is identified by the dimensions of the feed opening and the mantle diameter, a cone crusher is rated by the diameter of the cone lining. Cone crushers range in size from

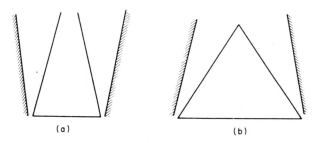

FIG. 6.11. Head and shell shapes of (a) gyratory, and (b) cone crushers.

559 mm to 3.1 m and have capacities up to 1100 t h^{-1} with a discharge setting of 19 mm, although two 3.1-m Symons cone crushers, each with capacities of 3000 t h^{-1}, have been installed in a South African iron-ore plant.[7]

The throw of cone crushers can be up to five times that of primary crushers, which must withstand heavier working stresses. They are also operated at much higher speeds. The material passing through the crusher is subjected to a series of hammer-like blows rather than being gradually compressed as by the slowly moving head of the gyratory.

The high-speed action allows particles to flow freely through the crusher, and the wide travel of the head creates a large opening between it and the bowl when in the fully open position. This permits the crushed fines to be rapidly discharged, making room for additional feed.

The fast discharge and non-choking characteristics of the cone crusher allow a reduction ratio in the range 3–7:1, but this can be higher in some cases.

The Symons cone crusher is the most widely used type of cone crusher. It is produced in two forms: the Standard for normal secondary crushing and the Short-head for fine, or tertiary duty (Figs. 6.12 and 6.13). They differ mainly in the shape of their crushing chambers. The Standard cone has "stepped" liners which allow a coarser feed than in the Short-head (Fig. 6.14). They deliver a product varying from 0.5 cm to 6.0 cm. The Short-head has a steeper head angle than the Standard, which helps to prevent choking from the much finer material being handled. It also has a narrower feed opening and a longer parallel section at the discharge, and delivers a product of 0.3–2.0 cm.

The parallel section between the liners at the discharge is a feature of all cone crushers and is incorporated to maintain a close control on product size. Material passing through the parallel zone receives more than one impact from the crushing members. The set on the cone crusher is thus the minimum discharge opening.

The distributing plate on the top of the cone helps to centralise the feed, distributing it at a uniform rate to all of the crushing chamber.

An important feature of the crusher is that the bowl is held down either by an annular arrangement of springs or by a hydraulic

FIG. 6.12. Standard cone crusher.

mechanism. These allow the bowl to yield if "tramp" material enters the crushing chamber, so permitting the offending object to pass. If the springs are continually "on the work", as may happen with ores containing many very tough particles, oversize material will be allowed to escape from the crusher. This is one of the reasons for using closed-circuit crushing in the final stages. It may be necessary to choose a screen for the circuit which has apertures slightly larger than the set of the crusher. This is to reduce the tendency for very tough particles, which are slightly oversize, to "spring" the crusher, causing an accumulation of such particles in the closed circuit and a build-up of pressure in the crushing throat.

The set on the crusher can easily be changed, or adjusted for liner

FIG. 6.13. Short-head cone crusher.

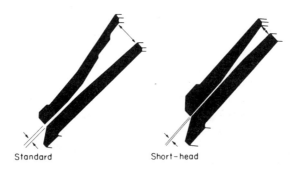

Standard   Short-head

FIG. 6.14. Liners of standard and short-head cone crushers.

wear, by screwing the bowl up or down by means of a capstan and chain arrangement or by adjusting the hydraulic setting, as on the "425 Vari-Cone" crusher manufactured by Hewitt-Robins, which allows the operator to change settings even if the equipment is operating under maximum load.[8] To close the setting, the operator opens a valve and presses a button starting a pump that adds hydraulic oil to the cylinder supporting the crusher head. To open the setting, another valve is opened allowing the oil to flow out of the cylinder. Efficiency is enhanced through automatic tramp iron clearing and reset. When tramp iron enters the crushing chamber, the crushing head will be forced down, causing hydraulic oil to flow into the accumulator. When the tramp iron has passed from the chamber, nitrogen pressure forces the hydraulic oil from the accumulator back into the supporting hydraulic cylinder, thus restoring the original setting.

The gyradisc crusher. This is a specialised form of cone crusher, used for producing very fine material, and such crushers have found application in the quarrying industry for the production of large quantities of sand at an economic cost.[9]

The main modification to the conventional cone crusher is that the machine has very short liners and a very flat angle for the lower liner (Fig. 6.15). Crushing is by *interparticle comminution* by the impact and attrition of a multi-layered mass of particles (Fig. 6.16).

The angle of the lower liner is less than the angle of repose of the ore, so that when the liner is at rest the material does not slide. Transfer through the crushing zone is by movement of the head. Each time the lower liner moves away from the upper liner, material enters the attrition chamber from the surge load above.

When reduction begins, material is picked up by the lower liner and is moved outward. Due to the slope of the liner it is carried to an advanced position and caught between the crushing members.

The length of stroke and the timing are such that after the initial stroke the lower liner is withdrawn faster than the previously crushed material falls by gravity. This permits the lower liner to recede and return to strike the previously crushed mass as it is falling, thus scattering it so that a new alignment of particles is obtained prior to

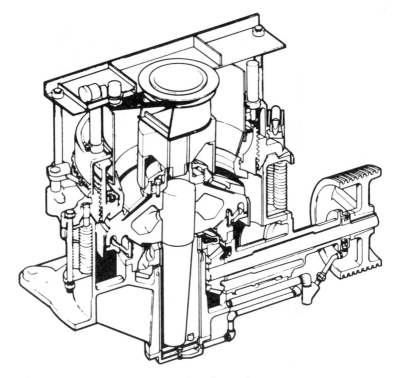

FIG. 6.15. Gyradisc crusher.

another impact. At each withdrawal of the head, the void is filled by particles from the surge chamber.

At no time does single-layer crushing occur, as with conventional crushers. Crushing is by particle on particle, so that the setting of the crusher is not as directly related to the size of product as it is on the cone crusher.

Their main use is in quarries, for producing sand and gravel. When used in open circuit they will produce a product of chippings from about 1 cm downwards, of good cubic shape, with a satisfactory amount of sand, which obviates the use of blending and rehandling. In closed circuit they are used to produce large quantities of sand. They may be used in open circuit on clean metalliferous ores with no primary

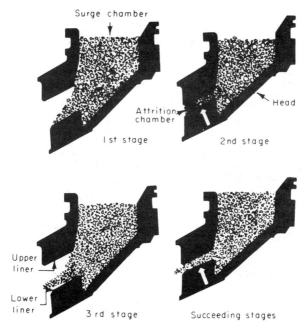

Surge chamber

Attrition chamber

Head

1st stage

2nd stage

Upper liner

Lower liner

3rd stage

Succeeding stages

FIG. 6.16. Action of gyradisc crusher.

slimes to produce an excellent ball-mill feed. Minus-19-mm material may be crushed to about 3 mm.[1]

Roll crushers. Roll crushers, or crushing rolls, are still used in some mills, although they have been replaced in most installations by cone crushers. They still have a useful application in handling friable, sticky, frozen, and less abrasive feeds, such as limestone, coal, chalk, gypsum, phosphate, and soft iron ores. Jaw and gyratory crushers have a tendency to choke near the discharge when crushing friable rock with a large proportion of maximum size pieces in the feed.

The mode of operation of roll crushers is extremely simple, the standard spring rolls (Fig. 6.17) consisting of two horizontal cylinders which revolve towards each other. The set is determined by shims

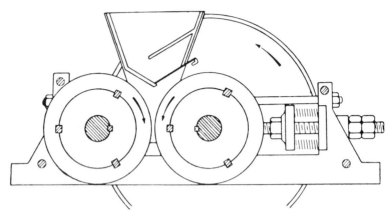

FIG. 6.17. Crushing rolls.

which cause the spring-loaded roll to be held back from the solidly mounted roll. Unlike jaw and gyratory crushers, where reduction is progressive by repeated pressure as the material passes down to the discharge point, the crushing process in rolls is one of single pressure.

Roll crushers are also manufactured with only one rotating cylinder, which revolves towards a fixed plate. Other roll crushers use three, four or six cylinders. In some crushers the diameters and speeds of the rolls may differ. The rolls may be gear driven, but this limits the distance adjustment between the rolls and modern rolls are driven by V-belts from separate motors.

Multi-roll machines may use rolls in pairs or in sets of three. Machines with more than two rolls are, however, rare in modern mills. The great disadvantage of roll crushers is that, in order for reasonable reduction ratios to be achieved, very large rolls are required in relation to the size of the feed particles. They therefore have the highest capital cost of all crushers.

Consider a spherical particle, of radius r, being crushed by a pair of rolls, of radius R, the gap between the rolls being $2a$ (Fig. 6.18). If μ is the coefficient of friction between the rolls and the particle, θ is the angle formed by the tangents to the roll surfaces at their points of contact with the particle (the angle of nip), and C is the compressive force exerted by the rolls, acting from the roll centres through the

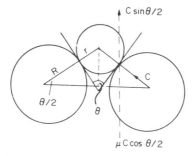

FIG. 6.18. Forces on a particle in crushing rolls.

particle centre, then for a particle to be *just* gripped by the rolls, equating vertically,

$$C \sin \theta/2 = \mu C \cos \theta/2. \tag{6.1}$$

Therefore
$$\mu = \tan \theta/2. \tag{6.2}$$

The coefficient of friction between steel and most ore particles is in the range 0.2–0.3, so that the value of the angle of nip θ should never exceed about 30°, or the particle will slip. It should also be noted that the value of the coefficient of friction decreases with speed, so that the speed of the rolls depends on the angle of nip, and the type of material being crushed. The larger the angle of nip (i.e. the coarser the feed), the slower the peripheral speed needs to be to allow the particle to be nipped. For smaller angles of nip (finer feed), the roll speed can be increased, so increasing the capacity. Peripheral speeds vary between about 1 m s⁻¹ for small rolls, up to about 15 m s⁻¹ for the largest sizes of 1800-mm diameter upwards.

The value of the coefficient of friction between a particle and moving rolls can be calculated from the equation

$$\mu_k = \left(\frac{1 + 1.12v}{1 + 6v}\right) \mu, \tag{6.3}$$

where μ_k is the kinetic coefficient of friction and v is the peripheral velocity of the rolls (m s⁻¹).

From Fig 6.18,

$$\cos \theta/2 = \frac{R + a}{R + r}. \tag{6.4}$$

Equation 6.4 can be used to determine the maximum size of rock gripped in relation to roll diameter and the reduction ratio (r/a) required. Table 6.1 lists such values for rolls crushing material where the angle of nip should be less than 20° in order for the particles to be gripped (in most practical cases the angle of nip should not exceed about 25°).

TABLE 6.1. MAXIMUM DIAMETER OF ROCK GRIPPED
IN CRUSHING ROLLS RELATIVE TO ROLL DIAMETER

Roll diameter (mm)	Maximum size of rock gripped (mm) Reduction ratio				
	2	3	4	5	6
200	6.2	4.6	4.1	3.8	3.7
400	12.4	9.2	8.2	7.6	7.3
600	18.6	13.8	12.2	11.5	11.0
800	24.8	18.4	16.3	15.3	14.7
1000	30.9	23.0	20.4	19.1	18.3
1200	37.1	27.6	24.5	22.9	22.0
1400	43.3	32.2	28.6	26.8	25.7

It can be seen that unless very large diameter rolls are used, the angle of nip limits the reduction ratio of the crusher, and since reduction ratios greater than 4:1 are rarely used, a flow-line may require coarse crushing rolls to be followed by fine rolls.

Smooth-surfaced rolls are usually used for fine crushing, whereas coarse crushing is often performed in rolls having corrugated surfaces, or with stub teeth arranged to present a chequered surface pattern. "Sledging" or "slugger" rolls have a series of intermeshing teeth, or slugs, protruding from the roll surfaces (Fig. 6.19). These dig into the rock so that the action is a combination of compression and ripping, and large pieces in relation to the roll diameter can be handled. Their main application is in the coarse crushing of soft or sticky iron ores,

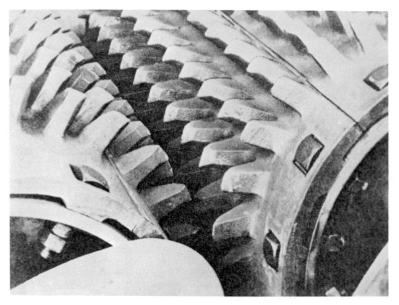

FIG. 6.19. Toothed crushing rolls.

friable limestone, coal, etc., rolls of 1-m diameter being used to crush material of top size 400 mm.

Wear on the roll surfaces is very high, and they often have a manganese steel tyre, which can be replaced when worn. The feed must be spread uniformly over the whole width of the rolls in order to give even wear. One simple method is to use a flat feed belt of the same width as the rolls.

Since there is no provision for the swelling of broken ore in the crushing chamber, roll crushers must be "starvation fed" if they are to be prevented from choking. Although the floating roll should only yield to an uncrushable body, choked crushing causes so much pressure that the springs are continually "on the work" during crushing, and some oversize escapes. Rolls should therefore be used in closed circuit with screens. Choked crushing also causes interparticle comminution, which leads to the production of material finer than the set of the crusher.

The capacity of the rolls can be calculated in terms of the ribbon of material that will pass the space between the rolls. Thus theoretical capacity is equal to

$$188.5 \, N \, D \, W \, sd \quad \text{kg h}^{-1} \tag{6.5}$$

where N is the speed of rolls (rev min^{-1}), D is the roll diameter (m), W is the roll width (m), s is the specific gravity of feed material (kg m^{-3}), and d is the distance between the rolls (m).

In practice, allowing for voids between the particles, loss of speed in gripping the feed, etc., the capacity is usually about 25% of the theoretical.

Impact Crushers

In this class of crusher, comminution is by impact rather than compression, by sharp blows applied at high speed to free-falling rock. The moving parts are beaters, which transfer some of their kinetic energy to the ore particles on contacting them. The internal stresses created in the particles are often large enough to cause them to shatter. These forces are increased by causing the particles to impact upon an anvil or breaker plate.

There is an important difference between the states of materials crushed by pressure and by impact. There are internal stresses in material broken by pressure which can cause later cracking. Impact causes immediate fracture with no residual stresses. This stress-free condition is particularly valuable in stone used for brick-making, building, and roadmaking, in which binding agents, such as bitumen, are subsequently added to the surface. Impact crushers, therefore, have a wider use in the quarrying industry than in the metal-mining industry. They may give trouble-free crushing on ores that tend to be plastic and pack when the crushing forces are applied slowly, as is the case in jaw and gyratory crushers. These types of ore tend to be brittle when the crushing force is applied instantaneously by impact crushers.[1]

Figure 6.20 shows a cross-section through a typical *hammer mill*. The hammers are made from manganese steel or, more recently,

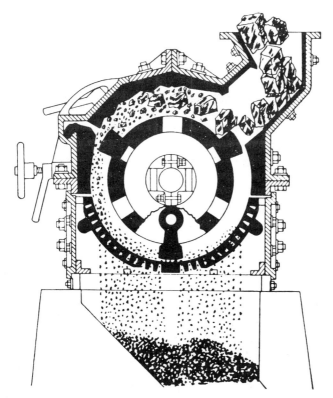

FIG. 6.20. Hammer mill.

nodular cast iron, containing chromium carbide, which is extremely abrasion resistant. The breaker plates are made of the same material.

The hammers are pivoted so that they can move out of the path of oversize material, or tramp metal, entering the crushing chamber. Pivoted hammers exert less force than they would if rigidly attached, so they tend to be used on smaller impact crushers or for crushing soft material. The exit from the mill is perforated, so that material which is not broken to the required size is retained and swept up again by the rotor for further impacting.

This type of machine is designed to give the particles velocities of the order of that of the hammers. Fracture is either due to the severity of impact with the hammers or to the subsequent impact with the casing or grid. Since the particles are given very high velocities, much of the size reduction is by attrition, i.e. breaking of particle on particle, and this leads to little control on product size and a much higher proportion of fines than with compressive crushers.

The hammers can weigh over 100 kg and can work on feed up to 20 cm. The speed of the rotor varies between 500 and 3000 rev min^{-1}.

Due to the high rate of wear on these machines (wear can be taken up by moving the hammers on the pins) they are limited in use to relatively non-abrasive materials. They have extensive use in limestone quarrying and in the crushing of coal. A great advantage in quarrying is in the fact that they produce a very good cubic product.

For much coarser crushing, the fixed hammer *impact mill* is often used (Fig. 6.21). In these machines the material falls tangentially on to a rotor, running at 250–500 rev min^{-1}, receiving a glancing impulse, which sends it spinning towards the impact plates. The velocity imparted is deliberately restricted to a fraction of the velocity of the

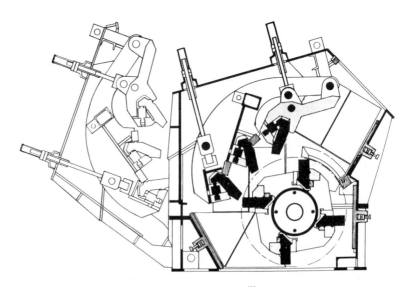

FIG. 6.21. Impact mill.

rotor to avoid enormous stress and probable failure of the rotor bearings.

The fractured pieces which can pass between the clearances of the rotor and breaker plate enter a second chamber created by another breaker plate, where the clearance is smaller, and then into a third smaller chamber. This is the *grinding path* which is designed to reduce flakiness and gives very good cubic particles.

The rotary impact mill gives a much better control of product size than does the hammer mill, since there is less attrition. The product shape is much more easily controlled and energy is saved by the removal of particles once they have reached the size required.

The blow bars are reversible to even out wear, and can easily be removed and replaced.

Large impact crushers will reduce 1.5-m top size run-of-mine ore to 20 cm, at capacities of around 1500 t h^{-1}, although crushers with capacities of 3000 t h^{-1} have been manufactured. Since they depend on high velocities for crushing, wear is greater than for jaw or gyratory crushers. Hence impact crushers should not be used on ores containing over 15% silica.[1] However, they are a good choice for primary crushing when high reduction ratios are required—the ratio can be as high as 40:1—and a high percentage of fines, and the ore is relatively non-abrasive.

Rotary Coal Breakers

Where large tonnages of coal are treated, the rotary coal breaker is often used (Fig. 6.22).

This is very similar in operation to the cylindrical trommel screen (Chapter 8), consisting of a cylinder of 1.8–3.6 m in diameter and length of about 1½ to 2½ times the diameter, revolving at a speed of about 12–18 rev min^{-1}. The machine is massively constructed, with perforated walls, the size of the perforations being the size to which the coal is to be broken. The run-of-mine coal is fed into the rotating cylinder, at up to 1500 t h^{-1} in the larger machines. The machine utilises differential breakage, the coal being much more friable than the associated stones and shales, and rubbish such as wood, steel, etc.,

FIG. 6.22. Rotary coal breaker.

from the mine. The small particles of coal and shale quickly fall through the holes, while the larger lumps are retained, and are lifted by longitudinal lifters within the cylinder until they reach a point where they slide off the lifters and fall to the bottom of the cylinder, breaking by their own impact, and fall through the holes. The lifters are inclined to give the coal a forward motion through the breaker. Large pieces of shale and stone do not break as easily, and are usually discharged from the end of the breaker, which thus cleans the coal to a certain degree and, as the broken coal is quickly removed from the breaker, produces few fines. Although the rotary breaker is an expensive piece of equipment, maintenance costs are relatively low, and it produces positive control of top size of product.[10]

Crushing Circuits and Control

In recent years, efforts have been made to improve crusher

efficiency in order to reduce capital and operating costs. Automatic control of crushing circuits is increasingly used, larger crushers have been constructed; and mobile crushing units have been used, which allow relatively cheap ore transportation by conveyor belts rather than by trucks to a fixed crushing station. [11–14] A mobile crusher is a completely self-contained unit, mounted on a frame that is moved by means of a transport mechanism in the open pit as mining progresses. Mobile units typically use jaw, hammer, or roll crushers fed directly, or by apron feeders, at rates of up to 1000 t h^{-1}. Some units employ large gyratory crushers with capacities of up to 6000 t h^{-1}. [15]

A typical flowsheet for a crushing plant producing ball mill feed is shown in Figure 6.23. [16] The circuit is typical of current practice in that the secondary product is screened and conveyed to a storage bin, rather than feeding the tertiary crushers directly. The intermediate bins allow good mixing of the secondary screen oversize with the circulating load, and regulation of the tertiary crusher feed, providing more efficient crushing. The circuit is also more readily adaptable to automatic feed control to maintain maximum power utilisation. [17]

In some cases, the crushing circuit is designed not only to produce mill feed, but also to provide media for autogenous grinding. Figure

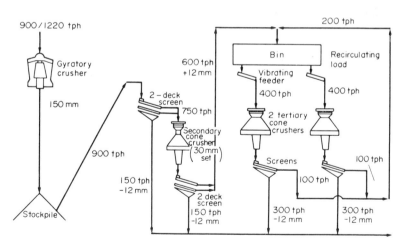

FIG. 6.23. Three-stage crushing circuit for ball mill feed (after Motz[16]).

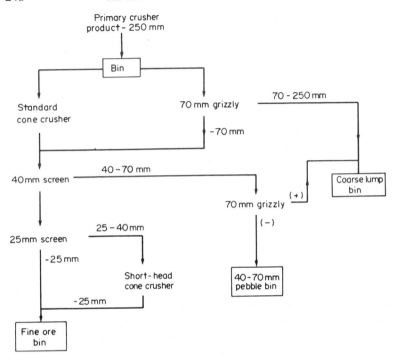

FIG. 6.24. Pyhasalmi crushing circuit.

6.24 shows the flowsheet for the crushing circuit at Pyhasalmi in Finland, where grinding is performed in lump mills, followed by pebble mills.[18]

Primary crushing is carried out underground and the product is hoisted to the fine crushing plant on top of the fine ore bin. This plant consists of two parallel crushing lines, one including a standard Symons cone crusher as the first stage, and the other commencing with a vibrating grizzly of 70 mm set. The grizzly oversize of 70–250 mm is conveyed directly to a coarse lump bin, while the undersize joins the cone crusher product from the other line. This material is screened on two double deck vibrating screens, to produce a fine ore product of −25 mm, an intermediate product of 25–40 mm, which is crushed to

−25 mm in an open circuit short-head Symons cone crusher, and a +40 mm fraction which is transferred to a 40–70 mm pebble bin via a 70 mm scalping grizzly.

Recent advances in instrumentation and process control hardware have made the implementation of computer control more common in crushing circuits. Instrumentation employed includes ore level detectors, oil flow sensors, power measurement devices, belt scales, variable speed belt drives and feeders, blocked chute detectors, and particle size measurement devices.[19] The importance of automatic control is exemplified by the crushing plant at Mount Isa in Australia, where the output increased by over 15% after controls were introduced.[20]

Supervisory control systems are not usually applied to primary crushers, the instrumentation basically being used to protect them. Thus lubrication flow indicators and bearing temperature detectors may be fitted, together with high and low level alarms in the chamber under the crusher.

The operating and process control objectives for secondary and tertiary crushing circuits differ from one plant to the next, but usually the main objective is to maximise crusher throughput at some specified product size. Numerous variables affect the performance of a crusher, but only three-ore feed rate, crusher opening and, in some cases, feed size, can be adjusted. Lynch[21] has described case studies of automatic control systems for various applications. When the purpose of the crushing plant is to produce feed for the grinding circuit, the most important objective of the control system is to ensure a supply of crushed ore at the rate required by the grinding plant. The fineness of the crusher product is maintained by the selection of screens of the appropriate aperture in the final closed circuit loop. The most effective way of maximising throughput is to maintain the highest possible crusher power draw, and this has been used to control many plants. There is an optimum closed side setting for crushers operating in closed circuit which provides the highest tonnage of finished screen product for a particular power or circulating load limit, although the actual feed tonnage to the crusher increases at larger closed side settings. The power draw can be maintained by the use of a variable speed belt feeding the crusher. Changes in ore hardness and size distribution are compensated for by a change in belt speed. Operations under such choked conditions also require sensing of upper and lower levels of feed in the crusher by mechanical, nuclear, sonic or proximity switches. Operation of the crusher at high power draw leads to increased fines production, such that if the increased throughput provided by the control system cannot be accommodated by the grinding plant, then the higher average power draw can be used to produce a finer product. This can be done by using screens with smaller apertures in the closed circuit, thus increasing the circulating load and hence the total crusher feed. In most cases, high screen loading decreases screening efficiency, particularly that of the particles close to the screen aperture size. This has the effect of reducing the effective "cut-size" of the screen, producing a finer product. Thus a possible control scheme during periods of excess closed circuit crushing capacity or reduced throughput requirement is to increase the circulating load by reducing the number of screens used, leading to a finer product. The implementation of this type of control loop requires accurate knowledge of the behaviour of the plant under various conditions.

In those circuits where the crushers produce a final saleable product (e.g. road-stone quarries), the control objective is usually to maximise the production of certain size fractions from each tonne of feed. Since screen efficiency decreases as circulating load increases, producing a finer product size, circulating load can be used to control the product size. This can be affected by control of the crusher setting, as on the Allis–Chalmers hydrocone crusher,[3,22,23] which has a hydraulic setting adjustment system which can be controlled automatically to optimise the crusher parameters. The required variation in crusher setting can be determined by the use of accurate mathematical models of crusher performance,[21] from empirical data, or by measuring product size on-line. The latter approach is receiving much attention, research into optical imaging techniques being carried out, while Foxboro are manufacturing a device that scans and classifies the surface of the product on the discharge belt, this method being prone to error if the material is segregrated.

Additional loops are normally required in crushing circuits to control levels in surge bins between different stages. For instance, the crusher product surge bins can be monitored such that at high level, screen feed is increased to draw down the bins.

In most crushing plants there are long process delays, and standard PI controllers can be inadequate for the control loops. It is therefore desirable to provide a model of the process which incorporates the actual time delays, and use this model to adjust the controller inputs.[24] Such a dynamically compensated controller was developed for the Mount Isa crusher circuit.[20] The variation in operating conditions and significant circuit delays would seem to be well suited to control strategies based on modern model-based control techniques (see Chapter 3). In order to apply modern control theory to crushing, it is first necessary to develop dynamic models that are sufficiently detailed to reproduce the essential dynamic characteristics of crushing. For a cone crusher, this would involve accurately predicting the discharge size distribution, throughput and power consumption as a function of time. Then optimal estimation using a Kalman filter has to be developed for the dynamic model. Finally, an optimal control algorithm is employed to determine the values of the manipulated variables which optimise the chosen control objectives. The optimal control algorithm will also supervise the set-points of the regulatory control loops. Herbst and Oblad[25] have used a dynamic model of a cone crusher in conjunction with an extended Kalman filter to optimally estimate the size distributions of the material within the crusher as well as the measured output variables. In addition, estimation of the specific crushing rates has been implemented for the purpose of detecting ore hardness changes. The dynamic model and estimator are incorporated in a simulator as a first step in developing an optimal control scheme for crushing circuits. An overview of some of the recent work on crushing and screening models and a description of general policies for the control of crushing plants has been given by Whiten.[26]

References

1. Lewis, F. M., Coburn, J. L., and Bhappu, R. B., Comminution: a guide to size-reduction system design. *Min. Engng.* **28,** 29 (Sept. 1976).

2. Grieco, F. W., and Grieco, J. P., Manufacturing and refurbishing of jaw crushers, *CIM Bulletin,* **78,** 38 (Oct. 1985).
3. Anon., Crushers, *Min. Mag.* 94 (Aug. 1981).
4. McQuiston, F. W., and Shoemaker, R. S., *Primary Crushing Plant Design,* AIMME, New York (1978).
5. Taggart, A. F., *Handbook of Mineral Dressing,* Wiley, New York, 1945.
6. Broman, J., Optimising capacity and economy in jaw and gyratory crushers, *Engng. & Min. J.,* 69 (June 1984).
7. White, L., Processing responding to new demands, *Engng Min. J.* 219 (June 1976).
8. Anon., Rugged roller-bearing crusher, *Mining Mag.,* 240 (Sept. 1985).
9. Anon., A new concept in the production of fine material, *Quarry Managers' J.* (June 1967).
10. *Coal Preparation Course,* Vol. 1, South African Coal Processing Soc., Johannesburg, 1977.
11. Woody, R., Cutting crushing costs, *World Mining,* 76 (Oct. 1982).
12. Kok, H. G., Use of mobile crushers in the minerals industry, *Min. Engng.* **34,** 1584 (Nov. 1982).
13. Gresshaber, H. E., Crushing and grinding: design considerations, *World Mining,* **36,** 41 (Oct. 1983).
14. Frizzel, E. M., Mobile in-pit crushing-product of evolutionary change, *Mining Engng.* **37,** (June 1985).
15. Rixen, W., Energy saving ideas for open pit mining, *World Mining,* 85 (June 1981).
16. Motz, J. C., Crushing, in *Mineral Processing Plant Design* (eds A. L. Mular and R. B. Bhappu), p. 203, A.I.M.M.E., New York (1978).
17. Mollick, L., Crushing, *Eng. & Min. J.* **181,** 96 (June 1980).
18. Wills, B. A., Pyhasalmi and Vihanti concentrators, *Min. Mag.* 174 (Sept. 1983).
19. Horst, W. E., and Enochs, R. C., Instrumentation and process control, *Eng. & Min. J.* **181,** 70 (June 1980).
20. Manlapig, E. V., and Watsford, R. M. S., Computer control of the lead-zinc concentrator crushing plant operations of Mount Isa Mines Ltd, *Proc. 4th IFAC Symp. on Automation in Mining,* Helsinki (Aug. 1983).
21. Lynch, A. J., *Mineral Crushing and Grinding Circuits,* Elsevier, Amsterdam (1977).
22. Flavel, M. D., Scientific methods to design crushing and screening plants, *Min. Engng.* **29,** 65 (July 1977).
23. Flavel, M. D., Control of crushing circuits will reduce capital and operating costs, *Min. Mag.* 207 (March 1978).
24. McKee, D. J., and Thornton, A. J., Emerging automatic control approaches in mineral processing, in *Mineral Processing at a Crossroads—Problems and Prospects,* eds. B. A. Wills and R. W. Barley, p. 117, Martinus Nijhoff, Dordrecht, Netherlands, 1986.
25. Herbst, J. A., and Oblad, A. E., Modern control theory applied to crushing, Part 1: development of a dynamic model for a cone crusher and optimal estimation of crusher operating variables, in *Automation for Mineral Resource Development,* p. 301, Pergamon, Oxford, 1986.
26. Whiten, W. J., Models and control techniques for crushing plants, in *Control '84 Minerals/Metallurgical Processing,* ed. J. A. Herbst, p. 217, AIME, New York, 1984.

CHAPTER 7

GRINDING MILLS

Introduction

Grinding is the last stage in the process of comminution; in this stage the particles are reduced in size by a combination of impact and abrasion, either dry or in suspension in water. It is performed in rotating cylindrical steel vessels known as *tumbling mills*. These contain a charge of loose crushing bodies—the grinding medium—which is free to move inside the mill, thus comminuting the ore particles. The grinding medium may be steel rods, or balls, hard rock, or, in some cases, the ore itself. In the grinding process, particles between 5 and 250 mm are reduced in size to between 10 and 300 μm.

All ores have an economic optimum mesh of grind (Chapter 1) which will depend on many factors, including the extent to which the values are dispersed in the gangue, and the subsequent separation process to be used. It is the purpose of the grinding section to exercise close control on this product size and, for this reason, correct grinding is often said to be the key to good mineral processing. Undergrind of the ore will, of course, result in a product which is too coarse, with a degree of liberation too low for economic separation; poor recovery and enrichment ratio will be achieved in the concentration stage. Overgrinding needlessly reduces the particle size of the substantially liberated major constituent (usually the gangue) and may reduce the particle size of the minor constituent (usually the mineral value) below the size required for most efficient separation. Much expensive energy is wasted in the process. It is important to realise that grinding is the most energy intensive operation in mineral processing. It has been estimated that 50% of the 10^{10}-kWh energy consumption in US mills is used in comminution.[1] On a survey of the energy consumed in a number of Canadian copper concentrators it was shown that the average energy consumption in kWh t^{-1} was 2.2 for crushing, 11.6 for grinding, and 2.6 for flotation.[2] Since grinding is the greatest single operating cost, the ore should not be ground any finer than is justified economically. Finer grinding should not be carried out beyond the point where the net smelter return for the increased recovery becomes less than the added operating cost.[3] It can be shown, using Bond's equation (5.2), that 19% extra power must be consumed in grinding one screen size finer on a $\sqrt{2}$ screen series.

Although tumbling mills have been developed to a high degree of mechanical efficiency and reliability, they are extremely wasteful in terms of energy expended since the ore is mostly broken as a result of repeated, random impacts, which break liberated as well as unliberated particles. At present there is no way that these impacts can be directed at the interfaces between the mineral grains, which would produce optimum liberation, although various ideas have been postulated.[4]

Although the correct degree of liberation is the principal purpose of grinding in mineral processing, this treatment is sometimes used to increase the surface area of the valuable minerals even though they may already be essentially liberated from the gangue. This is very important in processes where grinding is followed by hydrometallurgical methods of treatment. Thus in gold-ore treatment, leaching with cyanide solution follows, and in some cases takes place during, the grinding process. Leaching is much more efficient on particles which have large surface areas in relation to their mass and in such cases overgrinding may not be a disadvantage, as the increase in power consumption may be offset by the increased gold recovery. This may not, however, be the case in the leaching of base metal ores, where a relatively coarse grind may be required in order to merely expose the valuable minerals to the lixivant, any further increase in metal extraction by finer grinding not being justified.

Grinding within a tumbling mill is influenced by the size, quantity, the type of motion, and the spaces between the individual pieces of the medium in the mill. As opposed to crushing, which takes place between relatively rigid surfaces, grinding is a more random process, and is subject to the laws of probability. The degree of grinding of an ore particle depends on the probability of it entering a zone between the medium shapes and the probability of some occurrence taking place after entry. Grinding can take place by several mechanisms, including impact or compression, due to forces applied almost normally to the particle surface; chipping due to oblique forces; and abrasion due to forces acting parallel to the surfaces (Fig. 7.1). These mechanisms distort the particles and change their shape beyond certain limits determined by their degree of elasticity, which causes them to break.

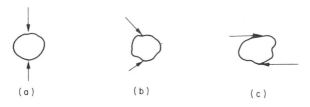

FIG. 7.1. Mechanisms of breakage: (a) impact or compression, (b) chipping, (c) abrasion.

Grinding is usually performed wet, although in certain applications dry grinding is used. When the mill is rotated the mixture of medium, ore, and water, known as the *mill charge*, is intimately mixed, the medium comminuting the particles by any of the above methods depending on the speed of rotation of the mill. Most of the kinetic energy of the tumbling load is dissipated as heat, noise, and other losses, only a small fraction being expended in actually breaking the particles.

Apart from laboratory testing, grinding in mineral processing is a continuous process, material being fed at a controlled rate from storage bins into one end of the mill and overflowing at the other end after a suitable dwell time. Control of product size is exercised by the type of medium used, the speed of rotation of the mill, the nature of the ore feed, and the type of circuit used.

The Motion of the Charge in a Tumbling Mill

The distinctive feature of tumbling mills is the use of loose crushing bodies, which are large, hard, and heavy in relation to the ore particles, but small in relation to the volume of the mill, and which occupy slightly less than half the volume of the mill.

Due to the rotation and friction of the mill shell, the grinding medium is lifted along the rising side of the mill until a position of dynamic equilibrium is reached, when the bodies cascade and cataract down the free surface of the other bodies, about a dead zone where little movement occurs, down to the *toe* of the mill charge (Fig. 7.2).

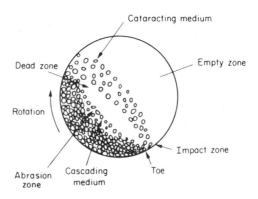

FIG. 7.2. Motion of charge in a tumbling mill.

The speed at which a mill is run is important, since it governs the nature of the product and the amount of wear on the shell liners. For instance, a practical knowledge of the trajectories followed by the steel balls in a mill determines the speed at which it must be run in order that the descending balls shall fall on to the toe of the charge, and not on to the liner, which could lead to rapid liner wear.

The driving force of the mill is transmitted via the liner to the charge. At relatively low speeds, or with smooth liners, the medium tends to roll down to the toe of the mill and essentially abrasive comminution occurs. This *cascading* leads to finer grinding, with increased slimes production and increased liner wear. At higher speeds the medium is projected clear of the charge to describe a series of parabolas before landing on the toe of the charge. This *cataracting* leads to comminution by impact and a coarser end product with reduced liner wear. At the *critical speed* of the mill the theoretical trajectory of the medium is such that it would fall outside the shell. In practice, *centrifuging* occurs and the medium is carried around in an essentially fixed position against the shell.

In travelling around inside the mill the medium (and the large lumps of ore) follows a path which has two parts. The lifting section near to the shell liners is circular while the drop back to the toe of the mill charge is parabolic (Fig. 7.3(a)).

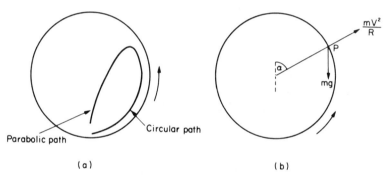

(a)   (b)

FIG. 7.3. (a) Trajectory of grinding medium in tumbling mill, (b) forces acting on the medium.

Consider a ball, or rod, which is lifted up the shell of a mill of radius R metres, revolving at N rev min^{-1}. The rod abandons its circular path for a parabolic path at point P (Fig. 7.3(b)), when the weight of the rod is just balanced by the centrifugal force, i.e. when

$$\frac{mV^2}{R} = mg \cos \alpha,$$

where m is the mass of the rod or ball (kg), V is the linear velocity of the rod (m s^{-1}), and g is the acceleration due to gravity (m s^{-2}).

Since

$$V = \frac{2\pi RN}{60},$$

$$\cos \alpha = \frac{4\pi^2 N^2 R}{60^2 g}$$

$$= 0.0011 N^2 R.$$

When the diameter of the rod, or ball, is taken into account, the radius of the outermost path is $(D - d)/2$ where D is the mill diameter and d the rod or ball diameter in metres.

Thus

$$\cos \alpha = 0.0011 N^2 \frac{(D - d)}{2} . \tag{7.1}$$

The critical speed of the mill occurs when $\alpha = 0$, i.e. the medium abandons its circular path at the highest vertical point. At this point, $\cos \alpha = 1$.

Therefore

$$N_c = \frac{42.3}{\sqrt{(D - d)}} \quad \text{rev min}^{-1}, \tag{7.2}$$

where N_c is the critical speed of the mill.

Equation 7.2 assumes that there is no slip between the medium and the shell liner and, to allow for a margin of error, it has been common practice to increase the value of the calculated critical speed by as much as 20%.[5] It is questionable, however, whether with modern liners maintained in reasonable condition this increase in the value is necessary or desirable.

Mills are driven in practice at speeds of 50–90% of critical speed, the choice being influenced by economic considerations. Increase in speed increases capacity, but there is little increase in efficiency (i.e. kWh t^{-1}) above about 40–50% of the critical speed. Very low speeds are sometimes used when full mill capacity cannot be attained. High speeds are used for high-capacity coarse grinding. Cataracting at high speeds converts the potential energy of the medium into kinetic energy of impact on the toe of the charge and does not produce as much very fine material as the abrasive grinding produced by cascading at lower speeds. It is essential, however, that the cataracting medium should fall well inside the mill charge and not directly onto the liner, thus excessively increasing steel consumption. Most of the grinding in the mill takes place at the toe of the charge, where not only is there direct impact of the cataracting medium on to the charge, but also the ore packed between the cascading medium receives the shock transmitted.

At the extreme toe of the load the descending liner continuously underruns the churning mass, and moves some of it into the main mill charge. The medium and ore particles in contact with the liners are held with more firmness than the rest of the charge due to the extra weight bearing down on them. The larger the ore particle, rod, or ball the less likely it is to penetrate the interior of the charge and the more likely it is to be carried to the breakway point by the liners. The cataracting effect should thus be applied in terms of the rod or ball of largest diameter.

Tumbling Mills

Tumbling mills are of three basic types: rod, ball, and autogenous.[6,7] Structurally, each type of mill consists of a horizontal cylindrical shell, provided with renewable wearing liners and a charge of grinding medium. The drum is supported so as to rotate on its axis on hollow *trunnions* attached to the end walls (Fig. 7.4). The diameter of the mill determines the pressure that can be exerted by the medium on the ore particles and, in general, the larger the feed size the larger needs to be the mill diameter. The length of the mill, in conjunction with the diameter, determines the volume, and hence the capacity of the mill.

The feed material is usually fed to the mill continuously through one end trunnion, the ground product leaving via the other trunnion, although in certain applications the product may leave the mill through a number of ports spaced around the periphery of the shell. All types of mill can be used for wet or dry grinding by modification of feed and discharge equipment.

FIG. 7.4. Tumbling mill.

Construction of Mills

Shell.   Mill shells are designed to sustain impact and heavy loading, and are constructed from rolled mild steel plates, butt-welded together. Holes are drilled to take the bolts for holding the liners. Normally one or two access manholes are provided. For attachment of the trunnion heads, heavy flanges of fabricated or cast steel are usually welded or bolted to the ends of the plate shells, planed with parallel faces which are grooved to receive a corresponding spigot on the head, and drilled for bolting to the head.

Mill ends.   The mill ends, or trunnion heads, may be of nodular or grey cast iron for diameters less than about 1 m. Larger heads are constructed from cast steel, which is relatively light, and can be welded. The heads are ribbed for reinforcement and may be flat, slightly conical, or dished. They are machined and drilled to fit shell flanges (Fig. 7.5).

FIG. 7.5. Tube mill end and trunnion.

Trunnions and bearings. The trunnions are made from cast iron or steel and are spigoted and bolted to the end plates, although in small mills they may be integral with the end plates. They are highly polished to reduce bearing friction. Most trunnion bearings are rigid high-grade iron castings with a 120–180° lining of white metal in the bearing area, surrounded by a fabricated mild steel housing, which is bolted into the concrete foundations (Fig. 7.6).

The bearings in smaller mills may be grease lubricated, but oil lubrication is favoured in large mills, via motor-driven oil pumps. The effectiveness of normal lubrication protection is reduced when the mill is shut down for any length of time, and many mills are fitted with manually operated hydraulic starting lubricators which force oil between the trunnion and trunnion bearing, preventing friction damage to the bearing surface, on starting, by re-establishing the protecting film of oil (Fig. 7.7).

Some manufacturers are installing large roller bearings, which can withstand higher forces than plain metal bearings (Fig. 7.8).

FIG. 7.6. 180° oil-lubricated trunnion bearing.

Drive. Tumbling mills are most commonly rotated by a pinion meshing with a girth gear ring bolted to one end of the machine (Fig. 7.9).

The pinion shaft is driven from the prime mover through vee-belts, in small mills of less than about 180 kW. For larger mills the shaft is coupled directly to the output shaft of a slow-speed synchronous motor, or to the output shaft of a motor-driven helical or double helical gear reducer. In some mills thyristors and DC motors are being used to give variable speed control. Very large mills driven by girth gears require two to four pinions, and complex load sharing systems must be incorporated.

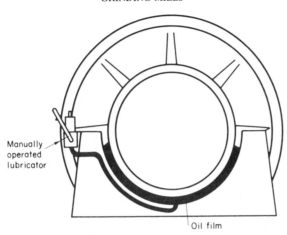

FIG. 7.7. Hydraulic starting lubricator.

FIG. 7.8. Trunnion with roller-type bearings.

FIG. 7.9. Gear/pinion assembly on ball mill.

Large mills can be rotated by a central trunnion drive, which has the advantage of requiring no expensive ring gear, the drive being from one or two motors, with the inclusion of two- or three-speed gearing.[6]

At Sydvaranger, in Norway, a 9.65 m long by 6.5 m diameter ball mill is driven by a gearless ring, or "wraparound" 8.1 MW motor. The drive motor rotor is built into the shell, and the stator is mounted to surround it.[8,9] The stator outer diameter is 11.8 m, and the rotor outer diameter is 8.85 m (Fig. 7.10). The stator has a closed circuit cooling system for the windings, and the mill speed is infinitely variable by frequency control, allowing automatic adjustments to the mill throughput as ore grindability changes. A 5.5-m diameter × 7.3-m gearless-drive mill was commissioned at Bougainville Copper, Papua New Guinea, in June 1985.[10] The mill is powered by a 6-mW cycloconvertor-fed synchronous ring motor, and is believed to be only the second application of a ring motor on a wet grinding mill, although the technology is relatively common in the European cement industry,

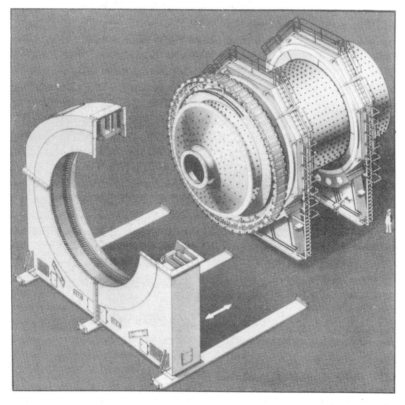

Fig. 7.10. Ball mill with gearless ring motor.

the mill being developed due to the limitations in the strength of ring gears on very large mills.

A G.E.C. mill utilising *tyres* to rotate and support the mill shell was installed at South Crofty tin mine, Cornwall, in 1982. The mill has a stepless variable speed hydraulic drive, and when integrated with on-stream size analysis can be controlled automatically to maintain a preset product size under varying feed conditions. Tyre driven mills are manufactured by G.E.C. in sizes up to 3.0 m diameter and 7.6 m long, but correct tyre selection and drive design is reported to require substantial engineering effort.[6]

Liners. The internal working faces of mills consist of renewable liners, which must withstand impact, be wear-resistant, and promote the most favourable motion of the charge. Rod mill ends have plain, flat liners, slightly coned to encourage the self-centring and straight-line action of rods. They are made usually from manganese or chrome-molybdenum steels, having high impact strength. Ball-mill ends usually have ribs to lift the charge with the mill rotation. These prevent excessive slipping and increase liner life. They can be made from white cast iron, alloyed with nickel (Ni-hard), other wear-resistant materials, and rubber. Trunnion liners are designed for each application and can be conical, plain, with advancing or retarding spirals. They are manufactured from hard cast iron or cast alloy steel, a rubber lining often being bonded to the inner surface for increased life.

Shell liners have an endless variety of lifter shapes. Smooth linings result in much abrasion, and hence a fine grind, but with associated high metal wear. The liners are therefore generally shaped to provide lifting action and to add impact and crushing, the most common shapes being wave, Lorain, stepped, and ship-lap (Fig. 7.11). The liners are attached to the mill shell and ends by forged steel countersunk liner bolts.

Rod-mill liners are also generally of alloyed steel or cast-iron—and of the wave type, although Ni-hard step liners may be used with rods up to 4 cm in diameter. Lorain liners are extensively used for coarse grinding in rod and ball mills, and consist of high carbon rolled steel plates held in place by manganese or hard alloy steel lifter bars.

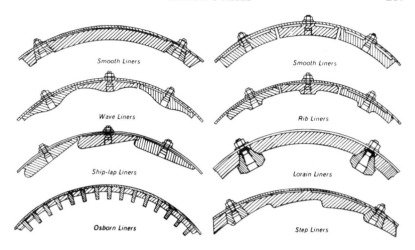

FIG. 7.11. Mill shell liners.

Ball-mill liners may be made of hard cast iron when balls of up to 5 cm in diameter are used, but otherwise cast manganese steel, cast chromium steel, or Ni-hard are used.

Mill liners are a major cost in mill operation, and efforts to prolong liner life are constantly being made. There are at least ten wear-resistant alloys used for ball mill linings, the more abrasion resistant alloys containing large amounts of chromium, molybdenum and nickel being the most expensive.[11] However, with steadily increasing labour costs for replacing liners, the trend is towards selecting liners which have the best service life regardless of cost.[12]

Rubber liners and lifters have supplanted steel in some operations, and have been found to be longer lasting, easier and faster to install, and their use results in a significant reduction of noise level. However, increased medium consumption has been reported using rubber liners rather than Ni-hard liners. Rubber lining may also have drawbacks in processes requiring the addition of flotation reagents directly into the mill, or temperatures exceeding 80°C.[13] A concept which has found recent application for ball mills is the "angular spiral lining". The circular cross section of a conventional mill is changed to a square cross section with rounded corners by the addition of rubber-lined, flanged frames, which are offset to spiral in a direction opposite to the mill rotation. Double wave liner plates are fitted to these frames, and a sequential lifting of the charge down the length of the mill results, which increases the grinding ball to pulp mixing through axial motion of the grinding charge, along with the normal cascading motion. Substantial increases in throughput, along with reductions in power and grinding medium consumptions, have been reported.[14]

A revolutionary new liner was installed in a secondary pebble mill at the LKAB Kiruna concentrator in Sweden in 1980. The Orebed Magnetic Lining manufactured by Trelleborg[15] is built from permanent ceramic magnets, vulcanised into rubber or any suitable polymer. It is thinner than conventional liners, and the magnets keep the lining in contact with the steel shell, while any magnetic material in the charge, either magnetic minerals or mill scale, is attracted to the liner. This magnetic material forms the wear-resistant layer, which is continuously renewed as it wears. Energy consumption at Kiruna decreased by 11.4%, and pebble consumption decreased by more than 30%. After 5000 hours operation, no visible liner wear could be detected.

Mill feeders.   The type of feeding arrangement used on the mill depends on whether the grinding is done in open or closed circuit and whether it is done wet or dry. The size and rate of feed are also important. Dry mills are usually fed by some sort of vibratory feeder. Three types of feeder are in use in wet-grinding mills. The simplest form is the *spout feeder* (Fig. 7.12), consisting of a cylindrical or elliptical chute supported independently of the mill, and projecting directly into the trunnion liner. Material is fed by gravity through the spout to feed the mills. They are often used for feeding rod mills operating in open circuit or mills in closed circuit with hydrocyclone classifiers. *Drum feeders* (Fig. 7.13) may be used in lieu of a spout feeder when headroom is limited. The entire mill feed enters the drum via a chute or spout and an internal spiral carries it into the trunnion liner. The drum also provides a convenient method of adding grinding balls to a mill.

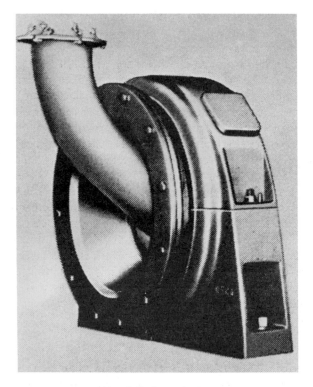

FIG. 7.12. Spout feeder.

Combination drum–scoop feeders (Fig. 7.14) are generally used for wet grinding in closed circuit with a spiral or rake classifier. New material is fed directly into the drum, while the scoop picks up the classifier sands for regrinding. Either a single or double scoop can be used, the latter providing an increased feed rate and more uniform flow of material into the mill; the counter-balancing effect of the double-scoop design serves to smooth out power fluctuation and it is normally incorporated in large-diameter mills. *Scoop feeders* are sometimes used in place of the drum–scoop combination when mill feed is in the fine-size range.

FIG. 7.13. Drum feeder on ball mill.

Types of Mill

Rod mills.   These may be considered as either fine crushers or coarse grinding machines. They are capable of taking feed as large as 50 mm and making a product as fine as 300 μm, reduction ratios normally being in the range 15–20:1. They are often preferred to fine crushing machines when the ore is "clayey" or damp, thus tending to choke crushers.

The distinctive feature of a rod mill is that the length of the cylindrical shell is between 1.5 and 2.5 times its diameter (Fig. 7.15). This ratio is important because the rods, which are only a few centimetres shorter than the length of the shell, must be prevented from turning so that they become wedged across the diameter of the cylinder. The ratio must not, however, be so large for the maximum diameter of shell in use that the rods deform and break. Since rods longer than about 6 m will bend, this establishes the maximum length of mill. Thus with a mill 6.4 m long the diameter should not be over

Fig. 7.14. Drum-scoop feeder.

4.57 m. Rod mills of up to 4.57 m in diameter by 6.4 m in length are in use, run by 1640 kW motors.[16] Rod and other grinding mills are rated by power rather than capacity, since the capacity is determined by many factors, such as the grindability, determined by laboratory testing (Chapter 5), and the reduction in size required. The power required for a certain required capacity may be estimated by the use of Bond's equation:

$$W = \frac{10 \ W_i}{\sqrt{P}} - \frac{10 \ W_i}{\sqrt{P}}. \tag{5.2}$$

FIG. 7.15. Rod mill.

The calculated power requirement is adjusted by utilising efficiency factors dependent on the size of mill, size and type of media, type of grinding circuit, etc., to give the operating power requirement.[17]

Rod mills are classed according to the nature of the discharge. A general statement can be made that the closer the discharge is to the periphery of the shell, the quicker the material will pass through and less overgrinding will take place.

Centre peripheral discharge mills (Fig. 7.16) are fed at both ends through the trunnions and discharge the ground product through circumferential ports at the centre of the shell. The short path and steep gradient give a coarse grind with a minimum of fines, but the reduction ratio is limited. This mill can be used for wet or dry grinding and has found its greatest use in the preparation of specification sands, where high tonnage rates and an extremely coarse product are required.

End peripheral discharge mills (Fig. 7.17) are fed at one end through the trunnion, discharging the ground product from the other end of the mill by means of several peripheral apertures into a close-fitting

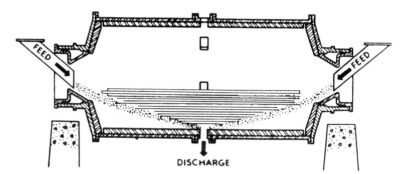

FIG. 7.16. Central peripheral discharge mill.

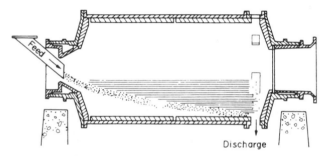

FIG. 7.17. End peripheral discharge mill.

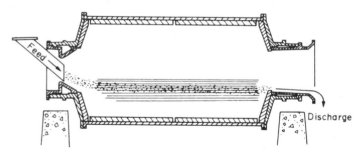

FIG. 7.18. Overflow mill.

circumferential chute. This type of mill is used mainly for dry and damp grinding, where moderately coarse products are involved.

The most widely used type of rod mill in the mining industry is the *trunnion overflow* (Fig. 7.18), in which the feed is introduced through one trunnion and discharges through the other. This type of mill is used only for wet grinding and its principal function is to convert crushing-plant product into ball-mill feed. A flow gradient is provided by making the overflow trunnion diameter 10–20 cm larger than that of the feed opening. The discharge trunnion is often fitted with a spiral screen to remove tramp material.

Rod mills are charged initially with a selection of rods of assorted diameters, the proportion of each size being calculated to provide maximum grinding surface and to approximate to a seasoned or equilibrium charge. A seasoned charge will contain rods of varying diameters ranging from fresh replacements to those which have worn down to such a size as to warrant removal. Actual diameters in use range from 25 to 150 mm. The smaller the rods the larger is the total surface area and hence the greater the grinding efficiency. The largest diameter should be no greater than that required to break the largest particle in the feed. A coarse feed or product normally requires larger rods. Generally, rods should be removed when they are worn down to about 25 mm in diameter or less, depending on the application, as small ones tend to bend or break. High carbon steel rods are used as they are hard and break rather than warp when worn, so do not entangle with other rods. Optimum grinding rates are obtained with new rods when the volume is 35% of that of the shell. This reduces to 20–30% with wear and is maintained at this figure by substitution of new rods for worn ones. This proportion means that with normal voidage, about 45% of the mill volume is occupied. Overcharging results in inefficient grinding and increased liner and rod consumption. Rod consumption varies widely with the characteristics of the mill feed, mill speed, rod length, and product size; it is normally in the range 0.1–1.0 kg of steel per tonne of ore for wet grinding, being less for dry grinding.

Rod mills are normally run at between 50–65% of the critical speed, so that the rods cascade rather than cataract; many operating mills have been sped up to close to 80% of critical speed without any reports of excessive wear.[18] The feed pulp density is usually between 65 and 85% solids by weight, finer feeds requiring lower pulp densities. The grinding action results from line contact of the rods on the ore particles; the rods tumble in essentially a parallel alignment, and also spin, thus acting rather like a series of crushing rolls. The coarse feed tends to spread the rods at the feed end, so producing a wedge or cone-shaped array. This increases the tendency for grinding to take place preferentially on the larger particles, thereby producing a minimum amount of extremely fine material (Fig. 7.19). This selective grinding gives a product of relatively narrow size range, with little oversize or slimes. Rod mills are therefore suitable for preparation of feed to gravity concentrators, certain flotation processes with slime problems, magnetic cobbing, and ball mills. They are nearly always run in open circuit because of this controlled size reduction.

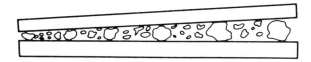

FIG. 7.19. Grinding action of rods.

Ball mills. The final stages of comminution are performed in tumbling mills using steel balls as the grinding medium and so designated ball mills.

Since balls have a greater surface area per unit weight than rods, they are better suited for fine finishing. The term ball mill is restricted to those having a length to diameter ratio of 1.5 to 1 and less. Ball mills in which the length to diameter ratio is between 3 and 5 are designated *tube mills.* These are sometimes divided into several longitudinal compartments, each having a different charge composition; the charges can be steel balls or rods, or pebbles, and they are often used dry to grind cement clinker, gypsum, and phosphate. Tube mills having only one compartment and a charge of hard, screened ore particles as the grinding medium are known as *pebble mills.* They are widely used in the South African gold mines. Since the weight of pebbles per unit volume is 35–55% of that of steel balls, and as the power input is directly proportional to the volume weight of the grinding medium, the power input and capacity of pebble mills are correspondingly lower. Thus in a given grinding circuit, for a certain feed rate, a pebble mill would be much larger than a ball mill, with correspondingly higher capital cost. However, it is claimed that the increment in capital cost can be justified economically by a reduction in operating cost attributed to the lower cost of the grinding medium. This may, however, be partially offset by higher power cost per tonne of finished product.[16]

Ball mills are also classified by the nature of the discharge. They may be simple trunnion overflow mills, operated in open or closed circuit, or *grate discharge* (low-level discharge) mills. The latter type is fitted with discharge grates between the cylindrical mill body and the discharge trunnion. The pulp can flow freely through the openings in the grate and is then lifted up to the level of the discharge trunnion (Fig. 7.20). These mills have a lower pulp level than overflow mills, thus reducing the dwell time of particles in the mill. Very little overgrinding takes place and the product contains a large fraction of coarse material, which is returned to the mill by some form of classifying device. Closed-circuit grinding (Fig. 7.21), with high circulating loads, produces a closely sized end product and a high output per unit volume compared with open circuit grinding. Grate discharge mills usually take a coarser feed than overflow mills and are

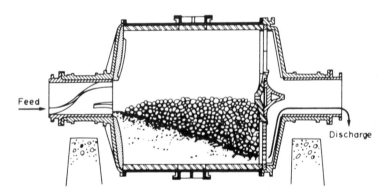

FIG. 7.20. Grate discharge mill.

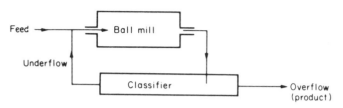

FIG. 7.21. Simple closed-grinding circuit.

not required to grind so finely, the main reason being that with many small balls forming the charge the grate open area plugs very quickly. The trunnion overflow mill is the simplest to operate and is used for most ball-mill applications, especially for fine grinding and regrinding. Energy consumption is said to be about 15% less than that of a grate discharge mill of the same size, although the grinding efficiencies of the two mills are the same.[16]

Ball mills are rated by power rather than capacity. Apart from the single 6.5 m diameter mill in operation at Sydvaranger (Fig. 7.10), the largest currently operating ball mills are 5.5 m in diameter by 7.3 m in length and are driven by 4000-kW motors. The trend in recent years has been to use larger mills, but there are several cases where large ball mills have not obtained design capacities predicted from small scale results with the application of conventional scale-up procedures. For instance, the performance of the 5.5-m diameter by 6.4-m ball mills at Bougainville Copper Ltd did not measure up to the design expectations, being particularly inefficient in grinding coarse material. As most of the copper losses were in the coarse size fractions of the flotation tailings, it was concluded that the most economical way of reducing these losses was to scalp the coarse material from flotation feed and regrind it, a 5.5-m diameter gearless-drive mill being installed to do this.[10]

Ball mill scale-up studies have been conducted in Australia and the United States, the results emphasising that there are limitations to conventional procedures for estimating large mill requirements from small-scale results.[19-22] These constraints may limit ball mill diameters to dimensions not far from the present maxima. Mill capacities can still be increased by increasing mill lengths, but without the gain in specific capacity associated with diameter increases.

Grinding in a ball mill is effected by point contact of balls and ore particles and, given time, any degree of fineness can be achieved. The process is completely random—the probability of a fine particle being struck by a ball is the same as that of a coarse particle. The product from an open-circuit ball mill therefore exhibits a wide range of particle size, and overgrinding of at least some of the charge becomes a problem. Closed-circuit grinding in mills providing low residence time for the particles is almost always used in the last stages to overcome this.

Several factors influence the efficiency of ball-mill grinding. The pulp density of the feed should be as high as possible, consistent with ease of flow through the mill. It is essential that the balls are coated with a layer of ore; too dilute a pulp increases metal-to-metal contact, giving increased steel consumption and reduced efficiency. Ball mills should operate between 65 and 80% solids by weight, depending on the ore. The viscosity of the pulp increases with the fineness of the particles, therefore fine-grinding circuits may need lower pulp densities.

The efficiency of grinding depends on the surface area of the grinding medium. Thus balls should be as small as possible and the charge should be graded such that the largest balls are just heavy enough to grind the largest and hardest particles in the feed. A seasoned charge will consist of a wide range of ball sizes and new balls added to the mill are usually of the largest size required. Undersize balls leave the mill with the ore product and can be removed by passing the discharge over screens. Various formulae have been proposed for the required ratio of ball size to ore size,[23] none of which are entirely satisfactory. They can be summarised as

$$d = kD^{0.5-1}, \qquad (7.3)$$

where d is the ball diameter, D is the feed size, and k is a constant, varying from 55 for chert and 35 for dolomite.

Such formulae give a good basis for determining the ball-mill charge sizes, but thereafter the correct size should be determined by trial and error. Primary grinding usually requires a graded charge of 10-cm to 5-cm diameter balls, while secondary grinding requires 5 cm to 2 cm. Segregation of the ball charge within the mill is achieved in the

Hardinge mill (Fig. 7.22). The conventional drum shape is modified by fitting a conical section, the angle of the cone being about 30°. Due to the centrifugal force generated, the balls are segregated so that the largest are at the feed end of the cone, i.e. the largest diameter and greatest centrifugal force, and the smallest are at the discharge. By this means, a regular gradation of ball size and of size reduction is produced.

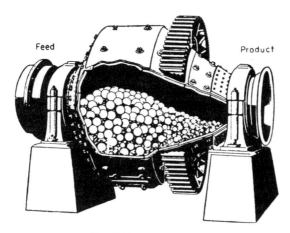

Feed   Product

FIG. 7.22. Hardinge mill.

Grinding balls are usually made of forged or rolled high-carbon or alloy steel, or cast steel, and consumption varies between 0.1 to as much as 1 kg per tonne of ore depending on hardness of ore, fineness of grind and medium quality. Medium consumption can be a very high proportion, sometimes as much as 40%, of the total milling cost, so is an area that often warrants special attention. Good quality grinding media may be more expensive, but may be economic due to lower wear rates. Very hard media, however, may lead to lower grinding efficiencies due to slippage, and this also should be taken into account. Finer grinding may lead to improved metallurgical efficiency, but at the expense of higher grinding energy and media consumption. Therefore, particularly with ores of low value, where milling costs are crucial, the economic limit of grinding has to be carefully assessed.

Apart from direct removal of metal from the grinding media surface by abrasive wear, additional corrosive wear is apparent during wet grinding, in which less resistant corrosion product films are abraded away.[24] However, data collected in the United States show that in the primary grinding of molybdenite ore abrasion is the major cause of metal loss, corrosion representing less than 10% of the total loss.[25]

The charge volume is about 40–50% of the internal volume of the mill, about 40% of this being void space. The energy input to a mill increases with the ball charge, and reaches a maximum at a charge volume of approximately 50% (Fig. 7.23), but for a number of reasons, 40–50% is seldom exceeded. The efficiency curve is in any case quite flat about the maximum. In overflow mills the charge volume is usually 40%, but there is a greater choice in the case of grate discharge mills. The optimum mill speed increases with charge volume, as the increased weight of charge reduces the amount of cataracting taking place.

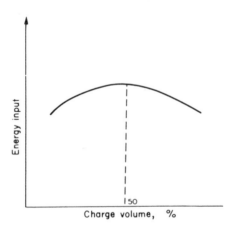

FIG. 7.23.   Energy input versus charge volume.

Ball mills are often operated at higher speeds than rod mills, so that the larger balls cataract and impact on the ore particles. The work input to a mill increases in proportion to the speed, and ball mills are run at as high a speed as is possible without centrifuging. Normally this is 70–80% of the critical speed, the higher speeds often being used to increase the amount of cataracting taking place in order to break hard or coarse feeds.

Autogenous Mills

There has been increasing emphasis during recent years towards the use of autogenous and semi-autogenous grinding (SAG) circuits. On suitable ores, the advantages over conventional circuits include lower capital cost, the ability to handle wet and sticky material, relatively simple flowsheets, the large size of available equipment, lower manpower requirements, and minimal grinding media expense.

The autogenous mill uses tumbling to effect comminution, but instead of utilising media such as steel rods or balls, comminution is achieved by the action of the ore particles on each other. Semi-autogenous milling refers to grinding methods using a combination of the ore and a reduced load of steel rods or balls as the medium.[26] Experience indicates that the ball charges used in semi-autogenous grinding have generally been most effective in the range of 6–10% of the mill volume, including voids.

Where the proportion of coarse fraction in the feed is too low, pebble milling is sometimes used. In this method the coarsest fraction of the feed is separated, by screening, and the remainder is crushed in conventional machinery to a considerably smaller size. This crushed material and the coarse fraction are then put into pebble mills for completion of fine grinding.

Autogenous milling may be performed wet or dry. Dry mills have more environmental problems, do not handle materials containing clay well, and are more difficult to control than wet mills. However, in certain applications, involving grinding of minerals such as asbestos, talc, and mica, dry semi-autogenous milling is used exclusively. On the other hand, few completely dry autogenous mills are operated because of the relatively low throughput.[16]

Although Scandinavian and South African practice favours length to diameter ratios of 1–2, in North America primary autogenous mills are distinguished by their very large diameters relative to their lengths.

The largest in present use weigh over 1100 tonnes, are 11 m in diameter by 4.3 m long, and are driven by 9000-kW motors. The Cascade mill (Fig. 7.24), manufactured by Koppers–Hardinge, is typical of this category in that the length is one-third of the diameter. It is lined with wearing plates held by lift bars bolted to the shell. Lift bars are essential to reduce slippage of the mill load, which causes rapid wear of the liners and also impairs the grinding action.

Two 9.75-m diameter autogenous mills were installed at Palabora Mining Co. in South Africa in 1977 to produce copper flotation feed from primary crushed ore. Each mill has the capacity to grind up to 800 t h^{-1}, reducing the ore from -180 mm to 50% -300 μm. Each mill has a maximum power consumption of 7000 kW. Milling costs have been reduced by over 20% compared with conventional grinding.

In comparison with such large diameter mills, which are usually operated at 75–80% of critical speed, autogenous mills in the gold

Fig. 7.24. Cascade mills in operation.

mines of South Africa have been limited in size to 4.88 m diameter and up to 12.2 m in length, and are normally operated at very high speeds, up to 90% critical. The difference in practice is attributed to the very high hardness of the gold ores, limitations on manufacturing facilities for large diameter mills, and a long history of pebble milling.

Primary autogenous mills can achieve size reductions from 25 cm to 0.1 mm in one piece of equipment. The particle-size distribution of the product depends on the characteristics and structure of the ore being ground. Fractures in rock being reduced autogenously are principally at the grain or crystal boundaries, due to the relatively gentle action compared with steel-grinding media. Thus the product sizing is predominantly around the region of grain or crystal size. This is generally desirable for minerals treatment as the wanted minerals are released with minimal overgrinding. This type of mill is widely used in North America to grind specular hematite in such a way as to release the crystals from the quartzitic gangue with minimum degradation, and to grind sandstone so as to release the discrete silica grains. It is usually thought that autogenous grinding produces less extreme fines than conventional grinding. This is in fact true for materials having a distinct grain structure, but with materials, such as low-grade copper ores, which lack a well-defined grain structure, more fines and greater surface area are produced.[27]

Autogenous primary mills cannot usually be selected from bench scale grinding tests as they require more extensive testing than rod or ball mills. While grinding rods and balls can be obtained in the required sizes and quantities and their actions during milling can be reasonably predicted, in an autogenous mill the grinding medium is also the material to be ground and consequently is itself a variable.

Although primary autogenous and semi-autogenous mills have a definite role in comminution, the growth in autogenous milling came during a period of low energy costs. Although steel consumption is reduced over conventional milling, unit power costs can be higher than conventional crushing and grinding by between 25 and 100%.

However, secondary autogenous grinding (pebble milling) and intermediate autogenous grinding (lump milling) feature unit power costs very comparable to conventional milling, and are widely used in Scandinavia and South Africa. Lump milling uses 150–200 mm rocks, and pebble milling uses 30–70 mm pebbles, usually screened from the crushing circuit (see Chapter 6).

Vibratory mills.   Vibratory mills are designed for continuous, or batch, grinding to give a very fine end product from a wide variety of materials, the operation being performed either wet or dry.

Two tubes functioning as vibrating grinding cylinders are located one above the other in a plane inclined at 30° to the perpendicular (Fig. 7.25). Between them lies an eccentrically supported weight connected by a flexible universal joint to a 1000–1500 rev min^{-1} motor. Rotation

FIG. 7.25. Vibratory mill.

of the eccentric vibrates the tubes to produce an oscillation circle of a few millimetres. The cylinders are filled to about 60–70% with grinding medium, usually steel balls of diameter 10–50 mm. The material being ground passes longitudinally through the cylinders like a fluid, in a complex spinning helix, thus allowing the grinding medium to reduce it by attrition. The material is fed and discharged through flexible bellow-type hoses.

The outstanding features of correctly designed vibratory ball mills are their small size and lower power consumption relative to throughput when compared with other types of mill. High-energy vibration mills can grind materials to surface areas of around 500 m^2 g^{-1}, a degree of fineness which is impossible in a conventional mill.[28] Vibratory mills are made with capacities up to 15 t h^{-1}, although units of greater capacity than about 5 t h^{-1} involve considerable engineering problems. The processed material size range is between approximately 30-mm feed and −10-μm end product, although the mill product size is very sensitive to feed-size variation, which limits its use for controlled grinding. They have been used to regrind tin concentrates prior to cleaning, their function being to remove the waste from the cassiterite crystals without destroying the crystals, which often happens in the conventional ball mill environment.

Centrifugal mills.   The concept of centrifugal grinding is an old one, and although a patent of 1896 describes this process, it has so far not gained full-scale industrial application. Conventional mills have critical speeds above which centrifugal force prevents the mill charge mixing and tumbling to its best advantage. Throughput can therefore be increased only by increasing the size of the mill, particularly its diameter, and there are serious design and engineering problems associated with this.

In centrifugal milling, the forces on the charge inside the mill are increased by operating the mill in a centrifugal, rather than a gravitational field. Comminution is more rapid, and the size of machine needed for a given grinding duty is thus reduced.

The Chamber of Mines of South Africa studied centrifugal milling concepts in detail, which has led to a co-operative development programme with Lurgi of West Germany.[29,30] Operation of a prototype 1 m diameter, 1 m long, 1 MW mill over an extended period at Western Deep Levels gold mine has proved it to be the equivalent of a conventional 4 m × 6 m ball mill in every respect.

Tower mills.   An alternative to ball mills for very fine grinding operations is the *tower mill* (Fig. 7.26). Steel balls or pebbles are placed in a vertical grinding chamber in which an internal screw flight provides medium agitation. The feed enters at the top, with mill water, and is reduced in size by attrition and abrasion as it falls, the finely ground particles being carried upwards by pumped liquid and overflowing to a classifier. Oversize particles are returned to the bottom of the chamber, and final classification is by hydrocyclones, which return underflow to the mill sump for further grinding. According to the manufacturer's claims, advantages are a small installation area, low noise levels, efficient energy usage, minimal overgrinding and low capital and operating costs. Product sizes may be 1–100 μ at capacities of up to 100 t/h or more. The Japanese have been the primary users of these mills, grinding a variety of materials including limestone, silica, rock salt, coal and copper concentrates. A tower mill was installed at Quintana Minerals, New Mexico, in 1980 to regrind molybdenum concentrates, and a 200 kW mill is being used at Macassa, Ontario, for the simultaneous grinding and leaching of gold tailings.[31] At Cominco's Polaris Mine in Canada, the zinc concentrate had difficulty in meeting the magnesia specification in concentrates. The installation of a tower mill in the zinc circuit raised the recovery and also improved magnesia rejection.[32]

Roller mills.   These mills are often used for the dry fine grinding of medium soft materials of up to 4–5 mohs hardness.[33] Above this hardness, excessive wear offsets the advantage of lower energy consumption compared with conventional mills.

Table and roller mills are used to grind medium hard materials such as coal, limestone, phosphate rock and gypsum. Two or three rollers,

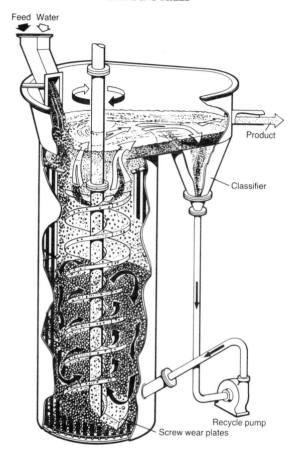

Fig. 7.26. Tower mill.

operating against coiled springs, grind material which is fed onto the centre of a rotating grinding table (Fig. 7.27). Ground material spilling over the edge of the table is air-swept into a classifier mounted on the mill casing, coarse particles being returned for further grinding.

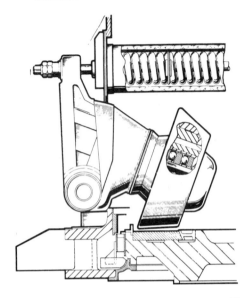

FIG. 7.27. Section through table and roller mill.

Pendulum roller mills are used for fine grinding non-metallic minerals such as barytes and limestone. Material is reduced by the centrifugal action of suspended rollers running against a stationary grinding ring (Fig. 7.28). The rollers are pivoted on a spider support fitted to a gear-driven shaft. Feed material falls onto the mill floor, to be scooped up by ploughs into the "angle of nip" between the rolls and the grinding ring. Ground material is air-swept from the mill into a classifier, oversize material being returned.

Grinding Circuits

The feed can be wet or dry, depending on the subsequent process and the nature of the product. Dry grinding is necessary with some materials due to physical or chemical changes which occur if water is added. It causes less wear on the liners and grinding media and there is

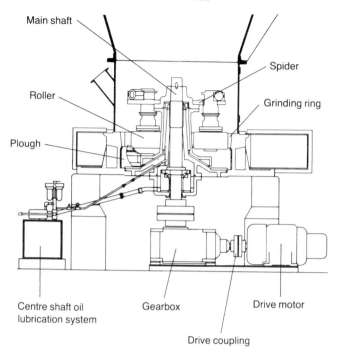

FIG. 7.28. Section through pendulum roller mill.

a higher proportion of fines in the product, which may be desirable in some cases.

Wet grinding is generally used in mineral processing operations because of the overall economies of operation.

The advantages of wet grinding are:

(1) Lower power consumption per tonne of product.
(2) Higher capacity per unit mill volume.
(3) Makes possible the use of wet screening or classification for close product control.
(4) Elimination of the dust problem.
(5) Makes possible the use of simple handling and transport methods such as pumps, pipes, and launders.

The type of mill for a particular grind, and the circuit in which it is to be used, must be considered simultaneously. Circuits are divided into two broad classifications: *open* and *closed*. In open circuit the material is fed into the mill at a rate calculated to produce the correct product in one pass (Fig. 7.29). This type of circuit is rarely used in mineral processing applications as there is no control on product size distribution. The feed rate must be low enough to ensure that every particle spends enough time in the mill to be broken down to product size. As a result, many particles in the product are overground, which consumes power needlessly, and the product may subsequently be difficult to treat.

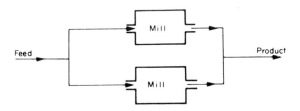

FIG. 7.29. Single-stage open circuit.

Grinding in the mining industry is almost always in closed circuit (Fig. 7.21), in which material of the required size is removed by a classifier, which returns oversize to the mill. Although virtually every grinding circuit has some form of classification, individual mills in the circuit can be open or closed.

In closed-circuit operation, no effort is made to effect all the size reduction in a single pass. Instead, every effort is made to remove material from the circuit as soon as it reaches the required size. When grinding to a specified size, an increase in capacity of up to 35% can be obtained by closed-circuit operation.

The material returned to the mill by the classifier is known as the *circulating load*, and its weight is expressed as a percentage of the weight of new feed.

Closed-circuit grinding reduces the residence time of particles in each pass, and so the proportion of finished sizes in the mill, compared with open-circuit grinding. This decreases overgrinding and increases the energy available for useful grinding as long as there is an ample supply of unfinished material present. As the tonnage of new feed increases, so the tonnage of circulating load increases, since the classifier underflow becomes coarser, but the *composite* mill feed becomes finer due to this increase in circulating load. Due to the reduced residence time in the mill, the mill discharge correspondingly becomes coarser, so that the difference in mean size between the composite feed and the discharge decreases. The capacity of a mill increases with decreasing ball diameter due to the increase in grinding surface, to the point where the angle of nip between contacting balls and particles is exceeded. Consequently, the more near-finished material there is in the composite feed, the higher is the proportion of favourable nip angles, and the finer the composite feed the smaller need be the mean ball diameter. Within limits, therefore, the larger the circulating load the greater is the useful capacity of the mill. The increase is most rapid in the first 100% of circulating load, then continues up to a limit, depending on the circuit, just before chokage of the mill occurs. The optimum circulating load for a particular circuit depends on the classifier capacity and the cost of transporting the load to the mill. It is usually in the range 100–350%, although it can be as high as 600%.

Rod mills are generally operated in open circuit because of their grinding action, especially when preparing feed for ball mills. The parallel grinding surfaces simulate a slotted screen and tend to retard the larger particles until they are broken. Smaller particles slip through the spaces between the rods and are discharged without appreciable reduction. Ball mills, however, are virtually always operated in closed circuit with some form of classifier.

Various types of classifying device can be used to close the circuit, mechanical classifiers being used in many of the older mills. They are robust, easy to control, and smooth out and tolerate surges well. They can handle very coarse sands products and are still used on many coarse-grinding circuits. They suffer from the severe disadvantage of classifying by gravitational force, which restricts their capacity when dealing with extremely fine material, hence reducing the circulating load possible.

Hydrocyclones (Chapter 9) classify by centrifugal action, which speeds up the classification of fine particles, giving much sharper separations and increasing the optimum circulating load. They occupy much less floor space than mechanical classifiers of the same capacity and have lower capital and installation costs. Due to their much quicker action, the grinding circuit can rapidly be brought into balance if changes are made and the reduced dwell time of particles in the return load gives less time for oxidation to occur, which is important with sulphide minerals which are to be subsequently floated. They have therefore become widely used in fine grinding circuits preceding flotation operations.

The action of all classifiers in grinding circuits is dependent on the differential settling rates of particles in a fluid, which means that particles are classified not only by size but also by specific gravity. A small dense particle may therefore behave in a similar way to a large, low-density particle. Thus, when an ore containing a heavy valuable mineral is ground, overgrinding of this material is likely to occur, due to it being returned in the circulating load, even though it is below the required product size.

Selective grinding of heavy sulphides in this manner prior to flotation can allow a coarser overall grind, the light gangue reporting to the classifier overflow, while the heavy particles containing valuable mineral are selectively reground. It can, however, pose problems with gravity and magnetic separation circuits, where particles should be as coarse and as closely sized as possible in order to achieve maximum separation efficiency. Such circuits are often closed by screens rather than classifiers. Fine screens have the disadvantage, however, of being relatively inefficient and delicate, and often a combination of classification and screening is used to reduce the load on the screens.

Such a circuit was used at Wheal Jane Ltd., Truro, for the primary grinding of the complex copper–zinc–tin ore. The secondary crusher product was fed on to a wedge-wire screen, which routed the oversize into the primary ball mill; the mill discharge was pumped to a cyclone, and the underflow passed on to the screen to remove any fine heavy minerals before being returned to the ball mill (Fig. 7.30).

It is universal in gold recovery plants, where coarse free gold is present, to incorporate some form of gravity concentrator into the grinding circuit. Native gold is extremely dense, and is invariably

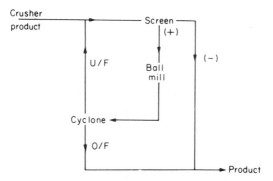

FIG. 7.30. Cyclone and screen in the closed circuit.

returned to the mill via the classifier. Being extremely malleable, once liberated it is merely deformed in the mill with no further breakage, so is continually recycled in the circuit.

At the Western Deep Levels Gold Mine in South Africa, the mill feed is first classified to remove fines which are sent to the cyanidation plant in order to leach out the gold. The coarse fraction is fed to tube mills, the discharge being classified, and the fines returned to the primary cyclone. The classifier oversize, which contains any free gold present, is then treated by gravity concentrators, the tailings being returned to the primary cyclone, while the gravity concentrate is sent

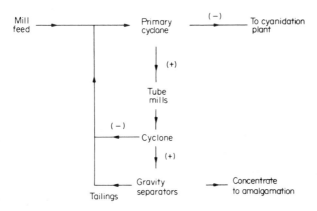

FIG. 7.31. Typical gold recovery grinding circuit.

for further treatment by amalgamation (Fig. 7.31). A similar system is used at the Afton copper concentrator in Canada, where jigs are used to remove coarse native copper from the primary grinding circuit.[34]

Grinding circuits are fed at a uniform rate from bins holding the crusher plant product. There may be a number of ball mills in parallel, each circuit being closed by its own classifier, and taking a definite fraction of the feed (Fig. 7.32).

Parallel mill circuits increase circuit flexibility, since individual units can be shut down or the feed rate can be changed, with little effect on the flowsheet. Fewer mills are, however, easier to control and capital and installation costs are lower, so the optimum number of mills must be decided at the design stage.

Multi-stage grinding in which mills are arranged in series can be used to produce successively finer products (Fig. 7.33), but the present trend is toward large single-stage primary ball mills, which greatly reduces capital and operating costs, and facilitates automatic control. The disadvantage of single-stage milling is that if too high a reduction ratio is produced, then relatively large balls are required for coarse feed, which may not efficiently grind the finer particles.

FIG. 7.32. Closed-circuit ball mills.

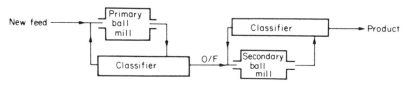

FIG. 7.33. Two-stage grinding circuit.

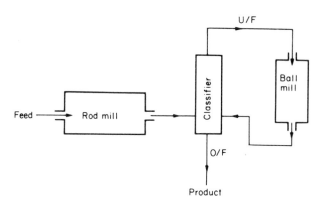

FIG. 7.34. Rod-ball mill grinding circuits.

Two-stage grinding is used where rod milling is substituted for tertiary crushing (Fig. 7.34).

The crusher product is fed to the rod mills, which operate in open circuit, producing ball-mill feed. This should, if capacity permits, be fed to the classifier rather than directly to the ball mill, as any finished material is immediately removed, and the rod-mill product, which may need diluting prior to pumping, is thickened before being fed to the ball mill.

Multi-stage autogenous milling is typical Scandinavian practice, and there are some interesting variations of this method. At Vihanti in Finland pebble milling follows rod milling, while at Pyhasalmi, autogenous grinding takes place in three stages, in a primary lump mill, two secondary mills, and one tertiary pebble mill.[35] The grinding circuit is shown in Figure 7.35, and is particularly interesting as it incorporates a heavy medium separation (Chapter 11) stage as an

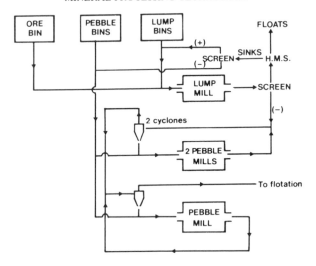

FIG. 7.35. Pyhasalmi grinding circuit.

integral part of the flowsheet. The secondary and tertiary mills are in closed circuit with 500 mm hydrocyclones, which produce a flotation feed of 60%—74 μm. Coarse material is discharged from the 3.2 m diameter × 4.5 m long primary lump mill via large grate openings measuring 75 mm × 120 mm, and is washed and screened to remove fines, before being passed to the HMS drum separator, which operates at a density of about 2.8. The HMS circuit reduces the build-up of coarse, hard gangue pebbles which reduce grinding efficiency and capacity. About 10% of the HMS feed is rejected as floats, while the sinks product, consisting of the heavy massive sulphides, is returned to the mill circuit.

At the Stekenjokk lead-zinc mine and the Aitik copper mine operated by Boliden in Sweden, two-stage autogenous milling is used, in which the primary mill provides pebbles for a secondary pebble mill[36] (Fig. 7.36). The primary mill, operating with coarse crushed material, is placed on a slightly higher level than the secondary pebble mill. Although most of the end grate sectors on the primary mill have normal openings for pulp, there are a number of large openings for pebbles, which are discharged to a classifier in the mill discharge

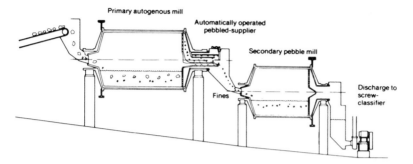

FIG. 7.36. Boliden two-stage grinding system.

trunnion. This screens the pebbles from the slurry, and feeds them to a gate which is controlled by the power draft of the secondary mill, pebbles not being required in secondary grinding, being diverted back to the primary mill. A similar system is in use at the Mount Pleasant tungsten mine in Canada.[37]

This concept has recently been taken a stage further with the objective of increasing mill throughput by removing waste pebbles from the mill discharge.[38] The majority of South African gold reefs are very narrow and mining constraints make the extraction of substantial quantities of barren foot and hanging wall inevitable. The gold-bearing reefs are conglomerates, which are more friable than the quartzites present in the foot and hanging walls. The conglomerate reefs are therefore ground preferentially and it was surmised that the larger pebbles left in the mill would be essentially barren. If these hard waste pebbles could be removed at this stage, mill capacity would be increased. Test work showed that the small percentage of the pebbles containing payable gold could be removed by a combination of photometric and radiometric sorting (Chapter 14). The experiment shows considerable promise, not only for the South African gold and platinum mines, both of which work narrow reefs which are more friable than the hanging and foot walls, but also for other mines exploiting deposits where the boundaries and hardness between ore and waste are very marked.

In recent years autogenous and semi-autogenous circuits have contributed towards substantial savings in capital and operating costs compared to conventional circuits, particularly for large-scale copper and molybdenum circuits in North and South America, where single and two-stage circuits are in operation. Primary circuits operate autogenously or semi-autogenously in either closed or open circuit. Critical sizes are extracted and crushed for recycle in some cases. Secondary grinding is conducted using ball or pebble mills. Descriptions of plant practice in North and South America can be found elsewhere.[39-42]

Control of the Grinding Circuit

The purpose of milling is to reduce the size of the ore particles to the point where economic liberation of the valuable minerals takes place. Thus it is essential that not only should a mill accept a certain tonnage of ore per day, but that it should also yield a product that is of known and controllable particle size. The principal variables which can affect control are: changes in the new feed rate and the circulating load, size distribution and hardness of the ore, and the rate of water addition to the circuit. Also important are interruptions in the operation of the circuit, such as stoppages for such reasons as to feed new grinding medium or to clear a choked cyclone. For the purposes of stabilising control, feed rate, densities and circulating loads can be maintained at values which experience has shown will produce the required product, but this method fails when disturbances cause deviations from normal operation. Fluctuation in feed size and hardness are probably the most significant factors disturbing the balance of a grinding circuit. Such fluctuations may arise from differences in composition, mineralisation, grain size, and crystallisation of the ore from different parts of the mine, from changes in the crusher settings, often due to wear, and from damage to screens in the crushing circuit. Minor fluctuations can be smoothed out by blending material excavated at different points in space and time. Ore storage tends to smooth out variations providing that segregation does not take place in the bins, and the amount of storage capacity available depends on the nature of the ore (such as its tendency to oxidise) and on the mill economics.

Increase in feed size or hardness produces a coarser mill product unless the feed rate is correspondingly reduced. Conversely, a decrease in feed size or hardness may allow an increase in mill throughput. A coarser mill product results in there being a greater circulating load from the classifier, thus increasing the volumetric flowrate. Since the product size from a hydrocyclone is affected by flowrate (Chapter 9), the size distribution of the circuit product will change. Control of the circulating load is therefore important in control of particle size. A high circulating load at a fixed product size means lower energy consumption, but excessive circulating loads can result in mill, classifier or pump overloading, hence there is an upper limit to mill feed rate.

Measurement of the circulating load can be made by sampling various pulp streams in the circuit. It may not be practicable to physically weigh the tonnage of circulating material, although the weight of new feed to each mill may be known from weightometers or other weighing methods used on the feeders.

In the simple ball mill-classifier circuit shown in Figure 7.21, suppose the new feed rate is F t h^{-1} and the circulating load (i.e. classifier underflow) is C t h^{-1}, then:

$$\text{circulating load ratio} = C/F.$$

A mass balance on the mill classifier gives:

$$\text{ball mill discharge} = \text{circulating load} + \text{product}$$

or

$$F + C = C + F.$$

If samples of ball mill discharge, circulating load and classifier overflow (circuit product) are taken, and screen analyses are performed, then if a, b and c are the percentage weights in any specific size fraction in the mill product, circulating load, and classifier overflow respectively, then a mass balance on the classifier in terms of such sized material is:

$$(F + C)a = Fc + Cb$$

or $$C/F = (a - c)/(b - a). \tag{7.4}$$

Using all the size analysis data available, the "best-fit" value of the circulating load ratio can be determined by the method of least-squares (Chapter 3).

a, b, and c may also represent the water/solid ratio of the products, since the relationship is then a mass balance of the weight of water in the circuit; since a, b, and c can be measured on-line by nuclear density gauges, the circuit circulating load can be continuously monitored.

Since grinding is extremely energy intensive, and the product from grinding affects subsequent processes, the need for close control is extremely important, and it is now generally accepted that some form of automatic control is required to maintain efficient performance. Automatic control of grinding circuits is now the most advanced and most successful area of the application of process control in the minerals industry. Excellent reviews of the state of the art in grinding circuit control have been made by Lynch,[43] and by Herbst and Rajamani.[44]

In implementing instrumentation and process control for grinding circuits the control objective must first be defined, which may be:

(1) To maintain a constant product size at maximum throughput.
(2) To maintain a constant feed rate within a limited range of product size.
(3) To maximise production per unit time in conjunction with downstream circuit performance (e.g. flotation recovery).

The most important variables associated with a conventional grinding circuit are shown in Figure 7.37.

Of the variables shown, only the ore feed rate and the water addition rates can be varied independently, as the other variables depend upon and respond to changes in these items. It is therefore ore feed rate and water addition to the mill and classifier which are the major variables used to control grinding circuits. The use of a variable speed pump feeding the classifiers introduces another independent variable, and has a significant effect on the stability of the circuit. Pump speed should, however, be viewed as a variable that provides the conditions under which particular control objectives can be achieved, rather than as a variable that actually achieves them.[45]

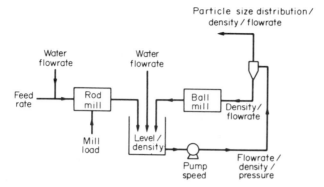

FIG. 7.37. Grinding circuit control variables.

Control of the feed rate is essential, and devices such as variable speed feeder belts in conjunction with weightometers are often used. Control of the grinding medium charge can be made by continuous monitoring of the mill power consumption. A fall in power consumption to a certain level necessitates charging of fresh grinding medium.

Continuous power monitoring finds its greatest use in autogenous grinding, the power drawn being used to control the mill feed rate. Control of power draft has been mainly made by load cells which measure directly the mass of material in the mill, but a dual microphone system has recently been tested and has been found to have considerable advantages over load cells.[46] Microphones have been used before, but mainly on air-swept mills, the sound intensity providing the set-point for feed control. The new system differs in that it uses two microphones, placed above and below the normal level of impact of the charge on the mill shell. As the mill load increases, the point of impact of the charge on the shell moves towards the upper microphone, and conversely down towards the lower microphone as it decreases. By comparing the output from the two microphones, it becomes possible to determine whether the charge level is rising or falling. This information is correlated with the power draw of the mill and used to calculate the rate of addition of the new feed. An unexpected benefit was that there proved to be a strong correlation between the sound intensity of the lower microphone and the pulp density in the mill. At the low pulp densities the fluidity of the charge permits stronger media-liner impacts and therefore more noise, while at high pulp density the viscosity tends to cushion the collisions, muffling the sound output. By using the output from the lower microphone to control water addition, a constant pulp density can be maintained. The use of mill noise measurements as indicators of mill pulp density and viscosity has also been demonstrated on a laboratory ball mill.[47] The results suggest that as pulp density increases, the change in mill noise can be used to identify the pulp rheological regime and that knowledge of the rheology can be advantageous in optimising grinding.

Flowrates and densities are measured continuously mainly by the use of magnetic flowmeters and nuclear density gauges respectively, sump level is commonly indicated by bubble tube, capacitance-type detectors or other electronic devices, and product particle size can either be measured directly by the use of on-line monitors (Chapter 4), or inferred by the use of mathematical models (Chapter 9).

There is an important difference in the dynamic response of the circuit to changes in ore feed rate and to changes in water addition rate. Changes in ore feed rate initiate a slow progressive change in which the final equilibrium state represents the maximum product response, while changes in classifier water addition initiate an immediate maximum response, the equilibrium product response being relatively small.[43] Increase in water addition rate also results in a simultaneous increase in circulating load and sump level, confirming the necessity of using a large capacity sump and variable speed pump to maintain effective control.

If the requirement of the control system is constant product size at constant feed rate, then the only manipulated variable is the classifier water, and the circuit must therefore tolerate variations in the cyclone overflow density and volumetric flowrate when ore characteristics change. Large variations in circulating load will also occur.

In many applications, the control objective is maximum throughput at constant particle size, which allows manipulation of both ore feedrate and classifier water addition. Allowing for the fact that the circuit has a limited capacity, this objective can be stated as a fixed product particle size set-point at a circulating load set-point corresponding to a value just below the maximum tonnage constraint. The circulating load is either calculated from measured values, measured directly, or inferred.

Since both ore feed rate and classifier water addition rate can be varied independently, two control strategies are available. In the first system product size is controlled by ore feed rate, and circulating load by classifier water addition rate, while in the second system product size is controlled by the classifier water, the circulating load being controlled by the ore feed rate. The choice of control strategy depends on which of the control loops, the particle size loop or the circulating load loop, is required to respond the faster, this depending on many factors, such as the ability of the grinding and concentration circuits to handle variations in flow, the sensitivity of the concentration process to deviations from the optimum particle size, time lags within the grinding circuit, and the number of grinding stages. If the particle size response must be fast, then this loop is controlled by the classifier water, whereas if a fast mill throughput response is more important, then the product size is controlled by the ore feed rate.

At Vihanti in Finland,[35] the latter strategy is employed, as the flotation circuit can tolerate short-term changes in particle size. The flowsheet, and instrumentation for the No. 2 grinding circuit, are shown in Figure 7.38. The control concentrates on keeping the particle size constant by regulating the crushed ore feed to the rod mill, and on stabilising the cyclone feed density by regulating the water addition to the cyclone pump sump.

The rod mill feed is measured by means of an electric belt weigher, and is kept constant by regulating the speed of the belt feeder. Water addition to the mill is controlled according to the feed rate set-point to maintain a constant slurry density. The rod mill discharge is fed to a sump where it joins the discharge from grinding circuit No. 1. The sump level is monitored by means of a pressure transducer and controlled by a variable speed pump, the slurry density being stabilised by water addition to the sump. The slurry is pumped to a 500 mm cyclone, the flowrate and density in the feedline being monitored. The cyclone underflow is fed to a 1.6 m diameter Hukki cone classifier,[48] the overflow from this joining the cyclone overflow and providing feed

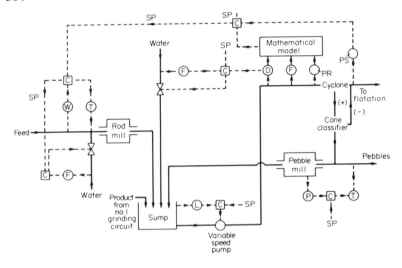

FIG. 7.38. Vihanti No. 2 Grinding Circuit.
F = Flowrate; W = Weight (belt scale); T = Thyristor control; D = Density;
PR = Pressure; PS = Particle size; P = Power; L = Level; SP = Set point;
C = Control.

to the flotation plant. Such two-stage classification is reported to increase the sharpness of separation.[49] The particle size was originally inferred from the cyclone feed data by means of an empirical mathematical model, but it is now measured directly by an Autometrics PSM-200 system, the particle size being controlled at 60%—75 μm. The coarse product from the cone classifier is fed to a pebble mill, the pebble feed being controlled according to the power consumed.

The grinding circuit at Amax Lead Co.'s Buick concentrator, which controls product size by classifier water addition rate, is shown in Figure 7.39.[50]

The plant treats 1.6 m t a^{-1} of a complex copper–lead–zinc ore by flotation, the circuit being extremely sensitive to particle size. The control system uses a weightometer to measure the rod mill feed rate, orifice plate measurement of feed water and sump water flow rates, and sonic detection of sump level. An Autometrics PSM-100 particle

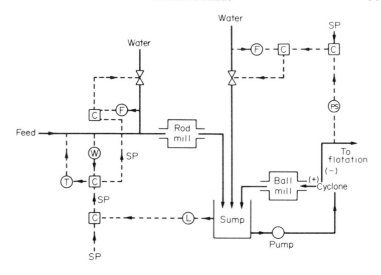

FIG. 7.39. Amax Buick Concentrator—Zinc Circuit.
F = Flowrate; W = Weight; T = Thyristor control; L = Level; PS = Particle size;
SP = Set point; C = Controller.

size analyser measures the percent passing 75 μm and the percentage solids in the cyclone overflow.

The feed control loop matches the actual rod mill feed, indicated by the belt weightometer, with the feed set-point, which is cascaded from the cyclone sump level detector. The feed water control loop maintains a constant preset ratio between rod mill feed rate and rod mill water addition to provide a constant slurry density in the rod mill. The particle size monitor signal is compared with the particle size set-point, the controller adjusting the set-point of water addition to the sump accordingly, in order to maintain a product which is 59%—75 μm in the cyclone overflow. Circulating load and sump level are controlled without the use of mass flow measurements or a variable speed pump, the sump level being a good indication of the circulating load, since the largest volume of flow into the sump comes from the ball mill discharge. The mill throughput loop therefore maintains as high a sump level as possible by manipulating the ore feed rate to the rod mill.

The major limitation of the control strategies discussed above is that control of the numerous plant variables by independent controllers to their optimum values is made difficult by the interaction of these variables. For instance, in the example quoted above, particle size is controlled by water addition to the sump, but increase in sump water also increases the circulating load. Similarly, circulating load is controlled by the ore feed rate loop, but increase in ore feed rate also increases the particle size of the product. It is also apparent that if the circulating load (in this case the sump level) is above its set-point value *and* the product size is also above the set-point value, then the water addition loop will increase the sump water addition to reduce the particle size, this action increasing the circulating load faster than the slower acting feed loop can reduce it. In practice, one or more control loops are de-tuned to minimise such interactions, which results in a general slow down in control response.

Alternative approaches to grinding circuit control based on multi-variable control systems have been investigated although the use of such methods in industrial grinding has so far been fairly limited.[51,52] Two approaches can be used, the first involving the application of multivariable frequency domain methods to design decouplers, using the Inverse Nyquist array method.[53] Once the decouplers have been designed, each loop behaves without affecting other loops and single loop controllers can be applied to each in turn. The second approach applies multi-variable model-based optimal control techniques.[54], in which the "states" of the system are estimated, typically by a Kalman filter (Chapter 3). A third method, "dynamic matrix control", has been used to control a simulated grinding circuit.[55]

References

1. Hartley, J. N., *et al.*, Chemical additives for ore grinding: how effective are they? *Engng. Min. J.* 105 (Oct. 1978).
2. Joe, E. G., Energy consumption in Canadian mills. *CIM Bulletin* **72**, 147 (June 1979).
3. Steane, H. A., Coarser grind may mean lower metal recovery but higher profits, *Can. Min. J.* **97**, 44 (May 1976).
4. Wills, B. A., Deyelopments and research needs in mineral processing, *Proc. 1st Int. Min. Proc. Symp.*, **1**, 1, Izmir (1986).
5. Rose, H. E., and Sullivan, R. M. E., *Ball, Tube and Rod Mills*, Constable, London, 1957.

6. Anon, Grinding mills-rod, ball and autogenous, *Min. Mag.* 197 (Sept. 1982).
7. Anon, Milling and grinding, *Mine & Quarry,* **15,** 21 (May 1986).
8. Sidery, D., Sydvaranger's new mill, *Min. Mag.* 30 (Jan. 1982).
9. Meintrup, W., and Kleiner, F., World's largest ore grinder without gears, *Min. Engng.* 1328 (Sept. 1982).
10. Tilyard, P. A., Process developments at Bougainville Copper Ltd, *Bull. Proc. Australas. Inst. Min. Metall.,* **291,** 33 (March 1986).
11. De Richemond, A. L., Choosing liner materials for ball mills, *Pit & Quarry,* 62, (Sept. 1982).
12. Malghan, S. G., Methods to reduce steel wear in grinding mills, *Min. Engng.* 684, (June 1982).
13. Norman, T. E., *Symposium Materials for the Mining Industry,* p. 207, Colorado, 1975.
14. Korpi, P. A., and Dopson, G. W., Angular spiral lining systems in wet grinding grate discharge ball mills, *Min. Engng.* **34,** 57 (Jan. 1982).
15. Anon, New mill lining wears less, *Min. Equipment Int.* **7,** 21 (Feb. 1983).
16. Lewis, F. M., Coburn, J. L., and Bhappu, R. B., Comminution: a guide to size reduction system design, *Min. Engng.* **28,** 29 (Sept. 1976).
17. Rowland, C. A., and Kjos, D. M., Rod and ball mills, in *Mineral Processing Plant Design* (ed. Mular and Bhappu), AIMME, New York, 1978.
18. McIvor, R. E., and Finch, J. A., The effects of design and operating variables on rod mill performance, *CIM Bulletin,* **79,** 39 (Nov. 1986).
19. Arbiter, N., and Harris, C. C., Scale-up problems with large ball mills, *Minerals and Metallurgical Processing,* **1,** 4 (May 1984).
20. Whiten, W. J., and Kavetsky, A., Studies on scale-up of ball mills, *Minerals and Metallurgical Processing,* **1,** 23 (May 1984).
21. Rowland, C. A., and Erickson, M. T., Large ball mill scale-up factors to be studied relative to grinding efficiency, *Minerals and Metallurgical Processing,* **1,** 165 (Aug. 1984).
22. Rowland, C. A., Ball mill scale-up-diameter factors, in *Advances in Mineral Processing,* ed. P. Somasundaran, p. 605, SME Inc., Littleton, 1986.
23. Lowrison, G. C., *Crushing and Grinding,* Butterworths, London, 1974.
24. Gangopadhyay, A. K., and Moore, J. J., Assessment of wear mechanisms in grinding media, *Minerals and Metallurgical Processing,* **2,** 145 (Aug. 1985).
25. Dodd, J., *et al.,* Relative importance of abrasion and corrosion in metal loss in ball milling, *Minerals and Metallurgical Processing,* **2,** 212 (Nov. 1985).
26. Barratt, D. J., Semi-autogenous grinding: a comparison with the conventional route, *CIM Bulletin* **72,** 74 (Nov. 1979).
27. Rowland, C. A., Jr., and Kjos, D. M., Autogenous and semi-autogenous mill selection and design, *Aust. Mining* 21 (Sept. 1975).
28. Smith, E. A., Grinding very hard and very soft materials, *Processing* 16 (Nov. 1974).
29. Lloyd, P. J. D., *et al.,* Centrifugal grinding on a commercial scale, *Engng. Min. J.* 49 (Dec. 1982).
30. Kitschen, L. P., Lloyd, P. J. D., and Hartman, R., The centrifugal mill: experience with a new grinding system and its applications, *Proc. XIV Int. Min. Proc. Cong.,* paper no. 1–9, CIM, Toronto, Canada, Oct. 1982.
31. Ramsey, T. L., Reducing grinding costs: the potential is great, *World Mining,* 78 (Oct. 1982).
32. Hopkins, J. L., Beyond Resolute, *Can. Min. J.* **107,** 12 (Aug. 1986).

33. Hilton, W., Comminution and classification of barytes, *Trans. Inst. Min. Metall.*, **93**, A145 (1984).
34. Anon, Afton—new Canadian copper mine, *World Mining* **31**, 42 (April 1978).
35. Wills, B. A., Pyhasalmi and Vihanti concentrators, *Min. Mag.* 176 (Sept. 1983).
36. Maripuu, R., Marklund, O., and Nystroem, L. S., Evolution of a highly successful grinding circuit, *World Mining* **36**, 48 (Oct. 1983).
37. Anon, Mount Pleasant Mine, *Min. Mag.* 16 (July 1983).
38. Freer, J. S., *et al.*, The effects of pebble mill removal on mill performance with a view to waste sorting in closed circuit with run-of-mine milling, *Proc. Gold 100 Conf.*, SAIMM, Joburg, 1986.
39. Vanderbeek, J. L., and Pennington, R. I., Investigation of alternative SAG mill recycle treatment options, *Minerals and Metallurgical Processing*, **3**, 108 (May 1986).
40. Vesely, M., and Fernandez, O., Crusher for critical size material at Los Bronces semiautogenous grinding plant, Chile, *Minerals and Metallurgical Processing*, **3**, 104 (May 1986).
41. Bymark, J. V., and von Bitter, R. W., Autogenous grinding at Hibbing Taconite Co., *Minerals and Metallurgical Processing*, **3**, 87 (May 1986).
42. Jerez, J., *et al.*, Automatic control of semi-autogenous grinding at Los Bronces, in *Automation for Mineral Resource Development*, eds. A. W. Norrie and D. R. Turner, p. 275, Pergamon Press, Oxford, 1986.
43. Lynch, A. J., *Mineral Crushing and Grinding Circuits*, Elsevier, Amsterdam, 1977.
44. Herbst, J. A., and Rajamani, K., Control of grinding circuits, in *Computer Methods for the 80's in the Mineral Industry* ed. A. Weiss, AIMME. New York, 1979.
45. Gault, G. A., *et al.*, Automatic control of grinding circuits: theory and practice, *World Mining* **32**, 59 (Nov. 1979).
46. Jaspan, R. K., *et al.*, ROM mill power control using multiple microphones to determine mill load, *Proc. Gold 100 Conf.*, SAIMM, Joburg, 1986.
47. Watson, J. L., and Morrison, S. D., Estimation of pulp viscosity and grinding mill performance by means of mill noise measurements, *Minerals and Metallurgical Processing*, **3**, 216 (Nov. 1986).
48. Hukki, R. T., A new way to grind and recover minerals, *Engng Min. J.* **178**, 66 (April 1977).
49. Heiskanen, K., Two-stage classification, *World Mining* **32**, 44 (June 1979).
50. Perkins, T. E., and Marnewecke, L., Automatic grind control at Amax Lead Co., *Mining Engng.* **30**, 166 (Feb. 1978).
51. Niemi, A. J., *et al.*, Experiences in multivariable control of sulphide ore grinding in Vuonos concentrator, *Proc. XIV Int. Min. Proc. Cong.*, paper III-5, CIM Toronto, Canada, Oct. 1982.
52. Pauw, O. G., *et al.*, The control of pebble mills at Buffelsfontein Gold Mine by use of a multivariable peak-seeking controller. *J. S. Afr. Inst. Min. Metall.*, **85**, 89 (1985).
53. Rosenbrock, H. H., *Computer-Aided Control System Design*, Academic Press, 1974.
54. Herbst, J. A., and Rajamani, K., The application of modern control theory to mineral processing operations. in *Proc. 12th CMMI Cong.*, ed. H. W. Glenn, South Afr. I.M.M., Jo'burg, 1982.
55. Lee, P. L., and Newell, R. B., Multivariable control of a grinding circuit, in *Automation for Mineral Resource Development*, eds. A. W. Norrie and D. R. Turner, Pergamon, Oxford, 1986.

CHAPTER 8

INDUSTRIAL SCREENING

Introduction

Industrial sizing of materials is extensively used, and the types of equipment are many and varied. Screening is generally carried out on relatively coarse material, as the efficiency decreases rapidly with fineness. Fine screens are very fragile and expensive and tend to become blocked rather easily with retained particles ("blind"). Screening is generally limited to material above about 250 μm in size, finer sizing normally being undertaken by classification (Chapter 9), although the boundary between the two methods will in practice depend on many factors, such as the type of ore, the plant throughput, etc. The trend towards finer crushing prior to grinding has also necessitated the use of screens which are capable of recovering the fine products at high efficiency and capacity.

There is, of course, a wide range of purposes for screening. The main purposes in the minerals industry are:

(a) to prevent the entry of undersize into crushing machines, so increasing their capacity and efficiency;
(b) to prevent oversize material from passing to the next stage in closed-circuit fine crushing and grinding operations;
(c) to prepare a closely sized feed to certain gravity concentration processes;
(d) to produce a closely sized end product. This is important in quarrying, where the final product size is an important part of the specification.

Performance of Screens

In its simplest form, the screen is a surface having a multiplicity of apertures of given dimensions. Material of mixed size presented to that surface will either pass through or be retained, according to whether the particles are smaller or larger than the governing dimensions of the aperture. The efficiency of screening is determined by the degree of perfection of separation of the material into size fractions above or below the governing dimensions of the aperture.

There has been no universally accepted method of defining screen performance and a number of widely used methods are employed. The most common screen performance criteria are those which state an efficiency based on the recovery of material at a given size, or on the mass of misplaced material in each product. This immediately leads to a range of possibilities, such as undersize in the overscreen product, oversize in the through-screen product, or a combination of the two.

An effficiency equation can be calculated from a mass balance across a screen as follows:

Consider a screen (Fig. 8.1) on to which there is a feed of F t h^{-1}; C t h^{-1} overflows from the screen, and U t h^{-1} is the rate of underflow.

Let f be the fraction of material above the cut point size in the feed; c be the fraction of material above the cut point size in the overflow; and

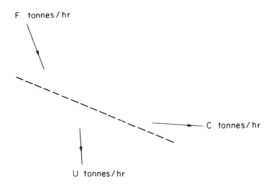

FIG 8.1. Mass balance on a screen.

u be the fraction of material above the cut point size in the underflow. f, c, and u can be determined by sieving a representative sample of each of the fractions on a laboratory screen of the same aperture size as the industrial screen and assuming this to be 100% efficient.

The mass balance on the screen is

$$F = C + U.$$

The mass balance of the oversize material is

$$Ff = Cc + Uu$$

and the mass balance of the undersize material is

$$F(1 - f) = C(1 - c) + U(1 - u).$$

Hence

$$\frac{C}{F} = \frac{f - u}{c - u}$$

and

$$\frac{U}{F} = \frac{c - f}{c - u}.$$

The recovery of oversize material into the screen overflow

$$= \frac{Cc}{Ff} = \frac{c(f - u)}{f(c - u)} \tag{8.1}$$

and the corresponding recovery of undersize material in the screen underflow

$$= \frac{U(1 - u)}{F(1 - f)}$$

$$= \frac{(1 - u)(c - f)}{(1 - f)(c - u)}. \tag{8.2}$$

These two relationships, (8.1) and (8.2), measure the effectiveness of the screen in separating the coarse material from the underflow and the fine material from the overflow.

A combined effectiveness, or overall efficiency, E is then obtained by multiplying the two equations together:

$$E = \frac{c(f - u)(1 - u)(c - f)}{f(c - u)^2(1 - f)}. \tag{8.3}$$

If there are no broken or deformed apertures, the amount of coarse material in the underflow is usually very low and a simplification of equation 8.3 can be obtained by assuming that it is, in fact, zero (i.e. $u = 0$), in which case the formula for fines recovery and that for the overall efficiency both reduce to

$$E = \frac{c - f}{c \, (1 - f)} \, . \tag{8.4}$$

This formula is widely used and implies that recovery of the coarse material in the overflow is 100%.[1]

Formulae such as the one derived are acceptable for assessing the efficiency of a screen under different conditions, operating on the *same feed*.

They do not, however, give an absolute value of the efficiency, as no allowance is made for the difficulty of the separation. A feed composed mainly of particles of a size near to that of the screen aperture—"near mesh" material—presents a more difficult separation than a feed composed mainly of very coarse and very fine particles with a screen aperture intermediate between them.

An efficiency or *partition curve* for a screen is drawn by plotting the partition coefficient, defined as the percentage of the feed reporting to the oversize product, against the geometric mean size on a logarithmic scale. (For particles in the range, say, $-125 + 63 \, \mu$m, the geometric mean size is $\sqrt{(125 \times 63)}$.) Figure 8.2 shows ideal and real partition curves.

The separation size, or *cut-point*, is obtained at 50% probability, i.e. the size at which a particle has equal chance of reporting to the undersize or oversize product. This is unlikely to correspond with the size of the screen apertures, the cut-point usually being less than the aperture size.

The efficiency of separation is assessed from the steepness of the curve (see Chapter 9). The efficiency curve effectively models the screen, and can be used for simulation and design purposes.[2]

Factors Affecting Screen Performance

Screen effectiveness must always be coupled with capacity as it is

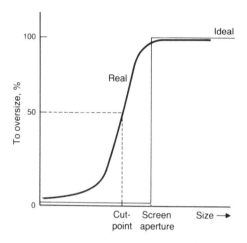

FIG. 8.2. Partition curve.

often possible by the use of a low feed-rate and a very long screening time to effect an almost complete separation. In practice economics dictates that relatively high feed-rates should be used, which reduces particle dwell time on the screen and often produces a thick bed of material through which the fines must travel to the screen surface. The net effect is reduced efficiency. High capacity and high efficiency are opposing requirements for any given separation, and a compromise is necessary to achieve the optimum result. At a given capacity, the effectiveness depends on the nature of the screening operation, i.e. on the overall chance of a particle passing through the screen once it has reached it.

The overall probability of passage of one particle is a product of the number of times that a particle strikes the screen and the probability of its passage through the screen during a single contact.[1]

As well as feed rate, an important factor dictating the number of strikes that a particle makes on the screen is the rate of vibration of the screen. Screens are vibrated in order to increase their efficiency, as blinding is reduced and segregation of the feed material is induced, which allows the fines to work through the layer of feed to the screen surface. Too high a vibration rate may, however, reduce the efficiency, as the particles may bounce from the screen wires and be thrown so far from the surface that there are very few strikes. Higher vibration rates can, in general, be used with higher feed-rates, as the deeper bed of material has a "cushioning" effect which inhibits particle bounce.

There are many factors affecting the chance of an individual passage through the screen once the particle has reached it. The angle of approach and particle orientation to the screen are important; the closer the perpendicular is to the angle of approach, the higher the chance of passage. If the particle shape is non-spherical, then some orientations will present a small cross-section for passage and some a large cross-section; mica, for instance, screens extremely badly, its flat, plate-like crystals tending to "ride" over the screen apertures.

The chance of passing through the aperture is proportional to the percentage of *open area* in the screen material, which is defined as the ratio of the net area of the apertures to the whole area of the screening surface. The smaller the area occupied by the screen material, the greater the chance of a particle reaching an aperture.

Open area decreases with the fineness of the aperture of screens with wires of the same diameter and in order to increase the open area of a fine screen, very thin wires must be used, which makes the screen fragile. This is the main reason for classifiers taking over from screens at fine aperture sizes. Due to the greater open area of coarse screens, the capacity is greater than that of fine screens and, in general, the maximum permissible rate of feeding is roughly proportional to the diameter of the apertures.

Probably the most important factor determining screen performance is the nature of the feed material. The efficiency is markedly reduced by the presence of particles close to the aperture size; these particles tend to "blind" the apertures, reducing the available open area and often report into the oversize fraction. This is often a problem with screens run in closed circuit with crushers, where a build-up of "near mesh" material occurs which progressively reduces efficiency. Screens are often used in these cases which have apertures slightly greater than the set of the crusher, thus allowing this material to pass out into the screen undersize.

Taggart[3] gives some probabilities of passage related to the particle size, which are shown in Table 8.1. The figures relate the probable chance per thousand of unrestricted passage through a square aperture

TABLE 8.1. PROBABILITY OF PASSAGE

Ratio of particle to aperture size	Chance of passage per 1000	Number of apertures required in path
0.001	998	1
0.01	980	2
0.1	810	2
0.2	640	2
0.3	490	2
0.4	360	3
0.5	250	4
0.6	140	7
0.7	82	12
0.8	40	25
0.9	9.8	100
0.95	2.0	500
0.99	0.1	10^4
0.999	0.001	10^6

of a spherical particle and gives the probable number of apertures in series in the path of the particle necessary to ensure its passage through the screen. It can be seen that as the particle size approaches that of the aperture, the chance of passage falls off very rapidly.

The amount of moisture present in the feed has a marked effect on screening efficiency, as does the presence of clays and other sticky materials which should be removed early in the treatment route (Chapter 2). Damp feeds screen very poorly as they tend to agglomerate and blind the screen apertures. Screening must always be performed on perfectly dry, or wet, material, but never on damp material. If screening efficiency is the only criterion, then wet screening is always superior; finer sizes may be processed, adherent fines are washed off large particles, and the screen is cleaned by the flow of pulp. There is no dust problem. There is, however, the cost of dewatering and drying the products, and in many instances this is so high that dry screening, with its deficiencies, is preferred.

Screen Types

There are many different types of industrial screens, which may be classified as either stationary or moving screens.

Stationary Screens

The grizzly. Very coarse material is usually screened on a grizzly, which, in its simplest form, consists of a series of heavy parallel bars set in a frame. Some grizzlies employ chains instead of bars and some are shaken or vibrated mechanically to help the sizing and to aid in the removal of the oversize (Fig. 8.3). The most common use of grizzlies in mineral processing is for sizing the feed to primary crushers. If the primary crusher has a 10-cm product, then the feed is passed over a grizzly with a 10-cm spacing between the bars, in order to "scalp", or remove, the undersize.

The bars of a grizzly are inclined usually at an angle of between 20-50°, the greater the incline, the greater is the throughput, but the lower is the efficiency. Feed flows down the grizzly in the direction of the bars to assist flow and to reduce clogging. Clogging is a major

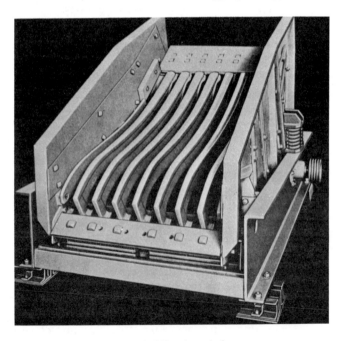

Fig. 8.3. Vibrating grizzly.

problem, especially on the cross-members which hold the bars together. The bars are usually tapered in cross-section to minimise clogging once particles have entered between them.

The size of particle screened on grizzlies can be as large as 300 mm, or as small as 20 mm. The capacity, which can be up to 1000 t h^{-1}, is proportional to the area.

Sieve bends. Static screens, such as the Dutch State Mines (DSM) sieve bend, and its later version, the Dorr Rapifine, have gained wide acceptance in the minerals industry for very fine wet screening purposes.

The sieve bend (Fig. 8.4) has a curved screen composed of horizontal wedge bars. Feed slurry enters the upper surface of the screen tangentially and flows down the surface in a direction perpendicular to the openings between the wedge bars. As the stream

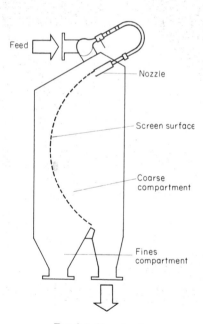

FIG. 8.4. Sieve bend.

of slurry passes each opening a thin layer is peeled off and directed to the underside of the screen. According to Fontein, [7] particles roughly twice the thickness of this layer are dragged along with the undersize fraction; particles larger than this size pass across the openings as their greatest part projects into the liquid flowing over the slot. The thickness of the layer peeled off is primarily a function of the space between the bars, and investigation has shown that the thickness of this layer is in the order of 25% of the slot width. [8] In general, therefore, a separation is produced at a size roughly equivalent to half the bar spacing and so very little plugging of the apertures should take place. Separation can be undertaken down to 50 μm and screen capacities are up to 180 m³ h⁻¹.

One of the problems associated with the DSM sieve bend is that whilst separation occurs on the face of the screen, the thin layer of pulp, having passed through the apertures, tends to continue down the back (convex side) of the cloth, being held on by a "wall effect". The Dorr Rapifine incorporates a periodic rapping to the screen to dislodge adhering particles, while the Bartles CTS screen [9] has a series of crimps on the back of the cloth to divert pulp clinging to the back of the cloth. A later version uses a bonded plastic strip for this purpose. The unit has a primary stage, followed by a scavenging stage (Fig. 8.5), which has been found to produce a cleaner oversize product. The capacity, at a separation size of 150 μm, is about 26 m³ h⁻¹ for 1-m width of screen.

Sieve bends have found an important application in closed-circuit grinding of heavy mineral ores. Overgrinding of the heavy minerals, which can occur when conventional classification is used, can be greatly reduced by the combination of classifiers and sieve bends. [10]

The Hukki screen (Fig. 8.6) employs a combination of classification and screening and has been used in grinding circuits. [11] It consists of an open stationary vessel with a cylindrical top section and a conical base. The top section includes a cylindrical screen. Feed enters centrally at the top and is distributed inside the cylindrical screen by means of a low-speed rotating mechanism. Wash water is introduced into the conical section. The undersize passes through the cylindrical screen, which consists of wedge bars, into a collecting launder, whilst the oversize fraction is discharged through the apex.

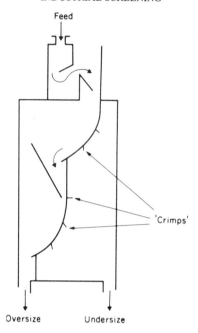

Feed

'Crimps'

Oversize   Undersize

FIG. 8.5. Bartles CTS screen.

Moving Screens

Revolving screens. One of the oldest screening devices is the *trommel* (Fig. 8.7), which is a slightly inclined, rotating cylindrical screen, which can be used wet or dry.

Material is fed in at one end of the drum, undersize material passing through the screen apertures and the oversize material coming off at the opposite end.

Trommels, such as the one shown, can be made to deliver several sized products. The main problem is that the fine screen wears quickly, as the whole of the feed must be fed on to it.

Trommels may be arranged in series, with the coarsest discharging its undersize into consecutively finer trommels. This method, however, requires a great deal of floor space.

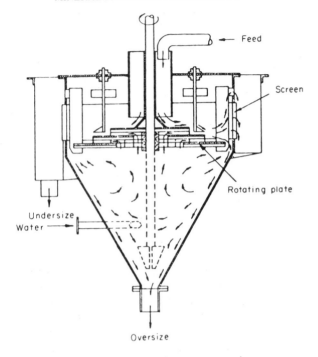

FIG. 8.6. Hukki screen.

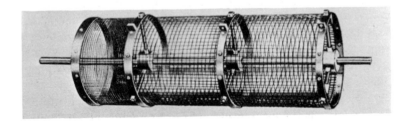

FIG. 8.7. Trommel screen.

The problem of undue wear on the fine-screen trommel is overcome in the *compound trommel* which has a series of concentric cylinders, with the coarsest screen at the centre, such that the coarsest fraction is removed first. It suffers from the disadvantage that failure of the inner screens is difficult to observe and they are difficult to replace when worn. It therefore has very limited use.

Trommels can handle material from 55 mm down to 6 mm, and even smaller sizes can be handled under wet screening conditions. They are widely used in the grading of sand and ballast and are still seen in a few ore-washing plants.

Although trommels are cheap, vibration-free, and robust, they have poor capacities since only part of the screen surface is in use at any one time and they "blind" very easily. They are tending to be replaced by one or other of the range of shaking or vibrating screens.

Shaking screens. Shaking screens have a reciprocating movement mechanically induced in the horizontal direction and are mounted either horizontally or with a gentle slope. They operate in the range of 60–800 strokes per minute, and find their greatest use as grading screens for relatively large sized feeds, down to about 12 mm. They are widely used dry in coal preparation, but find little use on abrasive metalliferous ores.

Reciprocating screens. Reciprocating screens employ a horizontal gyratory motion to the feed end of a rectangular screen by means of an unbalanced rotating shaft, rotating at about 1000 rev min^{-1}. The horizontal circular motion at the feed end gradually diminishes through the length of the machine to an elliptical movement, and finally to an approximate straight line reciprocating motion at the discharge end (Fig. 8.8).

The circular motion at the feed end immediately spreads the material across the full width of the screen surface, even though it is fed from a single point. This horizontal circular motion also stratifies the material, causing the fines to sink down against the screen mesh. As the material travels along the screen surface it enters the area of

FIG. 8.8. Plan view of reciprocating screen.

gradually diminishing screening motion at the discharge end. This reduced action aids the screening out of the "near mesh" particles. The screen is slightly inclined, and some commercial screens have ball trays below the screen; resilient balls confined in these trays are deflected against bevel strips by the machine motion and bounce continuously against the undersize of the screen mesh, thus reducing blinding.

These screens are used for fine separation, mainly dry, of light materials in the range 10 mm to 250 μm, and sometimes down to 40 μm.

Gyratory screens. This type of screen, which imparts gyratory motion throughout the whole screen cloth, is becoming widely used for fine-screening applications, wet or dry, down to 40 μm.[1] The basic components consist of a nest of sieves supported on a table which is mounted on springs on a base; suspended from beneath the table is a motor with double-shaft extensions, which drives eccentric weights and in doing so effects horizontal gyratory motion (Fig. 8.9). Vertical motion is imparted by the bottom weights, which swing the mobile mass about its centre of gravity, producing a circular tipping motion to the screen, the top weights producing the horizontal gyratory motion.

As in the reciprocating screen, ball trays may be fitted below a screen assembly to reduce blinding.

Figure 8.10 shows gyratory screens sizing zinc oxide in a Belgian plant.

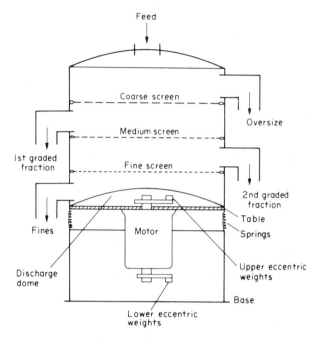

FIG. 8.9. Gyratory screen.

Vibrating screens. Vibrating screens are the most important screening machines for mineral processing applications.[5,6,12] They handle material up to 25 cm in size down to 250 μm. Their main application is in crushing circuits where they are required to handle material ranging, in general, from 25 cm to 5 mm in size.

Vibration is induced vertically either by the rotation of a mechanical reciprocating device applied to the casing or by electrical devices operating directly on the screen.

They can work at low slopes and need little headroom. In multiple-deck systems the feed is introduced to the top coarse screen, the undersize falling through to the finer screens, thus producing a range of sized fractions (Fig. 8.11).

Electrically vibrated screens, such as the Hummer (Fig. 8.12), operate with a high-frequency motion of very small throw, created by a

FIG. 8.10. Gyratory screens in operation.

moving magnet activated by alternating current. The electromagnetic vibrator is mounted above and connected directly to the screening surface.

Electric screens are often used to dewater finely crushed ores, such as washing-plant classifier sands products (Chapter 2). For dry screening they are limited to materials below about 12 mm in size. The most widely used screens for coarser sizing are mechanically vibrated. There are two main methods of producing vibration. Eccentric motion is often preferred for feeds coarser than about 4 cm, while below this size, unbalanced pulleys are used.

Coarse-screening machines of the type shown in Fig. 8.11 are known as "full-floating screens", since the moving parts of the screen float entirely on rubber mountings, allowing absolute freedom to develop maximum screening action. This freedom allows the screen to adjust itself to varying load conditions so that the whole action is concentrated on the efficient separation of the material.

Vibration in the finer screening ranges is often produced by unbalanced weights or flywheels attached to the drive shaft.

FIG. 8.11. Double-deck vibrating screen.

In the system shown in Fig. 8.13 the shaft is balanced, but the flywheels at each end of the shaft are unbalanced. The amplitude of throw can be adjusted by adding or removing weight elements bolted to the flywheels inside the rims. The vibrator generates an elliptical motion, slanting forward at the feed end; a circular motion at the centre; and an elliptical motion, slanting backwards at the discharge end (Fig. 8.14). Forward motion at the feed end serves to move oversize material rapidly out of the feed zone to keep the bed as thin as possible. This action facilitates passage of fines which should be completely removed in the first one-third of the screen length. As the oversize bed thins down, near the centre of the screen, the motion gradually changes to the circular pattern to slow down the rate of travel of the solid. At the discharge end, the oversize and remaining near-size materials are subjected to the increasingly retarding effect of the backward elliptical motion. This allows the near-size material more time to find openings in the screen cloth.

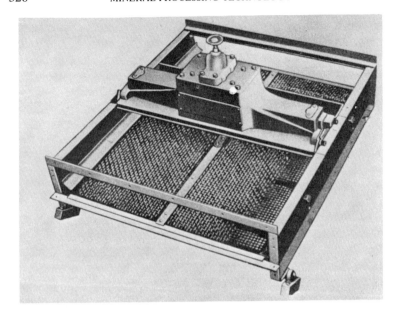

FIG. 8.12. Hummer screen.

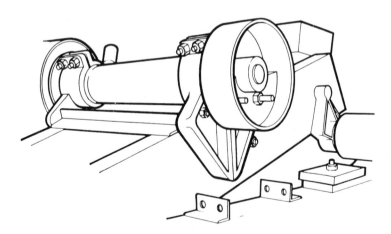

FIG. 8.13. Unbalanced-flywheel unit.

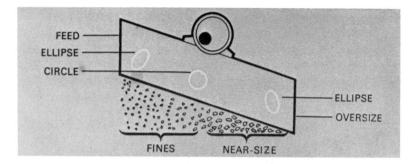

FIG. 8.14. Motion of vibrating screen.

In most screens, energy is wasted in changing the direction of motion. In the *resonance screen* the screen frame is arranged so that it is freely vibrating between rubber buffers connected to a heavy weighted balancing frame, having a natural resonance frequency which is the same as that of the vibrating screen. Movement is imparted to the screen by an eccentric drive and the rubber buffers restrict the movement of the screen and store up energy, which is reimparted to the live frame. Any movement given to the screen is transmitted to the balancing frame which stands on rubber pads. The motion sets up resonance vibrations, which, instead of being wasted, are transmitted back to the screening frame, making energy losses minimal. In addition to storing energy, the sharp return of motion of the deck produced by the buffers imparts a lively action to the deck and promotes good screening.

There are two main requirements which must be satisfied for efficient screening of fine materials. The first is the conveyance of particles to mesh apertures and their presentation in as many attitudes as possible. The second is to ensure that particles pass mesh apertures without blinding. In practice these are conflicting requirements, the former needing vibration frequencies around 20–50 Hz, while for prevention of blinding, frequencies of 10–30 kHz are necessary.[13] Several dual-frequency vibrating screens have been introduced which are claimed to produce higher efficiencies and throughputs.[14,15,16] The screens are vibrated by two motors, one mounted at the feed end and running at low speed and high amplitude, with the other at the discharge end running at high speed and low amplitude.

The Mogensen sizer operates on the principle that there is a definite and quantifiable probability that a particle will pass through an aperture larger than the maximum diameter of that particle.[17] In contrast to the conventional concept of screening, the material is allowed to fall freely through a system of oscillating and sloping screens all of which have a mesh size larger than the biggest particle to be treated (Fig. 8.15). The capacity is such that a particular screening duty can be met with a machine occupying a fraction of the floor space required by a conventional screen and blinding and wear is greatly reduced.

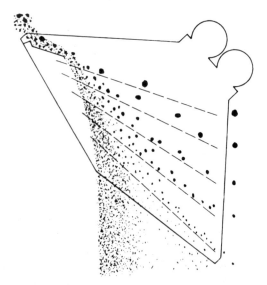

FIG. 8.15. Principle of Mogensen sizer.

The *rotating probability screen* (Fig. 8.16) has been developed for extracting fines from damp raw coal of which, in recent years, there has been a general increase due mainly to the increased use of water sprays to meet more stringent underground dust regulations. This has caused

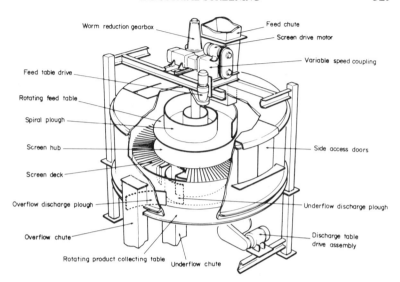

FIG. 8.16. Rotating probability screen.

considerable difficulty in separating fines by conventional means. Standard vibrating screens with woven wire-mesh apertures of 6 mm and below rapidly blind due to the build-up of damp fines, and developments in screen design, deck construction, and screening aids have had only limited success in overcoming this problem.

The rotating probability screen, developed by the NCB's Mining Research and Development Establishment, avoids the use of fine-aperture meshes and overcomes the problem of deck blinding.[18]

It consists of a horizontal, circular screen deck mounted on a vertical rotating shaft provided with a variable-speed drive. The screening surface is comprised entirely of small-diameter stainless steel rods radiating from a central hub. A uniform circular distribution of the feed material to the screen deck is achieved by a rotating feed table with a stationary spiral plough mounted above it.

The finer particles pass between the rotating rods while the coarser material is discharged around the periphery of the deck by centrifugal action. Both products are collected on a second rotating table, where they are kept separate by a cylindrical division and discharged by ploughs. The overall design of the screen is such that the material progresses either in free fall or by positive displacement over a smooth horizontal surface. The principle of its operation is based on the greater probability of the finer particles passing through the relatively large apertures between the rotating rods.

An important feature of the screen is that the effective screening size depends on the rotational speed which can be regulated either manually or automatically.

The main application is in situations where the ability to handle difficult, high moisture material is of more practical importance than accurate size separation. Such is the case with dry fines extraction in coal treatment. The coarse fraction is usually treated by gravity separation to remove relatively heavy incombustible minerals and shales, while the fine fraction is eventually blended with the treated fraction to give a product of the required ash content. The screen provides a means of continuously controlling the amount of fines extracted according to the ash content of the raw coal, so that the treated and untreated constituents of blended coal can be made available in the required proportions. The capacity of a 2.4-m deck-diameter screen is around 100 t h^{-1} when handling −19-mm raw coal, and separating at about 4 mm.

The *Delkor Linear Screen* is a new screening machine which is being installed in several South African mines, particularly for removing wood chips and fibre from the ore stream feeding carbon-in-pulp systems.[19] The machine (Fig. 8.17) comprises a coarse multi-filament screen cloth supported on cloth rollers and driven by a head pulley coupled to a variable speed drive unit. Slurry enters through a distributor on to the moving cloth, the undersize draining through the cloth by gravity and being collected in the underpan. The wood fibre and oversize material retained on the screen is discharged at the drive pulley using a belt wash spray, the spray ensuring a continuous, self-cleaning screen. By adjusting the weave configurations of the screen cloth, the cut-point for oversize removal can be varied. As the screen is not vibrated, its power consumption is much less than that required for vibrating screens, and noise and stress cracking are eliminated.

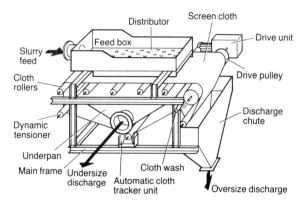

FIG. 8.17. Delkor linear screen.

Screening Surfaces

The type of screening surface chosen for a particular duty will depend on the aperture required and the nature of the work.

For very heavy-duty work, with the grizzly, parallel iron or steel bars or rails are set at a fixed distance apart.

Punched plate is used for many purposes. Heavy-duty "scalping" screens often use plate with circular or square holes. Punched plate with slotted openings is sometimes used for fine work.

Rod-deck screens are often used for heavy-duty work in the range 3–25 mm, especially in removing undersize from heavy tonnage crusher feeds in closed and open circuit. The tempered steel rods are sprung into place and are held firmly by moulded rubber spacers. The smaller-sized rods can be replaced individually by hand, while rods greater than 8 mm in diameter can be replaced by the aid of a tool supplied by the manufacturer.

Wedge wire screens are employed in many fine screening machines, such as sieve bends. Such screens are strong and have relatively large open areas. The wedge profile (Fig. 8.18) minimises particle blinding.

Woven-wire cloths, usually constructed from steel, stainless steel, copper, or bronze, are by far the most widely used screening surfaces, especially in the range encountered in crushing circuits. Various

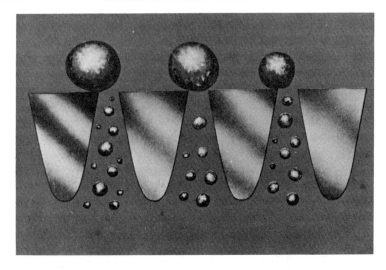

FIG. 8.18. Wedge wire screen.

shapes of aperture and types of weave are available, although square mesh is usually used for fairly coarse screening and rectangular for fine screening. Rectangular screen apertures have a greater open area than square-mesh screens of the same wire diameter. The wire diameter chosen depends on the nature of the work and the capacity required. Fine screens can have the same or greater open areas than coarse screens, but the wires used must be thinner and hence more fragile (Fig. 8.19).

Increasing the wire thickness increases their strength, but decreases open area and hence capacity.

Rectangular aperture screen cloths should be used with the long side of the mesh set across the flow for maximum capacity. They are often used on material that tends to "flake" into long thin fragments.

Various non-metallic screen surfaces, which greatly increase screen life due to reduced wear, are now used. Polyurethane or rubber offer exceptional resistance to abrasion and impact, while effectively reducing noise, and are lighter in weight than wire cloth.[20] Polyurethane offers more open area than other non-metallic screens and is manufactured with tapered holes, wider at the bottom than the

Fig. 8.19. Woven-wire cloth of different apertures but similar open area.

top, to reduce blinding. Renison Ltd., Tasmania, replaced the stainless steel wedge wire screens of their primary grinding circuit with polyurethane wedge-bar screen panels and found that the efficiency was significantly better, and the screen life about five times greater than before.[10]

References

1. Records, F. A., Sieving practice and the gyratory screen, *Process Technology* **18**, 47 (1973).
2. Lynch, A. J., and Narayanan, S. S., Simulation—the design tool for the future, in

Mineral Processing at a Crossroads, eds. B. A. Wills, and R. W. Barley, p. 89, Martinus Nijhoff Publishers, Dordrecht, 1986.

3. Taggart, A. F., *Handbook of Mineral Dressing*, Wiley, New York, 1945.
4. Anon., Screens—new ideas and new machines, *Mining Mag.*, 299 (Oct. 1980).
5. Pritchard, A. N., Vibrating screens in the mining industry, *Mine & Quarry* **9**, 46 (Oct. 1980).
6. Pritchard, A. N., Choosing the right vibrating screen, *Mine & Quarry* **9**, 36 (Dec. 1980).
7. Fontein, F. J., The D.S.M. sievebend, new tool for wet screening on fine sizes, application in coal washeries, *Second Int. Coal Prep. Congress*, Essen, 1954.
8. Stavenger, P. L., and Reynolds, V. R., Applications of the DSM screen, *Min. Cong. J.* 48 (July 1958).
9. Burt, R. O., Fine sizing of minerals, *Min. Mag.* **128**, 6 (June 1973).
10. Woodcock, J. T., Mineral processing progress in Australia, 1974, *Aust, Min.* 17 (Aug. 1975).
11. De Kok, S. K., Fine size in milling circuits. *Jl. S. Afr. I.M.M.* Spec. Issue, 86 (Oct. 1975).
12. Crissman, H., Vibrating screen selection, *Pit & Quarry*, **78**, 39 (June 1986) and **79**, 46 (Nov. 1986).
13. Adorjan, L. A., Mineral processing, *Min. Ann. Rev.* 225 (1975).
14. Anon., New concept in new unit, *Min. J.* 443 (Nov. 22, 1974).
15. Chalk, P. H., Screening on inclined decks. *Processing* 6 (Oct. 1974).
16. Adorjan, L. A., Mineral processing, *Min. Ann. Rev.* 221 (1977).
17. Mogensen, F. A., A new method of screening granular materials, *Quarry Managers J.* 409 (Oct. 1965).
18. Shaw, S. R., The Rotating Probability Screen—a new concept in screening, *Mine & Quarry*, **12**, 29 (Oct. 1983).
19. Anon, New linear screen offers wide applications, *Engng. & Min. J.*, **187**, 79 (Sept. 1986).
20. Anon., New wear-resistant surface from U.S. company, *Min. J.* 29 (Jan. 14, 1977).

CHAPTER 9

CLASSIFICATION

Introduction

Classification is a method of separating mixtures of minerals into two or more products on the basis of the velocity with which the grains fall through a fluid medium. In mineral processing, this is usually water, and wet classification is generally applied to mineral particles which are considered too fine to be sorted efficiently by screening. Since the velocity of particles in a fluid medium is dependent not only on the size, but also on the specific gravity and shape of the particles, the principles of classification are important in mineral separations utilising gravity concentrators.

Principles of Classification

When a solid particle falls freely in a vacuum, it is subject to constant acceleration and its velocity increases indefinitely, being independent of size and density. Thus a lump of lead and a feather fall at exactly the same rate.

In a viscous medium, such as air or water, there is resistance to this movement and the value increases with velocity. When equilibrium is attained between the gravitational and fluid resistances forces, the body reaches its *terminal velocity* and thereafter falls at a uniform rate.

The nature of the resistance depends on the velocity of descent. At low velocities motion is smooth because the layer of fluid in contact with the body moves with it, while the fluid a short distance away is motionless. Between these two positions is a zone of intense shear in the fluid all around the descending particle. Effectively all resistance to motion is due to the shear forces or viscosity of the fluid and is hence called *viscous resistance*. At high velocities the main resistance is due to the displacement of fluid by the body, and viscous resistance is relatively small; this is known as *turbulent resistance*.

Whether viscous or turbulent resistance predominates, the acceleration of particles in a fluid rapidly decreases and the terminal velocity is quickly reached.

Classifiers consist essentially of a *sorting column* in which a fluid is rising at a uniform rate (Fig. 9.1). Particles introduced into the sorting column either sink or rise according to whether their terminal velocities are greater or less than the upward velocity of the fluid. The sorting column therefore separates the feed into two products—an *overflow* consisting of particles with terminal velocities less than the velocity of the fluid and an *underflow* or *spigot product* of particles with terminal velocities greater than the rising velocity.

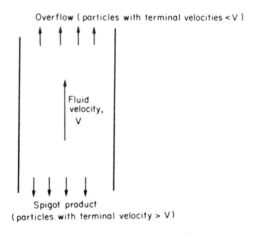

FIG. 9.1. Classifier sorting column.

Free Settling

Free settling refers to the sinking of particles in a volume of fluid which is large with respect to the total volume of particles, hence particle crowding is negligible. For well-dispersed ore pulps, free settling predominates when the percentage of solids by weight is less than about 15.[1]

Consider a spherical particle of diameter d and density Ds falling under gravity in a viscous fluid of density Df under free-settling conditions, i.e. ideally in a fluid of infinite extent. The particle is acted upon by three forces: a gravitational force acting downwards, an upward buoyant force due to the displaced fluid, and a drag force D acting upwards. The equation of motion of the particle is

$$mg - m'g - D = m\frac{dx}{dt} , \qquad (9.1)$$

where m is the mass of the particle, m' is the mass of displaced fluid, x is the particle velocity, and g is the acceleration due to gravity.

When the terminal velocity is reached, $dx/dt = 0$, and hence $D = (m - m')g$.

Therefore $$D = \frac{\pi}{6}gd^3(Ds - Df) . \qquad (9.2)$$

Stokes[2] assumed the drag force on a spherical particle to be entirely due to viscous resistance and deduced the expression

$$D = 3\pi d\eta\upsilon, \qquad (9.3)$$

where η is the fluid viscosity and υ is the terminal velocity.

Hence, substituting in equation 9.2,

$$3\pi d\eta\upsilon = \frac{\pi}{6}gd^3(Ds - Df)$$

and

$$\upsilon = \frac{gd^2(Ds - Df)}{18\eta} . \qquad (9.4)$$

This expression is known as *Stokes' law*.

Newton assumed that the drag force was entirely due to turbulent resistance, and deduced:

$$D = 0.055\pi d^2\upsilon^2 Df. \qquad (9.5)$$

Substituting in equation 9.2 gives

$$v = \left(\frac{3gd(Ds - Df)}{Df} \right)^{1/2}. \tag{9.6}$$

This is *Newton's law* for turbulent resistance.

Stokes' law is valid for particles below about 50 μm in diameter. The upper size limit is determined by the dimensionless Reynolds number (Chapter 4). Newton's law holds for particles larger than about 0.5 cm in diameter. There is, therefore, an intermediate range of particle size, which corresponds to the range in which most wet classification is performed, in which neither law fits experimental data.

Stokes' law (9.4) for a particular fluid can be simplified to

$$v = k_1 d^2 (Ds - Df) \tag{9.7}$$

and Newton's law (9.6) can be simplified to

$$v = k_2 [d(Ds - Df)]^{1/2}, \tag{9.8}$$

where k_1 and k_2 are constants, and $(Ds - Df)$ is known as the *effective density* of a particle of density Ds in a fluid of density Df.

Both laws show that the terminal velocity of a particle in a particular fluid is a function only of the particle size and density. It can be seen that:

(1) If two particles have the same density, then the particle with the larger diameter has the higher terminal velocity.

(2) If two particles have the same diameter, then the heavier particle has the higher terminal velocity.

Consider two mineral particles of densities Da and Db and diameters da and db respectively, falling in a fluid of density Df, at exactly the same settling rate. Their terminal velocities must be the same, and hence from Stokes' law (9.7):

$$da^2(Da - Df) = db^2(Db - Df)$$

or

$$\frac{da}{db} = \left(\frac{Db - Df}{Da - Df} \right)^{1/2}. \tag{9.9}$$

This expression is known as the *free-settling ratio* of the two minerals, i.e. the ratio of particle size required for the two minerals to fall at equal rates.

Similarly from Newton's law (9.8), the free settling ratio of large particles is

$$\frac{da}{db} = \left(\frac{Db - Df}{Da - Df} \right).$$ (9.10)

Consider a mixture of galena (density 7.5) and quartz (density 2.65) particles classifying in water. For small particles, obeying Stokes' law, the free settling ratio (equation 9.9) is

$$\left(\frac{7.5 - 1}{2.65 - 1} \right)^{1/2} = 1.99 \ ,$$

i.e. a small particle of galena will settle at the same rate as a small particle of quartz which has a diameter 1.99 times as large.

For large particles obeying Newton's law, the free-settling ratio (equation 9.10) is

$$\left(\frac{7.5 - 1}{2.65 - 1} \right) = 3.94 \ .$$

The free-settling ratio is therefore larger for coarse particles obeying Newton's law than for fine particles obeying Stokes' law. This means that *the density difference between the mineral particles has a more pronounced effect on classification at coarser size ranges.* This is important where gravity concentration is being utilised. Overgrinding of the ore must be avoided, such that particles are fed to the separator in as coarse a state as possible, so that a rapid separation can be made, exploiting the enhanced effect of specific gravity difference. The enhanced gravity effect does, however, mean that fine heavy minerals are more likely to be overground in conventional ball mill-classifier circuits, so it is preferable where possible to use rod mills for the primary coarse grind.

The general expression for free-settling ratio can be deduced from equations 9.9 and 9.10 as

$$\frac{da}{db} = \left(\frac{Db - Df}{Da - Df} \right)^{n} \ ,$$ (9.11)

where $n = 0.5$ for small particles obeying Stokes' law and $n = 1$ for large particles obeying Newton's law.

The value of n lies in the range 0.5–1 for particles in the intermediate size range of 50 μm–0.5 cm.

Hindered Settling

As the proportion of solids in the pulp increases, the effect of particle crowding becomes more apparent and the falling rate of the particles begins to decrease. The system begins to behave as a heavy liquid whose density is that of the pulp rather than that of the carrier liquid; *hindered-settling* conditions now prevail. Because of the high density and viscosity of the slurry through which a particle must fall in a separation by hindered settling, the resistance to fall is mainly due to the turbulence created,[3] and a modified form of Newton's law (9.8) can be used to determine the approximate falling rate of the particles

$$v = k[d(Ds - Dp)]^{1/2}, \qquad (9.12)$$

where Dp is the *pulp density*.

The lower the density of the particle, the more marked is the effect of reduction of the effective density, $Ds - Dp$, and the greater is the reduction in falling velocity. Similarly, the larger the particle, the greater is the reduction in falling rate as the pulp density increases.

This is important in classifier design; in effect, *hindered-settling reduces the effect of size, while increasing the effect of density on classification.*

This is illustrated by considering a mixture of quartz and galena particles settling in a pulp of density 1.5. The *hindered-settling ratio* can be derived from equation 9.12 as

$$\frac{da}{db} = \left(\frac{Db - Dp}{Da - Dp}\right). \qquad (9.13)$$

Therefore, in this system,

$$\frac{da}{db} = \frac{7.5 - 1.5}{2.65 - 1.5} = 5.22 .$$

A particle of galena will thus fall in the pulp at the same rate as a particle of quartz which has a diameter 5.22 times as large. This compares with the free-settling ratio, calculated as 3.94 for turbulent resistance.

The *hindered-settling ratio is always greater than the free-settling ratio*, and the denser the pulp, the greater is the ratio of the diameter of equal settling particles. For quartz and galena, the greatest hindered settling ratio that we can attain practically is about 7.5. Hindered-settling classifiers are used to increase the effect of density on the separation, whereas free-settling classifiers use relatively dilute suspensions to increase the effect of size on the separation (Fig. 9.2). Relatively dense slurries are fed to certain gravity concentrators, particularly those treating heavy alluvial sands. This allows high tonnages to be treated, and enhances the effect of specific gravity difference on the separation. The efficiency of separation, however, may be reduced since the viscosity of a slurry increases with density. For separations involving feeds with a high proportion of particles close to the required density of separation, lower slurry densities may be necessary, even though the density difference effect is reduced.

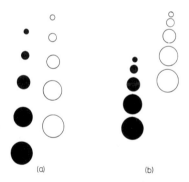

FIG. 9.2. Classification by (a) free settling, (b) hindered settling.

As the pulp density increases, a point is reached where each mineral particle is covered only with a thin film of water. This condition is known as a *quicksand*, and because of surface tension, the mixture is a perfect suspension and does not tend to separate. The solids are in a condition of *full teeter*, which means that each grain is free to move, but is unable to do so without colliding with other grains and as a result stays in place. The mass acts as a viscous liquid and can be penetrated by solids with a higher specific gravity than that of the mass, which will then move at a velocity impeded by the viscosity of the mass.

A condition of teeter can be produced in a classifier sorting column by putting a constriction in the column, either by tapering the column or by inserting a grid into the base (Fig. 9.3).

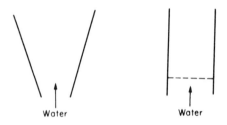

FIG. 9.3. Teeter chambers.

Such hindered-settling sorting columns are known as *teeter chambers*. Due to the constriction, the velocity of the introduced water current is greatest at the bottom of the column. A particle falls until it reaches a point where its falling velocity equals that of the rising current. The particle can now fall no further. Many particles reach this condition, and as a result, a mass of particles becomes trapped above the constriction and pressure builds up in the mass. Particles move upward along the path of least resistance, which is usually the centre of the column, until they reach a region of lower pressure at or near the top of the settled mass; here, under conditions in which they previously fell, they fall again. As particles from the bottom rise at the centre, those from the sides fall into the resulting void. A general circulation is built up, the particles being said to *teeter*. The constant jostling of teetering particles has a scouring effect which removes any entrained or adhering slimes particles, which then leave the teeter chamber and pass out through the classifier overflow. Cleaner separations can therefore be made in such classifiers.

Types of Classifier

Many different types of classifier have been designed and built. They may be grouped, however, into two broad classes depending on the direction of flow of the carrying current. Horizontal current classifiers such as mechanical classifiers are essentially of the free-settling type and accentuate the sizing function; vertical current or hydraulic classifiers are usually hindered settling types and so increase the effect of density on the separation.

A useful guide to some of the major manufacturers of classification equipment used in mineral processing can be found elsewhere.[4]

Hydraulic Classifiers

These are characterised by the use of water additional to that of the feed pulp, introduced so that its direction of flow opposes that of the settling particles. They normally consist of a series of sorting columns through each of which a vertical current of water is rising and particles are settling out (Fig. 9.4).

The rising currents are graded from a relatively high velocity in the first sorting column, to a relatively low velocity in the last, so that a series of spigot products can be obtained, with the coarser, denser particles in the first spigot and the fines in the latter spigots. Very fine slimes overflow the final sorting column of the classifier. The size of each successive vessel is increased, partly because the amount of liquid

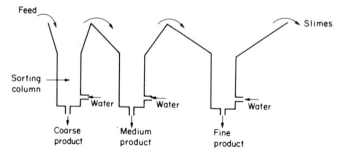

Fig. 9.4. Principle of hydraulic classifier.

to be handled includes all the water used for classifying in the previous vessels and partly because it is desired to reduce, in stages, the surface velocity of the fluid flowing from one vessel to the next.

Hydraulic classifiers may be free- or hindered-settling types. The former are rarely used; they are simple and have high capacities, but are inefficient in sizing and sorting. They are characterised by the fact that each sorting column is of the same cross-sectional area throughout its length.

The greatest use for hydraulic classifiers in the mineral industry is for sorting the feed to certain gravity concentration processes (Chapter 10). Such classifiers are of the hindered-settling type. These differ from the free-settling classifiers in that the sorting column is constricted at the bottom in order to produce a teeter chamber (Fig. 9.3). The hindered-settling classifier uses much less water than the free-settling type, and is more selective in its action, due to the scouring action in the teeter chamber, and the buoyancy effect of the pulp, as a whole, on those particles which are to be rejected. Since the ratio of sizes of equally falling particles is high, the classifier is capable of performing a concentrating effect, and the first spigot product is normally of higher grade than the other products (Fig. 9.5).

This is known as the *added increment* of the classifier and the first spigot product may in some cases be rich enough to be classed as a concentrate.

During classification the teeter bed tends to grow, as it is easier for particles to become entangled in the bed, rather than leave it. This tends to alter the character of the spigot discharge, as the density builds up. In modern multi-spigot *hydrosizers* the teeter bed composition is automatically controlled. The Stokes hydrosizer (Fig. 9.6) is commonly used to sort the feed to gravity concentrators.

Each teeter chamber is provided at its bottom with a supply of water under constant head which is used for maintaining a teetering condition in the solids that find their way down against the interstitial rising flow of water. Each teeter chamber is fitted with a discharge spigot which is, in turn, connected to a pressure-sensitive valve so that the classifying conditions set by the operator can be accurately controlled (Fig. 9.7).

The valve may be hydraulically or electrically operated; in operation

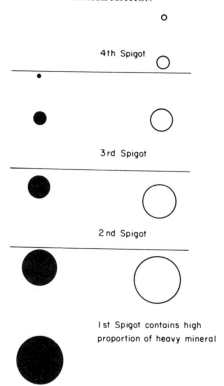

FIG. 9.5. Added increment of hindered-settling classifier.

it is adjusted to balance the pressure set up by the teetering material. The concentration of solids in a particular compartment can be held extremely steady in spite of the normal variations in feed rate taking place from time to time. The rate of discharge from each spigot will, of course, change in sympathy with these variations, but since these changing tendencies are always being balanced by the valve, the discharge will take place at a nearly constant density. For a quartz sand this is usually about 64% solids by weight, but is higher for heavier minerals.

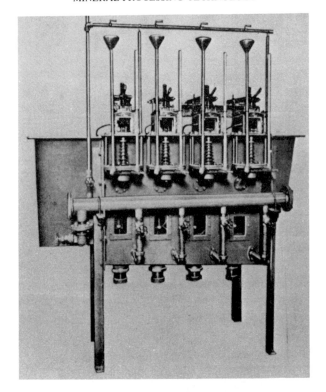

Fɪɢ. 9.6. Stokes multi-spigot hydrosizer.

Horizontal Current Classifiers

Settling cones. These are the simplest form of classifier in which there is little attempt to do more than separate the solids from the liquid, i.e. they are sometimes used as dewatering units in small-scale operations. They are often used in the aggregate industry to de-slime coarse sands products. The principle of the settling cone is shown in Fig. 9.8. The pulp is fed into the tank as a distributed stream at F, with the spigot discharge S initially closed. When the tank is full, overflow of water and slimes commences, and a bed of settled sand

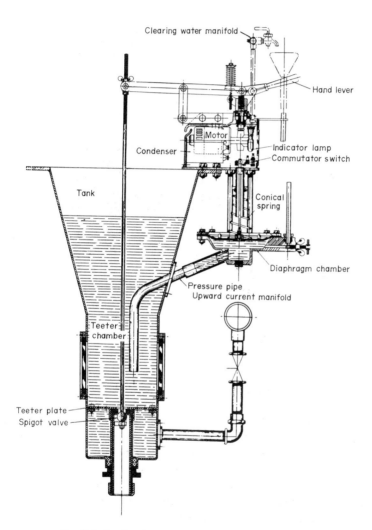

FIG. 9.7. Section through sorting column of hydrosizer.

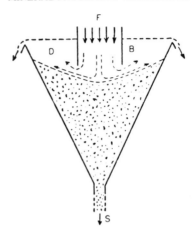

FIG. 9.8. Settling-cone operation.

builds up until it reaches the level shown. If the spigot valve is now opened and sand discharge maintained at a rate equal to that of the input, classification by horizontal current action takes place radially across zone D from the feed cylinder B to the overflow lip. The main difficulty in operation of such a device is the balancing of the sand discharge and deposition; it is virtually impossible to maintain a regular discharge of sand through an open pipe under the influence of gravity. Many different designs of cone have been introduced to overcome this problem.[1]

In the "Floatex" separator, which consists essentially of a hindered-settling classifier over a dewatering cone, automatic control of the coarse lower discharge is governed by the specific gravity of the teeter column. The use of the machine as a desliming unit and in upgrading coal and mica, as well as its possible application in closed-circuit classification of metalliferous ores, is discussed by Littler.[5]

Mechanical classifiers. Several forms of classifier exist in which the material of lower settling velocity is carried away in a liquid overflow, and the material of higher settling velocity is deposited on the bottom of the equipment and is dragged upwards against the flow of liquid, by some mechanical means.

Mechanical classifiers have widespread use in closed-circuit grinding operations and in the classification of products from ore-washing plants (Chapter 2). In washing plants they act more or less as sizing devices, as the particles are essentially unliberated, so are of similar density. In closed-circuit grinding they have a tendency to return small dense particles to the mill, causing overgrinding (Chapter 7).

The principle of the mechanical classifier is shown in Fig. 9.9.

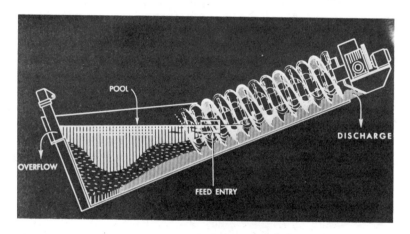

FIG. 9.9. Principle of mechanical classifier.

The pulp feed is introduced into the inclined trough and forms a settling pool in which particles of high falling velocity quickly fall to the bottom of the trough. Above this coarse sand is a quicksand zone where essentially hindered settling takes place. The depth and shape of this zone depends on the classifier action, and on the feed pulp density. Above the quicksand is a zone of essentially free settling material, comprising a stream of pulp flowing horizontally across the top of the quicksand zone from the feed inlet to the overflow weir, where the fines are removed.

The settled sands are conveyed up the inclined trough by mechanical rakes or by a helical screw. The conveying mechanism also serves to keep fine particles in suspension in the pool by gentle agitation and when the sands leave the pool they are slowly turned over by the raking action, thus releasing entrained slimes and water, increasing the efficiency of the separation. Washing sprays are often directed on the emergent sands to wash the released slimes back into the pool.

The *rake classifier* (Fig. 9.10) utilises rakes actuated by an eccentric motion, which cause them to dip into the settled material and to move it up the incline for a short distance. The rakes are then withdrawn, and return to the starting-point, where the cycle is repeated; the settled material is thus slowly moved up the incline to the discharge.

In the *duplex* type shown, one set of rakes is moving up, while the other set returns, Simplex and quadruplex machines are also made in which there are one or four raking assemblies.

FIG. 9.10. Rake classifier.

Spiral classifiers (Fig. 9.11) use a continuously revolving spiral to move the sands up the slope. They can be operated at steeper slopes than the rake classifier, in which the sands tend to slip back when the

FIG. 9.11. Spiral classifier.

rakes are removed. Steeper slopes aid the drainage of sands, giving a cleaner, drier product. Agitation in the pool is less than in the rake classifier which is important in separations of very fine material.

The size at which the separation is made and the quality of the separation depend on a number of factors.

Increasing the feed rate increases the horizontal carrying velocity and thus increases the size of particle leaving in the overflow. The feed should not be introduced directly into the pool, as this causes agitation and releases coarse material from the hindered-settling zone, which may report to the overflow. The feed stream should be slowed down by spreading it on an apron, partially submerged in the pool, and sloped towards the sand discharge end, so that most of the kinetic energy is absorbed in the part of the pool furthest from the overflow.

The speed of the rakes or spiral determines the degree of agitation of the pulp and the tonnage rate of sand removal. For coarse separations, a high degree of agitation may be necessary to keep the coarse particles in suspension in the pool, whereas for finer separations, less agitation and thus lower raking speeds are required. It is essential, however, that the speed is high enough to transport the sands up the slope.

The height of the overflow weir is an operating variable in some mechanical classifiers. Increasing the weir height increases the pool volume, and hence allows more settling time and decreases the surface agitation, thus reducing the pulp density at overflow level, where the final separation is made. High weirs are thus used for fine separations.

Dilution of the pulp is the most important variable in the operation of mechanical classifiers. In closed-circuit grinding operations, ball mills rarely discharge at less than 65% solids by weight, whereas mechanical classifiers never operate at more than about 50% solids. Water to control dilution is added in the feed launder, or into the sand near the vee of the pool. Water addition determines the rate of settling of the particles; increased dilution reduces the density of the weir overflow, and increases free settling, allowing finer particles to settle out of the influence of the horizontal current. Finer separations are thus produced, providing that the overflow pulp density is above a value known as the *critical dilution*, which is normally about 10% solids. Below this density, the effect of increasing rising velocity with dilution becomes more important than the increase in particle-settling rates produced by decrease of pulp density. The overflow therefore becomes coarser with increasing dilution (Fig. 9.12). In mineral processing applications, however, very rarely is the overflow density less than the critical dilution.

One of the major disadvantages of the mechanical classifier is its inability to produce overflows of very fine particle size at reasonable pulp densities. To produce such separations, the pulp may have to be diluted to such an extent to increase particle-settling rates that it becomes too thin for subsequent operations. It may therefore require thickening before concentration can take place. This is undesirable as, apart from the capital cost and floor space of the thickener, oxidation of liberated particles may occur in the thickener, which may affect subsequent processes, especially froth flotation.

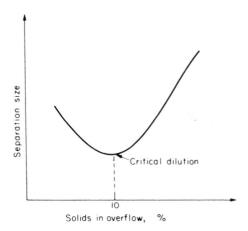

FIG. 9.12. Effect of dilution of overflow on classifier separation.

The Hydrocyclone

This is a continuously operating classifying device that utilises centrifugal force to accelerate the settling rate of particles. It is one of the most important devices used in the minerals industry, and there are over 50 hydrocyclone manufacturers in the world.[6]

Its main use in mineral processing is as a classifier, which has proved extremely efficient at fine separation sizes. It is used increasingly in closed-circuit grinding operations[7] but has found many other uses, such as de-sliming, de-gritting, and thickening. It has also found wide acceptance recently for the washing of fine coal.[8]

A typical hydrocyclone (Fig. 9.13) consists of a conically shaped vessel, open at its apex, or underflow, joined to a cylindrical section, which has a tangential feed inlet. The top of the cylindrical section is closed with a plate through which passes an axially mounted overflow pipe. The pipe is extended into the body of the cyclone by a short, removable section known as the *vortex finder*, which prevents short-circuiting of feed directly into the overflow.

The feed is introduced under pressure through the tangential entry which imparts a swirling motion to the pulp. This generates a vortex in

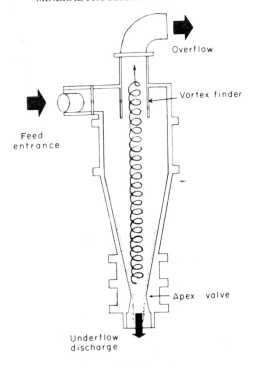

FIG. 9.13. Hydrocyclone.

the cyclone, with a low-pressure zone along the vertical axis. An air core develops along the axis, normally connected to the atmosphere through the apex opening, but in part created by dissolved air coming out of solution in the zone of low pressure.

The classical theory of hydrocyclone action is that particles within the flow pattern are subjected to two opposing forces—an outward centrifugal force and an inwardly acting drag (Fig. 9.14). The centrifugal force developed accelerates the settling rate of the particles (there is evidence to show that Stokes' law applies with reasonable accuracy to separations in cyclones of conventional design), thereby separating particles according to size and specific gravity. Faster settling particles move to the wall of the cyclone, where the velocity is lowest, and migrate to the apex opening. Due to the action of the drag

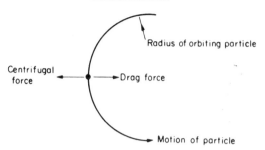

FIG. 9.14. Forces acting on an orbiting particle in the hydrocyclone.

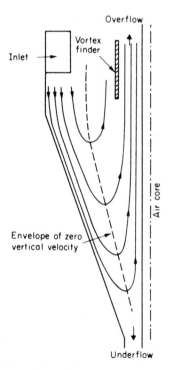

FIG. 9.15. Distribution of the vertical and radial components of velocity in a hydrocyclone.

force, the slower-settling particles move towards the zone of low pressure along the axis and are carried upward through the vortex-finder to the overflow.

The existence of an outer region of downward flow and an inner region of upward flow necessitates a position at which there is no vertical velocity. This applies throughout the greater part of the cyclone body, and an envelope of zero vertical velocity should exist throughout the body of the cyclone (Fig. 9.15). Particles thrown outside the envelope of zero vertical velocity by the greater centrifugal force exit via the underflow, while particles swept to the centre by the greater drag force leave in the overflow. Particles lying on the envelope of zero velocity are acted upon by equal centrifugal and drag forces and have an equal chance of reporting either to the underflow or overflow.

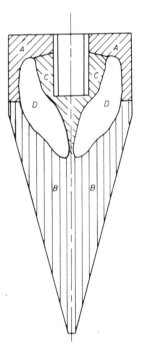

FIG. 9.16. Regions of similar size distribution within cyclone.

Experimental work carried out by Renner and Cohen[9] has shown that classification does not take place throughout the whole body of the cyclone as the classical model postulates. Using a high-speed probe, samples were taken from several selected positions within a 150-mm diameter cyclone, and were subjected to size analysis. The results showed that the interior of the cyclone may be divided into four regions that contain distinctively different size distributions (Fig. 9.16).

Essentially unclassified feed exists in a narrow region A adjacent to the cylinder wall and roof of the cyclone. Region B occupies a very large part of the cone of the cyclone and contains fully classified coarse material, i.e. the size distribution is practically uniform and resembles that of the coarse underflow product. Similarly, fully classified fine material is contained in region C, a narrow region surrounding the vortex finder and extending below the latter along the cyclone axis. Only in the toroid-shaped region D does classification appear to be taking place. Across this region, size fractions are radially distributed, so that decreasing sizes show maxima at decreasing radial distances from the axis.

Hydrocyclones have replaced mechanical classifiers in many modern grinding plants (Fig. 9.17), as they are more efficient, especially in the finer size ranges, and they require less floor space. Due to the relatively short residence time of particles within the cyclone, the mill circuit can rapidly be brought into balance if any changes are made and oxidation of particles within the circuit is reduced. They are almost universally used for classification between 150 and 5 μm, although coarser separations are possible.

Cyclone Efficiency

The commonest method of representing cyclone efficiency is by a *performance* or *partition* curve (Fig 9.18), which relates the weight fraction, or percentage, of each particle size in the feed which reports to the apex, or underflow, to the particle size. The *cut point*, or separation size, of the cyclone is often defined as that point on the partition curve for which 50% of particles in the feed of that size report

FIG. 9.17. Ball mill in closed circuit with hydrocyclones.

to the underflow, i.e. particles of this size have an equal chance of going either with the overflow or underflow.[10] This point is usually referred to as the d_{50} size. The sharpness of the cut depends on the slope of the central section of the partition curve; the closer to vertical is the slope, the higher is the efficiency. The slope of the curve can be expressed by taking the points at which 75% and 25% of the feed particles report to the underflow. These are the d_{75} and d_{25} sizes,

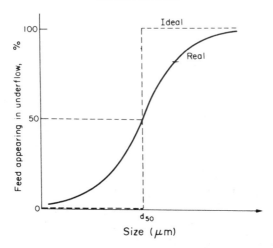

FIG. 9.18. Partition curve for hydrocyclone.

respectively. The efficiency of separation, or the so-called *imperfection* I, is then given by

$$I = \frac{d_{75} - d_{25}}{2d_{50}} . \tag{9.14}$$

Many mathematical models of hydrocyclones include the term "corrected d_{50}" taken from the "corrected" classification curve. It is assumed that in all classifiers, solids of all sizes are entrained in the coarse product liquid by short-circuiting in direct proportion to the fraction of feed water reporting to the underflow.

For example, if the feed contains 16 t h^{-1} of material of a certain size, and 12 t h^{-1} report to the underflow, then the percentage of this size reporting to the underflow, and plotted on the normal partition curve, is 75%.

However, if, say, 25% of the feed water reports to the underflow, then 25% of the feed material will short-circuit with it; therefore, 4 t h^{-1} of the size fraction will short-circuit to the underflow, and only

$8 \, t \, h^{-1}$ leave in the underflow due to classification. The corrected recovery of the size fraction is thus

$$\frac{12 - 4}{16 - 4} \times 100 = 67\%.$$

The uncorrected partition curve can therefore be corrected by utilising the equation

$$y' = \frac{y - R}{1 - R}, \tag{9.15}$$

where y' is the corrected mass fraction of a particular size reporting to underflow, y is the actual mass fraction of a particular size reporting to the underflow, and R is the fraction of the feed liquid which is recovered in the coarse product stream. Figure 9.19 shows uncorrected and corrected classification curves.

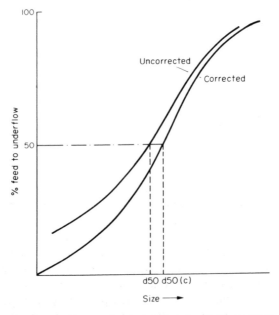

FIG. 9.19. Uncorrected and corrected classification curves.

The method of construction of the partition curve can be illustrated by means of an example. Suppose a cyclone is being fed with quartz (density 2700 kg m^{-3}) in the form of a slurry, of pulp density 1670 kg m^{-3}. The cyclone underflow density is 1890 kg m^{-3}, and the overflow 1460 kg m^{-3}.

Using equation 3.6, the percentage solids by weight in the cyclone feed is 63.7%. Therefore, the dilution ratio (water–solids ratio) of feed is

$$\frac{36.3}{63.7} = 0.57.$$

Similarly, the underflow and overflow dilution ratios can be calculated to be 0.34 and 1.00 respectively.

If the cyclone is fed at the rate of F t h^{-1} of dry solids and the underflow and overflow mass flow-rates are U and V t h^{-1} respectively, then, since the total amount of water entering the cyclone must equal the amount leaving in unit time:

$$0.57 \, F = 0.34U + V$$

or

$$0.57 \, F = 0.34U + (F - U).$$

Therefore

$$\frac{U}{F} = 0.652.$$

The underflow is thus 65.2% of the total feed weight and the overflow is 34.8% of the feed.

The performance curve for the cyclone can now be prepared by tabulating the data as in Table 9.1. Columns 1, 2 and 3 represent the screen analyses of the overflow and underflow, and columns 4 and 5 relate these results in relation to the feed material. Column 4, for example, is prepared by multiplying the results of column 2 by 0.652. Adding column 4 to column 5 produces column 6, the reconstituted size analysis of the feed material. Column 8 is determined by dividing each weight in column 4 by the corresponding weight in column 6. Plotting column 8 against column 7, the arithmetic mean of the sieve

TABLE 9.1.

(1)	(2)	(3)	(4)	(5)	(6)	(7)	(8)
Size (μm)	Wt.%		Wt.% of feed		Reconstituted feed	Nominal size (arithmetic mean)	% of feed to U/F
	U/F	O/F	U/F	O/F			
+1168	14.7	—	9.6	—	9.6	—	100.0
589–1168	21.8	—	14.2	—	14.2	878.5	100.0
295–589	25.0	5.9	16.3	2.1	18.4	442.0	88.6
208–295	7.4	9.0	4.8	3.1	7.9	251.5	60.8
147–208	6.3	11.7	4.1	4.1	8.2	177.5	50.0
104–147	4.8	11.2	3.1	3.9	7.0	125.5	44.3
74–104	2.9	7.9	1.9	2.7	4.6	89.0	41.3
−74	17.1	54.3	11.2	18.9	30.1	—	37.2
Total	100.0	100.0	65.2	34.8	100.0		

size ranges, produces the partition curve, from which the d_{50} (177.5 μm) can be determined. The partition curve can be corrected by utilising equation 9.15. The value of R in this example is

$$\frac{65.2 \times 0.34}{100 \times 0.57} = 0.39.$$

The assumption that the fraction of feed material which is not classified is proportional to the volume fraction of water in the underflow has been reported to be subject to error, and a mathematical procedure to extrapolate the partition curve towards finer sizes has been developed.[11]

Lynch[12] describes the application of the "reduced efficiency curve", which is obtained by plotting corrected weight percentage of particles reporting to the underflow against the actual size divided by the corrected d_{50} (Fig. 9.20), and suggests that it can be used to derive the actual performance curve after any changes in operating conditions, the curve being independent of hydrocyclone diameter, outlet dimensions, or operating conditions.

Although partition curves are extremely useful in assessing classifier performance, the minerals engineer is usually more interested in knowing fineness of grind (i.e. cyclone overflow size analysis) than the

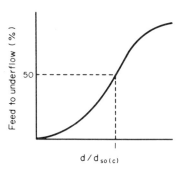

FIG. 9.20. Reduced efficiency curve.

cyclone d_{50}. A simple fundamental relationship between fineness of grind and the efficiency curve of a hydrocyclone has been developed by Kawatra and Seitz.[13]

Factors Affecting Cyclone Performance

The effects of changing operating and design parameters in cyclones are very complex in that all parameters are interrelated. It is almost impossible to select a cyclone to give the precise separation required and it is nearly always necessary to adjust feed inlet, vortex finder, apex opening and pulp pressure, and dilution. Designers, therefore, tend to specify cyclones capable of handling the flow rates required, with provision for fitting suitable ranges of feed, overflow, and underflow openings. There are a number of empirical relationships which are used by designers in predicting performance and designing cyclones.

Bradley[14] lists eight different equations to calculate the cut-point d_{50}. One of the oldest is that of Dahlstrom:[15]

$$d_{50} = \frac{13.7(D_o D_i)^{0.68}}{Q^{0.53}(S - L)^{0.5}} \qquad (9.16)$$

where d_{50} is the cut-point (μm), D_o is the overflow diameter (cm), D_i is the inlet diameter (cm), Q is the total flow rate ($m^3 \ h^{-1}$), S is the sp. gr. of solids, and L is the sp. gr. of liquid.

Such equations are not, however, directly applicable to industrial scale cyclones, as most of the work was carried out on dilute slurries using very small diameter cyclones.

Plitt[16] has developed a mathematical model which gives reasonable predictions of performance of large diameter cyclones, operating at high solids content, over a wide range of operating conditions. The model has been successfully applied to the development of automatic control systems in the comminution circuits of various Australian mines. Four fundamental parameters were determined in terms of the operating and design variables, these being the cut-size, the flow split between underflow and overflow, the sharpness of separation, and the capacity in terms of the pressure drop across the cyclone. The model formulated enables the complete cyclone performance to be calculated with reasonable accuracy without requiring experimental data.

The equation for the cut-size is:

$$d_{50(c)} = \frac{14.8 D_c^{0.46} D_i^{0.6} D_o^{1.21} \exp(0.063V)}{D_u^{0.71} h^{0.38} Q^{0.45} (S - L)^{0.5}} \ , \quad (9.17)$$

where $d_{50(c)}$ is "corrected" d_{50} (μm); D_c, D_i,* D_o, D_u are inside diameters of hydrocyclone, inlet, vortex finder, and apex, respectively (cm); V is the volumetric percentage of solids in feed; h is the distance from bottom of the vortex finder to top of underflow orifice (cm); Q is the flow rate of feed slurry ($m^3 \ h^{-1}$); and S, L are density of solids, density of liquid, respectively ($g \ cm^{-3}$).

The equation for the volumetric flow-rate of slurry to the cyclone is

$$Q = \frac{0.021 P^{0.56} D_c^{0.21} D_i^{0.53} h^{0.16} (D_u^2 + D_o^2)^{0.49}}{\exp (0.0031V)} \quad (9.18)$$

where P is the pressure drop across the cyclone in kPa (1 psi = 6.895 kPa).

For preliminary design purposes, Mular and Jull[17] have developed expressions, from results obtained on Krebs cyclones,[18] relating d_{50} to

*For non-circular inlets, $D_i = \sqrt{4A/\pi}$, where A cm^2 is the cross-sectional area of cyclone inlet.

the operating variables for "typical" cyclones, of varying inside diameter. A "typical" cyclone has an inlet area of about 7% of the cross-sectional area of the feed chamber, a vortex finder of diameter 35–40% of the cyclone diameter, and an apex diameter normally not less than 25% of the vortex-finder diameter.

The equation for the cyclone cut-point is:

$$d_{50(c)} = \frac{0.77 D_c^{1.875} \exp(-0.301 + 0.0945V - 0.00356V^2 + 0.0000684V^3)}{Q^{0.6}(S-1)^{0.5}}.$$

(9.19)

The maximum volume of slurry that the cyclone can handle is given by:

$$Q = 9.4 \times 10^{-3} \sqrt{P} D_c^2, \text{ m}^3 \text{ h}^{-1}.$$

(9.20)

Equations such as these have been used in computer controlled grinding circuits to infer cut-points from measured data, but their use in this respect is declining with the increased application of on-line particle size monitors (Chapter 4). Their great value, however, is in the design and optimisation of circuits employing cyclones by the use of computer simulation. For instance, Krebs Engineers applied mathematical models to a grinding-classification circuit and predicted that two-stage cycloning, as opposed to the more common single stage, would allow a 6 percent increase in grinding circuit throughput. Without the aid of cyclone modelling, such work would be costly and time-consuming.[19,20]

The equations can also be very valuable in the selection of cyclones for a particular duty, the final control of cut-point and capacity made by adjusting the size of inlet, vortex-finder, and apex.

For example, consider a primary grinding mill, fed with ore (s.g. 3.7 g cm^{-3}) at the rate of 201.5 t h^{-1}. The mill is to be in closed circuit with cyclones, to produce a circulating load of 300% and a cut-point of 74 μm.

The total cyclone feed is therefore 806 t h^{-1}. Assuming 50% solids in the cyclone feed, then the slurry density (equation 3.6) is 1.574 kg l^{-1}, and the volumetric flowrate to the cyclones (equation 3.7) is 1024 m^3 h^{-1}.

The volumetric % solids in the feed is $1.574 \times 50/3.7 = 21.3\%$. Combining equations 9.19 and 9.20:

$$d_{50(c)} = \frac{12.67 D_c^{0.675} \exp(-0.301 + 0.0945V - 0.00356V^2 + 0.0000684V^3)}{P^{0.3}(S - 1)^{0.5}}$$

(9.21)

For a cyclone cutting at 74 μm ($d_{50(c)}$) operating at a pressure of, say, 12 psi (82.74 kPa), $D_c = 66$ cm. Therefore, 660 mm cyclones, or their nearest manufactured equivalent, would probably be chosen, final adjustments to the cut-size being made by changing the vortex-finder, spigot openings, pressure, etc.

The maximum volume handled by 660 mm cyclones at 12 psi is 372.5 m^3 (equation 9.20), so three cyclones would be required to handle the total flow-rate.

A computer program (CYCLONE) incorporating equations 9.19–9.21 is listed in Appendix 3. Conway[21] has developed a more comprehensive program, from published work and from data supplied by hydrocyclone manufacturers, for the prediction of cyclone performance, parameters, and product size distributions. The user inputs the desired d_{50} size, the relative densities of the solids and pulp, and the volumetric flow-rate of the pulp. The program then outputs either the cyclone size that is optimum for the inputed data, or all the cyclone sizes that will satisfy the inputed data at the user's choice. If the size distribution of the feed is available, the program calculates the size distributions, mass and volume splits, and relative densities of the overflow and underflow pulp. The imperfection is also calculated.

Since the operating variables have an important effect on the cyclone performance, it is necessary to avoid fluctuation in flow rate, etc., during operation. Pump surging should be eliminated either by automatic control of level in the sump, or by a self-regulating sump, and adequate surge capacity should be installed to eliminate flow-rate fluctuations.

The feed flow-rate and the pressure drop across the cyclone are closely related (equation 9.20). The value of the pressure drop is required to permit design of the pumping system for a given capacity or to determine the capacity for a given installation. Usually the pressure drop is determined from a feed-pressure gauge located on the inlet line some distance upstream from the cyclone. Within limits, an increase in feed flow-rate will improve efficiency by increasing the centrifugal force on the particles. All other variables being constant, this can only be achieved by an increase in pressure and a corresponding increase in power, since this is directly related to the product of pressure drop and capacity. Since increase in feed rate, or pressure drop, increases the centrifugal force effect, finer particles are carried to the underflow, and d_{50} is decreased, but the change has to be large to have a significant effect. Figure 9.21 shows the effect of pressure on the capacity and cut-point of cyclones.

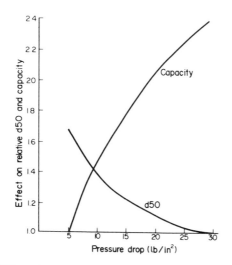

FIG. 9.21. Effect of pressure on capacity and cut-point of hydrocyclones.

The effect of increase in feed-pulp density is complex, as the effective pulp viscosity and degree of hindered settling is increased within the cyclone. The sharpness of the separation decreases with increasing pulp density and the cut-point rises due to the greater resistance to the swirling motion within the cyclone, which reduces the effective pressure drop. Separation at finer sizes can only be achieved with feeds of low solids content and large pressure drop. Normally, the

MPT—M

feed concentration is no greater than about 30% solids by weight, but for closed-circuit grinding operations, where relatively coarse separations are often required, high feed concentrations of up to 60% solids by weight are often used, combined with low-pressure drops, often less than 10 psi (68.9 kPa). Figure 9.22 shows that feed concentration has an important effect on the cut-size at high pulp densities.

The shape of the particles in the feed is also an important factor in separation, very flat particles such as mica often reporting to the overflow, even though they are relatively coarse.

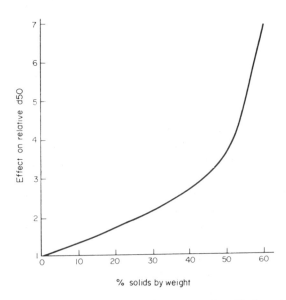

FIG. 9.22. Effect of solids concentration on cut-point of hydrocyclones.

In practice, the cut-point is mainly controlled by the cyclone design variables, such as inlet, vortex-finder, and apex openings, and most cyclones are designed such that these are easily changed.

The area of the inlet determines the entrance velocity and an increase in area increases the flow-rate. Also important is the geometry of the feed inlet. In most cyclones the shape of the entry is developed from circular cross-section to rectangular cross-section at the entrance to the cylindrical section of the cyclone. This helps to "spread" the flow along the wall of the chamber. The inlet is normally tangential, but involuted feed entries are also common (Fig. 9.23). Involuted entries are said to minimise turbulence and reduce wear.

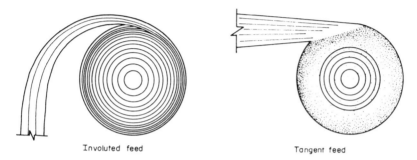

Involuted feed   Tangent feed

FIG. 9.23. Involuted and tangential feed entries.

The diameter of the vortex finder is a very important variable. At a given pressure drop across the cyclone, an increase in the diameter of the vortex finder will result in a coarser cut-point and an increase in capacity.

The size of the apex, or spigot opening, determines the underflow density, and must be large enough to discharge the coarse solids that are being separated by the cyclone. The orifice must also permit the entry of air along the axis of the cyclone in order to establish the air vortex. Cyclones should be operated at the highest possible under-flow density, since unclassified material leaves the underflow in proportion to the fraction of feed water leaving via the underflow. Under correct operating conditions, the discharge should form a hollow cone spray with a 20–30° included angle (Fig. 9.24). Air can then enter the cyclone, the classified coarse particles will discharge freely, and solids concentrations greater than 50% by weight can be achieved. Too small an apex opening can lead to the condition known as "roping", where an extremely thick pulp stream of the same diameter as the apex is formed, and the air vortex may be lost, the separation efficency will fall, and oversize material will discharge

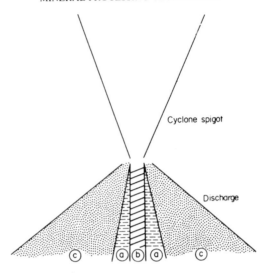

F<small>IG</small>. 9.24. Effect of spigot size on cyclone underflow: zone (a), correct operation; zone (b), "roping"—spigot too small; zone (c), excessively dilute—spigot too large.

through the vortex finder. Too large an apex orifice results in the larger hollow cone pattern seen in Fig. 9.24. The underflow will be excessively dilute and the additional water will carry unclassified fine solids that would otherwise report to the overflow.

Several investigators have concluded that the cyclone diameter has no effect on the cut-point and that for geometrically similar cyclones the efficiency curve is a function only of the feed material characteristics; the inlet and outlet diameters are the critical design variables, the cyclone diameter merely being the size of the housing required to accommodate these apertures.[22-24] However, from theoretical considerations, it is the cyclone diameter which controls the radius of orbit and thus the centrifugal force acting on the particles. As there is a strong interdependence between the aperture sizes and cyclone diameter, it is difficult to distinguish the true effect, and Plitt[16] concludes that the cyclone diameter has an independent effect on separation size.

For geometrically similar cyclones at constant flow rate, $d_{50} \propto$ diameter^x, but the value of x is open to much debate. The value of x

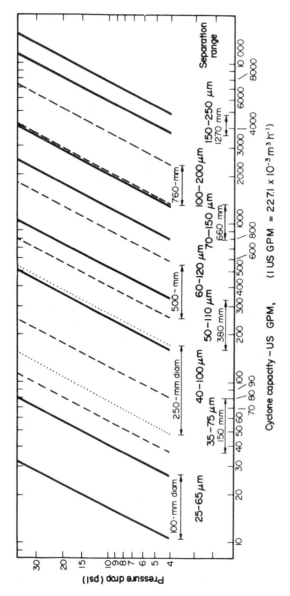

FIG. 9.25. Hydrocyclone performance chart (Krebs).

FIG. 9.26. Bank of cyclones in parallel.

using the Krebs–Mular–Jull model is 1.875, for Plitt's model it is 1.18, and Bradley[14] concluded that x varies from 1.36 to 1.52.

In practice, the cut-point is determined to a large extent by the cyclone size. The size required for a particular application can be estimated from the empirical models developed, but these tend to become unreliable with extremely large cyclones due to the increased turbulence within the cyclone, and it is therefore more common to choose the required model by referring to manufacturer's charts, which show capacity and separation size range in terms of cyclone size. A typical performance chart is shown in Fig. 9.25. This is for Krebs cyclones, operating at less than 30% feed solids by weight, and with solids density in the range 2.5–3.2 kg^{-1}.

Since fine separations require small cyclones, which have only small capacity, several have to be connected in parallel if high capacity is required (Fig. 9.26). Cyclones used for de-sliming duties are usually of very small diameter, and a large number may be required if substantial flow-rates must be handled. The practical problems of distributing the

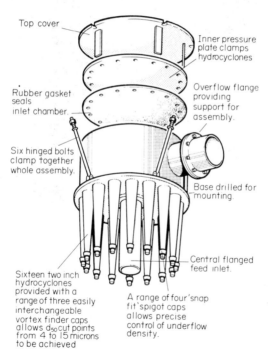

Top cover

Inner pressure plate clamps hydrocyclones

Rubber gasket seals inlet chamber.

Overflow flange providing support for assembly.

Six hinged bolts clamp together whole assembly.

Base drilled for mounting.

Central flanged feed inlet.

Sixteen two inch hydrocyclones provided with a range of three easily interchangeable vortex finder caps allows d$_{50}$ cut points from 4 to 15 microns to be achieved

A range of four 'snap fit' spigot caps allows precise control of underflow density.

FIG. 9.27. Mozely 16 × 44 mm cyclone assembly.

FIG. 9.28. Interior of Mozely cyclone assembly.

feed evenly and minimising blockages have been largely overcome by
the use of Mozely cyclone assemblies.[25] A 16 × 44 mm assembly is
shown in Fig. 9.27. The feed is introduced into a housing via a
central feed inlet at pressures of up to 50 psi (344.8 kPa). The housing
contains an assembly of 16 × 44 mm cyclones, the feed being forced
through a trash screen and into each cyclone without the need for
separate distributing ports (Fig. 9.28). The overflow from each cyclone
leaves via the inner pressure plate, and leaves the housing through the
single overflow pipe on the side. The assembly design reduces
maintenance to a minimum, the removal of the top cover allowing
easy access so that individual cyclones can be removed without
disconnecting feed or overflow pipework.

Since separation at large particle size requires large diameter
cyclones with consequent high capacities, in many cases, where coarse
separations are required, cyclones cannot be utilised, as the plant
throughput is not high enough. This is often a problem in pilot plants

where simulation of the full-size plant cannot be achieved, as scaling down the size of cyclones to allow for smaller capacity also reduces the cut-point produced.

References

1. Taggart, A. F., *Handbook of Mineral Dressing*, Wiley, New York, 1945.
2. Stokes, Sir G. G., *Mathematical and Physical Paper III*, Cambridge University Press, 1891.
3. Golson, C. E., *Modern Classification Methods*, Wemco Bulletin G7-B31.
4. Anon., Classifiers Part 2: Some of the major manufacturers of classification equipment used in mineral processing, *Mining Mag.*, 40 (July 1984).
5. Littler, A., Automatic hindered-settling classifier for hydraulic sizing and mineral beneficiation, *Trans. Instn. Min. Metall.*, **95**, C133 (Sept. 1986).
6. Edmiston, K. J., International guide to hydrocyclones, *World Mining* **36**, 61 (April 1983).
7. Barber, S. P., *et al.*, Cyclones as classifiers in ore grinding circuits: a case study, in *Hydrocyclones* (ed. G. Priestley and H. S. Stephens). BHRA Fluid Engineering, Cranfield (1980).
8. Draeger, E. A., and Collins, J. W., Efficient use of water only cyclones, *Mining Engng* **32**, 1215 (Aug. 1980).
9. Renner, V. G., and Cohen, H. E., Measurement and interpretation of size distribution of particles within a hydrocyclone, *Trans. IMM.*, Sec. C, **87**, 139 (June 1978).
10. Svarovsky, L., *Hydrocyclones*, Holt, Rinehart & Winston Ltd, Eastbourne, 1984.
11. Austin, L. G., and Klimpel, R. R., An improved method for analysing classifier data, *Powder Technology* **29**, 227 (July/Aug. 1981).
12. Lynch, A. J., *Mineral Crushing and Grinding Circuits*, Elsevier, Amsterdam (1977).
13. Kawatra, S. K., and Seitz, R. A., Calculating the particle size distribution in a hydrocyclone product for simulation purposes, *Minerals and Metallurgical Processing*, **2**, 152 (Aug. 1985).
14. Bradley, D., *The Hydrocyclone*, Pergamon Press, Oxford, 1965.
15. Dahlstrom, D. A., Fundamentals and applications of the liquid cyclone, *Chem. Engng Prog. Symp. Series No. 15*, **50**, 41 (1954).
16. Plitt, L. R., A mathematical model of the hydrocyclone classifier, *CIM Bull.* **69**, 114 (Dec. 1976).
17. Mular, A. L., and Jull, N. A., The selection of cyclone classifiers, pumps and pump boxes for grinding circuits, in *Mineral Processing Plant Design*, AIMME, New York, 1978.
18. Anon., *The Sizing of Hydrocyclones*, Krebs Engineers, California (1977).
19. Edmiston, K. J., Sizing operations are key to comminution circuits, *World Mining* 82 (Oct. 1982).

20. Apling, A. C., *et al.*, Hydrocyclone models in an ore grinding context, in *Hydrocyclones* (ed. G. Priestley and H. S. Stephens), BHRA Fluid Engineering, Cranfield (1980).
21. Conway, T. M., A computer program for the prediction of hydrocyclone performance, parameters, and product-size distributions, *Mintek Report No. M233*, Randburg, South Africa (Dec. 1985).
22. Lynch, A. J., Rao, T. C., and Prisbrey, K. A., The influence of hydrocyclone diameter on reduced efficiency curves, *Int. J. Min. Proc.* **1,** 173 (May 1974).
23. Lynch, A. J., Rao, T. C., and Bailey, C. W., The influence of design and operating variables on the capacities of hydrocyclone classifiers, *Int. J. Min. Proc.* **2,** 29 (Mar. 1975).
24. Rao, T. C., Nageswararao, K., and Lynch, A. J., Influence of feed inlet diameter on the hydrocyclone behaviour, *Int. J. Min. Proc.* **3,** 357 (Dec. 1976).
25. Anon., Small diameter hydrocyclones Richard Mozley prominent in UK development, *Ind. Minerals* 33 (Jan. 1983).

CHAPTER 10

GRAVITY CONCENTRATION

Introduction

Gravity methods of separation are used to treat a great variety of materials, ranging from heavy metal sulphides such as galena (sp. gr. 7.5) to coal (sp. gr. 1.3), at particle sizes in some cases below 50 μm.

These methods declined in importance in the first half of the century due to the development of the froth-flotation process, which allows the selective treatment of low-grade complex ores. They remained, however, the main concentrating methods for iron and tungsten ores and are used extensively for treating tin ores. Whenever gravity methods are chosen in preference to flotation, it is usually because relative costs favour the application. Minerals which are liberated at sizes above the normal flotation range may be concentrated even more economically using gravity methods.

In recent years, many companies have re-evaluated gravity systems due to increasing costs of flotation reagents, the relative simplicity of gravity processes, and the fact that they produce comparatively little environmental pollution. Modern gravity techniques have proved efficient for concentration of minerals having particle sizes in the 50–10-μm range and when coupled with improved pumping technology and instrumentation, have been incorporated in high-capacity plants.[1,2] In many cases a high proportion of the mineral in an ore-body can at least be pre-concentrated effectively by cheap and ecologically acceptable gravity systems; the amount of reagents and fuel used can be cut significantly when the more expensive methods are restricted to the processing of gravity concentrate. Gravity separation of minerals at coarser sizes as soon as liberation is achieved can also have significant advantages for later treatment stages due to decreased surface area, more efficient dewatering, and the absence of adhering chemicals which could interfere with further processing.

Gravity techniques to recover residual valuable heavy minerals in flotation tailings are being increasingly used. Apart from current production, there are many large tailings dumps which could be excavated cheaply and processed to give high value concentrates using recently developed technology.

Principles of Gravity Concentration

Gravity concentration methods separate minerals of different specific gravity by their relative movement in response to gravity and one or more other forces, the latter often being the resistance to motion offered by a viscous fluid, such as water or air.

It is essential for effective separation that a marked density difference exists between the mineral and the gangue. Some idea of the type of separation possible can be gained from the *concentration criterion*

$$\frac{D_h - D_f}{D_l - D_f} , \qquad (10.1)$$

where D_h is the specific gravity of the heavy mineral, D_l is the specific gravity of the light mineral, and D_f is the specific gravity of the fluid medium.

In very general terms, when the quotient is greater than 2.5, whether positive or negative, then gravity separation is relatively easy. As the value of the quotient decreases, so the efficiency of separation decreases, and below about 1.25 gravity concentration is not generally commercially feasible.[3]

The motion of a particle in a fluid is dependent not only on its specific gravity, but also on its size (Chapter 9); large particles will be affected more than smaller ones. The efficiency of gravity processes therefore increases with particle size, and the particles should be sufficiently coarse to move in accordance with Newton's law (equation 9.6). Particles which are so small that their movement is dominated mainly by surface friction respond relatively poorly to commercial high-capacity gravity methods. In practice, close size control of feeds to gravity processes is required in order to reduce the size effect and make the relative motion of the particles specific gravity dependent.

Gravity Separators

Many different machines have been designed and built in the past to affect separation of minerals by gravity, and they are comprehensively reviewed by Burt.[4] Many gravity devices have become obsolete, and only equipment that is used in modern mills will be described in this chapter. A list of manufacturers of gravity concentrators, together with a brief description of their products, can be found elsewhere.[5]

A classification of the more commonly used gravity separators on the basis of feed size range is shown in Fig. 1.8.

The heavy medium separation (HMS) process is widely used to preconcentrate crushed material prior to grinding and will be considered separately in the next chapter.

It is essential for the efficient operation of all gravity separators that the feed is carefully prepared. Grinding is particularly important in that the feed particles should be as coarse as possible consistent with adequate liberation; successive regrinding of middlings is required in most operations. Primary grinding should be performed where possible in open-circuit rod mills, but if fine grinding is required, closed-circuit ball milling should be used, preferably with screens closing the circuit rather than hydrocyclones in order to reduce selective overgrinding of heavy friable valuable minerals.

Gravity separators are extremely sensitive to the presence of *slimes* (ultra-fine particles), which increase the viscosity of the slurry and hence reduce the sharpness of separation, and obscure visual cut-points. It is common practice in most gravity concentrators to remove particles less than about 10 μm from the feed, and divert this fraction to the tailings, and this can account for considerable loss of values. De-sliming is often achieved by the use of hydrocyclones, although if hydraulic classifiers are used to prepare the feed, it may be preferable to de-slime at this stage, since the high shear forces produced in hydrocyclones tend to cause degradation of friable minerals.

The feed to jigs, cones, and spirals should, if possible, be screened before separation takes place, each fraction being treated separately. In most cases, however, removal of the oversize by screening, in conjunction with de-sliming, is adequate. Processes which utilise flowing-film separation, such as shaking tables and tilting frames, should always be preceded by good hydraulic classification in multi-spigot hydrosizers.

Although most slurry transportation is achieved by centrifugal pumps and pipelines, as much use as possible should be made of natural gravity flow; many old gravity concentrators were built on hillsides to achieve this. Reduction of slurry pumping to a minimum not only reduces energy consumption, but also reduces slimes production in the circuit. To minimise degradation of friable minerals, slurry pumping velocities should be as low as possible, consistent with maintaining the solids in suspension.

One of the most important aspects of gravity circuit operations is correct water-balance within the plant. Almost all gravity concentrators have an optimum feed pulp-density, and relatively little deviation from this density causes a rapid decline in efficiency. Accurate pulp-density control is therefore essential, and this is most important on the raw feed. Automatic density control should be used where possible, and the best way of achieving this is by the use of nucleonic density gauges (Chapter 3) controlling the water addition to the new feed. Although such instrumentation is expensive, it is usually economic in the long term. Control of pulp density within the circuit can be made by the use of settling cones preceding the gravity device. These thicken the pulp, but the overflow often contains solids, and should be directed to a central large sump or thickener. For substantial increase in pulp density, hydrocyclones or thickeners may be used. The latter are the more expensive, but produce less particle degradation and also provide substantial surge capacity. It is usually necessary to recycle water in most plants, so adequate thickener or cyclone capacity should be provided, and slimes build-up in the recycled water must be minimised.

If the ore contains an appreciable amount of sulphide minerals then, if the primary grind is finer than about 300 μm, these should be removed by froth flotation prior to gravity concentration, as they reduce the performance of spirals, tables, etc. If the primary grind is too coarse for effective sulphide flotation, then the gravity concentrate must be reground prior to removal of the sulphides. The sulphide flotation tailing is then usually cleaned by further gravity concentration.

The final gravity concentrate often needs cleaning by magnetic separation, leaching, or some other method, in order to remove other mineral contaminants. For instance, at the South Crofty tin mine in Cornwall, the gravity concentrate was subjected to cleaning by magnetic separators, which removed wolframite from the cassiterite product.

A comprehensive review of gravity separation circuits has been made by Ottley.[6]

Jigs

Jigging is one of the oldest methods of gravity concentration, yet even today the basic principles are not completely understood. The jig is normally used to concentrate relatively coarse material and, if the feed is fairly closed sized (e.g. 3–10 mm), it is not difficult to achieve good separation of a fairly narrow specific gravity range in minerals in the feed (e.g. fluorite, sp. gr. 3.2, from quartz, sp. gr. 2.7). When the specific gravity difference is large, good concentration is possible with a wider size range. Many large jig circuits are still operated in the coal, cassiterite, tungsten, gold, barytes, and iron-ore industries. They have a relatively high unit capacity on classified feed and can achieve good recovery of values down to 150 μm and acceptable recoveries often down to 75 μm. High proportions of fine sand and slime interfere with performance and the fines content should be controlled to provide optimum bed conditions.[1]

In the jig the separation of minerals of different specific gravity is accomplished in a bed which is rendered fluid by a pulsating current of water so as to produce stratification. The aim is to dilate the bed of material being treated and to control the dilation so that the heavier, smaller particles penetrate the interstices of the bed and the larger high specific gravity particles fall under a condition probably similar to hindered settling.

On the pulsion stroke the bed is normally lifted as a mass, then as the velocity decreases it tends to dilate, the bottom particles falling first until the whole bed is loosened. On the suction stroke it then closes slowly again and this is repeated at every stroke, the frequency usually varying between 55–330 c min^{-1}. Fine particles tend to pass through the interstices after the large ones have become immobile. The motion can be obtained either by using a fixed sieve jig, and pulsating the water, or by employing a moving sieve, as in the simple hand-jig. (Fig 10.1).

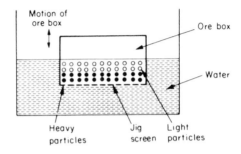

FIG. 10.1. Hand jig.

The jigging action. It was shown in Chapter 9 that the equation of motion of a particle settling in a viscous fluid is

$$m\,\frac{\mathrm{d}x}{\mathrm{d}t} = mg - m'g - D, \qquad (9.1)$$

where m is the mass of the mineral grain, $\mathrm{d}x/\mathrm{d}t$ is the acceleration, g is the acceleration due to gravity, m' is the mass of displaced fluid, and D is the fluid resistance due to the particle movement.

At the beginning of the particle movement, since the velocity x is very small, D can be ignored as it is a function of velocity.

Therefore

$$\frac{\mathrm{d}x}{\mathrm{d}t} = \left(\frac{m - m'}{m}\right)g \qquad (10.2)$$

and since the particle and the displaced fluid are of equal volume,

$$\frac{dx}{dt} = \left(\frac{D_s - D_f}{D_s}\right) g = \left(1 - \frac{D_f}{D_s}\right) g, \qquad (10.3)$$

where D_s and D_f are the respective specific gravities of the solid and the fluid.

The initial acceleration of the mineral grains is thus independent of size and dependent only on the densities of the solid and the fluid. Theoretically, if the duration of fall is short enough and the repetition of fall frequent enough, the total distance travelled by the particles will be affected more by the differential initial acceleration, and therefore by density, than by their terminal velocities and therefore by size. In other words, to separate small heavy mineral particles from large light particles a short jigging cycle is necessary. Although relatively short fast strokes are used to separate fine materials, more control and better stratification can be achieved by using longer, slower strokes, especially with the coarser particle sizes. It is therefore good practice to screen the feed to jigs into different size ranges and treat these separately. The effect of differential initial acceleration is shown in Fig. 10.2.

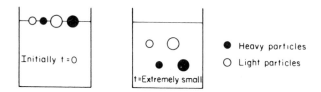

FIG. 10.2. Differential initial acceleration.

If the mineral particles are examined after a longer time they will have attained their terminal velocities and will be moving at a rate dependent on their specific gravity and size. Since the bed is really a loosely packed mass with interstitial water providing a very thick suspension of high density, hindered-settling conditions prevail, and the settling ratio of heavy to light minerals is higher than that for free settling (Chapter 9). Figure 10.3 shows the effect of hindered settling on the separation.

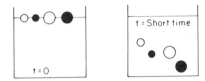

FIG. 10.3. Hindered settling.

The upward flow can be adjusted so that it overcomes the downward velocity of the fine light particles and carries them away, thus achieving separation. It can be increased further so that only large heavy particles settle, but it is apparent that it will not be possible to separate the small heavy and large light particles of similar terminal velocity.

Hindered settling has a marked effect on the separation of coarse minerals, for which longer, slower strokes should be used, although in practice, with coarser feeds, it is improbable that the larger particles have time to reach their terminal velocities.

At the end of a pulsion stroke, as the bed begins to compact, the larger particles interlock, allowing the smaller grains to move downwards through the interstices under the influence of gravity. The fine grains may not settle as rapidly during this *consolidation trickling* phase (Fig. 10.4) as during the initial acceleration or suspension, but if consolidation trickling can be made to last long enough, the effect, especially in the recovery of the fine heavy minerals, can be considerable.

Figure 10.5 shows an idealised jigging process by the described phenomena.

In the jig the pulsating water currents are caused by a piston having a movement which is a harmonic waveform (Fig. 10.6).

FIG. 10.4. Consolidation trickling.

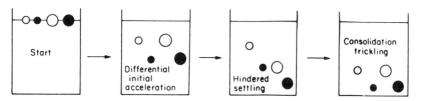

FIG. 10.5. Ideal jigging process.

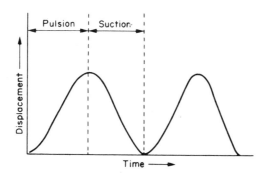

FIG. 10.6. Movement of the piston in a jig.

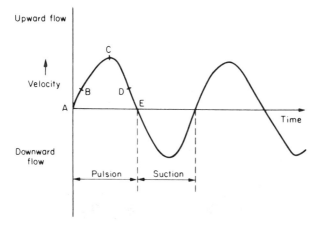

FIG. 10.7. Speed of flow through bed during jig cycle.

The vertical speed of flow through the bed is proportional to the speed of the piston. When this speed is greatest, the speed of flow through the bed is also greatest (Fig. 10.7).

The upward speed of flow increases after point A, the beginning of the cycle. As the speed increases, the grains will be loosened and the bed will be forced open, or dilated. At, say, point B, the grains are in the phase of hindered settling in an upward flow, and since the speed of flow from B to C still increases, the fine grains are pushed upwards by the flow. The chance of them being carried along with the top flow into the tailings is then at its greatest. In the vicinity of D, first the coarser grains and later on the remaining fine grains will fall back. Due to the combination of initial acceleration and hindered settling, it is mainly the coarser grains that will lie at the bottom of the bed.

At the point of transition between the pulsion and suction stroke, at point E, the bed will be completely compacted. Consolidation trickling can now occur to a limited extent. In a closely sized ore the heavy grains can now penetrate only with difficulty through the bed and may be lost to the tailings. Severe compaction of the bed can be reduced by the addition of *hutch water*, a constant volume of water, which creates a constant upward flow through the bed. This flow, coupled with the varying flow caused by the piston, is shown in Fig. 10.8. Thus suction is reduced by hutch-water addition, and is reduced in duration; by adding a large quantity of water, the suction may be entirely

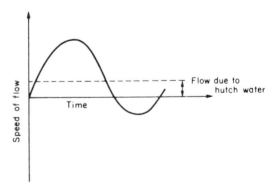

Fig. 10.8. Effect of hutch water on flow through bed.

eliminated. The coarse ore then penetrates the bed more easily and the horizontal transport of the feed over the jig is also improved. However, fines losses will increase, partly because of the longer duration of the pulsion stroke, and partly because the added water increases the speed of the top flow.

A mathematical model, which expresses the kinetics of jigging, has been developed by Karantzavelos and Frangiscos.[7]

Types of jig. Essentially the jig is an open tank filled with water, with a horizontal jig screen at the top, and provided with a spigot in the bottom, or *hutch* compartment, for concentrate removal (Fig. 10.9). The jig bed consists of a layer of coarse, heavy particles, or *ragging*, placed on the jig screen on to which the slurry is fed. The feed flows across the ragging and the separation takes place in the jig bed so that grains with a high specific gravity penetrate through the ragging and screen to be drawn off as concentrate, while the light grains are carried away by the cross-flow to be discarded as tailings. The harmonic motion produced by the eccentric drive is supplemented by a large amount of continuously supplied hutch water, which enhances the upward and diminishes the downward velocity of the water (Fig. 10.8).

One of the oldest types of jig is the *Harz* (Fig 10.10) in which the plunger moves up and down vertically in a separate compartment. Up to four successive compartments are placed in series in the hutch. A high-grade concentrate is produced in the first compartment,

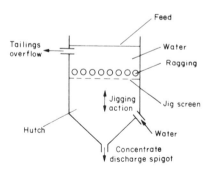

FIG. 10.9. Basic jig construction.

successively lower grades being produced in the other compartments, tailings overflowing the final compartment. If the feed particles are larger than the apertures of the screen, jigging "over the screen" is used, and the concentrate grade is partly governed by the thickness of the bottom layer, determined by the rate of withdrawal through the concentrate discharge port.

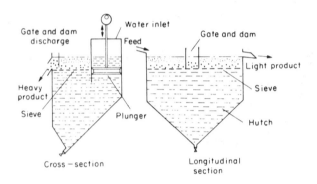

FIG. 10.10. Harz Jig.

The *Denver mineral jig* (Fig 10.11) is widely used, especially for removing heavy minerals from closed grinding circuits, thus preventing overgrinding. The rotary water valve can be adjusted so as to open at any desired part of the jig cycle, synchronisation between the valve and the plungers being achieved by a rubber timing belt. By suitable adjustment of the valve, any desired variation can be achieved, from complete neutralisation of the suction stroke with hydraulic water to a full balance between suction and pulsion.

Conventional mineral jigs consist of square or rectangular tanks, sometimes combined to form two, three or four cells in series. In order to compensate for the increase in cross-flow velocity over the jig bed, caused by the addition of hutch water, trapezoidal-shaped jigs were developed. By arranging these as sectors of a circle, the modular *circular*, or *radial* jig, was introduced, in which the feed enters in the centre and flows radially over the jig bed towards the tailings discharge at the circumference (Fig. 10.12).

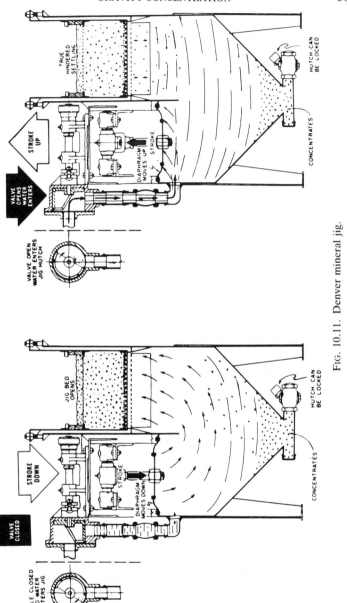

FIG. 10.11. Denver mineral jig.

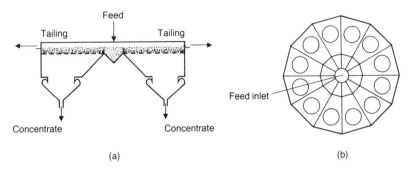

FIG. 10.12. (a) Outline of circular jig; (b) radial jig made up of twelve modules.

The main advantage of the circular jig is its very large capacity, and *IHC Radial Jigs* (Fig. 10.13) have been installed on most newly built tin dredges in Malaysia and Thailand since their development in 1970. They are also in use for the treatment of gold, diamonds, iron ore, etc., the largest, of 7.5 m in diameter, being capable of treating up to 300 $m^3 h^{-1}$ of feed with a maximum particle size of 25 mm. In the IHC jig, the harmonic motion of the conventional eccentric-driven jig is replaced

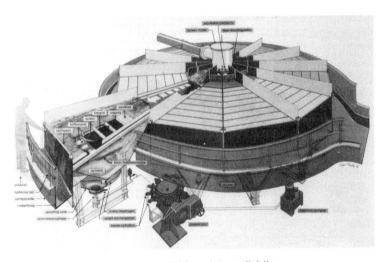

FIG. 10.13. IHC modular radial jig.

by an asymmetrical "saw-tooth" movement of the diaphragm, with a rapid upward, followed by a slow downward, stroke (Fig. 10.14). This produces a much larger and more constant suction stroke, giving the finer particles more time to settle in the bed, thus reducing their loss to tailings, the jig being capable of accepting particles as fine as 60 microns.

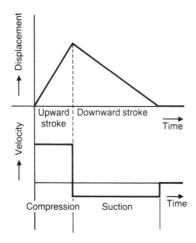

FIG. 10.14. IHC jig drive characteristics.

Coal jigs. Jigs are still the most widely used coal-cleaning device, and are preferred to the more expensive heavy medium process when the coal has relatively little middlings, or "near-gravity" material, as is often the case with British coals. No feed preparation is required, as is necessary with HMS, and for coals which are easily washed, i.e. those consisting predominantly of liberated coal and denser rock particles, the lack of close density control is not a disadvantage.

Two types of air-pulsated jig — *Baum* and *Batac* — are used in the coal industry. The standard Baum jig (Fig. 10.15), with some design modifications,[8,9] has been used for nearly 100 years, and is still the dominant device. Air under pressure is forced into a large air chamber on one side of the jig vessel causing pulsations and suction to the jig

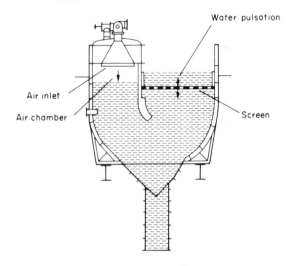

FIG. 10.15. Baum jig.

water, which in turn causes pulsations and suction through the screen plates upon which the raw coal is fed, thus causing stratification. Various methods are used to continuously separate the refuse from the lighter coal product, and all modern Baum jigs are fitted with some form of automatic refuse extraction.[10] One form of control incorporates a float immersed in the bed of material. The float is suitably weighted to settle on the dense layer of refuse moving across the screen plates. An increase in the depth of refuse raises the float, which automatically controls the refuse discharge, either by adjusting the height of a moving gate, or by controlling the pulsating water which lifts the rejects over a fixed weir plate.[11] This system is reported to respond quickly and accurately.

In Britain it is now commonplace for the automatic control system to determine the variations in refuse bed thickness by measuring the differences in water pressure under the screen plates arising from the resistance offered to pulsation. This is reportedly more sensitive to the changes in bed weight than the float method.[12] In the Radar Operated Shale Extraction ("Rose") system a vertical standpipe is arranged so as to be in communication with the wash-water beneath the perforated bed plates surmounting the compartment adjacent to an extractor.[13] Under the influence of pulsation, the column of water in the standpipe rises and falls in synchronisation with rotor valve timing, the peak height attained during each cycle being dependent on the resistance of the bed, which in turn is dependent on bed thickness. As bed thickness increases, so the volume of water forced into the standpipe during the pulsion phase of each jig cycle increases and the peak level rises pro-rata. A number of devices have been used to measure peak height, such as floats, electric probes, photo-cells and pneumatic instruments, but the Rose system makes use of the principles of radar distance measurement, the system being capable of resolving peak level changes to one millimetre difference between successive pulsation cycles, thereby automatically matching its response to the jig characteristics and to changes in the incoming feed rate. The system has the advantage of having no moving parts, except for control valve actuators, and does not require any physical contact between the measurement transducers and the wash water or bed.

In many situations the Baum jig still performs satisfactorily, with its ability to handle large tonnages (up to 1000 t h^{-1}) of coal of a wide size range. However, the distribution of the stratification force, being on one side of the jig, tends to cause unequal force along the width of jig screen and therefore uneven stratification and some loss in the efficiency of separation of the coal from its heavier impurities. This tendency is not so important in relatively narrow jigs, and in the United States multiple float and gate mechanisms have been used to counteract the effects. The *Batac jig*[14] is also pneumatically operated (Fig. 10.16), but has no side air chamber like the Baum jig. Instead, it is designed with a series of multiple air chambers, usually two to a cell, extending under the jig for its full width, thus giving uniform air distribution. The jig uses electronically controlled air valves which provide a sharp cut-off of the air input and exhaust. Both inlet and outlet valves are infinitely variable with regard to speed and length of stroke, allowing for the desired variation in pulsation and suction by which proper stratification of the bed may be achieved for differing raw coal characteristics. As a result, the Batac jig can wash both coarse and fine sizes well.[15]

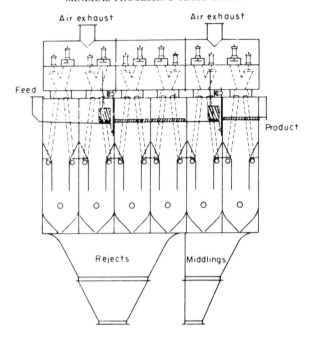

FIG. 10.16. Batac jig.

Pinched Sluices and Cones

Pinched sluices of various forms have been used for heavy mineral separations for centuries. In its simplest form (Fig. 10.17) it is an inclined launder about 1 m long, narrowing from about 200 mm in width at the feed end to about 25 mm at the discharge. Pulp of between 50 and 65% solids enters gently and stratifies as it descends; at the discharge end these strata are separated by various means, such as by splitters, or some type of tray.[16] Figure 10.18 shows pinched sluices in operation on an Australian heavy mineral sand concentrator. A computer model of the pinched sluice has been developed, which predicts the underflow grade, pulp density and flowrate in terms of the feed and operating conditions.[17] The model can also be used to

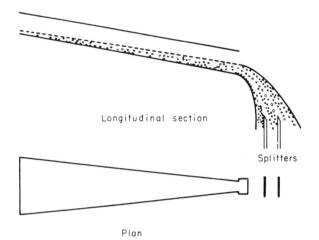

FIG. 10.17. Pinched sluice.

determine the effect of a change in operating conditions, and for the optimisation of rougher, cleaner and scavenger circuits.

The *Reichert cone* is a wet gravity concentrating device designed for high-capacity applications. Its principle of operation is similar to that of a pinched sluice, but the pulp flow is not restricted or influenced by side-wall effect, which is somewhat detrimental to pinched-sluice operation.

The Reichert cone concentrator was developed in Australia in the early 1960s primarily to treat titanium-bearing beach sands, and the success of cone circuits has led to their application in many other fields.

The single unit comprises several cone sections stacked vertically (Fig. 10.19), so as to permit several stages of upgrading. The cones are made of fibreglass and are mounted in circular frames over 6 m high. Each cone is 2 m in diameter and there are no moving parts in the unit. A cross section through a Reichert cone system is shown in Fig. 10.20. The system shown is one of many possible systems using double and single cones, together with trays, which direct heavy mineral fractions from the centre draw-off areas of the cones to external collection boxes and also serve to further upgrade the fraction, acting as a sort of pinched sluice.

FIG. 10.18. Pinched sluices in operation.

The feed pulp is distributed evenly around the periphery of the cone. As it flows towards the centre of the cone the heavy particles separate to the bottom of the film. This concentrate is removed by an annular slot in the bottom of the concentrating cone; the part of the film flowing over the slot is the tailings. The efficiency of this separation process is relatively low and is repeated a number of times within a single machine to achieve effective performance. A typical machine system for rougher concentration duties will consist of four double single-cone stages in series, each retreating the tailings of the preceding stage. This machine will produce a concentrate from the upper three stages and the product from the fourth stage as middlings.[18] A cone concentration system can be set to produce a marked reverse classification effect, in that the slots tend to reject coarser, lighter particles to tailings, while retaining finer, heavier particles in the concentrate flows.

Reichert cones have a high capacity, operating normally in the range 65–90 t h^{-1}, but in exceptional cases this can be from 40 to 100 t h^{-1},

FIG. 10.19. Reichert cones.

with a feed density of between 55–70% solids by weight. They accept
feeds of up to 3 mm in size and can treat material as fine as 30 μm,
although they are most efficient in the 100–600-μm size range.[19]

The success of cone circuits in the Australian mineral sand industry
has led to their application in other fields. Preconcentration of tin and

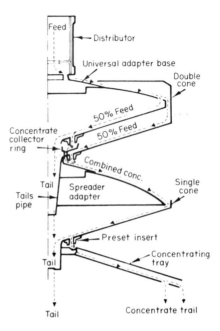

FIG. 10.20. Cross-section through Reichert cone concentrator system.

gold, the recovery of tungsten, and the concentration of magnetite are all recent successful applications. In many of these applications, cones, due to their high capacities and low operating costs, have replaced spirals and shaking tables.

One of the largest single installations is operated at Palabora Mining Co. in South Africa. Sixty-eight Reichert cones treat 34 000 t d^{-1} of flotation tailings, after preliminary de-sliming and low intensity magnetic separation for removal of magnetite. A complex circuit consists of forty-eight rougher-scavenger units, each with a six-cone configuration, and twenty cleaner-recleaners with eight-cone configurations.

The cone section of the plant produces an upgrading of about 200:1, with a recovery of 85% of the +45-μm material. The concentrate is further upgraded on shaking tables to produce final concentrates of uranothorite and baddeleyite.

Spirals

Spiral concentrators have, over numerous years, found many varied applications in mineral processing, but perhaps their most extensive usage has been in the treatment of heavy mineral sand deposits, such as those carrying ilmenite, rutile, zircon, and monazite (see Chapter 13).

The Humphreys spiral (Fig. 10.21) was introduced in 1943, its first

FIG. 10.21. Humphreys spiral concentrators.

commercial application being on chrome-bearing sands.[20] It is composed of a helical conduit of modified semicircular cross-section. Feed pulp of between 15–45% solids by weight and in the size range 3 mm to 75 μm is introduced at the top of the spiral and, as it flows spirally downwards, the particles stratify due to the combined effect of centrifugal force, the differential settling rates of the particles, and the effect of interstitial trickling through the flowing particle bed. These mechanisms are complex, being much influenced by the slurry density and particle size. Some workers[21] have reported that the main separation effect is due to hindered settling, with the largest, densest particles reporting preferentially to the concentrate, which forms in a band along the inner edge of the stream (Fig. 10.22). Bonsu,[22] however, reported that the net effect is reverse classification, the smaller, denser particles preferentially entering the concentrate band.

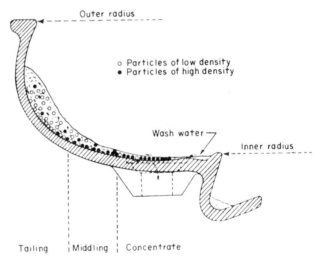

FIG. 10.22. Cross-section of spiral stream.

Ports for the removal of the higher specific-gravity particles are located at the lowest points in the cross-section. Wash water, added at the inner edge of the stream, flows outwardly across the concentrate band. The width of concentrate band removed at the ports is controlled by adjustable splitters. The grade of concentrate taken from descending ports progressively decreases, tailings being discharged from the lower end of the spiral conduit.

Until relatively recently, all spirals were very similar, based upon the original Humphreys design. However, in the last few years there have been considerable developments in spiral technology, and a wide range of devices are now available. The main areas of development have been in the introduction of spirals with only one concentrate take-off, at the bottom of the spiral, and the use of spirals without added wash water. Wash water-less spirals reportedly offer lower cost, easy operation and simple maintenance, and have been installed at several gold and tin processing plants. For example, these spirals form the major recovery stage in North America's first large-scale tin mine — the East Kemptville mine in Nova Scotia, Canada.[23] The various types of spiral available are described by Burt and Yashin,[24] who also outline the principles of spiral concentration, and give examples of spiral application in the treatment of a variety of minerals. The principles and application of spirals are also discussed by Sivamohan and Forssberg,[25] and a mathematical model has been developed for spiral concentrators.[26]

Improved spiral concentrator design has offered an efficient and economic alternative to the Reichert cone concentrator. Goodman *et al.*[27] have reviewed the development of new spiral models and describe the mechanism of separation and the effects of operating parameters. Some of the traditional areas of spiral application are described, together with examples of new applications such as treatment of fine alluvials and tailings, separation of liberated minerals at coarse sizes, and fine coal recovery. One of the most important developments in fine coal washing in the 1980s has been the introduction of spiral separators specifically designed for coal. It is common practice to separate down to 0.5 mm coal using heavy medium cyclones (Chapter 11), and below this by froth flotation. Spiral circuits have been installed to process the size range which is least effectively treated by these two methods, typically 0.2–1 mm.[28,29]

Double-spiral concentrators, with two starts integrated into the one space around a common column, have been used in Australia for many years and have also been accepted elsewhere. At Mount Wright in

Canada 4000 double-start spirals are upgrading specular hematite ore at 5000 t h^{-1} at 90%recovery.

Spirals are made with slopes of varying steepness, the angle effecting the specific gravity of separation, but having little effect on the concentrate grade and recovery. Shallow angles are used, for example, to separate coal from shale, while steeper angles are used for normal heavy mineral–silica separations. The steepest angles are used to separate heavy minerals from heavy waste minerals, for example zircon (s.g. 4.7) from kyanite and staurolite (s.g. 3.6). Capacity ranges from 1 to 3 t h^{-1} on low slope spirals to about double this for the steeper units. Spiral length is usually five or more turns for roughing duty and three turns in some cleaning units. Because treatment by spiral separators involves a multiplicity of units, the separation efficiency is very sensitive to the pulp-distribution system employed. Lack of uniformity in feeding results in substantial falls in operating efficiency and can lead to severe losses in recovery.

The range of spiral and cone separators now available offers process options that can profitably complement, and in certain cases replace, more expensive treatment methods. Wellings[30] has reviewed recent cone and spiral separator developments, and discusses typical flowsheets where these devices are succesfully exploited.

Shaking Tables

When a flowing film of water flows over a flat, inclined surface the water closest to the surface is retarded by the friction of the water absorbed on the surface; the velocity increases towards the water surface. If mineral particles are introduced into the film, small particles will not move as rapidly as large particles, since they will be submerged in the slower-moving portion of the film. Particles of high specific gravity will move more slowly than lighter particles, and so a lateral displacement of the material will be produced (Fig. 10.23).

The flowing film effectively separates coarse light particles from small dense particles, and this mechanism is utilised to some extent in the shaking-table concentrator (Fig 10.24), which is perhaps the most metallurgically efficient form of gravity concentrator, being used to

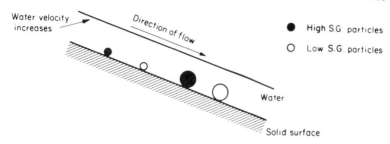

FIG. 10.23. Action in a flowing film.

treat the smaller, more difficult flow-streams, and to produce finished
concentrates from the products of other forms of gravity system.

It consists of a slightly inclined deck A on to which feed, at about
25% solids by weight, is introduced at the feed box and is distributed
along C; wash water is distributed along the balance of the feed side
from launder D. The table is vibrated longitudinally, by the
mechanism B, using a slow forward stroke and a rapid return, which
causes the mineral particles to "crawl" along the deck parallel to the
direction of motion. The minerals are thus subjected to two
forces—that due to the table motion and that, at right angles to it, due
to the flowing film of water. The net effect is that the particles move
diagonally across the deck from the feed end and, since the effect of the
flowing film depends upon the size and density of the particles, they
will fan out on the table, the smaller, denser particles riding highest
towards the concentrate launder at the far end, while the larger lighter

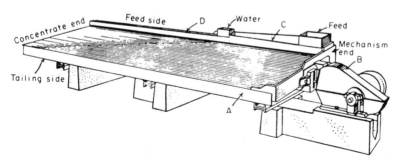

FIG. 10.24. Shaking table.

particles are washed into the tailings launder, which runs along the length of the table. Figure 10.25 shows an idealised diagram of the distribution of table products. An adjustable splitter at the concentrate end is often used to separate this product into two fractions—a high-grade concentrate and a middlings fraction.

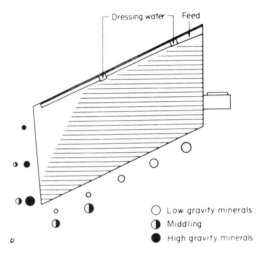

FIG. 10.25. Distribution of table products.

Although true flowing film concentration requires a single layer of feed, in practice a multi-layered feed is introduced on to the table, enabling much larger tonnages to be dealt with. Vertical stratification due to shaking action takes place behind the *riffles*, which generally run parallel with the long axis of the table and are tapered from a maximum height on the feed side, till they die out near the opposite side, part of which is left smooth (Fig. 10.25). In the protected pockets behind the riffles the particles stratify so that the finest and heaviest particles are at the bottom and the coarsest and lightest particles are at the top (Fig. 10.26). Layers of particles are moved across the riffles by the crowding action of new feed and by the flowing film of wash water. Due to the taper of the riffles, progressively finer sized and higher density particles are continuously being brought into contact with the

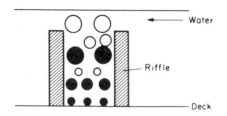

FIG. 10.26. Vertical stratification between riffles.

flowing film of water that tops the riffles. Final concentration takes place at the unriffled area at the end of the deck, where the layer of material is at this stage usually only one or two particles deep.

The significance of the many design and operating variables and their interactions have been reviewed by Sivamohan and Forssberg[31], and the development of a mathematical model of a shaking table is described by Manser et al.[32] The separation on a shaking table is controlled by a number of operating variables, such as wash water, feed pulp density, deck slope, amplitude, and feed rate, and the importance of these variables in the model development is discussed.

Many other factors, including particle shape and the type of deck, play an important part in the table separations. Flat particles, such as mica, although light, do not roll easily across the deck in the water film; such particles cling to the deck and are carried down to the concentrate discharge. Likewise, spherical dense particles may move easily in the film towards the tailings launder.

The table decks are usually constructed of wood, lined with materials with a high coefficient of friction, such as linoleum, rubber, and plastics. Decks made from fibreglass are also used which, although more expensive, are extremely hard wearing. The riffles on such decks are incorporated as part of the mould.

Particle size plays a very important role in the table separations; as the range of sizes in a table feed increases, the efficiency of separation decreases. If a table feed is made up of a wide range of particle sizes, some of these sizes will be cleaned inefficiently. As can be seen from Fig. 10.25, in an idealised separation, the middlings produced are not "true middlings", i.e. particles of associated mineral and gangue, but relatively coarse dense particles and fine light particles. If these particles are returned to the grinding circuit, together with the true middlings, then they will be needlessly reground.

Since the shaking table effectively separates coarse light from fine dense particles, it is common practice to classify the feed, since classifiers put such particles into the same product, on the basis of their equal settling rates. In order to feed as narrow a size range as possible on to the table, classification is usually performed in multi-spigot hydrosizers (Chapter 9), each spigot product, comprising a narrow range of equally settling particles, being fed to a separate set of shaking tables. A typical gravity concentrator employing shaking tables (Fig. 10.27) may have an initial grind in rod mills in order to liberate as much mineral at as coarse a size as possible to aid separation. The hydrosizer products feed separate sets of tables, the middlings being reground before being returned to the hydrosizer. Riffled tables, or *sand tables*, normally operate on feed sizes in the range 3 mm to 100 μm, and the hydrosizer overflow, consisting primarily of particles finer than this, is usually thickened and then distributed to tables whose decks have a series of planes rather than riffles and are designated *slimes tables*.

Dewatering of the hydrosizer overflow is often performed by hydrocyclones, which also remove particles in the overflow smaller than about 10 μm, which will not separate efficiently by gravity methods due to their extremely slow settling rates.

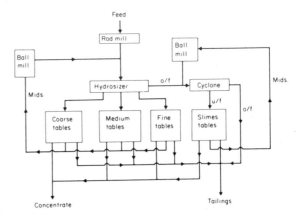

FIG. 10.27. Typical shaking-table concentrator flowsheet.

Successive stages of regrinding are a feature of many gravity concentrators. The mineral is separated at all stages in as coarse a state as possible in order to achieve reasonably fast separation and hence high throughputs.

The capacity of a table varies according to size of feed particles and the concentration criteria. Tables can handle up to 2 t h^{-1} of 1.5-mm sand and perhaps 1 t h^{-1} of fine sand. On 100–150 μm feed materials, table capacities may be as low as 0.5 t h^{-1}. On coal feeds, however, which are often tabled at sizes of up to 15 mm, much higher capacities are common. A normal −5-mm raw coal feed can be tabled with high efficiency at 12.5 t h^{-1} per deck, whilst tonnages as high as 15.0 t h^{-1} per deck are not uncommon when the top size in the feed is 15 mm.[33]

The introduction of double- and triple-deck units (Fig. 10.28) has improved the area/capacity ratio at the expense of some flexibility and control.

FIG. 10.28. Triple-deck table.

Separation can be influenced by the length of stroke, which can be altered by means of a handwheel on the vibrator, or head motion, and by the reciprocating speed. The length of stroke usually varies within the range of 10 mm to 25 mm or more, the speed being in the range 240–325 strokes per minute. Generally, a fine feed requires a higher speed and shorter stroke than a coarse feed. The head motion is designed to give a stroke which increases in speed as it goes forward until it is jerked to a halt before being sharply reversed, allowing the particles to slide forward during most of the backward stroke due to their built-up momentum. The head motion of a standard Wilfley table (Fig. 10.29) consists of two toggles driven by a pitman P. The back toggle B is seated against a fixed mounting C, whereas the front toggle A bears on a yoke D connected to the table deck. The system is held together by a spring E and driven by the eccentric and pitman. At the beginning of the forward stroke, the toggles are at their flattest and the spring in maximum compression. As the pitman rises, the toggles steepen out, being at their most acute angle at the top of the stroke when the table reaches its maximum speed. The pitman now descends, flattening the toggle angle and abruptly reversing the direction of the table; the decompression of the spring aids the forward stroke to some extent.

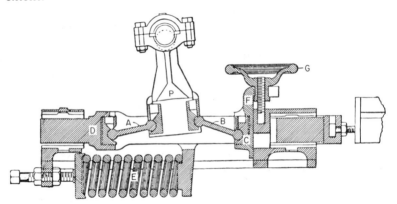

FIG. 10.29. Head motion of Wilfley table.

The quantity of water used in the feed pulp varies, but for ore tables normal feed dilution is 20–25% solids by weight, while for coal tables pulps of 33–40% solids are used. In addition to the water in the feed pulp, clear water flows over the table for final concentrate cleaning.

This varies from a few litres to almost 100 l min^{-1} according to the nature of the feed material.

Tables slope from the feed to the tailings discharge end and the correct angle of incline is obtained by means of a handwheel. In most cases the line of separation is clearly visible on the table so this adjustment is easily made.

The table is slightly elevated along the line of motion from the feed end to the concentrate end. This moderate slope, which the high-density particles climb more readily than the low-density minerals, greatly improves the separation, allowing much sharper cuts to be made between concentrate, middlings, and tailings. The correct amount of end elevation varies with feed size and is greatest for the coarsest and highest gravity feeds. The end elevation should never be less than the taper of the riffles, otherwise there is a tendency for water to flow out towards the riffle tips rather than across the riffles. Normal end elevations in ore tabling range from a maximum of 90 mm for a very heavy, coarse sand to as little as 6 mm for an extremely fine feed.

Ore-concentrating tables are used primarily for the concentration of minerals of tin, iron, tungsten, tantalum, mica, barium, titanium, zirconium, and, to a lesser extent, gold, silver, thorium, uranium, and others. By far the largest single use of shaking tables, however, is in the washing of coal. It is estimated that 45 million tonnes of metallurgical coal are cleaned annually by tables in the United States alone.[34]

Pneumatic Tables

Originally developed for seed separation, pneumatic or air tables have an important use in the treatment of heavy mineral sand deposits, and in the upgrading of asbestos, and in applications where water is at a premium. Pneumatic tables use a throwing motion to move the feed along a flat riffled deck, and blow air continuously up through a porous bed. The stratification produced is somewhat different from that of wet tables. Whereas in wet tabling the particle size increases and the density decreases from the top of the concentrate band to the tailings, on an air table both particle size and density decrease from the top down, the coarsest particles in the middlings band having the lowest density. Pneumatic tabling is therefore similar in effect to hydraulic classification. They are commonly used in combination with wet tables to clean ziron concentrates, one of the products obtained from heavy mineral sand deposits. Such concentrates are often contaminated with small amounts of fine silica, which can effectively be separated from the coarse zircon particles by air tabling. Some fine zircon may be lost in the tailings, and can be recovered by treatment on wet shaking tables.

Bartles–Mozley Tables

The main drawback of a true flowing film concentrator is that the sorting method can only be applied to an approximate monolayer of particles, which means that as particle size becomes very small, separator surface area must become very large. The Bartles–Mozley concentrator is used for the recovery of particles with an upper size of about 100 μm and a lower size of about 5 μm, or in the case of gold and platinum, down to 1 μm, and feed rates of over 5 t h^{-1} may be achieved.[35,36] The flowing-film thickness used is typically 0.5–1 mm, which is some 5–10 times the particle size for a 100 μm particle, and 100–200 times the particle size for a 5 μm particle. Separation is based on the fact that when a suspension of particles is subjected to continuous shear due to either the forward motion of the film across the surface or the movement of the surface, pressure develops across the plane of shear at right angles to the surface, the larger lighter particles moving towards the zone of least shear strain at the surface, and the smaller denser particles towards the zone of greater shear strain in the bed.[37] The concentrator (Fig. 10.30) uses orbital motion to develop shear in the bed, and consists of forty fibreglass decks, each 1.1 m wide by 1.5 m long, arranged in two sandwiches, each of twenty decks. Each deck surface is smooth and is attached to its neighbours by 12-mm-thick plastic formers, which also serve to define the pulp channel. The deck assemblies are supported on two suspension cables, at a 1–3° angle to the horizontal, within a free-standing steel framework.

An integral drive assembly is located between the two-deck

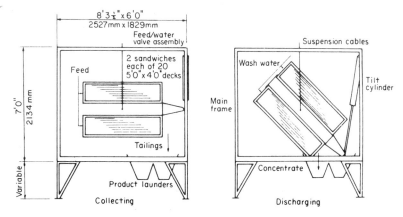

FIG. 10.30. Operation of Bartles–Mozley concentrator.

sandwiches and comprises an out-of-balance weight driven by a variable speed motor.

The feed slurry is evenly distributed to all forty decks through a simple piping system that permits each deck to be fed at four points across its width. In operating the table, which is essentially a semi-batch process, feed is run on to the decks for a period of up to 35 min, depending on the nature of the material being treated, after which the feed is automatically interrupted by a pneumatically operated pinch valve, with simultaneous tilting of the table. By means of an air cylinder, the table is first tilted to drain, and then further to about 45° for the washing cycle. Low-pressure water is fed to the decks through the pulp lines to remove the settled concentrate. The deck assembly is then returned automatically to the original position, the wash-water valve is shut off, and the slurry feed valve opened to commence the next cycle. Since the separator is a semi-continuous device, the concentrates are usually collected into a stirred surge tank prior to feeding to the next step in the process.

The Bartles–Mozley concentrator now has an established place in the industry and is operating on a variety of minerals. It was developed initially for the recovery of fine cassiterite from tin ores in Cornwall, but is also used profitably for the recovery of fine mineral from old mine dumps and tailings deposits.

Bartles Cross-belt Separator

The Bartles–Mozley concentrator is a roughing or scavenging device, its mode of operation precluding its use on feeds containing a large proportion of heavy minerals. The Bartles cross-belt separator, however, does not suffer from this constraint, and was introduced to upgrade Bartles–Mozley concentrates.[36] It consists (Fig. 10.31) of a 2.4-m-wide endless PVC belt, whose top surface is inclined from a central longitudinal ridge towards both side edges. The longitudinally horizontal belt moves slowly over drive and head pulleys, a rotating out-of-balance-weight imparting orbital motion to the belt. The belt assembly is freely suspended from the main frame by four wires. Feed

Fig. 10.31. Bartles cross-belt separator.

material is introduced along half of the length of the central ridge. Heavy mineral particles are deposited on the belt while light gangue material, held in suspension by the orbital shear, flows down the belt. The concentrate travels sideways with the belt from the feed zone through a cleaning zone where middlings particles are washed down to a middlings launder. Finally the clean concentrates are discharged over the head pulley. The unit was originally used at Geevor Tin Mines, Cornwall,[38] and there are several installations in the chromite industry in the Philippines and Europe.[37]

Duplex Concentrator

This machine (Fig. 10.32) was originally developed for the recovery of tin from low grade feeds, but has a wider application in the recovery

FIG. 10.32. Duplex concentrator.

of tungsten, tantalum, gold, chromite and platinum from fine feeds.[39] Two decks are used alternatively to provide continuous feeding, the feed slurry being fed onto one of the decks, the lower density minerals running off into the discharge launder, while the heavy minerals remain on the deck. The deck is washed with water after a preset time, in order to remove the gangue minerals, after which the deck is tilted and the concentrate is washed off. One table is always concentrating, while the other is being washed or is discharging concentrates. The concentrator has a capacity of up to $5 \, t \, h^{-1}$ of $-100 \, \mu m$ feed. producing enrichment ratios of between 20 and 500, and is available with various sizes and numbers of decks.

Mozley Laboratory Separator

This flowing film device, which uses orbital shear, is now used in many mineral processing laboratories, and is designed to treat small samples (100 g) of ore, allowing a relatively unskilled operator to obtain information for a recovery grade curve within a very short time.[40,41] It very closely simulates commercial shaking table performance, and in this respect has proven invaluable for ore evaluation.[42]

Gold Ore Concentrators

Although most of the gold from the South African gold mines is recovered by dissolution in cyanide solution, a proportion of the coarse ($+75 \, \mu m$) gold is recovered by gravity separators, followed by amalgamation of the gravity concentrates. It has been argued that separate treatment of the coarse gold in this way constitutes a security risk, increases costs, and (in the case of amalgamation) is a health hazard. It has been further argued that these disadvantages can be avoided if sufficient cyanidation time is allowed for the dissolution of the gold particles. However, the benefits and costs of gravity concentration are discussed by Loveday and Forbes[43] who show that it can improve the overall recovery.

The coarse gold must be concentrated in the grinding circuit to prevent it from being flattened into thin platelets (see Chapter 7). The primary concentrators are typically *Johnson drums*, the concentrate being cleaned either by shaking tables or *endless-belt strakes* to produce a high grade concentrate suitable for amalgamation.[44]

The Johnson concentrator consists of a continuously rotating cylindrical drum, which is tilted slightly and lined internally with riffled rubber sheeting. Pulp flows through the drum, and the concentrate collects in the riffles. As the drum turns, the concentrate is carried to the top and is washed into a launder. The endless-belt strake, which is usually used to clean the concentrate, consists of a flat, endless rubber belt, inclined slightly to the horizontal. The upper surface has saw-tooth riffles running across the belt. The belt moves continuously against the pulp stream and spray water washes the concentrate from the riffles after the belt passes over the head pulley.

The *plane table*[44] is a high capacity roughing device used on some of the South African mines. It consists of riffled rubber, covering a series of inclined smooth surfaces with steps between them. Pulp flows down the length of the riffles, which collect the concentrate, which is drawn off through slots at the steps. The faster flowing pulp is carried across the slots and is collected at the end of the table. The concentrates are then cleaned, usually by shaking tables.

Alluvial Tin Concentration

The bulk of the world's tin is mined in South-East Asia, Malaysia being the biggest producer. Although the deposits are very low grade (in some cases as low as 0.02% Sn), they are amenable to fairly simple and efficient gravity concentration, as the heavy minerals are essentially liberated and coarse-grained. The size range of the cassiterite varies from about 6 mm to minus 50 μm, but the greatest bulk lies in the range 100–250 μm, which presents little problem in processing. The bulk of Malaysian tin is mined by dredging (Fig. 10.33), buckets transferring the alluvium to washing trommels on board the dredge, which remove oversize rocks. The ore is then de-slimed and concentrated by jigs to produce a pre-concentrate which is transferred to a central concentrator for further cleaning on shaking

Fig. 10.33. Malaysian tin dredge.

tables, followed by magnetic and high-tension separation to recover other economic heavy minerals such as ilmenite, monazite, and zircon from the cassiterite (see Chapter 13). There are many mines which mine the alluvium by hydraulic monitoring, pumping the resultant slurry to simple sluices, or *palongs* (Fig. 10.34). As the ore flows down the palongs, the heavy minerals settle out behind riffles spaced at intervals across the slurry flow. Periodically the slurry is diverted, the riffles are raised, and the heavy minerals are washed from the sluices and pumped to jigs for further cleaning. The jig concentrate is then transported to the central concentrator.

References

1. Terril, I. J., and Villar, J. B., Elements of high capacity gravity separation, *CIM Bull.* **68**, 94 (May 1975).

Fig. 10.34. Palong concentrators.

2. Paterson, O. D., How many low grade ores are now being recovered by gravity concentration, *World Mining* **30,** 44 (July 1977).
3. Taggart, A. F., *Handbook of Mineral Dressing*, Wiley, New York, 1945.
4. Burt, R. O., *Gravity Concentration Technology*, Elsevier, Amsterdam, 1985.
5. Anon, Gravity concentration Part 2: some of the manufacturers, *Mining Mag.*, 335 (Oct. 1984)
6. Ottley, D. J., Technical, economic and other factors in the gravity concentration of tin, tungsten, columbium, and tantalum ores, *Minerals Sci. Engng.*, **11,** 99 (April 1979).
7. Karantzavelos, G. E., and Frangiscos, A. Z., Contribution to the modelling of the jigging process, in *Control '84 Minerals/Metallurgical Processing*, ed. J. A. Herbst, p. 97, AIME, New York, 1984.
8. Harrington, D., Practical design and operation of Baum jig installations, *Mine & Quarry Special Chinese Supplement*, 16 (July/Aug. 1986)
9. Green, P., Designers improve jig efficiency, *Coal Age*, **89,** 50 (Jan. 1984).
10. Adams, R. J., Control system increases jig performance, *Min. Equip. Int.*37 (Aug. 1983).
11. Wallace, W. M., Electronically controlled Baum jig washing, *Mine & Quarry* **8,** 43 (July/Aug. 1979).
12. Wallace, W., Practical aspects of Baum jig coal washing, *Mine & Quarry* **10,** 40 (Sept. 1981).

13. Anon, The Rose washbox, *Colliery Guardian*, **233**, 5 (Jan. 1984).
14. Zimmerman, R. E., Performance of the Batac jig for cleaning fine and coarse coal sizes, *Trans. Soc. Min. Engng AIME* **258**, 199 (Sept. 1975)
15. Chen, W. L., Batac jig in five U.S. plants, *Mining Engng.*, **32**, 1346 (Sept. 1980).
16. Sivamohan, R., and Forssberg, E., Principles of sluicing, *Int. J. Min. Proc.*, **15**, 157 (Oct. 1985).
17. Subasinghe, G. K. N. S., and Kelly, E. G., Modelling pinched sluice type concentrators, in *Control '84 Minerals/Metallurgical Processing*, ed. J. A. Herbst, p. 87, AIME, New York, 1984.
18. Anon., How Reichert cone concentration recovers minerals by gravity, *World Mining* **30**, 48 (July 1977).
19. Forssberg, K. S., and Sandström, E., Operational characteristics of the Reichert cone in ore processing, *13th IMPC, Warsaw*, **2**, 259 (1979).
20. Hubbard, J. S., Humphreys, I. B., and Brown, E. W., How Humphreys spiral concentrator is used in modern dressing practice, *Mining World* (May 1953).
21. Mills, C., Process design, scale-up and plant design for gravity concentration, in *Mineral Processing Plant Design*, eds. A. L. Mular and R. B. Bhappu, AIMME, New York, 1978.
22. Bonsu, A. K., Influence of pulp density and particle size on spiral concentration efficiency, M.Phil. Thesis, Camborne School of Mines, 1983.
23. Scales, M., Cassiterite connection, *Can. Min. J.*, **107**, 25 (Nov. 1986).
24. Burt, R. O., and Yashin, A. V., Spiral concentration: current trends in design and operation, in *Mineral Processing and Extractive Metallurgy*, ed. M. J. Jones and P. Gill, p. 117, IMM, London, 1984.
25. Sivamohan, R., and Forssberg, E., Principles of spiral concentration, *Int. J. Min. Proc.*, **15**, 173 (Oct. 1985).
26. Tucker, P., et al, A mathematical model of spiral concentration, as part of a generalized gravity-process simulation model, and its application at two Cornish tin operations, *Proc. XVth Int. Min. Proc. Cong.*, **3**, 3, Cannes (1985).
27. Goodman, R. H., et al, Advanced gravity concentrators for improved metallurgical performance, *Minerals and Metallurgical Processing*, **2**, 79 (May 1985).
28. Jackson, D., Fine coal spirals raise efficiency in preparation, *Coal Age*, **89**, 66 (June 1984).
29. Butcher, S. G., International coal preparation trends, *Colliery Guardian*, **234**, 284 (July 1986).
30. Wellings, D. E. A., Recent developments in cone and spiral separators, *World Mining Equip.*, **10**, 20 (June 1986).
31. Sivamohan, R., and Forssberg, E., Principles of tabling, *Int. J. Min. Proc.*, **15**, 281 (Nov. 1985).
32. Manser, R. J., et al, Development of a mathematical model of a shaking table concentrator, *Proc. 19th APCOM Symposium*, ed. R. V. Ramani, p. 631, AIME, New York, 1986.
33. Terry, R. L., Minerals concentration by wet tabling, *Minerals Processing* **15**, 14 (July/Aug. 1974).
34. Tiernon, C. H., Concentrating tables for fine coal cleaning, *Mining Engng* **32**, 1228 (Aug. 1980).
35. Burt, R. O., and Ottley, D. J., Fine gravity concentration using the Bartles–Mozley concentrator, *Int. J. Min. Proc.* **1**, 347 (1974).
36. Player, G. D., Operation and applications of fine-particle gravity separation

equipment, in *Mineral Processing and Extractive Metallurgy*, eds. M. J. Jones and P. Gill, p. 245, IMM, London, 1984.

37. Mills, C., and Burt, C. R., Thin film gravity concentrating devices and the Bartles–Mozley concentrator, *Min. Mag.* 32 (July 1979).
38. Burt, R. O., On stream testwork of the Bartles crossbelt concentrator, *Min. Mag*, 631 (Dec. 1977).
39. Anon, Duplex concentrator for gravity separation, *Min. J.* 57 (Jan. 28, 1983).
40. Anon, British developed separator aids small-scale mineral studies, *Min. Mag.* 45 (Jan. 1979).
41. Anon, Laboratory separator modification improves recovery of coarse-grained heavy minerals, *Min. Mag.* 158 (Aug. 1980).
42. Wills, B.A., Laboratory simulation of shaking table performance, *Min. Mag.* 489 (June 1981).
43. Loveday, B. K., and Forbes, J. E., Some considerations in the use of gravity concentration for the recovery of gold, *J. S. Afr. I.M.M.* 121 (May 1982).
44. Bath, M. D., Duncan, A. J., and Rudolph, E. R., Some factors influencing the recovery of gold by gravity concentration, *J. S. Afr. I.M.M.* **73,** 363 (June 1973).

CHAPTER 11

HEAVY MEDIUM SEPARATION (HMS)

Introduction

Heavy medium separation (or dense medium separation (DMS), or the sink-and-float process) is applied to the pre-concentration of minerals, i.e. the rejection of gangue prior to grinding for final liberation. It is also used in coal preparation to produce a commercially graded end-product, clean coal being separated from the heavier shale or high-ash coal.

In principle, it is the simplest of all gravity processes and has long been a standard laboratory method for separating minerals of different specific gravity. Heavy liquids of suitable density are used, so that those minerals lighter than the liquid float, while those denser than it sink (Fig. 11.1).

Since most of the liquids used in the laboratory are toxic, or give off harmful vapours, the heavy medium used in industrial separations is a thick suspension, or pulp, of some heavy solid in water, which behaves as a heavy liquid.

The process offers some advantages over other gravity processes. It has the ability to make sharp separations at any required density, with a high degree of efficiency even in the presence of high percentages of near-density material. The density of separation can be closely controlled, within a relative density of ± 0.005 kg l^{-1} and can be maintained, under normal conditions, for indefinite periods. The separating density can, however, be changed at will and fairly quickly, to meet varying requirements. The process is, however, rather expensive, mainly due to the ancillary equipment needed to clean the medium.

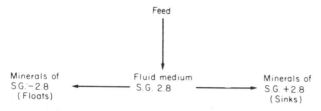

FIG. 11.1. Principle of heavy medium separation.

Heavy medium separation is applicable to any ore in which, after a suitable degree of liberation by crushing, there is enough difference in specific gravity between the particles to separate those which will repay the cost of further treatment from those which will not. The process is most widely applied when the density difference occurs at a coarse particle size, as separation efficiency decreases with size due to the slower rate of settling of the particles. Particles should preferably be larger than about 3 mm in diameter, in which case separation can be effective on a difference in specific gravity of 0.1 or less.

Separation down to 500 μm, and less, in size can, however, be made by the use of centrifugal separators. Providing a density difference exists, there is no upper size limit except that determined by the ability of the plant to handle the material.

Heavy medium separation is possible with ores in which the minerals are coarsely aggregated. If the values are finely disseminated throughout the host rock, then a suitable density difference between the crushed particles cannot be developed by coarse crushing.

Preconcentration is most often performed on metalliferous ores which are associated with relatively light country rock. Thus finely disseminated galena and sphalerite often occur with pyrite as replacement deposits in rocks such as limestone or dolomite. Similarly in some of the Cornish tin ores the cassiterite is found in lodes with some degree of banded structure in which it is associated with other high specific-gravity minerals such as the sulphides of iron, arsenic, and copper, as well as iron oxides. The lode fragments containing these minerals therefore have a greater density than the siliceous waste and will, therefore, allow early separation. Wall rock adjacent to the lode may likewise be disposed of and will in many cases form the majority of the waste generated, since the working of narrow lodes often involves the removal of waste rock from the walls to facilitate access. Problems may arise if the wall rock is mineralised with low-value, high-density minerals such as iron oxides and sulphides, a situation which is often encountered. A good example is a wolfram mine in France where the schist wall rock was found to contain pyrrhotite which raised its density to such an extent that heavy medium separation to preconcentrate the wolfram lode material was impossible. As a result, the whole of the run-of-mine ore had to be comminuted in order to obtain the wolfram.

The Heavy Medium

Liquids

Heavy liquids have wide use in the laboratory for the appraisal of gravity-separation techniques on ores. Heavy liquid testing may be performed to determine the feasibility of heavy medium separation on a particular ore, and to determine the economic separating density, or it may be used to assess the efficiency of an existing heavy medium circuit by carrying out tests on the sink and float products. The aim is to separate the ore samples into a series of fractions according to density, establishing the relationship between the high and low specific gravity minerals.

Tetrabromoethane (TBE), having a specific gravity of 2.96, is commonly used and may be diluted with white spirit or carbon tetrachloride (sp. gr. 1.58) to give a range of densities below 2.96.

Bromoform (sp. gr. 2.89) may be mixed with carbon tetrachloride to give densities in the range 1.58–2.89. For densities of up to 3.3, di-iodomethane is useful, diluted as required with triethyl orthophosphate. Aqueous solutions of sodium polytungstate have certain advantages over organic liquids, such as being virtually non-volatile and non-toxic, and densities of up to 3.1 can easily be achieved.[1] Clerici solution (thallium formate–thallium malonate solution) allows separation at densities up to 4.2 at 20°C, or 5.0 at 90°C. Separations of up to 12.0 kg l^{-1} can be achieved by the use of *magnetohydrostatics*, i.e. the utilization of the supplementary weighting force produced in a solution of a paramagnetic salt when situated in a magnetic field gradient. This type of separation is applicable primarily to non-magnetic minerals with a lower limiting particle size of about 50 μm.[2,3,4]

Many heavy liquids give off toxic fumes and must be used with adequate ventilation: the Clerici liquids are extremely poisonous and must be handled with exteme care. The use of liquids on a commercial scale has therefore not been found practicable, and industrial processes employ finely ground solids suspended in water.

Suspensions

Below a concentration of about 30% by volume, finely ground suspensions in water behave essentially as simple Newtonian fluids. Above this concentration, however, the suspension becomes non-Newtonian and a certain minimum stress, or yield stress, has to be applied before shear will occur and the movement of a particle can commence. Thus small particles, or those close to the medium density, are unable to overcome the rigidity offered by the medium before movement can be achieved. This can be overcome to some extent either by increasing the shearing forces on the particles, or by decreasing the apparent viscosity of the suspension. The shearing force may be increased by substituting centrifugal force for gravity. The viscous effect may be decreased by agitating the medium, which causes elements of liquid to be sheared relative to each other. In practice the medium is never static, as motion is imparted to it by paddles, air, etc., and also by the sinking material itself. All these factors, by reducing the yield stress, tend to bring the parting, or separating density as close as possible to the density of the medium in the bath.[5]

In order to produce a stable suspension of sufficiently high density, with a reasonably low viscosity, it is necessary to use fine, high specific-gravity solid particles, agitation being necessary to maintain the suspension and to lower the apparent viscosity. The solids comprising the medium must be hard, with no tendency to slime, as degradation increases the apparent viscosity by increasing the surface area of the medium. The medium must be easily removed from the mineral surfaces by washing, and must be easily recoverable from the fine-ore particles washed from the surfaces. It must not be affected by the constituents of the ore and must resist chemical attack, such as corrosion.

Galena was initially used as the medium and, when pure, it can give a bath density of about 4.0. Above this level, ore separation is slowed down by the viscous resistance. Froth flotation, which is an expensive process, was used to clean the contaminated medium, but the main disadvantage is that galena is fairly soft and tends to slime easily, and it also has a tendency to oxidise, which impairs the flotation efficiency. Galena is still used as the medium in a few lead-zinc concentrators, such as the Sullivan concentrator in Canada (Chapter 12).

The most widely used medium for metalliferous ores is now ferrosilicon, whilst magnetite is used in coal preparation. Recovery of medium in both cases is by magnetic separation.

Magnetite (sp. gr. 5.1) is relatively cheap, and is used to maintain bath densities of up to 2.5 kg l^{-1}.

Ferrosilicon (sp. gr. 6.7–6.9) is an alloy of iron and silicon which should contain not less than 82% Fe and 15–16% Si. If the silicon content is less than 15%, the alloy will tend to corrode, while if it is more than 16% the magnetic susceptibility will be greatly reduced. Losses of ferrosilicon from a heavy medium circuit vary widely, from as little as 0.1 to more than 2.5 kg/tonne of ore treated, the losses, apart from spillages, mainly occurring in magnetic separation and by the adhesion of medium to ore particles. Corrosion usually accounts for relatively small losses, and can be effectively prevented by maintaining the ferrosilicon in its passive state. This is normally achieved by atmospheric oxygen diffusing into the medium, or by the addition of small quantities of sodium nitrite.[6]

At 15% Si, the density of ferrosilicon is 6.8 and a medium of density 3.2 can be prepared. Various size grades are produced from 95%–150 μm to 95%– 40 μm, the finer material being employed for treatment of finer ore. Atomised ferrosilicon consists of rounded particles which produce lower apparent viscosities, and can be used to produce bath densities up to 3.4 kg l^{-1}.[7]

Separating Vessels

Several types of separating vessel are in use, and these may be classified into gravitational ('static-baths') and centrifugal (dynamic) vessels.

Gravitational vessels

Gravitational units comprise some form of vessel into which the feed and medium are introduced and the floats are removed by paddles, or merely by overflow. Removal of the sinks is the most difficult part of separator design. The aim is to discharge the sinks particles without removing sufficient of the medium to cause disturbing downward currents in the vessel.

The *Wemco cone separator* (Fig. 11.2) is widely used for ore treatment, since it has a relatively high sinks capacity. The cone, which

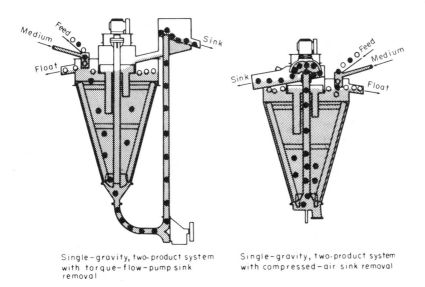

Single–gravity, two-product system   Single–gravity, two-product system
with torque–flow–pump sink   with compressed–air sink removal
removal

FIG. 11.2. Wemco cone separator.

has a diameter of up to 6 m, accommodates feed particles of up to 10 cm in diameter, with capacities of up to 500 t h^{-1}.

The feed is introduced on to the surface of the medium by free-fall, which allows it to plunge several centimetres into the medium. Gentle agitation by rakes mounted on the central shaft helps keep the medium in suspension. The float fraction simply overflows a weir, whilst the sinks are removed by pump or by external or internal air lift.

Drum separators (Fig. 11.3) are built in several sizes, up to 4.3 m diameter by 6 m long, with maximum capacities of 450 t h^{-1}, and can treat feeds of up to 30 cm in diameter. Separation is accomplished by the continuous removal of the sink product through the action of lifters fixed to the inside of the rotating drum. The lifters empty into the sink launder when they pass the horizontal position. The float product overflows a weir at the opposite end of the drum from the feed chute. Longitudinal partitions separate the float surface from the sink-discharge action of the revolving lifters.

The comparatively shallow pool depth in the drum compared with the cone separator minimises settling out of the medium particles giving a uniform gravity throughout the drum.

Where single-stage heavy-medium treatment is unable to produce the desired recovery, two-stage separation can be achieved in the *two-compartment drum separator* (Fig. 11.4), which is, in effect, two drum separators mounted integrally and rotating together, one feeding the other. The lighter medium in the first compartment separates a pure float product. The sink product is lifted and conveyed into the second compartment where the middlings and the true sinks are separated.

Although drum separators have very large sinks capacities, and are inherently more suited to the treatment of metallic ores, where the sinks product is normally 60–80% of the feed, rather than to coal, where the sinks product is only 5–20% of the feed,[8] they are very commonly used in the coal industry because of their simplicity, reliability, and relatively small maintenance needs.

The *Drewboy bath* is widely used in the UK coal industry because of its high floats capacity (Fig. 11.5). The raw coal is fed into the separator at one end, and the floats are discharged from the opposite end by a star-wheel with suspended rubber, or chain straps, while the sinks are

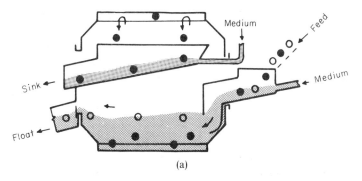

(a)

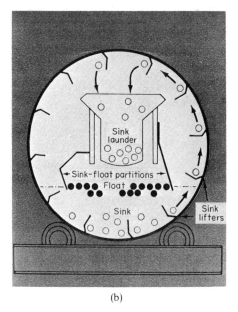

(b)

FIG. 11.3. Drum separator: (a) side view, (b) end view

lifted out from the bottom of the bath by a radial-vaned wheel
mounted on an inclined shaft. The medium is fed into the bath at two
points—at the bottom of the vessel, and with the raw coal, the
proportion being controlled by valves.

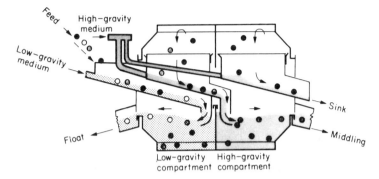

Fɪɢ. 11.4. Two-compartment drum separator.

The *Norwalt washer* was developed in South Africa, and most installations are to be found in that country.[9] Raw coal is introduced into the centre of the annular separating vessel, which is provided with stirring arms (Fig. 11.6). The floats are carried round by the stirrers, and are discharged over a weir on the other side of the vessel, being carried out of the vessel by the medium flow. The discard sinks to the bottom of the vessel and is dragged along by scrapers attached to the

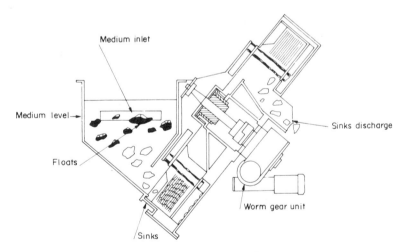

Fɪɢ. 11.5. Drewboy bath.

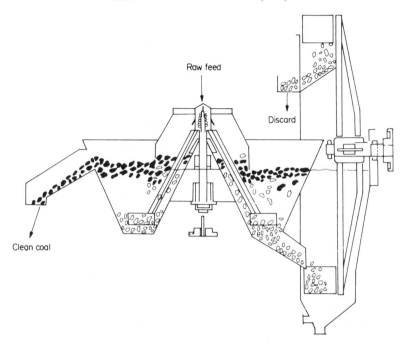

FIG. 11.6. Norwalt bath.

bottom of the stirring arms, and is discharged via a hole in the bottom of the bath into a sealed elevator, either of the wheel or bucket type, which continuously removes the sinks product.

Centrifugal Separators

Cyclone heavy medium separators have now become widely used in the treatment of ores and coal. They provide a high centrifugal force and a low viscosity in the medium, enabling much finer separations to be achieved than in gravitational separators. Feed to these devices is typically de-slimed at about 0.5 mm, to avoid contamination of the medium with slimes, and to minimise medium consumption. In recent years work has been carried out in many parts of the world to extend the range of particle size treated by centrifugal separators, particularly those operating in coal preparation plants, where advantages to be gained are elimination of de-sliming screens and froth flotation of the screen undersize, as well as more accurate separation of fine coal. Froth flotation has little effect on sulphur reduction, whereas pyrite can be removed and oxidised coal treated by HMS. The work has shown that good separations can be achieved for coal particles as fine as 0.1 mm, but that below this size separation efficiency is very low. Since a typical British coal can contain 10% material less than 0.1 mm, froth flotation must be retained to clean these finer fractions, although HMS with no de-sliming has been performed in the United States.[10] Tests on a lead–zinc ore have shown that good separations can be achieved down to 0.16 mm using a centrifugal separator.[11] These and similar results elsewhere, together with the progress made in automatic control of medium consistency, add to the growing evidence that HMS can be considered for finer material than had been thought economical or practical until recently. As the energy requirement for grinding, flotation and dewatering is often up to ten times that required for HMS, a steady increase of fines pre-concentration plants is likely.

The *DSM cyclone* (Fig. 11.7) was developed by the Dutch State Mines, and is used to treat ores and coal in the size range 40–0.5 mm. The principle of operation is very similar to that of the conventional hydrocyclone (Chapter 9). The ore is suspended in a very fine medium of ferrosilicon or magnetite, and is introduced tangentially to the cyclone under pressure, normally being gravity fed via a constant head. Gravity feeding reduces the degradation which can occur by pumping. The sinks product leaves the cyclone in the apex, while the floats product is discharged via the central vortex finder. Mathematical models of the DSM cyclone have been developed by King *et al.*[12] and Rao *et al.*[13]

The *Vorsyl separator* (Fig. 11.8) is used in many coal-preparation plants for the treatment of small coal sizes up to about 50 mm at feed rates of up to 120 t h^{-1}.[14,15] The feed to the separator, consisting of de-slimed raw coal, together with the separating medium of magnetite, is introduced tangentially, or more recently by an involute entry, at the top of the separating chamber, under pressure. Material of specific gravity less than that of the medium passes into the clean coal outlet

FIG. 11.7. DSM cyclones.

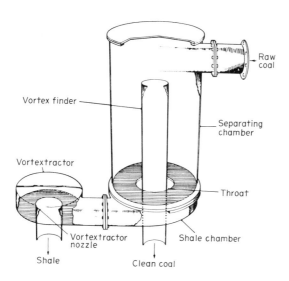

FIG. 11.8. Vorsyl separator.

via the vortex finder, while the near gravity material and the heavier shale particles move to the wall of the vessel due to the centrifugal acceleration induced. The particles move in a spiral path down the chamber towards the base of the vessel where the drag caused by the proximity of the orifice plate reduces the tangential velocity and creates a strong inward flow towards the throat. This carries the shale, and near gravity material, through zones of high centrifugal force, where a final precise separation is achieved. The shale, and a proportion of the medium, discharge through the throat into the shallow shale chamber, which is provided with a tangential outlet, and is connected by a short duct to a second shallow chamber known as the vortextractor. This is also a cylindrical vessel with a tangential inlet for the medium and reject and an axial outlet. An inward spiral flow to the outlet is induced, which dissipates the inlet pressure energy and permits the use of a large diameter outlet nozzle without the passing of an excessive quantity of medium.

The *LARCODEMS* (LARge COal DEnse Medium Separator) is a recent development for the treatment of raw coal of up to 100 mm in one vessel.[16] The unit (Fig. 11.9) consists of a cylindrical chamber which is inclined at approximately 30° to the horizontal. Feed medium at the required relative density is introduced under pressure, either by pump or static head, into the involute tangential inlet at the lower end. At the top end of the vessel is another involute tangential outlet connected to the vortextractor. Raw coal of 0.5–100 mm is fed into the separator by a chute connected to the top end, the clean coal after separation being removed through the bottom outlet. High relative density particles pass rapidly to the separator wall and are removed through the top involute outlet and the vortextractor.

A single unit is to be incorporated as the main processor in the 250 t h^{-1} coal preparation plant at Point of Ayr Colliery in the United Kingdom. As the 250 t h^{-1} LARCODEMS is only 1.2 m diameter by 3 m long, it could have a dramatic effect on the design and construction of future coal preparation plants.

The *Dyna Whirlpool* separator is similar to the LARCODEMS, and is also used for treating fine coal, particularly in the Southern Hemisphere, as well as diamonds, fluorspar, tin and lead–zinc ores, in the size range 0.5–30 mm.[17]

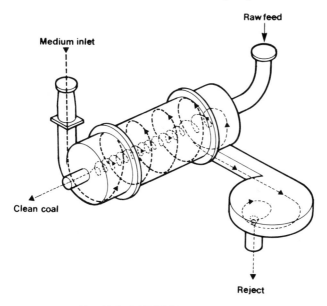

FIG. 11.9. LARCODEMS separator.

It consists of a cylinder of predetermined length (Fig. 11.10), having identical tangential inlet and outlet sections at either end. The unit is operated in an inclined position and medium of the required density is pumped under pressure into the lower outlet. The rotating medium creates a vortex throughout the length of the unit and leaves via the upper tangential discharge and the lower vortex outlet tube. Raw feed entering the upper vortex tube is sluiced into the unit by a small quantity of medium and a rotational motion is quickly imparted by the open vortex. Float material passes down the vortex and does not contact the outer walls of the unit, thus greatly reducing wear. The floats are discharged from the lower vortex outlet tube. The heavy sink particles of the feed penetrate the rising medium towards the outer wall of the unit and are discharged with medium through the sink discharge pipe. Since the sinks discharge is close to the feed inlet, the sinks are removed from the unit almost immediately, again reducing

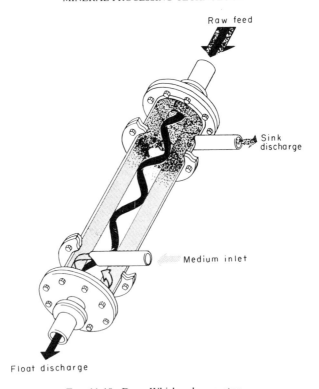

Raw feed

Sink
discharge

Medium inlet

Float discharge

FIG. 11.10. Dyna Whirlpool separator.

wear considerably. Only near-gravity particles which are separated further along the unit actually come into contact with the main cylindrical body. The tangential sink discharge outlet is connected to a flexible sink hose and the height of this hose may be used to adjust back pressure to finely control the cut-point. The capacity of the separator can be as high as 100 t h^{-1}, and it has several advantages over the DSM cyclone. Apart from the reduced wear, which not only decreases maintenance costs but also maintains performance of the unit, operating costs are lower, since only the medium is pumped. The unit has a much higher sinks capacity and can accept large fluctuations in sink/float ratios.[18]

The *Tri-Flo* separator (Fig. 11.11) can be regarded as two Dyna Whirlpool separators joined in series, and has been installed in a number of coal, metalliferous, and non-metallic ore treatment plants.[11,19] Involute medium inlets and sink outlets are used, which produce less turbulence than tangential inlets.

The device can be operated with two media of differing densities in order to produce sink products of individual controllable densities, as at the Prestavel Plant in Italy. The separator treats a $+1$ mm fluorspar–galena ore at separating densities of 2.75 kg l^{-1} and 3.2 kg l^{-1}. The fluorite concentrate from the No. 1 sink outlet contains 91.8% CaF_2 and 0.46% Pb, at 90% CaF_2 recovery, and the No. 2 sink concentrate contains 41.5% Pb and 40% CaF_2 with a 90.3% lead recovery. This concentrate and the -1 mm fines are ground and treated in the flotation plant.

Two-stage treatment using a single medium density produces a float and two sinks products with only slightly different separation densities. With metalliferous ores, the second sink product can be regarded as a scavenging stage for the dense minerals, thus increasing their recovery. This second product may be recrushed, and after de-sliming,

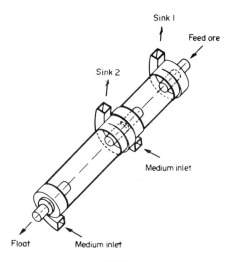

FIG. 11.11. Tri-Flo separator.

returned for retreatment. Where the separator is used for washing coal, the second stage cleans the floats to produce a higher grade product. Two stages of separation also increase the sharpness of separation. Tri-Flos are manufactured in sizes ranging from 250 mm to 500 mm with capacities of 15 t h^{-1} to 90 t h^{-1}.

HMS Circuits

The feed to a heavy medium circuit must be screened to remove fine ore, and slimes should be removed by washing, thus alleviating any tendency which such slime content may have for creating sharp increases in medium viscosity.

The greatest expense in any heavy medium circuit is the provision for reclaiming and cleaning the medium which leaves the separator with the sink and float products. A typical circuit is shown in Fig. 11.12,

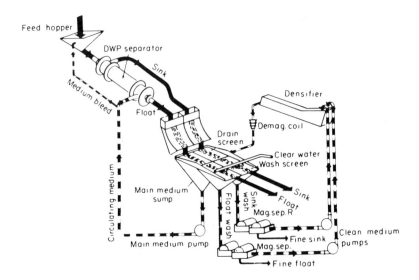

FIG. 11.12. Typical HMS circuit.

in which the separator is a Dyna Whirlpool, although similar circuits are used with other separators.

The sink and float fractions pass on to separate vibrating drainage screens where more than 90% of the medium in the separator products is recovered and pumped back via a sump into the separating vessel. The products then pass under washing sprays where substantially complete removal of medium and adhering fines is accomplished. The finished float and sink products are discharged from the screens for disposal or further treatment.

The undersize products from the washing screens, consisting of medium, wash water, and fines, are too dilute and contaminated to be returned directly as medium to the separatory vessel. They are treated individually as shown, or together, by magnetic separation, to recover the magnetic ferrosilicon, or magnetite, from the non-magnetic fines. Reclaimed, cleaned medium is thickened to the required density by a spiral classifier (densifier), which continuously returns it to the HMS circuit. The densified medium discharge passes through a demagnetising coil to assure a non-flocculated, uniform suspension in the separatory vessel.

Although dynamic heavy media vessels offer high efficiency, one major problem associated with their use has been the often high loss of medium. Ferrosilicon losses can account for 10-35% of the total operating cost of a heavy medium plant. In coal preparation, magnetite losses are typically in the range 0.5–1.5 kg t^{-1} of coal cleaned. Magnetite is usually ground to -50 microns and the finer the particles the more likely they are to be lost in the cleaning circuit. The problem with using coarser sizes of magnetite is that it tends to introduce an undesirable amount of non-magnetic coal and refuse into the recirculating media cycle, due to the coarser drainage screens required.

It is essential for efficient operation that the medium be free of fine ore particles, which increase the viscosity. If the amount of fines in the circuit reaches a high proportion due, say, to inefficient screening of the feed, it may be necessary to divert an increased amount of medium into the cleaning circuit. Many circuits have such a provision, allowing medium from the draining screen to be diverted into the washing screen undersize sump.

Typical Heavy Medium Separations

The most important use of HMS is in coal preparation where a relatively simple separation removes the low-ash coal from the heavier high-ash discard and associated shales and sandstones.

HMS is preferred to the cheaper Baum jig method of separation when washing coals with a relatively large proportion of middlings, or near-density material, since the separating density can be controlled to much closer limits.

British coals, in general, are relatively easy to wash, and jigs are used in many cases. Where HMS is preferred, Drum and Drewboys separators are most widely used for the coarser fractions, with DSM cyclones and Vorsyl separators being preferred for the fines. HMS is essential with most Southern Hemisphere coals, where a high middlings fraction is present. This is especially so with the large, low-grade coal deposits found in the South African Transvaal. Drums and Norwalt baths are the most common separators utilised to wash such coals, with DSM cyclones and Dyna Whirlpools being used to treat the finer fractions.

At Amcoal's Landau Colliery in the Transvaal, a two-density operation is carried out in order to produce two saleable products. After preliminary screening of the run-of-mine coal, the coarse (+7 mm) fraction is washed in Norwalt bath separators, utilising magnetite as the medium to give a separating density of 1.6. The sinks product from this operation, consisting predominantly of sand and shales, is discarded, and the floats product is routed to Norwalt baths operating at a lower density of 1.4. This separation stage produces a low-ash floats product, containing about 7.5% ash, which is used for metallurgical coke production, and a sinks product, which is the process middlings, containing about 15% ash, which is used as power-station fuel.

The fine (0.5–7 mm) fraction is treated in a similar two-stage manner utilising Dyna Whirlpool separators.

In metalliferous mining, HMS has found its greatest use in the preconcentration of lead–zinc ores, in which the disseminated sulphide minerals often associate together as replacement bandings in the light country rock, such that marked specific gravity differences between particles crushed to fairly coarse sizes can be exploited.

A heavy medium plant was incorporated into the lead–zinc circuit at Mount Isa Mines Ltd, Australia, in 1982 in order to increase the plant throughput by 50%. The ore, containing roughly 6.5% lead, 6.5% zinc and 200 ppm silver, consists of galena, sphalerite, pyrite, and other sulphides finely disseminated in distinct bands in quartz and dolomite (Fig. 11.13). Liberation of the ore into particles which are either sulphide-rich or predominantly gangue begins at around −50 mm, and becomes substantial below 18 mm.

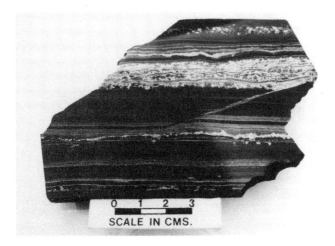

FIG. 11.13. Mount Isa ore. Bands of sulphide minerals in carbonaceous host rock.

The plant treats about 800 t h^{-1} of material, in the size range 1.7–13 mm by DSM cyclones, at a separating density of 3.05 kg l^{-1}, to reject 30–35% of the run-of-mine ore as tailings, with 96–97% recoveries of lead, zinc, and silver to the preconcentrate. The pre-concentrate has a 25% lower Bond Work Index, and is less abrasive because the lower specific gravity siliceous material mostly reports to the rejects. The rejects are used as a cheap source of fill for underground operations. The plant is extensively instrumented, the process control strategy being described elsewhere.[20]

HMS is also used to pre-concentrate iron, tin, and tungsten ores, and non-metallic ores such as fluorite, barite, etc. It has a very important use in the pre-concentration of diamond ores, prior to recovery of the diamonds by electronic sorting or grease-tabling. Diamonds are the lowest grade of all ores mined, and concentration ratios of several million to one must be achieved. HMS produces an initial enrichment of the ore in the order of 100–1000 to 1 by making use of the fact that diamonds have a fairly high specific gravity (3.5), and are relatively easily liberated from the ore, since they are loosely held in the parent rock. Gravity and centrifugal separators are utilised, with ferrosilicon as the medium, and separating densities of between 2.6 and 3.0 are used. Clays in the ore sometimes present a problem by increasing the medium viscosity, thus reducing separating efficiency and the recovery of diamonds to the sinks.

Laboratory Heavy Liquid Tests

Laboratory testing may be performed on ores in order to assess the suitability of heavy medium separation on the crushed material and to determine the economic separating density.

Liquids covering a range of densities in incremental steps are prepared, and the representative sample of crushed ore is introduced into the liquid of highest density. The floats product is removed and washed and placed in the liquid of next lower density, whose float product is then transferred to the next lower density and so on. The sinks product is finally drained, washed, and dried, and then weighed, together with the final floats product, to give the density distribution of the sample by weight (Fig. 11.14).

Care should be taken when evaluating ores of fine-particle size that sufficient time is given for the particles to settle into the appropriate fraction. Centrifuging is often carried out on fine materials to increase the settling time, but this should be done with care, as there is a tendency for the floats to become entrained in the sinks fraction.[21] Unsatisfactory results are often obtained with porous materials, such

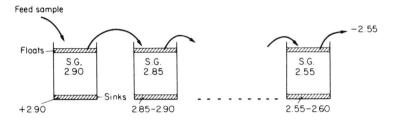

FIG. 11.14. Heavy liquid testing.

as magnesite ores, due to the entrainment of liquid in the pores, which changes the apparent density of the particles.[22]

After assaying the fractions for metal content, the distribution of material and metal in the density fractions of the sample can be tabulated. Table 11.1 shows such a distribution from tests performed on a tin ore. It can be seen from columns 3 and 6 of the table that if a separation density of 2.75 was chosen, then 68.48% of the material, being lighter than 2.75, would be discarded as a float product, and only 3.81% of the tin would be lost in this fraction. Similarly, 96.19% of the tin would be recovered into the sink product, which accounts for 31.52% of the original total feed weight.

TABLE 11.1. HEAVY LIQUID TEST RESULTS

1 Specific gravity fraction	2 % Weight	3 Cumulative % weight	4 Assay (% Sn)	5 Distribution (% Sn)	6 Cumulative distribution (% Sn)
−2.55	1.57	1.57	0.003	0.04	0.04
2.55–2.60	9.22	10.79	0.04	0.33	0.37
2.60–2.65	26.11	36.90	0.04	0.93	1.30
2.65–2.70	19.67	56.57	0.04	0.70	2.00
2.70–2.75	11.91	68.48	0.17	1.81	3.81
2.75–2.80	10.92	79.40	0.34	3.32	7.13
2.80–2.85	7.87	87.27	0.37	2.60	9.73
2.85–2.90	2.55	89.82	1.30	2.96	12.69
+2.90	10.18	100.00	9.60	87.31	100.00

For more precise evaluation of the results of laboratory tests the results should be plotted graphically.

Heavy liquid tests are important in coal preparation in order to determine the required density of separation and the expected yield of coal of the required ash content. The "ash content" refers to the amount of incombustible material in the coal. Since coal is lighter than the contained minerals, the higher the density of separation the higher is the *yield*

$$\left(\text{yield} = \frac{\text{weight of coal floats product} \times 100\%}{\text{total feed weight}} \right)$$

but the higher is the ash content. The ash content of each density fraction from heavy liquid testing is determined by taking about 1 g of the fraction, placing it in a cold well-ventilated furnace, and slowly raising the temperature to 815°C, maintaining the sample at this temperature until constant weight is obtained. The residue is cooled and then weighed. The ash content is the mass of ash expressed as a percentage of the initial sample weight taken.

Table 11.2 shows the results of heavy liquid tests performed on a coal sample. The coal was separated into the density fractions shown in column 1, and the weight fractions and ash contents are tabulated in columns 2 and 3 respectively. The weight per cent of each product is multiplied by the ash content to give the ash product (column 4).

The total floats and sinks products at the various separating densities shown in column 5 are tabulated in columns 6–11. To obtain the cumulative per cent for each gravity fraction, columns 2 and 4 are cumulated from top to bottom to give columns 6 and 7 respectively. Column 7 is then divided by column 6 to obtain the cumulative per cent ash (column 8). Cumulative sink ash is obtained in essentially the same manner, except that columns 2 and 4 are cumulated from bottom to top to give columns 9 and 10 respectively.

The results of Table 11.2 are plotted in Fig. 11.15 as typical *washability curves*.

Suppose an ash content of 12% is required in the coal product. It can be seen from the washability curves that such a coal would be produced at a yield of 55% (cumulative per cent floats), and the required density of separation is 1.465.

TABLE 11.2

(1) Sp. gr. fraction	(2) Wt. %	(3) Ash %	(4) Ash product	(5) Separating density	(6) Wt. %	(7) Cumulative float (Clean coal)		(8) Ash %	(9) Wt. %	(10) Cumulative sink (Discard)	(11) Ash %
						Ash product				Ash product	
−1.30	0.77	4.4	3.39	1.30	0.77	3.39		4.4	99.23	2213.16	22.3
1.30–1.32	0.73	5.6	4.09	1.32	1.50	7.48		5.0	98.50	2209.67	22.4
1.32–1.34	1.26	6.5	8.19	1.34	2.76	15.67		5.7	97.24	2201.48	22.6
1.34–1.36	4.01	7.2	28.87	1.36	6.77	44.54		6.6	93.23	2172.61	23.3
1.36–1.38	8.92	9.2	82.06	1.38	15.69	126.60		8.1	84.31	2090.55	24.8
1.38–1.40	10.33	11.0	113.63	1.40	26.02	240.23		9.2	73.98	1976.92	26.7
1.40–1.42	9.28	12.1	112.29	1.42	35.30	352.52		10.0	64.70	1864.63	28.8
1.42–1.44	9.00	14.1	126.90	1.44	44.30	479.42		10.8	55.70	1737.73	31.2
1.44–1.46	8.58	16.0	137.28	1.46	52.88	616.70		11.7	47.12	1600.45	34.0
1.46–1.48	7.79	17.9	139.44	1.48	60.67	756.14		12.5	39.33	1461.01	37.1
1.48–1.50	6.42	21.5	138.03	1.50	67.09	894.17		13.3	32.91	1322.98	40.2
+1.50	32.91	40.2	1322.98	—	100.00	2217.15		22.2	0.00	0.00	0.0
Total	100.0	22.2	2217.15								

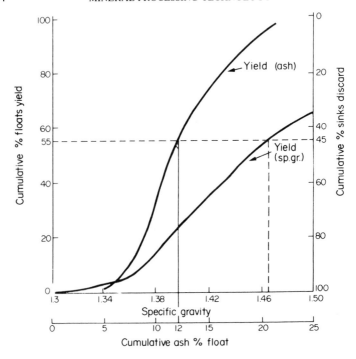

FIG. 11.15. Typical coal washability curves.

The difficulty of the separation in terms of operational control is dependent mainly on the amount of material present in the feed which is close to the required density of separation. For instance, if the feed were composed entirely of pure coal at a sp. gr. of 1.3 and shale at a density of 2.7, then the separation would be easily carried out over a wide range of operating densities. If, however, the feed consists of appreciable middlings, and much material is present very close to the chosen separating density, then only a small variation in this density will seriously affect the yield and ash content of the product.

The amount of near-gravity material present is sometimes regarded as being the weight of material in the range ±0.1 or ±0.05 kg l^{-1} of the separating density, and separations involving feeds with less than about 7% of ±0.1 near-gravity material are regarded by coal preparation engineers as being fairly easy to control. Such separations are often performed in Baum jigs, as these are cheaper than heavy medium plants, which require expensive media-cleaning facilities, and no feed preparation, i.e. removal of the fine particles by screening, is required. However, the density of separation in jigs is not easy to control to fine limits, as it is in heavy medium baths, and for near-gravity material much above 7%, heavy medium separation is preferred.

Heavy liquid tests can be used to evaluate any gravity separation process on any ore, and Table 11.3 can be used to indicate the type of separator which could effect the separation in practice.[23]

Table 11.3 takes no account of the particle size of the material and experience is therefore required in its application to heavy liquid results, although some idea of the effective particle size range of gravity separators can be gained from Fig. 1.8. The throughput of the plant must also be taken into account with respect to the type of separator chosen. For instance, if a throughput of only a few tonnes per hour is envisaged, there would be little point in installing Reichert cones, which are high-capacity units, operating most effectively at about 70 t h^{-1}.

TABLE 11.3

Wt. % within ±0.1 gravity of separation	Gravity process recommended	Type
0–7	Almost any process	Jigs, tables, spirals
7–10	Efficient process	Sluices, cones, HMS
10–15	Efficient process with good operation	
15–25	Very efficient process with expert operation	HMS
Above 25	Limited to a few exceptionally efficient processes with expert operation	HMS with close control

Efficiency of Heavy Medium Separation

Laboratory testing assumes perfect separation and, in such batch tests, conditions are indeed close to the ideal, as sufficient time can be taken to allow complete separation to take place.

When an ore sample is immersed in a heavy liquid, particles of high specific gravity relative to the liquid sink rapidly and particles of low specific gravity rise rapidly. However, particles with a specific gravity approaching that of the liquid move slowly, whilst those with the same specific gravity as the liquid come to rest anywhere.

In a continuously operating process, with constant discharge of float and sink, particles with specific gravity approaching that of the medium may not have time to reach the sink (or float) product and will be misplaced into the other product. Particles with a high or low specific gravity in comparison with the medium are least affected, but particles with the same specific gravity as the medium have an equal chance of reporting to the sink or float product. Efficiency of separation therefore varies depending on the specific gravity of the particle, ranging from 100%, where all the particles report to the correct product, i.e. those of highest and lowest specific gravity, to 50%, where the particles have the same density as the medium.

The difficulty, or ease, of separation is thus dependent on the amount of material present of specific gravity approaching that of the medium. Conversely, the efficiency of a particular separating process depends on its ability to separate material of specific gravity close to that of the medium.

The efficiency of separation can be represented by the slope of a *Partition*, or *Tromp curve*, which describes the separating efficiency for the separator, whatever the quality of the feed, and can be used for comparison and estimation of performance.

The partition curve relates the *partition coefficient*, i.e. the percentage of the feed material of a particular specific gravity which reports to the sinks product, against specific gravity (Fig. 11.16).

The ideal Tromp curve shows that all particles having a density higher than the separating density report to sinks, while those lighter report to floats.

The partition curve for a real separation shows that efficiency is highest for particles of density far from the operating density and decreases for particles approaching the operating density.

Almost all partition curves give a reasonable straight-line relationship between the distribution of 25% and 75%, and the slope of the line between these distributions is used to show the efficiency of the process.

The *probable error of separation* or the *Ecart probable* (*Ep*) is defined as half the difference between the density where 75% is recovered to sinks and that at which 25% is recovered to sinks, i.e. from Fig. 11.16,

$$Ep = \frac{A - B}{2} \ .$$

The density at which 50% of the particles report to sinks is shown as the *effective density of separation*, which may not be exactly the same as the operating density.

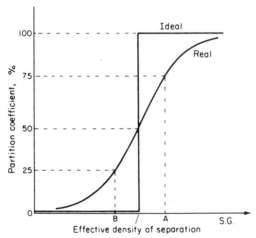

FIG. 11.16. Partition or Tromp curve.

The lower the *Ep*, the nearer to vertical is the line between 25% and 75% and the more efficient is the separation. An ideal separation has a vertical line with an *Ep* = 0, whereas in practice the *Ep* usually lies in the range 0.02–0.08.

The *Ep* is not commonly used as a method of assessing the efficiency of separation in units such as tables, spirals, cones etc., due to the many operating variables (wash water, table slope, speed, etc.) which can affect the separation efficiency. It is, however, ideally suited to the relatively simple and reproducible HMS process.

Construction of Partition Curves

The probable error of separation of an operating heavy medium vessel can be determined by sampling the sink and float products and performing heavy liquid tests to determine the amount of misplaced material in each of the products. The range of liquid densities applied must envelope the working density of the heavy medium unit. The results of heavy liquid tests on samples of floats and sinks from a vessel separating coal (floats) from shale (sinks) are shown in Table 11.4.

Columns 1 and 2 are the results of laboratory tests on the float and sink products and columns 3 and 4 relate these results to the total distribution of the feed material to floats and sinks which must be determined by weighing the products over a period of time. The weight fraction in columns 3 and 4 can be added together to produce the reconstituted feed weight distribution in each density fraction (column 5). Column 6 gives the nominal specific gravity of each density range, i.e. material in the density range 1.30–1.40 is assumed to have a specific gravity lying midway between these densities—1.35.

The partition coefficient (column 7) is the percentage of feed material of a certain nominal specific gravity which reports to sinks, i.e.

$$\frac{\text{column } 4}{\text{column } 5} \times 100\% \ .$$

The partition curve can therefore be constructed by plotting the partition coefficient against the nominal specific gravity, from which the probable error of separation of the vessel can be determined.

An alternative, rapid, method of determining the partition curve of a separator is to utilise density tracers. Specially developed colour-coded plastic tracers can be fed to the process, the partitioned products

TABLE 11.4. COAL-SHALE SEPARATION EVALUATION

Specific gravity fraction	(1) Floats analysis (wt. %)	(2) Sinks analysis (wt. %)	(3) Floats % of feed	(4) Sinks % of feed	(5) Reconstituted feed (%)	(6) Nominal sp. gr.	(7) Partition coefficient
−1.30	83.34	18.15	68.83	3.15	71.98	—	4.39
1.30–1.40	10.50	10.82	8.67	1.89	10.56	1.35	17.80
1.40–1.50	3.35	9.64	2.77	1.68	4.45	1.45	37.75
1.50–1.60	1.79	13.33	1.48	2.32	3.80	1.55	61.05
1.60–1.70	0.30	8.37	0.25	1.46	1.71	1.65	85.38
1.70–1.80	0.16	5.85	0.13	1.02	1.15	1.75	88.70
1.80–1.90	0.07	5.05	0.06	0.88	0.94	1.85	93.62
1.90–2.00	0.07	4.34	0.06	0.75	0.81	1.95	92.68
+2.00	0.42	24.45	0.35	4.25	4.60	—	92.39
Totals	100.00	100.00	82.60	17.40	100.00		

being collected and hand sorted by density (colour). It is then a simple matter to construct the Tromp curve directly by noting the proportion of each density of tracer reporting to either the sink or float product. Application of tracer methods has shown that considerable uncertainties can exist in experimentally determined Tromp curves unless an adequate number of tracers is used, and Napier-Munn[24] presents graphs that facilitate the selection of sample size and the calculation of confidence limits.

Partition curves can be used to predict the products that would be obtained if the feed or separation gravity were changed. The curves are *specific to the vessel for which they were established* and are not affected by the type of material fed to it, provided:

(a) The feed size range is the same—efficiency generally decreases with decrease in size; Fig. 11.17 show typical efficiencies of bath (drum, cone, etc.) and centrifugal (DSM, DWP, etc.) separators versus particle size. It can clearly be seen that, in general, below about 10 mm, centrifugal separators are better than baths;

(b) the separating gravity is in approximately the same range—the higher the effective separation density the greater the probable error, due to the increased medium viscosity. It has in fact been shown that the *Ep* is directly proportional to the separating density, all other factors being the same;[25]

(c) the feed rate is the same.

The partition curve for a vessel can be used to determine the amount of misplaced material which will report to the products for any particular feed material. For example, the distribution of the products from the tin ore, which was evaluated by heavy liquid tests (Table 11.1), can be determined for treatment in an operating separator. Figure 11.18 shows a partition curve for a separator having an *Ep* of 0.07

The curve can be shifted slightly along the abscissa until the effective density of separation corresponds to the laboratory evaluated separating density of 2.75. The distribution of material to sinks and floats can now be evaluated, e.g. at a nominal specific gravity of 2.725, 44.0% of the material reports to the sinks and 56.0% to the floats.

The performance is evaluated in Table 11.5. Columns 1, 2 and 3 show the results of the heavy liquid tests, which were tabulated in

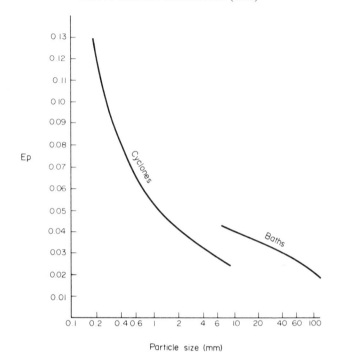

Fɪɢ. 11.17. Effect of particle size on efficiency of heavy media separators.

Table 11.1. Columns 4 and 5 are the distributions to sinks and floats of material at each of the nominal specific gravities. Columns 6 and 9 show the weights of each fraction of the ore reporting to sinks and floats respectively [column 6 = column 1 × column 4]. The assay of each fraction is assumed to be the same whether or not the material reports to sinks or floats.

The total distribution of the feed to sinks is the sum of all the fractions in column 6, i.e. 40.26%, while the recovery of tin into the sinks is the sum of the fractions in column 8, i.e. 95.29%. This compares with a distribution of 31.52% and a recovery of 96.19% of tin in the ideal separation.

This method of evaluating the performance of a separator on a particular feed is tedious, and is ideal for a computerised approach,

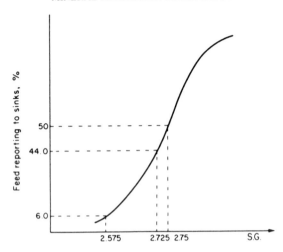

FIG. 11.18. Partition curve for $Ep = 0.07$.

providing that the equation for the partition curve is known. There are numerous functional representations for 'S' shaped curves that have been used in representing partition curves. Loveday[26] assumed that the quantity of misplaced material decreases exponentially with increasing difference in specific gravity from the separating density. The model implies that the efficiency curve is symmetrical, such that if the specific gravity of the component (x) is less than the separating density (x'), then:

$$y(x) = 100 - 50 \, \exp[(x - x')/z] \qquad (11.1)$$

where $y(x)$ is recovery of component to the floats fraction (%) and z = constant.

When $x > x'$:

$$y(x) = 50 \, \exp[(x' - x)/z] \qquad (11.2)$$

and when $x = x'$:

$$y(x) = 50. \qquad (11.3)$$

Equations 11.1–3 are useful for predicting the performance of a vessel

TABLE 11.5. Tin Ore Evaluation

Specific gravity fraction	Nominal sp. gr.	Feed			Distribution to (%)		Sinks			Floats		
		(1) Wt. %	(2) % Sn	(3) % Dist.	(4) Sinks	(5) Floats	(6) Wt. %	(7) % Sn	(8) % (feed) distribution	(9) Wt. %	(10) % Sn	(11) % (feed) distribution
−2.55	—	1.57	0.003	0.04	0.00	100.00	0.00	0.003	0.00	1.57	0.003	0.04
2.55–2.60	2.575	9.22	0.04	0.33	6.0	94.00	0.55	0.04	0.02	8.67	0.04	0.31
2.60–2.65	2.625	26.11	0.04	0.93	13.5	86.5	3.52	0.04	0.13	22.59	0.04	0.80
2.65–2.70	2.675	19.67	0.04	0.70	27.0	73.0	5.31	0.04	0.19	14.35	0.04	0.51
2.70–2.75	2.725	11.91	0.17	1.81	44.0	56.0	5.24	0.17	0.80	6.67	0.17	1.01
2.75–2.80	2.775	10.92	0.34	3.32	63.0	37.0	6.88	0.34	2.09	4.04	0.34	1.23
2.80–2.85	2.825	7.87	0.37	2.60	79.5	20.5	6.26	0.37	2.07	1.61	0.37	0.53
2.85–2.90	2.875	2.55	1.30	2.96	90.5	9.5	2.32	1.30	2.68	0.24	1.30	0.28
+2.90	—	10.18	9.60	87.31	100.00	0.00	10.18	9.60	87.31	0.00	9.60	0.00
Totals		100.00	1.12	100.00			40.26	2.65	95.29	59.74	0.09	4.71

of known Ep, since by definition, when $y(x) = 75\%$, $x' - x = Ep$, such that from equation 11.1:

$$z = -Ep/\ln 0.5. \tag{11.4}$$

Computer programs (GRAV and YIELDOP) for predicting the performance of heavy media circuits, and which utilise equations 11.1–4, can be found in Appendix 3.

A computer program developed by Wizzard et al.[27] uses separation data to calculate the points of the partition curve, and fits an exponential type equation to the curve, resulting in a mathematical expression that is used to determine various performance criteria. The mathematical equation is known as a *Weibull* function,[25,28] and is expressed as

$$y(x) = 100 \, (y_0 + a \exp[-(x - x_0)]^{b/c}) \tag{11.5}$$

where x_0, y_0, a, b and c are constants whose values must be determined for each partition curve.

The five constants are not independent, however, as when $y = 50\%$, $x = x'$ (the separating density). Thus:

$$x_0 = x' - [c \ln\{a/(0.5 - y_0)\}]^{1/b} \tag{11.6}$$

Values of x_0, y_0, a, b and c can be determined from the partition data and from equations 11.5 and 11.6 by means of an appropriate non-linear regression technique.[27] The Weibull equation is very flexible in that it can represent curves of many different shapes, including those which are asymmetrical about the specific gravity of separation.

Jowett[29] has also analysed the mathematical form of partition curves in order to help clarify procedures to be used in computer prediction of separation results. As an aid to improving understanding of the curves, the binomial expansion has been used to simulate the probable distribution of components of different densities arising from stratification in gravity devices. It is shown that this model leads to partition curves which give a near-linear plot on arithmetic probability paper, i.e. they have normal distribution form.

Organic Efficiency

The term organic efficiency is often used to express the efficiency of coal preparation plants. It is defined as the ratio (normally expressed as a percentage) between the actual yield of a desired product and the theoretical possible yield at the same ash content.

For instance, if the coal, whose washability data is plotted in Fig. 11.15, produced an operating yield of 51% at an ash content of 12%, then, since the theoretical yield at this ash content is 55%, the organic efficiency is equal to 51/55, or 92.7%.

Organic efficiency cannot be used to compare the efficiencies of different plants, as it is a dependent criterion, and is much influenced by the washability of the coal. It is possible, for example, to obtain a high organic efficiency on a coal containing little near-gravity material, even when the separating efficiency, as measured by partition data, is quite inefficient.

References

1. Anon., Sodium metatungstate, a new medium for binary and ternary density gradient centrifugation, *Makromol. Chem.*, **185**, 1429 (1984).
2. Walker, M. S., *et al.*, A new method for the commercial separation of particles of differing densities using magnetic fluids — the MC process, *Proc. XVth Int. Min. Proc Cong.*, **1**, 307, Cannes (1985).
3. Parsonage, P., Small-scale separation of minerals by use of paramagnetic liquids, *Trans. IMM*, Sec. B, **86**, 43 (Feb. 1977).
4. Parsonage, P., Factors that influence performance of pilot-plant paramagnetic liquid separator for dense particle fractionation, *Trans. IMM*, Sec. C, **89**, 166 (Dec. 1980).
5. Hill, N. W., A look at the dense medium process, *Mine & Quarry* **1**, 47 (April 1972).
6. Hunt, M. S., *et al.*, The influence of ferrosilicon properties on dense medium separation plant consumption, *Bull. Proc. Australas. Inst. Min. Metall.*, **291**, 73 (Oct. 1986).
7. Ferrara, G., and Schena, G. D., Influence of contamination and type of ferrosilicon on viscosity and stability of dense media, *Trans. Inst. Min. Metall.*, **95**, C211 (Dec. 1986).
8. Ruff, H., and Turner, J. F., Coal preparation and mineral processing—different approaches to common problems, *J. Camb. Sch. Mines* **76**, 30 (1976).
9. Anon., South African coal washer, *Colliery Engng.* 30 (Jan. 1961).
10. Anon., Feeding to zero: Island Creek's experience in Kentucky, *Coal Age*, **90**, 66 (Jan. 1985).

11. Ruff, H. J., New developments in dynamic dense-medium systems, *Mine & Quarry*, **13**, 24 (Dec. 1984).
12. King, R. P., *et al.*, Prediction of dense medium cyclone performance for beneficiation of fine coal, *Proc. XVth Int. Min. Proc Cong.*, **1**, 258, Cannes (1985).
13. Rao, T. C., *et al.*, Modelling of dense-medium cyclones treating coal, *Int. J. Min. Proc.*, **17**, 287 (July 1986).
14. Abbott, J., Bateman, K. W., and Shaw, S. R., The Vorsyl separator, *9th Commonwealth Mining and Metall. Cong.*, paper 33 (1969).
15. Shaw, S. R., The Vorsyl dense-medium separator: some recent developments, *Mine & Quarry* **13**, 28 (April 1984).
16. Jenkinson, D. E., Coal preparation into the 1990's, *Colliery Guardian*, **233**, 301 (July 1985).
17. Wills, B. A., and Lewis, P. A., Applications of the Dyna Whirlpool in the minerals industry, *Mining Mag.* 255 (Sept. 1980).
18. Hacioglu, E., and Turner, J. F., A study of the Dynawhirlpool, *Proc. XVth Int. Min. Proc Cong.*, **1**, 244, Cannes (1985).
19. Ruff, H. J., Operating experience with the Tri-Flo dense medium separator, *Trans. Ins. Min. Metall.*, **95**, C225 (Dec. 1986).
20. Munro, P. D., *et al.*, The design, construction and commissioning of a heavy medium plant for silver–lead–zinc ore treatment—Mount Isa Mines Ltd, *Proc. XIVth Int. Min. Proc. Cong.*, paper VI–6, Toronto (Oct. 1982).
21. Yakubu, N. A., and Wills, B. A., Heavy liquid testing at fine particle size, *J. Camb. Sch. Mines* **79**, 66 (1979).
22. Ignjatovic, R., Gravity concentration of porous magnesite fines, *Proceedings of the 10th International Mining Congress*, paper 18, London, 1973.
23. Mills, C., Process design, scale-up and plant design for gravity concentration, in *Mineral Processing Plant Design* (ed. Mular and Bhappu), AIMME, New York (1978).
24. Napier-Munn, T. J., Use of density tracers for determination of the Tromp curve for gravity separation processes, *Trans. Inst. Min. Metall.*, **94**, C45 (March 1985).
25. Gottfried, B. S., A generalisation of distribution data for characterizing the performance of float-sink coal cleaning devices, *Int. J. Min. Proc.* **5**, 1 (1978).
26. Loveday, B. K., Prediction of gravity or inertial separation efficiencies from heavy-liquid tests and Tromp-type efficiency curves: an improved method, *Trans. I.M.M. Sec. C*, **79**, 137 (June 1970).
27. Wizzard, J. T., Killmeyer, R. P., and Gottfried, B. S., Computer program for evaluating coal washer performance, *Min. Engng.* **35**, 252 (March 1983).
28. Fallon, N. E., and Gottfried, B. S., Statistical representation of generalized distribution data for float-sink coal-cleaning devices: Baum jigs, Batac jigs, Dynawhirlpools, *Int. J. Min. Proc.*, **15**, 231 (Oct. 1985).
29. Jowett, A., An appraisal of partition curves for coal-cleaning processes, *Int. J. Min. Proc.*, **16**, 75 (Jan. 1986).

CHAPTER 12

FROTH FLOTATION

Introduction

Flotation is undoubtedly the most important and versatile mineral-processing technique, and both use and application are being expanded to treat greater tonnages and to cover new areas.

Originally patented in 1906, flotation has permitted the mining of low-grade and complex ore bodies which would have otherwise been regarded as uneconomic. In earlier practice the tailings of many gravity plants were of a higher grade than ore treated in many modern flotation plants.

Flotation is a selective process and can be used to achieve specific separations from complex ores such as lead–zinc, copper–zinc, etc. Initially developed to treat the sulphides of copper, lead, and zinc, the field of flotation has now expanded to include oxides, such as hematite and cassiterite, oxidised minerals, such as malachite and cerussite, and non-metallic ores, such as fluorite, phosphates, and fine coal.

Principles of Flotation

The theory of froth flotation is complex and is not completely understood. The subject has been reviewed comprehensively by a number of authors,[1-8] and will only be dealt with briefly here.

Froth flotation utilises the differences in physico-chemical surface properties of particles of various minerals. After treatment with reagents, such differences in surface properties between the minerals within the flotation pulp become apparent and, for flotation to take place, an air-bubble must be able to attach itself to a particle, and lift it

457

to the water surface (Fig. 12.1). The process can only be applied to relatively fine particles, as if they are too large the adhesion between the particle and the bubble will be less than the particle weight and the bubble will therefore drop its load.

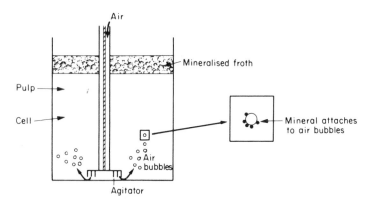

FIG. 12.1. Principle of froth flotation.

In flotation concentration, the mineral is usually transferred to the froth, or float fraction, leaving the gangue in the pulp or tailing. This is *direct flotation* as opposed to *reverse flotation*, in which the gangue is separated into the float fraction.

The air-bubbles can only stick to the mineral particles if they can displace water from the mineral surface, which can only happen if the mineral is to some extent water repellent or *hydrophobic*. Having reached the surface, the air bubbles can only continue to support the mineral particles if they can form a stable froth, otherwise they will burst and drop the mineral particles. To achieve these conditions it is necessary to use the numerous chemical reagents known as flotation reagents.[9,10]

The activity of a mineral surface in relation to flotation reagents in water depends on the forces which operate on that surface. The forces tending to separate a particle and a bubble are shown in Fig. 12.2. The tensile forces lead to the development of an angle between the mineral surface and the bubble surface. At equilibrium,

$$\gamma_{s/A} = \gamma_{s/w} + \gamma_{w/A} \cos \theta, \qquad (12.1)$$

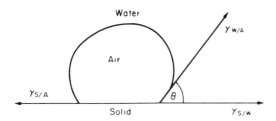

FIG. 12.2. Contact angle between bubble and particle in an aqueous medium.

where $\gamma_{s/A}$, $\gamma_{s/w}$ and $\gamma_{w/A}$ are the surface energies between solid–air, solid–water and water–air, respectively, and θ is the *contact angle* between the mineral surface and the bubble.

The force required to break the particle–bubble interface is called the *work of adhesion*, $W_{s/A}$, and is equal to the work required to separate the solid–air interface and produce separate air–water and solid–water interfaces, i.e.

$$W_{s/A} = \gamma_{w/A} + \gamma_{s/w} - \gamma_{s/A}. \tag{12.2}$$

Combining with equation 12.1 gives

$$W_{s/A} = \gamma_{w/A} (1 - \cos \theta). \tag{12.3}$$

It can be seen that the greater the contact angle the greater is the work of adhesion between particle and bubble and the more resilient the system is to disruptive forces. The *floatability* of a mineral therefore increases with the contact angle; minerals with a high contact angle are said to be *aerophilic*, i.e. they have a higher affinity for air than for water. Most minerals are not water repellent in their natural state and flotation reagents must be added to the pulp. The most important reagents are the *collectors*, which adsorb on mineral surfaces, rendering them hydrophobic (or aerophilic) and facilitating bubble attachment. The *frothers* help maintain a reasonably stable froth. *Regulators* are used to control the flotation process; these either activate or depress mineral attachment to air-bubbles and are also used to control the pH of the system.

Collectors

All minerals are classified into *non-polar* or *polar* types according to their surface characteristics. The surfaces of non-polar minerals are characterised by relatively weak molecular bonds. The minerals are composed of covalent molecules held together by van der Waals forces, and the non-polar surfaces do not readily attach to the water dipoles, and in consequence are hydrophobic. Minerals of this type, such as graphite, sulphur, molybdenite, diamond, coal, and talc, thus have high natural floatabilities with contact angles between 60° and 90°. Although it is possible to float these minerals without the aid of chemical agents, it is universal to increase their hydrophobicity by the addition of hydrocarbon oils or frothing agents. Creosote, for example, is widely used to increase the floatability of coal. Use is made of the natural hydrophobicity of diamond in *grease tabling*, a classical method of diamond recovery which is still used in some plants. The pre-concentrated diamond ore slurry is passed over inclined vibrating tables, which are covered in a thick layer of petroleum grease. The diamonds become embedded in the grease because of their water repellency, while the water-wetted gangue particles are washed off the table. The grease is skimmed off the table either periodically or continuously, and placed in perforated pots (Fig. 12.3), which are immersed in boiling water. The grease melts, and runs out through the perforations, and is collected and re-used, while the pot containing the diamonds is transported to the diamond-sorting section.

Minerals with strong covalent or ionic surface bonding are known as polar types, and exhibit high free energy values at the polar surface. The polar surfaces react strongly with water molecules, and these minerals are naturally hydrophilic.

The polar group of minerals have been subdivided into various classes depending on the magnitude of polarity,[11] which increases from groups 1 to 5 (Table 12.1). Minerals in group 3 have similar degrees of polarity, but those in group 3(a) can be rendered hydrophobic by sulphidisation of the mineral surface in an alkaline aqueous medium. Apart from the native metals, the minerals in group 1 are all sulphides, which are only weakly polar due to their covalent bonding, which is relatively weak compared to the ionic bonding of the

Fig. 12.3. Diamond recovery grease table.

carbonate and sulphate minerals. In general, therefore, the degree of polarity increases from sulphide minerals, through sulphates, to carbonates, halites, phosphates, etc., then oxides–hydroxides, and, finally, silicates and quartz.

Hydrophobicity has to be imparted to most minerals in order to float them. In order to achieve this, surfactants known as collectors are added to the pulp and time is allowed for adsorption during agitation in what is known as the conditioning period. Collectors are organic compounds which render selected minerals water-repellent by

TABLE 12.1. CLASSIFICATION OF POLAR MINERALS

Group 1	Group 2	Group 3(a)	Group 4	Group 5
Galena	Barite	Cerrusite	Hematite	Zircon
Covellite	Anhydrite	Malachite	Magnetite	Willemite
Bornite	Gypsum	Azurite	Gothite	Hemimorphite
Chalcocite	Anglesite	Wulfenite	Chromite	Beryl
Chalcopyrite			Ilmenite	Feldspar
Stibnite		Group 3(b)	Corundum	Sillimanite
Argentite		Fluorite	Pyrolusite	Garnet
Bismuthinite		Calcite	Limonite	Quartz
Millerite		Witherite	Borax	
Cobaltite		Magnesite	Wolframite	
Arsenopyrite		Dolomite	Columbite	
Pyrite		Apatite	Tantalite	
Sphalerite		Scheelite	Rutile	
Orpiment		Smithsonite	Cassiterite	
Pentlandite		Rhodochrosite		
Realgar		Siderite		
Native Au, Pt, Ag, Cu		Monazite		

adsorption of molecules or ions on to the mineral surface, reducing the stability of the hydrated layer separating the mineral surface from the air-bubble to such a level that attachment of the particle to the bubble can be made on contact.

Collector molecules may be ionising compounds, which dissociate into ions in water, or non-ionising compounds, which are practically insoluble, and render the mineral water-repellent by covering its surface with a thin film.

Ionising collectors have found very wide application in flotation. They have complex molecules which are asymmetric in structure and are *heteropolar*, i.e. the molecule contains a non-polar hydrocarbon group and a polar group which may be one of a number of types. The non-polar hydrocarbon radical has pronounced water-repellent properties, whereas the polar group reacts with water.

Ionising collectors are classed in accordance with the type of ion, anion or cation that produces the water-repellent effect in water. This classification is given in Fig. 12.4.

The structure of sodium oleate, an anionic collector in which the hydrocarbon radical, which does not react with water, constitutes the non-polar part of the molecule, is shown in Fig. 12.5.

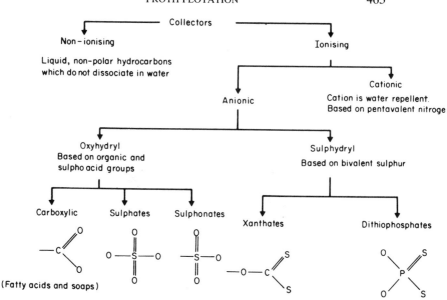

FIG. 12.4. Classification of collectors (after Glembotskii *et al.*[2]).

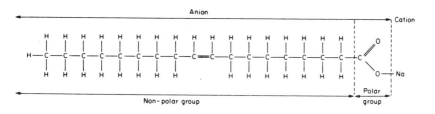

FIG. 12.5. Structure of sodium oleate.

Amphoteric collectors possess a cationic and an anionic function, depending on the working pH, and have been used to treat sedimentary phosphate deposits,[12] and to improve the selectivity of cassiterite flotation.[13]

Because of chemical, electrical, or physical attraction between the polar portions and surface sites, the collectors adsorb on the particles with their non-polar ends orientated towards the bulk solution, thereby imparting hydrophobicity to the particles (Fig. 12.6).

Generally speaking, collectors are used in small amounts, substantially those necessary to form a monomolecular layer on particle surfaces (starvation level), as increased concentration, apart from the cost, tends to float other minerals, reducing selectivity. It is always harder to eliminate a collector already adsorbed than to prevent its adsorption.

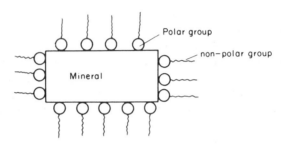

FIG. 12.6. Collector adsorption on mineral surface.

An excessive concentration of a collector can also have an adverse effect on the recovery of the valuable minerals, possibly due to the development of collector multi-layers on the particles, reducing the proportion of hydrocarbon radicals oriented into the bulk solution. The hydrophobicity of the particles is thus reduced, and hence their floatability. The flotation limit may be extended without loss of selectivity by using a collector with a longer hydrocarbon chain, thus producing greater water-repulsion, rather than by increasing the concentration of a shorter chain collector. However, chain length is usually limited to two to five carbon atoms, since the solubility of the collector in water rapidly diminishes with increasing chain length and, although there is a corresponding decrease in solubility of the collector products, which therefore adsorb very readily on the mineral surfaces, it is, of course, necessary for the collector to ionise in water for chemisorption to take place on the mineral surfaces. Not only the chain length, but also the chain structure, affects solubility; branched chains have higher solubility than straight chains.

It is common to add more than one collector to a flotation system. A selective collector may be used at the head of the circuit, to float the highly hydrophobic minerals, after which a more powerful, but less selective one is added to promote recovery of the slower floating minerals.

Anionic Collectors

These are the most widely used collectors in mineral flotation and may be classified into two types according to the structure of the polar group (Fig. 12.4). *Oxyhydryl* collectors have organic and sulpho-acid anions as their polar groups and, as with all anionic collectors, the cation takes no significant part in the reagent–mineral reaction.

Typically, oxyhydryl collectors are organic acids, or soaps. The carboxylates are known as fatty acids, and occur naturally in vegetable oils and animal fats from which they are extracted by distillation and crystallisation. The salts of oleic acid, such as sodium oleate (Fig. 12.5) and linoleic acid, are commonly used.[3] As with all ionic collectors, the longer the hydrocarbon chain length, the more powerful is the water-repulsion produced, but solubility decreases. Soaps (the salts of fatty acids), however, are soluble even if the chain length is long. The carboxylates are strong collectors, but have relatively low selectivity. They are used for the flotation of minerals of calcium, barium, strontium, and magnesium, the carbonates of non-ferrous metals, and the soluble salts of alkali metals and alkaline earth metals.

The sulphates and sulphonates are used more rarely. They possess similar properties to fatty acids, but have lower collecting power. However, they have greater selectivity and are used for floating barite,[14,15] kyanite, and scheelite.

The oxyhydryl collectors have been used to float cassiterite, but have now been largely replaced by other reagents such as arsonic and phosphonic acids and sulphosuccinamates.[3,13,16,17]

The most widely used collectors are of the *sulphydryl* type where the polar group contains bivalent sulphur (*thio* compounds). They are very powerful and selective in the flotation of sulphide minerals. The *mercaptans* (thiols) are the simplest of thio compounds, having the general formula RS^-Na (or $K)^+$, where R is the hydrocarbon group. They have been used as selective collectors for some of the more refractory sulphide minerals.[18] The most widely used thiol collectors are the *xanthogenates* (technically known as the *xanthates*) and the *dithiophosphates* (Aerofloat collectors). The xanthates are the most important for sulphide mineral flotation. They are prepared by reacting an alkali hydroxide, an alcohol and carbon disulphide:

$$ROH + CS_2 + KOH = RO.CS.SK + H_2O, \qquad (12.4)$$

where R is the hydrocarbon group and contains normally one to six carbon atoms, the most widely used xanthates being ethyl, isopropyl, isobutyl, amyl, and hexyl. Sodium ethyl xanthate is typical and has the structure shown in Fig. 12.7. The anion consists of a hydrocarbon non-polar radical and a connected polar group. Although the cation (sodium or potassium) plays no part in the reactions leading to mineral hydrophobicity, it has been shown[19] that the sodium form of the alkyl xanthates decreases in efficacy with age, probably due to water absorption from the atmosphere, whereas the potassium salts are not affected by this problem.

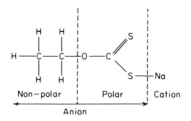

FIG. 12.7. Structure of sodium ethyl xanthate.

The dithiophosphates have pentavalent phosphorus in the polar group, rather than tetravalent carbon (Fig. 12.8).

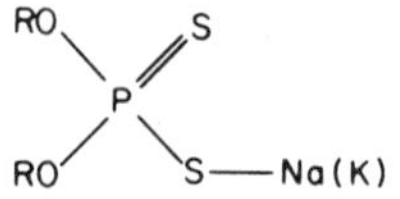

FIG. 12.8. Dithiophosphates.

The reaction between sulphide minerals and sulphydryl collectors is complex. Xanthates are assumed to adsorb on sulphide mineral surfaces due to chemical forces between the polar group and the surface, resulting in insoluble metal xanthates, which are strongly hydrophobic. Mechanisms involving the formation and adsorption of dixanthogen, xanthic acid, etc., have also been proposed. It has been established that the sulphide is not joined to the collector anions without the previous action of oxygen. The solubilities of sulphide minerals in water are very low, suggesting that sulphides should be relatively inert in aqueous solution. However, they are thermodynamically unstable in the presence of oxygen, and surface oxidation to S^{2-}, $S_2O_3^{2-}$, and SO_4^{2-} can occur, depending on the E_h–pH conditions. Figure 12.9 shows the E_h–pH (Pourbaix) diagram for galena. At cathodic potential, the surface of galena is converted to lead, and sulphide ions pass into solution. Under anodic conditions (i.e. when cathodic reduction of oxygen occurs, e.g.

$$\tfrac{1}{2}O_2 + H_2O + 2e \rightarrow 2OH^-)$$

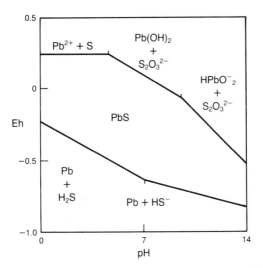

FIG. 12.9. E_h–pH diagram for galena (equilibrium lines correspond to dissolved species at 10^{-4}M) (after Woods[20]).

lead will dissolve or form oxidised metal species on the surface, depending on pH. The initial oxidation of sulphide leads to the formation of elemental sulphur, e.g. in acid solution:

$$MS \rightarrow M^{2+} + S^0 + 2e \qquad (12.5)$$

with its equivalent in neutral or alkaline solution:

$$MS + 2H_2O \rightarrow M(OH)_2 + S^0 + 2H^+ + 2e. \qquad (12.6)$$

The presence of elemental sulphur in the mineral surface can lead to hydrophobicity, and the mineral may be floated in the absence of collectors, although control of these redox conditions is difficult in practice. Usually the cathodic reduction of oxygen is strong enough to provide a sufficient electron sink for oxidation of the sulphide mineral surface to oxy-sulphur species, which are not hydrophobic, and so collectors are required to promote flotation. The oxidation products are more soluble than the sulphides, and the reaction of xanthates and other thiol collectors with these products by an ion-exchange process is the major adsorption mechanism for the flotation of sulphides.[21] For instance, if the sulphide surface oxidises to thiosulphate, the following reactions can occur:

$$2MS + 2O_2 + H_2O \rightarrow MS_2O_3 + M(OH)_2$$

and $$MS_2O_3 + 2ROCS_2^- \rightarrow M(ROCS_2)_2 + S_2O_3^{2-}$$

or

$$2MS + 4ROCS_2^- + 3H_2O \rightarrow 2M(ROCS_2)_2 + S_2O_3^{2-} + 6H^+ + 8e$$
$$\text{(anodic reaction)} \qquad (12.7)$$

The insoluble metal xanthate so formed renders the mineral surface hydrophobic. However, strong oxidising conditions can lead to the formation of sulphates, e.g.:

$$MS + 2ROCS_2^- + 4H_2O \rightarrow M(ROCS_2)_2 + SO_4^{2-} + 8H^+ + 8e$$
$$(12.8)$$

Although sulphates react strongly with xanthates, they are relatively soluble in aqueous solution, and so do not form stable hydrophobic surface products, the metal xanthates formed tending to scale off the mineral.

The solubilities of the hydrophobic xanthates of copper, lead, silver, and mercury are very low, but the xanthates of zinc and iron are much more soluble. Typically, ethyl xanthates are only weak collectors of pure sphalerite, but replacement of the crystal lattice zinc atoms by copper improves the flotation properties of the mineral. The alkali earth metal xanthates (calcium, barium, magnesium) are very soluble and xanthates have no collector action on the minerals of such metals, nor on oxides, silicates, or aluminosilicates, which permits extremely selective flotation of sulphides from gangue minerals.

Xanthates are used as collectors for oxidised minerals such as malachite, cerussite and anglesite, and for native metals such as gold, silver and copper. Comparatively high concentrations are necessary with the oxidised minerals, and often higher xanthates such as amyl are preferred.

Xanthates and similar compounds tend to oxidise fairly easily, which can lead to complications in flotation. After a few months storage, they develop a strong odour and a deeper colour due to the formation of "dixanthogen", e.g. with potassium ethyl xanthate:

$$2\left[C_2H_5-O-\underset{\underset{S-K}{|}}{C}=S \right] + \tfrac{1}{2}O_2 + CO_2 \rightarrow S = \underset{\underset{\underset{\underset{S = C-O-C_2H_5}{|}}{S}}{|}}{C}-O-C_2H_5 + K_2CO_3 \tag{12.9}$$

Dixanthogens and similar products of oxidation are themselves collectors,[22-24] and their formation can lead to loss of selectivity and control in complex flotation circuits.

Xanthates also form insoluble metal salts with ions of copper, lead, and other heavy metals which may be present in the slurry, which reduces the effectiveness of the collector. By using alkaline conditions, preferably as early as the grinding circuit, these heavy metal ions can be precipitated as relatively insoluble hydroxides. Alkaline conditions also inhibit xanthate breakdown, which proceeds more rapidly as the pH is lowered:

$$H^+ + ROCS^{2-} \rightleftharpoons HX \rightarrow ROH + CS_2. \tag{12.10}$$

With xanthic acid (HX) and xanthate ions in equilibrium, the unstable xanthic acid decomposes to alcohol and carbon disulphide.

Dithiophosphates are not as widely used as the xanthates, but are still important reagents in practice. They are comparatively weak collectors, but give good results in combination with xanthates. They are often used in the separation of copper from lead sulphides, as they are effective selective collectors for copper sulphide minerals.

It appears that the water repulsion imparted to the mineral surface is due to the formation of an oxidation product of the dithiophosphate collector which adsorbs on to the mineral surface. Thus, as with xanthates, the presence of oxygen, or another oxidising agent, is essential for flotation. Strong oxidising conditions destroy the hydrophobic substances and are thus undesirable, while oxidisation of the mineral surface itself may impede collector adsorption.

Various reviews of the interaction between xanthates, dithiophosphates and other thiol collectors with sulphide mineral surfaces have been made [25-27] and a list of the common thio collectors is given in Table 12.2, including references to papers providing more detail on these extremely important reagents.

Cationic Collectors

The characteristic property of this group of collectors is that the water repulsion is produced by the cation where the polar group is based on pentavalent nitrogen, the amines being the most common (Fig 12.10). The anions of such collectors are usually halides, or more rarely hydroxides, which take no active part in the reaction with minerals.

Unlike the xanthates, the amines are considered to adsorb on mineral surfaces primarily due to electrostatic attraction between the polar head of the collector and the charged electrical double layer on the mineral surface.[39] Such forces are not as strong or irreversible as the chemical forces characteristic of anionic collectors, so these collectors tend to be relatively weak in collecting power.

Cationic collectors are very sensitive to the pH of the medium, being most active in slightly acid solutions and inactive in strongly alkaline

TABLE 12.2. COMMON THIOL COLLECTORS AND THEIR USES

Reagent	Formula	pH range	Main uses	References
o-alkyl dithiocarbonates (Xanthates)	$R-O-\overset{\displaystyle S}{\overset{\|}{C}}-S^--K^+$ (or Na$^+$)	8–13	Flotation of sulphides, oxidised minerals such as malachite, cerussite, and elemental metals	3, 28–30
Dialkyl dithiophosphates (Aerofloats)	$R-O-\overset{\displaystyle S}{\overset{\|}{\underset{\|}{P}}}-S^--K^+$ (or Na$^+$)	4–12	Selective flotation of copper and zinc sulphides from galena	31–33
Dialkyl dithiocarbamate	$\overset{R}{\underset{R}{>}}N-\overset{\displaystyle S}{\overset{\|}{C}}-S^--K^+$ (or Na$^+$)	5–12	Similar properties to xanthates, but more expensive	33–35
Isopropyl thionocarbamate (Minerec 1661/ Z-200)	$(CH_3)_2CH-O-\overset{\displaystyle S}{\overset{\|}{C}}-N\overset{H}{\underset{C_2H_5}{<}}$	4–9	Selective flotation of copper sulphides from pyrite	33, 36, 37
Mercaptobenzothiazole (R404/425)	(benzothiazole ring) $\overset{N}{\underset{S}{>}}C-S-Na^+$	4–9	Flotation of tarnished or oxidised lead and copper minerals. Floats pyrite at pH 4–5	33, 38

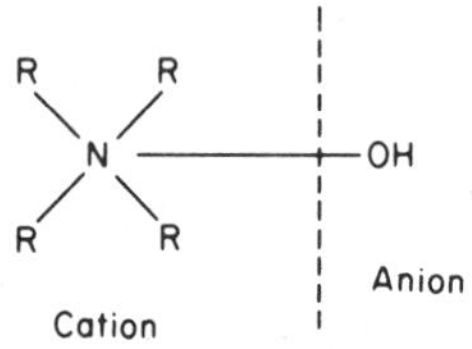

FIG. 12.10. Cationic amine collector.

and acid media. They are used for floating oxides, carbonates, silicates and alkali earth metals such as barite, carnallite, and sylvite. The primary amines (i.e. those where only one hydrocarbon group is present with two hydrogen atoms) are strong collectors of apatite and they can selectively float sedimentary phosphates from calcareous ores. The collector requirement can be reduced by adding a non-polar reagent such as kerosene, which is co-adsorbed on the mineral surface. Since the zeta potentials of both apatite and dolomite are negative in the relevant pH range, the selective flotation of the phosphate may not be interpreted solely by the electrostatic model of adsorption, and experimental evidence for chemical interaction has been presented.[40]

Frothers

When mineral surfaces have been rendered hydrophobic by the use of a collector, stability of bubble attachment, especially at the pulp surface, depends to a considerable extent on the efficiency of the frother.

Ideally the frother acts entirely in the liquid phase and does not influence the state of the mineral surface. In practice, however, interaction does occur between the frother, mineral, and other reagents, and the selection of a suitable frother for a given ore can only be made after extensive test work.

In sulphide mineral flotation it is common practice to employ at least two frothers and more than one collector. Specific frothers are chosen to provide adequate physical properties to the froth, while the second frother interacts with the collectors to control the dynamics of the flotation process.[41]

Frothers are in many respects chemically similar to ionic collectors, and, indeed, many of the collectors, such as the oleates, are powerful frothers, being, in fact, too powerful to be used as efficient frothers, since the froths which they produce can be too stable to allow efficient transport to further processing. Froth build-up on the surfaces of thickeners, and excessive frothing of flotation cells, are problems occurring in many mineral processing plants. A good frother should have negligible collecting power, and also produce a froth which is just stable enough to facilitate transfer of floated mineral from the cell surface to the collecting launder.

Frothers are generally heteropolar surface-active organic reagents, capable of being adsorbed on the air–water interface. When surface-active molecules react with water, the water dipoles combine readily with the polar groups and hydrate them, but there is practically no reaction with the non-polar hydrocarbon group, the tendency being to force the latter into the air phase. Thus the heteropolar structure of the frother molecule leads to its adsorption, i.e. the molecules concentrate in the surface layer with the non-polar groups oriented towards the air and the polar groups towards the water (Fig. 12.11).

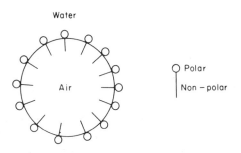

FIG. 12.11. Action of the frother.

Frothing action is thus due to the ability of the frother to adsorb on the air–water interface because of its surface activity and to reduce the surface tension, thus stabilising the air-bubble.

Frothers must be to some extent soluble in water, otherwise they would be distributed very unevenly in an aqueous solution and their surface-active properties would not be fully effective. The most effective frothers include in their composition one of the following groups:

Hydroxyl   —OH

Carboxyl $-C\overset{\displaystyle O}{\underset{\displaystyle OH}{\big\langle}}$

Carbonyl $= C = O$

Amino group $-NH_2$

Sulpho group $-OSO_2OH, -SO_2OH$

The acids, amines, and alcohols are the most soluble of the frothers. The alcohols (—OH) are the most widely used, since they have practically no collector properties, and in this respect are preferable to other frothers, such as the carboxyls, which are also powerful collectors; the presence of collecting and frothing properties in the same reagent may make selective flotation difficult. Frothers with an amino group, and certain sulpho group frothers, also have weak collector properties.

Pine oil, which contains aromatic alcohols, the most active frothing component being terpineol, $C_{10}H_{17}OH$, has been widely used as a frother. Cresol (cresylic acid), $CH_3C_6H_4OH$, has also had wide use.

A wide range of synthetic frothers, based mainly on high molecular-weight alcohols, are now in use in many plants. They have the important advantage over industrial products such as pine oil and cresol in that their compositions are much more stable, which makes it easier to control the flotation process and improves performance. A widely used synthetic alcohol frother is methyl isobutyl carbinol (MIBC). Another range of synthetic frothers are based on polyglycol ethers, and have been found to be very effective. They are marketed under various names, such as Dowfroth 250, Cyanamid R 65 and Union Carbide PG 400.

Although frothers are generally surface-active reagents, it has been shown that surface-inactive reagents, such as diacetone alcohol and ethyl acetal, behave as frothers in solid–liquid–air systems, although not in two-phase liquid–air systems.[42] Molecules of these reagents have two polar groups and are readily soluble in water. They adsorb on solid surfaces but do not appreciably change their hydrophobicity. When the mineral surface, on which the surface inactive frother is adsorbed, is approached by an air-bubble, the molecules reorientate and produce a sufficiently stable three-phase froth. Being surface inactive, these reagents do not reduce surface tension, and apart from the slight reduction due to collectors, the forces available for flotation are maintained at their maximum.

Regulators

Regulators, or modifiers, are used extensively in flotation to modify the action of the collector, either by intensifying or reducing its water-repellent effect on the mineral surface. They thus make collector action more selective towards certain minerals. Regulators can be classed as activators, depressants, or pH modifers.

Activators. These reagents alter the chemical nature of mineral surfaces so that they become hydrophobic due to the action of the collector. Activators are generally soluble salts which ionise in solution, the ions then reacting with the mineral surface.

A classical example is the activation of sphalerite by copper in solution. Sphalerite is not floated satisfactorily by a xanthate collector, since the collector products formed, such as zinc xanthate, are relatively soluble in water, and so do not provide a hydrophobic film around the mineral. Floatability can be improved by the use of large quantities of long-chain xanthates, but a more satisfactory method is to use copper sulphate as an activator, which is readily soluble and dissociates into copper ions in solution. Activation is due to the formation of molecules of copper sulphide at the mineral surface, due to the fact that copper is more electro-negative than zinc and therefore ionises less readily:

$$ZnS + Cu^{2+} \rightleftharpoons CuS + Zn^{2+} \qquad (12.11)$$

The copper sulphide deposited on the sphalerite surface reacts readily with xanthate to form insoluble copper xanthate, which renders the sphalerite surface hydrophobic. Recent work, however, suggests that this simple ion-exchange mechanism may be over-simplified, and an explanation based on galvanic effects is offered.[43]

The main use of copper sulphate as an activator is in the differential flotation of lead-zinc ores, where after lead flotation the sphalerite is activated and floated. To some extent, copper ions can also activate galena, calcite, and pyrite. When sphalerite is associated with pyrite or pyrrhotite, selectivity is usually ensured by the high alkalinity (pH 10.7–12) of the pulp, lime being added in conjunction with the copper sulphate activator.

Oxidised minerals of lead, zinc, and copper, such as cerussite, smithsonite, azurite, and malachite, float very inefficiently with sulphydryl collectors and require an extremely large amount, as heavy metal ions dissolved from the mineral lattice must be precipitated as metal xanthate before the collector interacts with the mineral. Such minerals are activated by the use of sodium sulphide or sodium hydrosulphide.[44,45] Large quantities of up to 10 kg t^{-1} of such "sulphidisers" may be required, due to the relatively high solubilities of the oxidised minerals.

In solution, sodium sulphide hydrolyses and then dissociates:

$$Na_2S + 2H_2O \rightleftharpoons 2NaOH + H_2S \qquad (12.12)$$

$$NaOH \rightleftharpoons Na^+ + OH^- \qquad (12.13)$$

$$H_2S \rightleftharpoons H^+ + HS^- \qquad (12.14)$$

$$HS^- \rightleftharpoons H^+ + S^{2-} \qquad (12.15)$$

Since the dissociation constants of equations 12.14 and 12.15 are extremely low and that of equation 12.13 is high, the concentration of OH^- ions increases at a faster rate than that of H^+ ions and the pulp becomes alkaline. Hydrolysis and dissociation of sodium sulphide release OH^-, S^{2-}, and HS^- ions into solution and these can react with and modify the mineral surfaces. Sulphidation causes sulphur ions to pass into the crystal lattice of the oxidised minerals, giving them a relatively insoluble pseudo-sulphide surface coating and allowing them to be floated by sulphydryl collectors. For example, in the sulphidisation of cerussite, the following reactions take place:[2]

$$Na_2S + H_2O \rightleftharpoons NaHS + NaOH \qquad (12.16)$$

$$PbCO_3 + 3NaOH = H_2O + Na_2CO_3 + NaHPbO_3 \quad (12.17)$$

$$NaHS + NaHPbO_2 = 2NaOH + PbS \qquad (12.18)$$

or

$$Na_2S + PbCO_3 = Na_2CO_3 + PbS \qquad (12.19)$$

The amount of sodium sulphide added to the pulp must be very strictly controlled, as it is a very powerful depressant for sulphide minerals and will, if in excess, depress the activated oxide minerals, preventing collector adsorption. The amount required is dependent on the pulp alkalinity, as an increase in pH causes equations 12.14 and 12.15 to proceed further to the right, producing more HS^- and S^{2-} ions. For this reason sodium hydrosulphide is often preferred to sodium sulphide, as the former does not hydrolyse and hence increase the pH. The amount of sulphidiser added should be sufficient only to produce a coherent film of sulphide on the mineral surface, such that xanthate can be adsorbed. With an increase in sulphidiser beyond that required for activation, concentrations of sulphide and hydrosulphide ions increase. The HS^- ions readily adsorb on the mineral surfaces, giving them a high negative charge, and preventing adsorption of the collector anions. Excess sodium sulphide also removes oxygen from the pulp:

$$Na_2S + 2O_2 = Na_2SO_4 \qquad (12.20)$$

Since oxygen is required in the pulp for the adsorption of sulphydryl collectors on sulphide surfaces, flotation efficiency is reduced.

In the flotation of mixed sulphide-oxidised ores, the sulphide minerals are always floated first, before sulphidisation of the oxidised surfaces. This prevents the depression of sulphides by sodium sulphide and the sulphidiser is subsequently added to the pulp in stages, in starvation levels.

Depressants. Depression is used to increase the selectivity of flotation by rendering certain minerals hydrophilic (water avid), thus preventing their flotation.

There are many types of depressant and their actions are complex and varied, and in most cases little understood, making depression more difficult to control than the application of other types of reagent.

Slime coating is an example of a naturally occurring form of depression. Slimes in a crushed and ground ore make flotation difficult, as they coat the mineral particles, retarding collector adsorption.[46] The particle size at which these effects become significant depends on the flotation system, but in general particles below 20 microns are potentially deleterious, and some form of de-sliming is usually carried out prior to flotation, resulting in an inevitable loss of slime values. Sometimes slimes can be removed from the mineral surfaces by vigorous agitation, or a slime dispersant may be used. Sodium silicate in solution increases the double-layer charge on particles, so that the slime layers which have formed readily disperse. The clean mineral surface can then interact with the collector. In this respect, therefore, sodium silicate is used as an activator, preventing depression by the slimes. Sodium silicate is also used as a depressant in some systems, being one of the most widely used regulating agents in the flotation of non-sulphide minerals, such as scheelite, calcite and fluorite. Sodium oleate is the major collector in the flotation of these minerals, but the selectivity in the separation of scheelite from calcite and fluorite is often inadequate. Sodium silicate has therefore been used to improve selectivity. Shin and Choi[47] have examined the mechanism of adsorption of sodium silicate and its interaction with these minerals.

Cyanides are widely used in the selective flotation of lead-copper-zinc and copper-zinc ores as depressants for sphalerite, pyrite and certain copper sulphides. Sphalerite rejection from copper concentrates is often of major concern, as zinc is a penalty element in copper smelting.

It is fairly well established that pure clean sphalerite does not adsorb short-chain xanthates until its surface is activated by copper ions (equation 12.11). However, copper ions resulting from very slight dissolution of copper minerals present in the ore may cause unintentional activation and prevent selective separation. Cyanide is added to the pulp to desorb the surface copper and to react with copper in solution forming soluble cyanide complexes. Sodium cyanide is most commonly used, which hydrolyses in aqueous solution to form free alkali, and relatively insoluble hydrogen cyanide:

$$NaCN + H_2O \rightleftharpoons HCN + NaOH \qquad (12.21)$$

The hydrogen cyanide then dissociates:

$$HCN \rightleftharpoons H^+ + CN^- \qquad (12.22)$$

The dissociation constant of equation 12.22 is extremely low compared with that of equation 12.21, so that an increase in pulp alkalinity reduces the amount of free HCN, but increases the concentration of CN^- ions. An alkaline pulp is essential, as free hydrogen cyanide is extremely dangerous. The major function of the alkali, however, is to control the concentration of cyanide ions available for dissolution of the copper to cupro-cyanide:

$$3CN^- + Cu^{2+} \rightleftharpoons [Cu(CN)_2]^- + \tfrac{1}{2}C_2N_2 \qquad (12.23)$$

Apart from the reactions of cyanide with metal ions in solution, it can react with metal xanthates to form soluble complexes, preventing xanthate adsorption on the mineral surface, although this cannot occur until the metal ions in solution have been complexed, according to equation 12.23. Hence if Cu^{2+} ions are in solution, the prevention of xanthate adsorption cannot occur unless the ratio of CN^- ions to Cu^{2+} ions is greater than 3:1. The greater the solubility of the metal xanthate in cyanide, the less stable is the attachment of the collector to the mineral. It has been shown that lead xanthates have very low solubilities in cyanide, copper xanthates are fairly soluble, while the xanthates of zinc, nickel, gold, and iron are highly soluble. Iron and zinc can, therefore, be very easily separated from lead in complex ores. In the separation of chalcopyrite from sphalerite and pyrite, very close control of cyanide ion concentration is needed. Cyanide should be added sufficient only to complex the heavy metal ions in solution, and to solubilise the zinc and iron xanthates. Excess cyanide forms soluble complexes with the slightly less soluble copper xanthate, depressing the chalcopyrite.

The depressive effect of cyanide depends on its concentration and on the concentration of the collector and the length of the hydrocarbon chain. The longer the chain, the greater is the stability of the metal xanthate in cyanide solutions and the higher the concentration of cyanide required to depress the mineral. Relatively low concentrations of xanthates with short hydrocarbon chains are therefore used for selective flotation where cyanides are used as depressants.

Cyanides are, of course, extremely toxic and must be handled with great care. They also have the disadvantage of being expensive and they depress and dissolve gold and silver, reducing the extraction of these metals into the froth product. Despite these disadvantages, they are widely used due to their high degree of selectivity in flotation. They also have the advantage of leaving the mineral surface relatively unaffected, so that subsequent activation is relatively simple, although residual cyanide ions in solution may interfere with the activator.

While many plants function efficiently with cyanide alone, in others an additional reagent, generally zinc sulphate, is added to ensure satisfactory depression of sphalerite. If copper ions are present, the introduction of zinc ions can prevent the copper depositing on the sphalerite surface by shifting equation 12.11 to the left.

However, other, more complex, reactions occur to assist depression and it is considered that cyanide reacts with zinc sulphate to form zinc cyanide, which is relatively insoluble, and precipitates on the sphalerite surface, rendering it hydrophilic and preventing collector adsorption:

$$ZnSO_4 + 2NaCN = Zn(CN)_2 + Na_2SO_4 \qquad (12.24)$$

In an alkaline pulp, zinc hydroxide, which adsorbs copper ions, is also formed and it is precipitated on to the sphalerite surface, preventing collector adsorption.

The use of zinc sulphate thus reduces cyanide consumption and cases have been known where depression of sphalerite has been achieved by the use of zinc sulphate alone.

Although cyanides and zinc sulphate are widely used, they do have many disadvantages, for example many concentrators being loath to use cyanides due to environmental problems. Zinc sulphate is effective only at high pH values, at which zinc hydroxide precipitates from solution. There is a need, therefore, for alternative selective depressants. Research on Pb–Zn ores in Yugoslavia has shown that sphalerite depression by zinc sulphate and sodium cyanide can be successfully replaced by a combination of ferrosulphate and sodium cyanide.[48] This has the advantage of reducing sodium cyanide consumption, with consequent economic and ecological advantages. Zinc bisulphite used with cyanide in alkaline conditions is being used to treat bulk copper–zinc–iron concentrates at the Cerro Colorado mill in Spain.[49] This reagent combination was found to give results which were more favourable than those obtained using standard depression techniques, such as cyanide–zinc sulphate, which was found to be very sensitive to variations in the chalcocite content of the ore.

Sphalerite activation can be prevented by eliminating copper ions from the flotation pulp, and in some plants precipitation with hydrogen sulphide or sodium sulphide is carried out.

Sulphur dioxide has developed into a most versatile and almost indispensable conditioning reagent for polysulphide ores. Although widely used as a galena depressant in copper-lead separations, it also deactivates zinc sulphides and enhances the flotation differential between zinc and other base metal sulphides. In copper cleaner and copper–lead separation circuits, very effective zinc rejection is attained through acidification of pulps by injection of SO_2. However, SO_2 cannot be employed when treating ores which contain the secondary copper minerals covellite or chalcocite, since they become soluble in the presence of sulphur dioxide and the dissolved copper ions activate zinc sulphides.[50] Sulphur dioxide does not appreciably depress chalcopyrite and other copper-bearing minerals. In fact, adsorption of xanthate on chalcopyrite is enhanced in the presence of SO_2, and the addition of SO_2 before xanthate results in effective sphalerite depression while increasing the floatability of chalcopyrite. The use of SO_2 in various Swedish concentrators is discussed by

Broman *et al.*[51] who point out that SO_2 has the advantage over cyanide in sphalerite depression in that there is little copper depression, and no dissolution of precious metals. However, it is pointed out that the use of SO_2 demands adaptation of the other reagent additions, and in some cases a change of collector type is required.

Potassium dichromate ($K_2Cr_2O_7$) is also used to depress galena in copper–lead separations. The depressive action is due to the chemical reaction between the galena surface and $CrO_4{}^2$ − anions, which produces insoluble lead dichromate which increases wettability and prevents flotation.

More than 40% of the Western world's molybdenum is produced as a by-product from porphyry copper ores. The small amount of molybdenite is collected along with copper in a bulk Cu–Mo concentrate. The two minerals are then separated, almost always by depressing the copper minerals and floating the molybdenite. Sodium hydrosulphide (or sodium sulphide) is used most extensively, though several other inorganic compounds, such as cyanides, and Noke's reagent (a product of the reaction of sodium hydroxide and phosporous pentasulphide), are also used.[52] Almost all the currently used depressants are inorganic. Numerous organic depressants have been developed over the years, but apart from sodium thioglycolate, none has been successfully commercialised.[53]

Organic reagents such as starch, tannin, quebracho, and dextrin do not ionise in solution, but form colloidal particles in the pulp which can be deposited on the mineral surfaces, preventing flotation in a similar manner to a slime coating. Large quantities of these reagents will depress all minerals and they are not as selective as the electrolytic depressants. They are used in small amounts to depress talc, graphite, and calcite. Starch is also used as a supplementary lead depressant in copper–lead separations.

The Importance of pH

It is evident from the foregoing that pulp alkalinity plays a very important role in flotation, and in practice, selectivity in complex separations is dependent on a delicate balance between reagent concentrations and pH.

Flotation where possible is carried out in an alkaline medium, as most collectors, including the xanthates, are stable under these conditions, and corrosion of cells, pipework etc., is minimised. Alkalinity is controlled by the addition of lime, sodium carbonate (soda ash), and to a lesser extent sodium hydroxide or ammonia. Sulphuric or sulphurous acids are used where a decrease in pH is required.

Lime, being cheap, is very widely used to regulate pulp alkalinity, and is used in the form of milk of lime, a suspension of calcium hydroxide particles in a saturated aqueous solution. Lime, or soda ash, is often added to the slurry prior to flotation to precipitate heavy metal ions from solution. In this sense, the alkali is acting as a "deactivator", as these heavy metal ions can activate sphalerite and pyrite and prevent their selective flotation from lead or copper minerals. Since the heavy metal salts precipitated by the alkali can dissociate to a limited extent and thus allow ions into solution, cyanide is often used with the alkali to complex them. Hydroxyl and hydrogen ions modify the electrical double layer and zeta potential (see Chapter 15) surrounding the mineral particles and hence the hydration of the surfaces and their floatability is affected. With xanthates as collectors, sufficient alkali will depress almost any sulphide mineral, and for any concentration of xanthate there is a pH value below which any given mineral will float, and above which it will not float. This *critical pH* value depends on the nature of the mineral, the particular collector and its concentration, and the temperature.[4] Figure 12.12 shows how the critical pH value for pyrite, galena and chalcopyrite depends on the concentration of sodium aerofloat collector.

It is evident from the curves that using 50 mg l^{-1} of sodium aerofloat, and a pH value of 8, chalcopyrite can be floated from galena and pyrite. On reducing the pH to 6, the galena can be floated from the pyrite.

Lime can also act as a strong depressant for pyrite and arsenopyrite when using xanthate collectors. Both the hydroxyl and calcium ions participate in the depressive effect on pyrite by the formation of mixed films of Fe(OH), FeO(OH), $CaSO_4$ and $CaCO_3$ on the surface, so

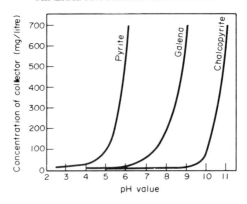

FIG. 12.12.   Relationship between concentration of sodium diethyl dithiophosphate and critical pH value (after Sutherland and Wark[4]).

reducing the adsorption of xanthate. Lime has no such effect with copper minerals, but does depress galena to some extent. In the flotation of galena, therefore, pH control is often affected by the use of soda ash, pyrite and sphalerite being depressed by cyanide. As was shown earlier, the effectiveness of sodium cyanide and sodium sulphide is governed to such a large extent by the value of pH that these reagents are of scarcely any value in the absence of alkalis. Since, where cyanide is used as a depressant, the function of the alkali is to control the cyanide ion concentration (equations 12.22 and 12.23), there is for each mineral and given concentration of collector a "critical cyanide ion concentration" above which flotation is impossible. Curves for several minerals are given in Fig. 12.13, and it can be seen that chalcopyrite can be floated from pyrite at pH 7.5 and 30 mg l^{-1} sodium cyanide. Since, of the copper minerals, chalcopyrite lies closest to pyrite relative to the influence of alkali and cyanide, all the copper minerals will float with the chalcopyrite. Thus by careful choice of pH value and cyanide concentration, excellent separations are theoretically possible, although in practice other variables serve to make the separation more difficult. Adsorption of xanthate by galena is uninfluenced by cyanide, the alkali alone acting as a depressant.

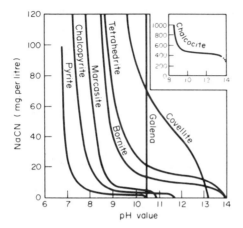

FIG. 12.13. Contact curves for several minerals (ethyl xanthate = 25 mg/1) (after Sutherland and Wark[4]).

The Importance of Pulp Potential

Recent work in Australia and the United States has shown that most sulphide minerals can, under certain conditions, be floated in the absence of collectors.[54] All these studies imply that, if not oxygen itself, then at least an oxidising potential is required for collectorless flotation. It has been established that sulphide minerals oxidise through a continuum of metal-deficient sulphides of decreasing metal content through to elemental sulphur (equations 12.5 and 12.6) by reactions of the type:

$$MS \rightarrow M_{1-x}S + xM^{2+} + 2xe \text{ (acid)} \qquad (12.25)$$

and

$$MS + xH_2O \rightarrow M_{1-x}S + xMO + 2xH^+ + 2xe \text{ (alkali)} \quad (12.26)$$

These sulphur-rich, metal-deficient zones can render the mineral hydrophobic, provided that the local conditions are such that the metal oxides/hydroxides formed by the reaction are solubilised. Excessive oxidation can produce thiosalts (equation 12.7), and ultimately, sulphate (equation 12.8), together with metal ions which may re-adsorb, as hydrolysis products, on to the mineral, producing hydrophilic surfaces.

Buckley et al.[55] studied the surface oxidation of galena, bornite, chalcopyrite and pyrrhotite, and found that for each mineral the initial oxidation reaction is the removal of a metal component from the surface region to leave a sulphide with similar structure to the original mineral but with lower metal content. Metal-deficient sulphide layers containing high sulphur–metal ratios are probably stabilised by the underlying mineral because they have the same sulphur lattices. The authors showed that flotation of the minerals could be accomplished without the aid of collectors when a metal-deficient sulphide, rather than elemental sulphur, is formed. Naturally hydrophobic sulphide minerals, such as molybdenite, have such layer structures, the behaviour of such minerals being explained in terms of the work of adhesion of water to the surface being largely determined by dispersion forces, with hydrogen bonding and ionic interactions being small. It is possible that a similar situation exists at the surface of other sulphides where a metal-deficient layer is formed. Although the metal is dissolved at low pH (equation 12.25), in neutral or alkaline conditions a hydroxy-oxide is formed (equation 12.26), which could be expected to be hydrophilic. However, collector-less flotation occurs under these conditions, and the authors conclude that the metal oxides are dissolved due to the turbulence in the flotation cells, or are abraded from the mineral surfaces.

The collectorless flotation process has also been tested on six different chalcopyrite ores while monitoring the potentials of the pulp.[56] The results confirmed that collectorless flotation is effective only under oxidising conditions. In addition, the flotation requires that the chalcopyrite surface be relatively free of hydrophilic oxidation products, which can be accomplished by treating the ore pulp with sodium sulphide. The role of sodium sulphide in collectorless flotation was initially thought to be one of sulphidising agent. However, the excess HS^-/S^{2-} ions that have not been consumed in sulphidisation may be oxidised to become elemental sulphur or polysulphides, depending on pH, which may deposit on the mineral surface. Thus the collectorless flotation process using sodium sulphide may provide an external source for these hydrophobic species that could enhance flotation. Collectorless flotation was also found to be pH dependent, becoming more favourable with decreasing pH.

However, as explained by Guy and Trahar,[57] the application of such findings to realistic separations is not straightforward, as the areas of floatability determined from experiments with single sulphides do not necessarily coincide with those determined from experiments with sulphide mixtures. Cations produced by sulphide oxidation may react in different ways in a given system. Apart from modifying the surfaces of some minerals by surface interactions, they may be precipitated as hydroxides which have a profound effect on sulphide floatability.

For instance, pyrite and pyrrhotite occur together in many important ores, and the galvanic interaction between these two minerals, and its effect on their floatabilities has been investigated.[58] The galvanic contact decreased the formation of hydroxide or oxide and sulphate species of iron on pyrrhotite, whereas such formations were increased on pyrite. The effect was to improve the floatability of pyrrhotite, while reducing that of pyrite.

The control of redox conditions is complicated not only by the galvanic interactions between the different minerals in the ore, but also by the interactions between the minerals and the steel grinding medium.[59] The reducing conditions at a sulphide mineral surface created by the oxidation of steel in a galvanic interaction can hinder the adsorption of the collector.

Learmont and Iwasaki[60] have studied the interaction between galena and steel media. They show that iron oxide, hydroxide or sulphate species form on the galena surface on contact with mild steel, decreasing the galena floatability. The time of contact and aeration conditions affect the severity of the flotation depression. Adam and Iwasaki[61] showed that the flotation response of pyrrhotite was similarly adversely affected. The formation of hydroxide or oxide and sulphate species of iron through the oxygen reduction reaction at the cathodically polarised surface of pyrrhotite was shown to be the mechanism responsible for the reduced floatability of pyrrhotite, the following reactions being proposed:

$$\tfrac{1}{2}O_2 + H_2O + 2e^- = 2OH^- \text{ (cathode)}$$

$$2H^+ + 2e^- = H_2 \text{ (cathode)}$$

$$FeS = Fe^{2+} + S^{2-} \text{ (dissociation)} \rightarrow Fe_2O(OH)_3 \text{ or}$$
$$FeOOH \rightarrow Fe(OH)SO_4$$

The formation of an iron hydroxide coating covering the mineral surface reduces mineral floatability.

Due to these many complex interactions, measurement of the pulp oxidation-reduction potential is difficult.[62] Electrodes which have different activities for the oxygen reduction reaction, such as platinum and gold, can give rise to different E_h values, and different sulphides can give rise to different E_h values in the same solution. Because of these complexities, on-line measurement of E_h to control redox conditions is still a control strategy of the future, although some concentrators do use such actions based on operating experience, and Outokumpu Oy is reportedly developing methods to control the electrochemical potentials of minerals directly in the ore pulps in order to attain the optimum combination of E_h and pH, as well as the optimum collector addition.[63]

Pre-flotation aeration of sulphide pulps has been practised in the Noranda Group (Canada) and other organisations for many years to help depress pyrite and pyrrhotite and promote chalcopyrite and galena.[64]

The introduction of talc pre-flotation at Woodlawns in Australia had a detrimental effect on copper circuit performance, due to the aeration provided by the talc cells.[65] The aeration promoted the flotation of the other sulphides, especially galena, relative to that of the chalcopyrite. Copper circuit performance was subsequently improved by the addition of a strong reducing agent, sodium sulphide, to the talc flotation tailing.

Nitrogen is used as the carrier gas in a few molybdenum flotation circuits, a reducing potential being used to minimise the consumption of the sulphide depressant which inhibits flotation of copper minerals.

Only recently has there been a revival of interest in studying the mechanism of depression of sulphides. The influence of the strongly reducing nature of sodium hydrosulphide on depressant action has been monitored by means of solution redox-potential measurements, and it would appear that the depressant activity is to some extent electrochemical, the HS^- ions, by virtue of their large negative E_h, destabilising the coating of thiol collector.[52] The oxidation-reduction effects in sulphide mineral depression have been reviewed by Chander.[66]

Laboratory Flotation Testing

In order to develop a flotation circuit for a specific ore, preliminary laboratory testwork must be undertaken in order to determine the choice of reagents and the size of plant for a given throughput as well as the flowsheet and peripheral data. Flotation testing is also carried out on ores in existing plants to improve procedures and for development of new reagents.[67]

It is essential that testwork is carried out on ore which is representative of that treated in the commercial plant. Samples for testwork must be representative, not only in chemical composition, but also relative to mineralogical composition and degree of dissemination. A mineralogical examination of drill cores or other individual samples should therefore be made before a representative sample is selected. Composite drill core samples are ideal for testing if drilling in the deposit has been extensive; the cores generally contain ore from points widely distributed over the area and in depth. It must be realised that ore-bodies are variable and that a representative sample will not apply equally well to all parts of the ore-body; it is used therefore for development of the *general* flotation procedure. Additional tests must be made on samples from various areas and depths to establish optimum conditions in each case and to give design data over the whole range of ore variation.

Having selected representative samples of the ore, it is necessary to prepare them for flotation testing, which involves comminution of the ore to its optimum particle size. Crushing must be carried out with care in order to avoid accidental contamination of the sample by grease or oil, or with other materials which have been previously crushed. Even in a commercial plant, a small amount of grease or oil can temporarily upset the flotation circuit. Samples are usually crushed with small jaw crushers or cone crushers to about 0.5 cm and then to about 1 mm with crushing rolls in closed circuit with a screen.

Storage of the crushed samples is important, since oxidation of the surfaces is to be avoided, especially with sulphide ores. Not only does oxidation inhibit collector adsorption, but it also facilitates the dissolution of heavy metal ions, which may interfere with the flotation process. Sulphides should be tested as soon as possible after obtaining the sample and ore samples must be shipped in sealed drums in as coarse a state as possible. Samples should be crushed as needed during the testwork, although a better solution is to crush all the samples and to store them in an inert atmosphere.

Wet grinding of the samples should always be undertaken immediately prior to flotation testing to avoid oxidation of the liberated mineral surfaces. Batch laboratory grinding, using ball mills, produces a flotation feed with a wider size distribution than that obtained in continuous closed-circuit grinding; to minimise this, batch rod mills are used which give products having a size distribution which approximates closely to that obtained in closed-circuit ball mills. True simulation is never really achieved, however, as overgrinding of high specific gravity minerals, which is a feature of closed-circuit grinding, is avoided in a batch rod mill.

A soft, dense mineral such as galena will be ground finer in closed circuit than predicted by the batch tests, and its losses due to production of ultra-fine particles may be substantial. Some sulphide minerals, such as sphalerite and pyrite, can be depressed more easily at the coarser sizes produced in batch grinding, but may be more difficult to depress at the finer sizes resulting from closed-circuit grinding. Predictions from laboratory tests can be improved if the mineral recovery from the batch tests is expressed as a function of *mineral* size rather than overall product size. The optimum mineral size can then be determined and the overall size estimated to give the optimum mesh of grind.[68] This method assumes that the same fineness of the valuable mineral will give the same flotation results both from closed-circuit and batch grinding, irrespective of the differences in size distributions of the other minerals.

It must be appreciated that the optimum grinding size of the particles depends not only on their grain size, but also on their floatability. Initial examination of the ore should be made to determine the degree of liberation in terms of particle size in order to estimate the required fineness of grind. Testwork should then be carried out over a range of grinding sizes in conjunction with flotation tests in order to determine the optimum mesh of grind. In certain cases, it may be necessary to overgrind the ore in order that the particles are small enough to be lifted by the air-bubbles. If the mineral is readily floatable a coarse grind may be utilised, the subsequent concentrate requiring regrinding to further free the mineral from the gangue, before further flotation is performed to produce a high-grade concentrate.

Initial floatability tests are often made on the liberated mineral particles, as a means of assessing a range of suitable collectors and regulators, and to determine the effective pH for flotation. A useful laboratory method is that of *contact angle measurement*, where a clean smooth surface of mineral is placed in distilled water, and a bubble of air from the end of a capillary tube is pressed down upon it. If after a short time no adhesion is visible on withdrawal of the bubble, the mineral surface is assumed to be clean, and the collector is then added. If the mineral surface now becomes hydrophobic, adherence of the introduced bubble to the surface results. The contact angle produced across the water phase (Fig. 12.2) is a measure of the floatability of the mineral. The method suffers from many disadvantages; it is extremely difficult to obtain a truly representative surface of the mineral of the required size (at least 0.5 cm^2). The mineral may not be representative of the naturally liberated surface after the intense polishing required to produce a completely clean, flat surface. The method is static, whereas true flotation is dynamic, particles adhering after impact with bubbles rising in the pulp.

In the *Hallimond tube* technique (Fig. 12.14), dynamic conditions prevail, similar to those in a commercial flotation cell. The mineral particles are held on a support of sintered glass inside the tube containing the distilled water and the collector under test. Air-bubbles are introduced through the sinter and any hydrophobic mineral particles are lifted by the bubbles, which burst at the water surface, allowing the particles to fall into the collecting tube. By treating a small weighed sample of mineral, the weight collected in the tube can be related to the floatability.

The bulk of laboratory testwork is carried out in batch flotation cells

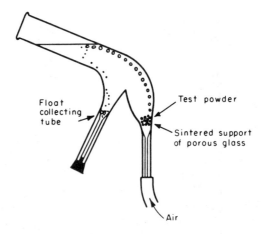

Float collecting tube

Test powder

Sintered support of porous glass

Air

FIG. 12.14. Hallimond tube.

(Fig. 12.15), usually with 500-g, 1-kg, or 2-kg samples of ore. The cells are mechanically agitated, the speed of rotation of the impellers being variable, and simulate the large-scale models commercially available. Introduction of air to the cell is normally via a hollow standpipe surrounding the impeller shaft. The action of the impeller draws air down the standpipe, the volume being controlled by a valve and by the speed of the impeller. The air stream is sheared into fine bubbles by the impeller, these bubbles then rising through the pulp to the surface, where any particles picked up are removed as a mineralised froth.

Batch tests are fairly straightforward in practice, but a few experimental points are worth noting.[67]

1. Agitation of the pulp must be vigorous enough to keep all the solids in suspension, without breaking up the mineralised froth column.

2. Conditioning of the pulp with the reagents is often required. This is a period of agitation, varying from a few seconds to 30 min before the air is turned on, which allows the surfaces of the mineral particles to react with the reagents.

3. Very small quantities of frother can have marked effects, and *stage additions* of frother are often needed to control the volume of

FIG. 12.15.   Laboratory flotation cell.

froth. The froth depth should be between 2 and 5 cm, as very shallow froths entail the risk of losing pulp into the concentrate container. Reduction of the amount of air is sometimes used to limit the amount of froth produced. This should be standardised for comparative tests in order to prevent the introduction of another variable.

4. As a matter of economics, flotation separations are carried out in as dense a pulp as possible consistent with good selectivity and operating conditions. The denser the pulp, the less cell volume is required in the commercial plant, and also less reagent is required, since the effectiveness of most reagents is a function of their concentrations in solution. The optimum pulp density is of great importance, as in general the more dilute the pulp, the cleaner the separation. Most commercial floats are in pulps of 25–40% solids by weight, although they can be as low as 8% and as high as 55%. It must be borne in mind that in batch flotation tests the pulp density varies continuously, from beginning to end, as solids are removed with the froth and water is added to maintain the cell pulp level. This continuous variation changes the concentration of reagents as well as the character of the froth.

5. As water contains dissolved chemicals which may affect flotation, water from the supply which will be used commercially should be used, rather than distilled water.

6. Normally only very small quantities of reagent are required for batch tests. In order to give accurate control of their addition rates, they may have to be diluted. Water-soluble reagents can be added as aqueous solutions by pipette, insoluble liquid reagents by graduated dropper or hyperdermic needle. Solids may either be emulsified or dissolved in organic solvents, providing the latter do not affect flotation.

7. Most commercial flotation operations include at least one cleaning stage, in which froth is refloated to increase its grade, the cleaner tailings often being recycled. Since cleaner tails are not recycled in batch tests, they do not always closely simulate commercial plants. If cleaning is critical, cycle tests may have to be undertaken. These are multiple-step flotation tests designed to measure the effect of circulating materials. The main objectives of cycle tests are to determine:

(a) The increase in recovery obtained by recirculating cleaner tailings.
(b) The variation in reagent requirements to compensate for the circulating load of reagents.

(c) The effect of build-up of slimes or other undesirables which may interfere with flotation.

(d) The froth handling problems.

Normally at least six cycles are required before the circuit reaches equilibrium and a complete material balance should be made on each cycle. Since the reagents are in solution, it is essential that liquids as well as solids recirculate, so any liquid used to adjust pulp density must be circuit liquid obtained from recent decantation or filtration steps.

Cycle tests are very laborious to carry out, and often the test fails to reach steady state. A method has been developed[69] whereby cycle test behaviour can be predicted from data developed from individual batch tests, and a computer program has been developed to arrive at a steady-state balance for a variety of simulated circuits.

Pilot Plant Testwork

Laboratory flotation tests provide the basis of design of the commercial plant. Prior to development of the plant, pilot scale testing is often carried out in order to:

(1) Provide continuous operating data for design. Laboratory tests do not closely simulate commercial plants, as they are batch processes.

(2) Prepare large samples of concentrate for survey by smelters, etc., in order to assess the possibility of penalties or bonuses for trace impurities.

(3) Compare costs with alternative process methods.

(4) Compare equipment performance.

(5) Demonstrate the feasibility of the process to non-technical investors.

Laboratory and pilot scale data should provide the optimum conditions for concentrating the ore and the effect of change of process variables. The most important data provided by testwork includes:[70]

(a) The optimum mesh of grind of the ore. This is the particle size at which the most economic recovery can be obtained. This depends not only on the grindability of the ore, but also on its floatability. Some readily floatable minerals can be floated at well above the liberating size of the mineral particles, the only upper limit to size being that at which the bubbles can no longer physically lift the particles to the surface. This upper size limit is normally around 300 μm. The lower size limit for flotation, at which problems of oxidation and other surface effects occur, is around 5 μm.

(b) Quantity of reagents required, and location of addition points.

(c) Pulp density; important in determining size and number of flotation cells.

(d) Flotation time; experimental data gives the time necessary to make the separation into concentrate and tailings. This depends on the particle size and the reagents used and is needed to determine the plant capacity.

(e) Pulp temperature, which affects the reaction rates. Water at room temperature is, however, used for most separations.

(f) The extent of uniformity of the ore; variations in hardness, grindability, mineral content and floatability should be investigated so that the variations can be accommodated in the design.

(g) Corrosion and erosion qualities of the pulp; this is important in determining the materials used to construct the plant.

(h) Type of circuit; many different types of circuit can be used, and laboratory tests should provide data for design of the best suited circuit. This should be as basic as possible at this stage. Many flow schemes used in operating plants have evolved over a long period, and attempted duplication in the laboratory is often difficult and misleading. The laboratory procedures should be kept as simple as possible so that the results can be interpreted into plant operation.

Basic Flotation Circuits

Commercial flotation is a continuous process. Cells are arranged in series forming a *bank* (Fig. 12.16).

Pulp enters the first cell of the bank and gives up some of its valuable mineral as a froth; the overflow from this cell passes to the second cell, where more mineralised froth is removed, and so on down the bank, until barren tailings overflow the last cell in the bank. The height of the froth column for each cell is determined by adjusting the height of the tailings overflow weir, the difference in height between this and the cell overflow lip determining the froth depth. New feed enters the first cell of a bank, the froth column in the first few cells being kept high, since there are plenty of hydrophobic mineral particles to sustain it. The pulp level is raised from cell to cell, as the pulp becomes depleted in floatable minerals, by progressively raising the cell tailings weir height. The last few cells in the bank contain relatively low-grade froths, comprising weakly aerophilic particles. These are the *scavengers*, usually comprising middlings particles, which are often recirculated to the head of the system. The scavenger cells, having little mineral to sustain a deep froth, have their tailings weirs raised so that pulp is almost overflowing the cell lip. This policy, which is used to remove all weakly floating material, is known as "pulling the cells hard" and ensures maximum recovery from the bank of cells. Excessive circulating loads should be avoided, however, as the rougher feed may

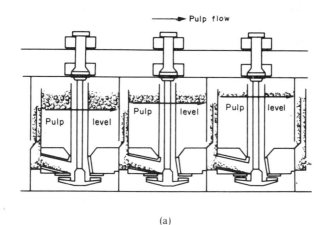

(a)

FIG. 12.16. (a) Bank of flotation cells.

FIG. 12.16. (b) Banks of cells.

be diluted, and the flotation time reduced. The flowsheet for this basic system is shown in Fig. 12.17. This flowsheet can be successfully operated only when the gangue is relatively unfloatable, and it requires extremely careful control to produce an even grade of concentrate if

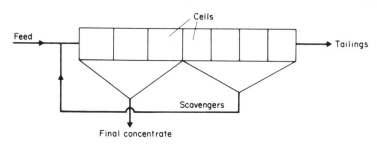

FIG. 12.17   Simple flotation circuit.

there are any fluctuations in the head grade. A preferable system is to
dilute the concentrate from the first few cells of the bank, known as
rougher concentrates, and refloat them in *cleaner* cells, where the weirs
are kept low to maintain a deep froth and produce a high-grade
concentrate. In this rougher–scavenger–cleaner system (Fig. 12.18),
the cleaning cells receive a comparatively high-grade feed, whilst the
scavenging section can be run with an excess of air so as to obtain
maximum recovery. Tailings from the cleaner cells, usually containing
aerophilic mineral particles, are generally recirculated to the rougher
cells, along with the scavengers. This type of circuit, besides being
useful for ores that need the maximum amount of aeration at the end of
the bank to produce profitable recovery, is often employed when the

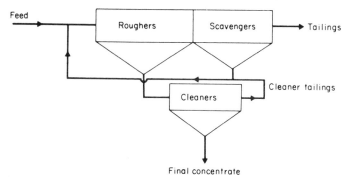

FIG. 12.18.   Rougher–scavenger–cleaner system.

gangue has a tendency to float and is difficult to separate from the mineral. In such cases, it may be necessary to utilise one or more recleaning banks of cells (Fig. 12.19).

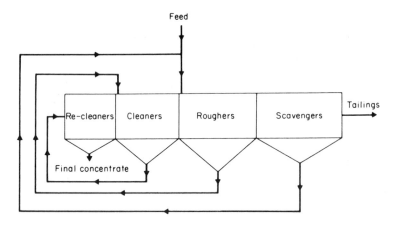

FIG. 12.19.   Circuit employing recleaning.

It is worth noting that the diluting water used to lower the pulp density of the cleaning bank passes to the roughing cells and dilutes the primary feed, which should therefore contain a correspondingly smaller proportion of water as it leaves the grinding section in order that the dilution of the cleaner tailing may bring it to the correct pulp ratio in the roughing cells.

Flowsheet Design

In designing a suitable flowsheet for a flotation plant, the primary mesh of grind is of major consideration. It can be estimated based on past experience and from mineralogical evaluation, but laboratory grind-flotation tests must be conducted to determine optimum conditions. The purpose of the primary grind is to promote *economic recovery* of the valuable ore minerals. Batch tests are performed, utilising various reagent combinations, on samples of ore ground to

various degrees. Incremental concentrate samples are weighed and
assayed, and the results plotted as recovery-time and recovery-grade
(Fig. 12.20) curves.

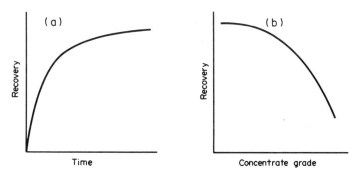

FIG. 12.20. (a) Recovery of metal to concentrate versus time, (b) Recovery versus
concentrate grade curve.

Initially, a mesh of grind should be chosen which gives a reasonable
rougher grade and recovery within an acceptable flotation time. If
grinding is too coarse, some of the valuable mineral, locked in
middlings grains, will not be floated. However, excessive flotation
times may eventually allow some of these particles to report to the
concentrate, reducing its grade. It is here that the flotation engineer
must use his experience in deciding what is, at this stage, a reasonable
concentrate grade and flotation time.

As grinding is invariably the greatest single operating cost, it should
not be carried out any finer than is justified economically. Later
testwork, having improved on the basic flotation scheme, will modify
the primary mesh of grind, taking into account the amount of
secondary grinding required to reach the specified concentrate grade,
and the number of cleaning stages required. Finer grinding should not
be performed beyond the point where the net smelter return for the
increment saved becomes less than the operating cost.[71]

After determining a suitable primary mesh of grind (which may be
modified in later testwork), further tests are performed to optimise

reagent additions, pH, pulp densities, etc. Having optimised flotation recovery, testwork is then aimed at producing the required concentrate *grade*, and determining the flowsheet which must be utilised to achieve this.

As Fig. 12.20 (a) indicates, most of the valuable mineral floats within a few minutes, whereas it takes much longer to float the residual small quantity. The rate equation for flotation can be expressed in a general way as follows:

$$v = -dW/dt = K_n W^n \qquad (12.27)$$

where v (weight/unit time) is the flotation rate, W is the weight of floatable mineral remaining in the pulp at time t, K_n is the rate constant, and n is the order of the reaction. The kinetics of flotation have been studied by many workers, the majority classifying flotation as a first order reaction ($n = 1$), others reporting second order kinetics.[72] Dowling et al.[73] applied thirteen rate models to batch copper flotation data and evaluated the results using statistical techniques. The flotation of the copper ore was shown to be essentially a first-order process, and all the models tested were found to give a reasonably good fit to the experimental data, though some models were clearly better than others.

The first order rate equation is usually expressed as:

$$R = 1 - \exp(-kt) \qquad (12.28)$$

where R is the cumulative recovery after time t,
          k is the first order rate constant (time^{-1}),
          t is the cumulative flotation time.

Plots of $\ln(1 - R)$ versus time should produce straight lines, but such plots are often concave upwards, which has led a number of workers to postulate the presence of fast and slow floating components. Agar[74] argues that such postulates are false, the non-linear plots resulting from the assumption that the maximum possible recovery is 100%, whereas in practice some floatable material is usually totally irrecoverable, as it may be encased in gangue. A modified first order rate equation of the form:

$$R = RI [1 - \exp(-kt)] \qquad (12.29)$$

is proposed, where RI is the maximum theoretical flotation recovery.

The flotation rate constant is dependent on the particle size and the degree of liberation of the mineral, the curve shown in Fig. 12.20 (a) being a summation of the flotation rates of all the particles within the ore. Figure 12.21 shows the variation of flotation rate constant of an ore as a function of the particle size. Extensive studies of the influence of particle size on flotation have been made.[75-77]

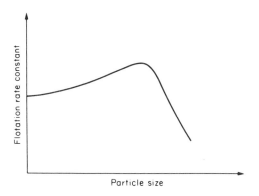

FIG. 12.21. Flotation-rate constant as a function of particle size.

It is clear that the flotation activity of an ore falls off slowly towards the range of fine particle size, mainly due to the increase in number of particles per unit weight and to the deteriorating conditions for bubble-particle contact and effects such as increased surface oxidation of the particles. Flotation activity falls off very rapidly above the optimum particle size, due to the lesser degree of liberation of the minerals and to the decreasing ability of the bubbles to lift the coarse particles. It can be seen that the floated material is composed of a fast floating fraction in the medium-size range and a more reluctant fraction comprising unliberated coarse material and fines. In a commercial flotation circuit the fast floating material will be recovered in the roughing section, while the more reluctant fraction is recovered by scavenging, certain losses into the tailings having to be accepted. Figure 12.22 relates the distribution in terms of the flotation rate constant.

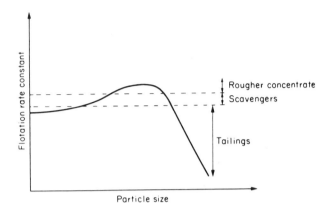

FIG. 12.22.   Rougher–scavenger system determined by flotation-rate constant.

The essential difference between the concentrates from the roughers and scavengers is that the latter comprises both coarse and fine particles while the rougher concentrate consists essentially of an intermediate size fraction.

The grade of the final cleaner concentrate is dependent on the grade of the rougher concentrate (Fig. 12.20(b)) and, in order to reach the specified optimum cleaning grade, it is necessary to keep the rougher grade at a predetermined value. The decision as to where the rougher–scavenger split should be can be made on the basis of batch tests where cumulative concentrate grade is plotted against time (Fig. 12.23), the time limit for rougher flotation then being fixed as that required to give a rougher concentrate with a high enough grade to produce the specified final concentrate grade with the chosen number of cleaning stages. The remaining flotation time (Fig. 12.20(a)) is for scavenging, this time sometimes being reduced by increasing the severity of the flotation conditions (i.e. increased aeration, addition of more powerful collector) after the removal of the rougher concentrate.

Agar et al.[78] have argued that the rougher–scavenger split should be made at the flotation time where separation efficiency (equation 1.1) is maximised. Separation efficiency (SE) reaches a maximum value when dSE/dt is zero, so that from equation 1.3:

$$\frac{dSE}{dt} = \frac{100m}{f(m-f)} \left[(c-f)\frac{dC}{dt} + \frac{Cdc}{dt} \right] \qquad (12.30)$$

= 0 at maximum separation efficiency.

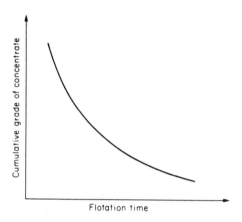

FIG. 12.23.   Cumulative grade in rougher flotation versus time.

If G is the grade of concentrate leaving the flotation cell lip at any given time, then:

$$\int_0^t Gdc = Cc. \qquad (12.31)$$

Therefore, differentiating equation 12.31:

$$Gdc = Cdc + cdC$$

or $$G = Cdc/dC + c. \qquad (12.32)$$

Multiplying equation 12.32 by dC/dt gives:

$$GdC/dt = Cdc/dt + cdC/dt \qquad (12.33)$$

and substituting equation 12.33 in equation 12.30:

$$\frac{dSE}{dt} = \frac{100m}{f(m-f)} \left[\frac{cdC}{dt} - \frac{fdC}{dt} + \frac{Gdc}{dt} - \frac{cdC}{dt} \right].$$

Therefore, at maximum separation efficiency, where $dSE/dt = 0$, $G = f$.

This means that at maximum separation efficiency, the grade of concentrate produced is equal to the flotation feed, and after this time the flotation system is no longer concentrating the valuable mineral.

Since separation efficiency = recovery of mineral − recovery of gangue (equation 1.1), separation efficiency is also maximised when:

$$\frac{d(Rm - Rg)}{dt} = 0, \text{ i.e. when } dRm/dt = dRg/dt.$$

Therefore, at maximum separation efficiency, the rate of flotation of valuable mineral is equal to that of the gangue, and above the optimum flotation time the gangue begins to float faster than the mineral. This optimum flotation time can be calculated from the first order rate equation (12.29). However, as shown by Agar,[74] this equation has to be modified for batch flotation tests to incorporate a correction factor for time. In batch flotation some hydrophobic solids will have air attached to them during the conditioning period, which causes them to float more rapidly than they would naturally. This causes a positive correction to time zero, as flotation actually started before the air flow was introduced. On the other hand, when air flow commences, several seconds elapse before a full depth of loaded froth is present in the cell, and this gives a negative correction to time zero. Agar's modified rate equation for batch flotation tests is:

$$R = RI[1 - \exp\{- k(t + b)\}] \tag{12.34}$$

where b is correction for time zero.

A plot of $\ln[(RI - R)/RI]$ versus $(t + b)$ should produce a line of slope k. However, RI and b are both unknown. Using experimental data, at the qth value of R and t:

$$\ln(RI - R_q)/RI] + k(t_q + b) = r_q$$

where r_q is the residual due to errors in the experimental data.

Hence

$$r_q^2 = \{\ln[(RI - R_q)/RI]\}^2$$
$$+ k^2(t_q + b)^2 + 2k(t_q + b).\ln[(RI - R_q)/RI].$$

Therefore, for n experimental data:

$$\sum_{q=1}^{n} r_q^2 = \sum_{q=1}^{n} [\ln\{(RI - R_q)/RI\}]^2 + k^2 \sum_{q=1}^{n} t_q^2 + nk^2b^2 + 2k^2b \sum_{q=1}^{n} t_q$$

$$+ 2k \sum_{q=1}^{n} [t_q.\ln\{(RI - R_q)/RI\}] + 2kb \sum_{q=1}^{n} \ln[(RI - R_q)/RI] \quad (12.35)$$

$\sum_{q=1}^{n} r^2$ is a minimum when $d \sum_{q=1}^{n} r^2/dk$ and $d \sum_{q=1}^{n} r^2/db$ are zero, i.e. when

$$d \sum_{q=1}^{n} r^2/dk = 2k \sum_{q=1}^{n} t^2 + 2nkb^2 + 4kb \sum_{q=1}^{n} t + 2 \sum_{q=1}^{n}$$

$$\quad (12.36)$$

$$[t.\ln\{(RI - R)/RI\}] + 2b \sum_{q=1}^{n} \ln[(RI - R)RI] = 0$$

and $d \sum_{q=1}^{n} r^2/db = 2nk^2b + 2k^2 \sum_{q=1}^{n} t + 2k \sum_{q=1}^{n} \ln[(RI - R)/RI] = 0.$

$$\quad (12.37)$$

Equations 12.36 and 12.37 can be solved to give:

$$\hat{k} = - \frac{\{n \sum_{q=1}^{n} t.\ln[(RI - R)/RI] - \sum_{q=1}^{n} \ln[(RI - R)/RI]. \sum_{q=1}^{n} t\}}{n \sum_{q=1}^{n} t^2 - \left(\sum_{q=1}^{n} t\right)^2} \quad (12.38)$$

and $\hat{b} = - \dfrac{\{k \sum_{q=1}^{n} t + \sum_{q=1}^{n} \ln[(RI - R)/RI]\}.}{nk} \quad (12.39)$

RI can initially be assigned a value of 100, and $\hat{k}$ and $\hat{b}$ calculated from equations 12.38 and 12.39. Using these values, $\sum_{q=1}^{n} r^2$ is then calculated from equation 12.35. RI is then decremented and the procedure repeated until values of $\hat{k}$, $\hat{b}$ and $\hat{RI}$ are found which minimise $\sum_{q=1}^{n} r^2$.

From equation 12.34:

$$dR/dt = RI.k \exp[-k(t + b)]$$

so that if the computations are performed for mineral (m) and gangue (g), then at optimum flotation time:

$$RI_m k_m \cdot \exp[-k_m(t + b_m)] = RI_g k_g \cdot \exp[-k_g(t + b_g)]$$

from which optimum flotation time =

$$\frac{\{\ln[RI_m k_m/(RI_g k_g)] - k_m b_m + k_g b_g\}}{k_m - k_g}. \qquad (12.40)$$

A computer program (KINETIC) for analysing the kinetics of batch flotation tests can be found in Appendix 3.

In a complex flotation circuit, the rougher flotation may be divided into stages, each delivering its concentrate into the cleaning circuit at a different location according to its grade. Thus the basic flowsheet consisting of cleaners and recleaners may be supplemented by a low-grade cleaning circuit (Fig. 12.24).

To ensure the recovery of the weakly aerophilic particles which are passed to the particular section of the cleaning plant, it is essential that the retention time in the cleaning stage is at least that of the corresponding roughing section.

Since the object of the scavenging section is to promote maximum recovery by minimising the losses to tailings, it is advisable to dimension the scavengers generously, to allow not only for the slow-floating character of the particles, but also to allow for fluctuations in the circuit. However, it is important to avoid excessive overloading of the system with large volumes of low-grade material, so a compromise must be made in designing the scavenger circuit.[79] It may be preferable to have a lower rougher concentrate grade (longer flotation time), and more cleaning stages, thus reducing the volume of scavenger concentrate produced.

This is particularly important in non-metallic flotation, where, due

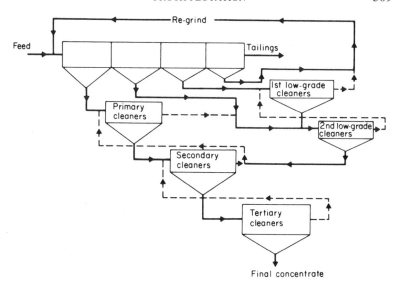

FIG. 12.24. Complex flotation circuit.

to the generally low ratio of concentration, large circulating loads are often produced. For instance, in the flotation of a low-grade metallic ore, the ratio of concentration may be as high as 50, so that only about 2% of the ore is removed as concentrate, and the circulating loads in the system are of this order of magnitude. Non-metallic ores, however, are often of high grade, and the ratio of concentration can be as low as 2, so that 50% of the ore is removed as concentrate, and very high circulating loads are produced. Control of such circuits can often be facilitated by the addition of a thickener, or agitator, which can act as surge capacity for large changes in circulating load which may arise when changes in ore grade occur.[70]

If cleaning does not give the required concentrate grade, regrinding of the rougher concentrate may be needed, it usually being necessary to at least regrind the scavenger concentrate, and sometimes the primary cleaner tailings, before recirculation to the rougher circuit. The purpose of primary grinding is to promote maximum recovery, by rendering most of the valuable mineral floatable so that the bulk of the gangue can be discarded, thus reducing the amount of material that must be further processed. In secondary grinding, or regrinding, the major consideration is the grade of the concentrate.

Regrinding of the middlings products is common practice in most flotation plants. Both the scavenger product and the cleaner tailings contain essentially a slow floating, fairly metal-rich fine fraction and a coarse product consisting mainly of unliberated mineral. These products are generally classified if the amount of fines is appreciable, after which the coarse product is reground and returned to the system with the new feed. The fine classified product is either recycled to the rougher circuit, or cleaned in a separate circuit to a grade high enough to be fed to the final concentrate or the main cleaner system.

Regrinding practice depends to a large extent on the ore mineralogy. In certain circumstances, particularly when the mineral is of high floatability, and is associated with an unfloatable gangue, it may be economic to grind at a relatively coarse size and regrind the rougher concentrate (Fig. 12.25). This is common practice with such minerals as molybdenite, which is readily floatable, when associated with hard, abrasive gangues. Removing the gangue as a tailings at a coarse particle size considerably reduces the power consumption in the grinding stage.

Figure 12.26 shows the flotation circuit of the Phoenix Copper Div. of Granby Mining Corp., Canada.[80] The main copper mineral is

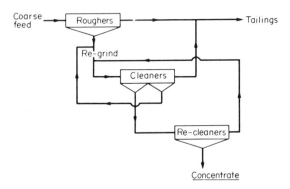

FIG. 12.25. Regrinding of rougher concentrates.

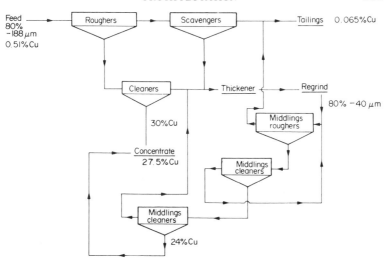

FIG. 12.26. Phoenix Copper Div., flotation circuit.

chalcopyrite, some of which is finely disseminated in the gangue, but it also occurs as complex grains with pyrite. The circuit removes the fairly coarse chalcopyrite early by one-stage rougher–scavenger–cleaning. The middlings from this stage, consisting essentially of the finely disseminated copper minerals, are reground and floated in a completely separate circuit utilising two cleaning stages, thus isolating the flotation of the coarse material ($80\% - 188$ μm) from the flotation of the very fine particles ($80\% - 40$ μm).

Selective flotation circuits, which concentrate two or more minerals, must incorporate substantial facilities for control. Where, for example, heavy sulphide ores are being treated, it is common for a bulk float to be initially removed. This isolates the sulphides from the non-sulphides, thus simplifying the subsequent selective separation of each sulphide component providing that this is not inhibited by the presence of reagents from the bulk float which are adsorbed on the mineral surfaces. If this is serious, direct selective flotation must be used, which is essentially two or more one-product circuits in series, although some plants treating difficult ores use a combination of bulk and selective flotation in the rougher operations. Mineral composition

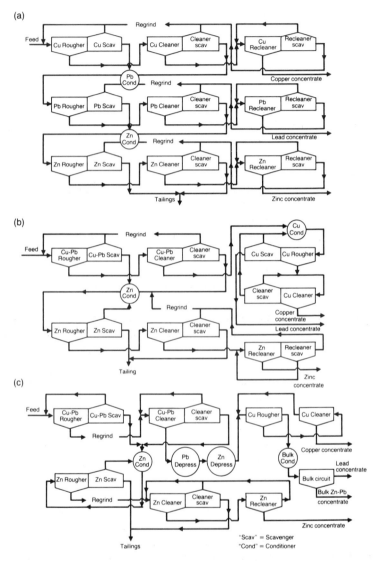

FIG. 12.27. Typical flotation flowsheets used for complex sulphide processing-production of three concentrates (after Barbery[81]).

and the degree of intergrowth of the valuable minerals are also important factors. Extremely fine intergrowth inhibits selective flotation separation, and there are some complex ores, containing copper, lead and zinc sulphides, which cause extreme difficulties in selective flotation. Figure 12.27 gives an outline of three flowsheets in use for such ores, from an "easy" coarse-grained ore (a), to a "difficult" fine-grained ore (c).

Some flotation plants are in operation where more than five concentrates are effectively recovered from a single feed and such operations demand considerable modification in the chemical nature of the feed pulp for each stage in the total treatment. The pH of the pulp may have to span a range from as low as 2.5 to as high as 10.5 to recover sulphide minerals alone, and further complications can arise if non-sulphide minerals, such as cassiterite, fluorite, barytes, etc., are to be recovered with sulphides.

Figure 12.28 shows a circuit which has been used to treat a complex ore containing copper, zinc, and iron sulphides, and cassiterite disseminated in a siliceous gangue.

A relatively coarse primary grind is used in order to recover as much cassiterite at as coarse a size as possible in the subsequent gravity process. After conditioning with copper sulphate to activate the sphalerite, the relatively large amount of sulphide minerals, which would interfere with the recovery of the cassiterite, is removed by bulk flotation at neutral pH. The bulk rougher concentrate is reground to release finely disseminated cassiterite, after which cleaning is undertaken, the cleaner tailings being recirculated to the head of the system. The bulk cleaner concentrate is conditioned with lime to about pH 11, which depresses the pyrite, and the copper and zinc sulphides are floated and cleaned, leaving the pyrite in the rougher tailings.

A significant problem in connection with flotation circuit design is that of transposing times from batch tests to flotation times in the continuous working circuit. The fundamental difference between a batch test and a continuous process is that every part of the flotation pulp in a batch test remains in the cell for the same length of time, whereas in a cell with continuous flow there is a spread, often quite considerable, in the retention times of different unit volumes. Some of the pulp takes a short cut and passes out of the cell relatively quickly,

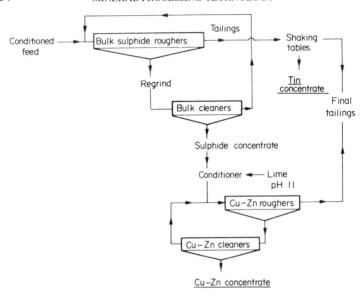

FIG. 12.28.   Copper–zinc–tin separation circuit.

with the result that flotation is incomplete. To reduce this problem the desired total flotation volume is divided into smaller portions.

The total cell volume required to give the specified nominal flotation time must, of course, be computed with allowance for the fact that only a part of the actual cell volume is available for the pulp. From the gross volume must be subtracted the volumes of the rotor machinery and stator, the froth layer and the air present in the pulp during the flotation process. Calculations indicate that the net volume in some cases is only about 50% of the gross cell volume. It must be remembered, however, that this factor only gives an adequate nominal retention time, without providing a safety margin for partial short-circuits in the flow as mentioned above or for the fluctuations that are liable to occur in the system. A factor of safety of two is usually applied to laboratory flotation times in order to determine the required cell volume of the full-scale plant.

It should be noted that although increased aeration results in faster flotation, it also results in shorter retention times, as a larger portion of the total volume is occupied by air. There is, therefore, an optimum rate of air supply to the cells, above which recovery may be reduced. This is not evident from the results of batch tests.

Circuit Flexibility

The decision having been reached to design a flotation circuit according to a certain scheme, it is necessary to provide for fluctuations in the flowrate of ore to the plant, both large and small, and for minor fluctuations in grade of incoming ore.

The simplest way of smoothing out grade fluctuations and of providing a smooth flow to the plant is by interposing a large storage agitator between the grinding section and the flotation plant:

$$\text{Grind} \rightarrow \frac{\text{Storage}}{\text{Agitator}} \rightarrow \frac{\text{Flotation}}{\text{Plant}}$$

Any minor variations in grade and tonnage are smoothed out by the agitator, from which material is pumped at a controlled rate to the flotation plant. The agitator can also be used as a conditioning tank, reagents being fed directly into it. It is essential to precondition the pulp sufficiently with the reagents before feeding to the flotation banks, otherwise the first few cells in the bank act as an extension of the conditioning system, and poor recoveries result.

Provision must be made to accommodate any major changes in flow rate which may occur; for example, if a number of mills have to be shut down for maintenance. This is achieved by splitting the feed into parallel banks of cells (Figs. 12.29 and 12.16(b)), each bank requiring an optimum flow rate for maximum recovery. Major reductions in flow rate below the designed maximum can then be accommodated by shutting off the feed to the required number of banks. The optimum number of banks required will depend on the ease of control of the particular circuit. More flexibility is built into the circuit by increasing the number of banks, but the problems of controlling large numbers of banks must be taken into account, and in plants that have installed very large unit processes, such as grinding mills, flotation machines, etc., in

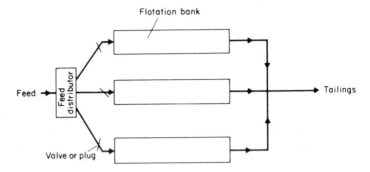

Fɪɢ. 12.29.   Parallel flotation banks.

order to reduce costs and facilitate automatic control, the need for many parallel banks has been reduced.

In designing each flotation bank, the number of cells required must be assessed; should a few large cells be incorporated or many small cells giving the same total capacity? This is determined by many factors. If a small cell in a bank containing many such cells has to be shut down, then its effect on production and efficiency is not as large as that of shutting down a large cell, in a bank consisting of only a few such cells. Maintenance costs, however, tend to be lower with large cells, since there are relatively fewer parts to change in a particular bank.

The desired residence time for maximum economic recovery, which is calculated from laboratory tests, assumes that every particle is given the chance to float in that time. This does not necessarily happen in a continuous process, as it is possible for particles to short-circuit immediately from one cell to the next. This becomes increasingly more serious the fewer cells there are in the bank. Designing a bank with many small cells gives particles which have short-circuited in one or more cells the chance to float in succeeding cells. The designer, therefore, must decide between small cells for greater flexibility and metallurgical performance and large cells, which have a smaller total capital cost, less floor area per unit volume and lower operating costs. In the USSR and Eastern Europe, it is common to install 30 or more machines in a single bank, while in the West, the trend is to install very large cells, especially in the roughing stage.

Many modern flotation plants use between eight and fourteen cells in the rougher banks to produce an optimum design, depending on the most economic layout of the plant. This has the effect of limiting the use of 28 m³ (1000 ft³) cells to mills with tonnage throughputs of 15 000 t d⁻¹ or higher, although some machine manufacturers, particularly Outokumpu, recommend using the largest cells possible, which reduces the number of mechanisms, in some cases to only two in a bank. There are reports that recovery is unimpaired, or even enhanced, at the same total retention time. As Young[82] observes, a clear difference of opinion has emerged, which requires further research.

Flexibility must be provided relative to the number of cells in a bank producing rougher and scavenger concentrates, in order to allow for changes in the grade of incoming ore. For instance, if the ore grade decreases, it may be necessary to reduce the number of cells producing rougher concentrate, in order to feed the cleaners with the required grade of material. A simple method of adjusting the "cell split" on a bank is shown in Fig. 12.30. If the bank shown has, say, twenty cells, each successive four cells feeding a common launder, then by plugging outlet B, twelve cells produce rougher concentrate. Similarly, by plugging outlet A, only eight cells produce rougher concentrates, and by leaving both outlets free, a ten–ten cell split is produced.

At North Broken Hill in Australia, the lead recleaner concentrate grade is automatically controlled by stabilising the mass flowrate of recleaner feed. An increase in flowrate increases the cleaner concentrate grade, due to the decreased residence time within the

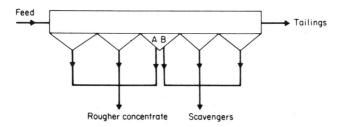

FIG. 12.30. Control of cell split.

cells. Automatically controlled froth diverter trays (Fig. 12.31) increase the number of cells producing concentrate to compensate for the increase in feed rate, and the number of cells producing middlings is correspondingly reduced (Fig. 12.32).

FIG. 12.31. Automatic froth diverter trays.

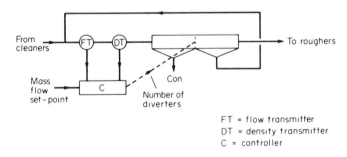

FT = flow transmitter
DT = density transmitter
C = controller

FIG. 12.32. Lead recleaning circuit. North Broken Hill, Australia.

Flotation Machines

Although many different machines are currently being manufactured and many more have been developed and discarded in the past, it is fair to state that two distinct groups have arisen: pneumatic, and mechanical machines. The type of machine is of great importance in designing a flotation plant and is frequently the characteristic causing most debate.

Pneumatic machines either use air entrained by turbulent pulp addition (cascade cells), or more commonly air either blown in or induced, in which case the air must be dispersed either by baffles or some form of permeable base within the cell. Generally pneumatic machines give a low-grade concentrate and little operating trouble. Since air is used not only to produce the froth and create aeration but also to maintain the suspension and to circulate it, an excessive amount is usually introduced and for this and other reasons they are now rarely used.

One of the developments in the pneumatic field is the *Davcra* cell (Fig. 12.33), which has been claimed to yield equivalent or better performance than a bank of mechanical machines.

The cell consists of a tank segmented by a vertical baffle. Air and feed slurry are injected into the tank through a cyclone-type dispersion nozzle, the energy of the jet of pulp being dissipated against the

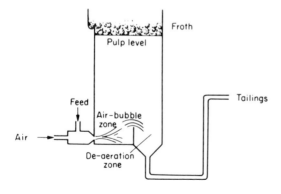

FIG. 12.33. Davcra cell.

vertical baffle. Dispersion of air and collection of particles by bubbles allegedly occurs in the highly agitated region of tank confined by the baffle. The pulp flows over the baffle into a quiescent region designed for bubble–pulp disengagement. The cell can be used for roughing or cleaning applications on a variety of minerals. Although not widely used, Davcra cells have replaced some mechanical cleaner machines at Chambishi copper mine in Zambia, with reported lower operating costs, reduced floor area and improved metallurgical performance. A four-compartment Davcra cell is in operation at Bougainville Copper Ltd in Papua New Guinea, as a tertiary cleaner in the copper circuit.[83]

A signficant development in recent years has been the increasing industrial use of *flotation columns*, primarily in Canada. Columns have been claimed to give better separations than cell-type machines, particularly when operating on fine materials. A typical configuration of a column is shown in Fig. 12.34. It consists of two distinct sections. In the section below the feed point (the recovery section), particles suspended in the descending water phase contact a rising swarm of air bubbles produced by a sparger in the column base. Floatable particles collide with and adhere to the bubbles and are transported to the

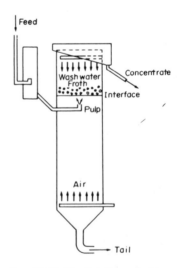

FIG. 12.34. The flotation column.

washing section above the feed point. Non-floatable material is removed from the base of the column as tailing. Gangue particles that are loosely attached to bubbles or are entrained in bubble slipstreams are washed back into the recovery section, hence reducing contamination of the concentrate. The wash water also serves to suppress the flow of feed slurry up the column toward the concentrate outlet. There is a downward liquid flow in all parts of the column preventing bulk flow of feed material into the concentrate. Two column flotation units were installed in the molybdenum circuit at Mines Gaspé, Canada, in 1980, and excellent results have been reported.[84] The units, of square cross section and with sides of length 91.5 cm and 46 cm respectively, replaced mechanical machines in the cleaner banks. Since then many of the copper–molybdenum producers in British Columbia have installed columns for molybdenum cleaning, and there now appears to be widespread interest in them. Instrumentation and some degree of automatic control is a necessity for column operation. Dobby et al.[85,86] describe extensive model and industrial-scale investigations, particularly at Gibraltar Mines Ltd., which have lead to the development of a control strategy for the columns.

The U.S. Bureau of Mines has recently compared column flotation with conventional flotation on a Montana chromite ore, the results showing that column flotation appears to be a physical improvement in the flotation separation process.[87] Because of the excellent results achieved, further studies of column flotation are underway on ores containing fluorite, manganese, platinum, palladium, titanium, and other minerals. The United Coal Co. in the United States has also pioneered the use of columns for the flotation of fine coal.[88] It is possible that, due to their froth washing capability, columns may find an increasing use in the future for treating ores that need extensive fine grinding, followed by de-sliming and multi-stage cleaning.

Froth separators were developed in the USSR in 1961 and had treated 8M tonnes of various ores by 1972.[89] The principle of the froth separator (Fig. 12.35) is that conditioned feed is discharged onto the top of a froth bed, the hydrophobic particles being retained, while the hydrophilic species pass through and are thereby separated. This method is particularly suited to the separation of coarse particles. The slurry is introduced at the top of the machine and descends over

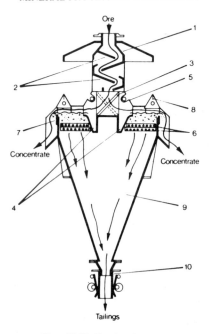

FIG. 12.35. Froth separator.

sloping baffles before entering an aeration trough, where it is strongly aerated before flowing horizontally onto the top of the froth bed. Water and solids which penetrate the froth bed pass between aerator pipes into the pyramidal tank. The aerators are rubber pipes with 40–60 fine holes per cubic centimetre, into which air is blown at 115 kPa. The machine, which has two froth discharge lips, each 1.6 m long, is capable of treating 50 t h^{-1} of solids at slurry densities of between 50 and 70% solids. Although little used in the West, they have great potential for treating coarse feeds at up to ten times the rate of mechanical machines. The upper size limit for flotation is increased to about 3 mm, but they are not suited to fines treatment, a typical feed size range being 75 μm to 2 mm. The rôle of flotation time is reversed, increasing flotation times reducing the recovery but increasing the concentrate grade.

Mechanical flotation machines are the most widely used, being characterised by a mechanically driven impeller which agitates the slurry and disperses the incoming air into small bubbles. The machines may be self-aerating, the depression created by the impeller inducing the air, or "supercharged" where air is introduced via an external blower. In a typical flotation bank, there are a number of such machines in series, "cell-to-cell" machines being separated by weirs between each impeller, whereas "open-flow" or "free-flow" machines allow virtually unrestricted flow of the slurry down the bank.

The most pronounced trend in recent years, particularly in the flotation of base metal ores, has been the move towards larger capacity flotation cells, with corresponding reduction in capital and operating costs, particularly where automatic control is incorporated. In the mid-60s, flotation cells were commonly 5.7 m^3 (200 ft^3) in volume,

FIG. 12.36. Mechanical flotation cells.

or less (Fig. 12.36), whereas 8.5 m^3 to 14.2 m^3 cells are now widely used (Fig. 12.37), with 28.3 m^3 cells, and larger being increasingly adopted. Manufacturers in the forefront of this industry include Denver Equipment (36.1 m^3), Galigher (42.5 m^3), Wemco (85 m^3), Outokumpu Oy (38 m^3) and Sala (44 m^3). Many other machines are manufactured, however, and are comprehensively reviewed by Harris[90] and Young.[82]

Most of the flotation machines now in use are of the "open-flow" type, as they are much better suited to high throughputs and are easier to maintain than cell-to-cell types. The Denver "Sub-A" is perhaps the most well-known cell-to-cell machine, having been used widely in the past in small plants, and in multi-stage cleaning circuits, where the pumping action of the impellers permitted the transfer of intermediate flows without external pumps. They are now manufactured with cell sizes of up to 14.2 m^3, and are used mostly as coal cleaning devices, where the users have reported a significant improvement in selectivity over open-flow designs.

FIG. 12.37. 14.2 m^3 Denver D–R flotation cells.

The flotation mechanism is suspended in an individual square cell separated from the adjoining cell by an adjustable weir (Fig. 12.38). A feed pipe conducts the flow of pulp from the weir of the preceding cell to the mechanism of the next cell, the flow being aided by the suction action of the impeller. The positive suction created by the impeller draws air down the hollow stand-pipe surrounding the shaft. This air stream is sheared into fine bubbles by the impeller action and is intimately mixed with the pulp which is drawn into the cell onto the rotating impeller. Directly above the impeller is a stationary hood, which prevents "sanding-up" of the impeller if the machine is shut down. Attached to this hood are four baffle vanes, which extend almost to the corners of the cell. These prevent agitation and swirling of the pulp above the impeller, thus producing a quiescent zone where bubbles can ascend with their mineral load without being subjected to

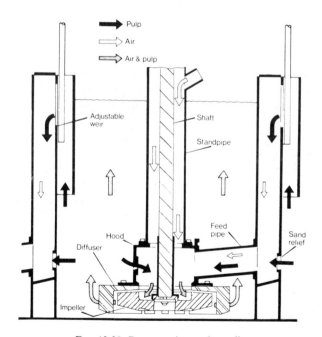

FIG. 12.38. Denver sub-aeration cell.

scouring which may cause them to drop it. The mineral-laden bubbles separate from the gangue in this zone and pass upward to form a froth. As the bubbles move to the pulp level, they are carried forward to the overflow lip by the crowding action of succeeding bubbles, and quick removal of froth is accomplished by froth paddles which aid the overflow.

Pulp from the cell flows over the adjustable tailings weir, and is drawn on to the impeller of the next cell where it is again subjected to intense agitation and aeration. Particles which are too heavy to flow over the tailings weir are by-passed through sand relief ports, which prevent the build-up of coarse material in the cell.

The amount of air introduced into the pulp depends upon the impeller speed, which is normally in the range of 7–10 m s^{-1} peripheral. More air may be obtained by increasing the impeller speed, but this may in certain circumstances overagitate the pulp, as well as increase impeller wear and power consumption. In such cases, supercharging may be applied by introducing additional air down the stand-pipe by means of an external blower.

Supercharging is required with the Denver D–R machine (Figs. 12.37 and 12.39), which ranges in size from 2.8 m^3 to 36.1 m^3, and which was developed as a result of the need for a machine to handle larger tonnages in bulk-flotation circuits. These units are characterised by the absence of intermediate partitions and weirs between cells. Individual cell feed pipes have been eliminated, and pulp is free to flow through the machine without interference. The pulp level is controlled by a single tailings weir at the end of the trough. Flotation efficiency is high, operation is simple and the need for operator attention is minimised. Most high-tonnage plants today use a free-flow type of flotation machine and many are equipped with automatic control of pulp level and other variable factors.

The most widely used flotation machine is the Wemco Fagergren (Figs 12.40 and 12.41) manufactured in sizes up to 85 m^3. The modern $1 + 1$ design consists of a rotor–disperser assembly, rather than an impeller, and the unit usually comprises a long rectangular trough, divided into sections, each containing a rotor–disperser assembly. The feed enters below the first partition, and tails go over partitions from one section to the next, the pulp level being adjusted at the end-tailings

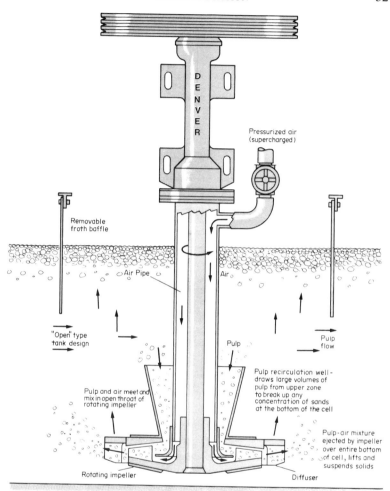

FIG. 12.39. Denver D–R flotation machine.

weir. Pulp passing through each cell, or compartment, is drawn upwards into the rotor by the suction created by the rotation. The rotor also draws air down the standpipe, no external blower being needed. The air is thoroughly mixed with the pulp before being broken into

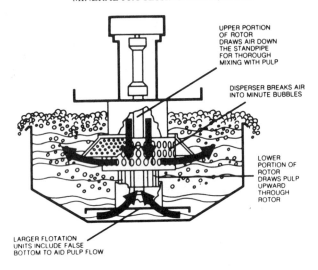

UPPER PORTION
OF ROTOR
DRAWS AIR DOWN
THE STANDPIPE
FOR THOROUGH
MIXING WITH PULP

DISPERSER BREAKS AIR
INTO MINUTE BUBBLES

LOWER
PORTION OF
ROTOR
DRAWS PULP
UPWARD
THROUGH
ROTOR

LARGER FLOTATION
UNITS INCLUDE FALSE
BOTTOM TO AID PULP FLOW

FIG. 12.40. Wemco Fagergen cell.

small, firm bubbles by the disperser, a stationary, ribbed, perforated band encompassing the rotor, by abruptly diverting the whirling motion of the pulp.

Perhaps the most well known of the supercharged machines is the Galigher Agitair[91] (Fig. 12.42).

This system again offers a straight-line flow of pulp through a suitably proportioned row of cells, flow being produced by a gravity head. The Agitair machine is often used in large-capacity plants. In each compartment, which may be up to 42.5 m^3 in volume, is a separate impeller rotating in a stationary baffle system. Air is blown into the pulp through the hollow standpipe surrounding the impeller shaft, and is sheared into fine bubbles, the volume of air being controlled separately for each compartment. Pulp depth is controlled by means of weir bars or dart valves at the discharge end of the bank, while the depth of froth in each cell can be controlled by varying the number and size of froth weir bars provided for each cell. Agitair machines produce copious froths and have found favour in mills

FIG. 12.41. Wemco 42.5 m³ flotation cell.

handling ores of poor floatability, which require large froth columns to help weakly aerophilic particles to overflow.

Outokumpu Oy is the largest mining company in Finland, and as well as operating several base metal mines and concentrators, it is particularly well known for its sophisticated products, including its range of OK flotation cells. It has been the policy at all Outokumpu concentrators to use very large flotation machines where possible, as not only does this reduce costs, but it also facilitates automatic control of such operations as level control, aeration, etc. In 1970, a working group was set up within the company to develop a new flotation mechanism. The initial studies were performed at the Pyhasalmi concentrator,[92] and in 1972 the Prototype OK-3 (3 m³ cell) was tested in the copper flotation circuit. During 1973 all the existing Agitair mechanisms in the copper rougher and scavenger flotation sections

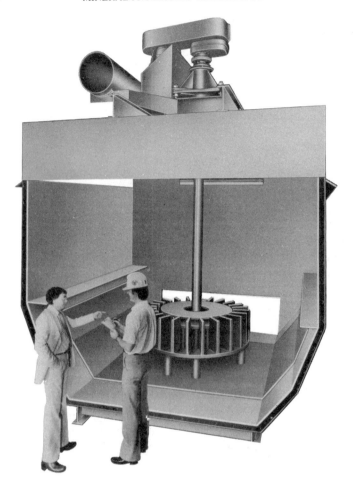

Fig. 12.42. 42.5 m³ Agitair flotation machine.

were replaced by OK-3 cells. Due to a decision to increase plant capacity, the zinc circuit was equipped with OK-16 machines in 1975, and in 1981 all the 56 small cells in pyrite flotation were replaced by one 2-cell OK-38 machine. In 1983, a prototype OK-60 cell was being tested in the circuit.

The OK impeller differs markedly from that of most other machines. It consists of a number of vertical slots which taper downwards, the top of the impeller being closed by a horizontal disc (Fig. 12.43). As it rotates, slurry is accelerated in the slots and expelled near the point of maximum diameter. Air is blown down the shaft and the slurry and air flows are brought into contact with each other in the rotor–stator clearance, the aerated slurry then leaving the mechanism to the surrounding cell volume. The slurry flow is replaced by fresh slurry which enters the slots near their base where the diameter and peripheral speed are less. Thus the impeller acts as a pump, drawing in slurry at the base of the cell, and expelling it outwards. The U-shaped cell profile minimises short-circuiting, as pulp flow is towards the bottom of the cell and the new feed entering is directed towards the mechanism due to the suction action of the rotor. It is because of this that banks containing only two large cells have been successfully operated, and OK cells are now in use in many of the world's concentrators.[93]

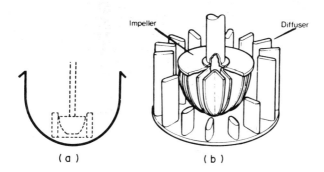

Impeller   Diffuser

(a)   (b)

Fig. 12.43. Outokumpu flotation machine: (a) cell profile, (b) impeller mechanism.

The Sala AS series of flotation cells, ranging in size from 1.2 to 44 m³, differs in design from the machines previously described. Most machines are designed to promote ideal mixing conditions, the vertical flows achieved maintaining solids in suspension. The Sala design (Fig. 12.44) minimises vertical circulation, the manufacturers claiming that the natural stratification in the slurry is beneficial to the process. The impeller is positioned under a stationary hood which extends out to, and supports, the stationary diffuser. The impeller is a flat disc with vertical radial blades on both surfaces, the upper blades expelling air which is blown down the standpipe, and the lower blades expelling slurry from the central base area of the tank, all slurry flow into the impeller being from below. The aerated slurry is then expelled through the conventional circular stator. Although the impeller has an unusually large diameter in relation to the rather shallow cell size, this preventing sanding in the corners, it is claimed that the air is dispersed into very closely sized fine bubbles, which are particularly suited to fines flotation. The machines are used to treat a variety of materials, including base metals, iron ore, coal and non-metallic minerals.

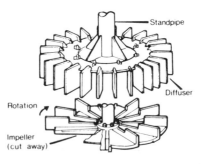

FIG. 12.44. Sala flotation mechanism.

Comparison of Flotation Machines

Selection of a particular type of flotation machine for a given circuit is usually the subject of great debate and controversy.

The main criteria in assessing cell performance are:[94]

(1) Metallurgical performance, i.e. product recovery and grade.
(2) Capacity in tonnes treated per unit volume.
(3) Power consumption/tonne.
(4) Economics, e.g. initial costs, operating and maintenance costs.

In addition to the above factors, less tangible factors, such as the ease of operation and previous experience of personnel with the equipment, may contribute.

Direct comparison of cells is by no means a simple matter. Although comparison of different cell types, such as mechanical against pneumatic, should be based on metallurgical performance when testing the same pulp in parallel streams, even here results are suspect; much depends on the operator's skill and prejudices, as an operator trained on one type of cell will prefer it to others.

In general the differences between mechanical machines are small and selection depends a lot on personal preference. One of the basic problems that hampers comparison of flotation cells is that a cell is required to perform more than selective collection operation; it is required to deliver the collected solids to a concentrate product with minimal entrainment of pulp. The observed rate of recovery from a cell can be dependent on the froth-removal rate, which in turn can be affected by such process variables as reagent additions, impeller speed, position of the pulp–froth interface and aeration. Research has shown that bubble size in mechanical machines of all designs is between 0.1 to 1 mm, the size being controlled mainly by the frother. The machine stator does not change the bubble size, but only the flow pattern in the cell.[90]

Mechanical and conventional pneumatic machines have been used for many decades, whereas froth separators and columns are fairly recent developments. Mechanical machines have been the dominant type, pneumatic machines, apart from the Davcra, now rarely being seen, except in a few old concentrators. Columns and froth separators have, as yet, to gain wide acceptance.

Although there is very little information on pneumatic machines, Arbiter and Harris[94] reported comparative tests on a number of mechanical and pneumatic cells showing the former to be generally superior. Gaudin[95] suggested that mechanical machines are better suited to difficult separations, particularly where fines are present. The impellers tend to have a scouring effect which removes slimes from particle surfaces.

An excellent analysis of the effectiveness of the various types of flotation machine in use today has been made by Young,[82] who discusses machine performance with regard to the basic objectives of flotation, which are the recovery of the hydrophobic species into the froth product, while achieving a high selectivity by retaining as much of the hydrophilic species as possible in the slurry. Achievement of recovery is dependent on the mechanism of particle–bubble attachment, which may be by "coursing bubble" contact between an ascending bubble and a falling particle, by precipitation of dissolved gas onto a particle surface, or by contact between a particle and an unstable "nascent" bubble in a pressure gradient. Coursing bubble attachment requires non-turbulent conditions, which is not found in mechanical or Davcra cells. A mechanical impeller can be compared to a turbine operating in a cavitating mode, air bubbles forming on the trailing, low pressure side of the impeller blades, while slurry flows are concentrated mainly upon the leading, high pressure side. The air and slurry flows are therefore separated to some extent, and the possibility of air precipitation on particles and for contact between particles and nascent bubbles is low. Mechanical impellers, therefore, do not appear to be the ideal device for particle–bubble contact, and the nozzle in the Davcra cell would appear to be much more efficient, which may explain why the Davcra cell can give the same recovery as a short bank of mechanical cells.

The particle–bubble contact in column machines is by coursing bubble only, and these are ideal displacement machines, whereas the mechanical cells are ideal mixers. The more favourable conditions for particle–bubble attachment, together with a lower tendency to break established bonds, may account for the high recoveries reported from flotation columns.

As selectivity is reduced by slurry turbulence, it is clear that the column machines, which also improve selectivity by froth washing, and to a lesser extent the Davcra, have an advantage over the mechanical machines. The froth separators, however, are not well suited to achieving selectivity from fine feeds, as the fine hydrophilic particles must descend through the total froth bed to report to the tailings, and this is difficult to achieve.

As Young concludes, mechanical flotation machines dominate the Western industry, but the reasons for this may be more due to commercial realities than to design excellence. The major Western manufacturers make only this type, and most flotation engineers are familiar with no other. However, in the future, the mechanical machines will no doubt encounter the increasing challenge of other types, and there is no reason why a number of different units could not be installed in concentrators for specific duties. For example, coarsely textured ores could be ground to -2 mm, and classified at about 200 μm, the coarse fraction being treated by froth separators, and the fine fractions in mechanical or Davcra machines. On the other hand, ores requiring extremely fine grinding and multi-stage cleaning could be processed by flotation columns.

Electroflotation

Industrial flotation is rarely applied to particles below 10 μm in size due to lack of control of air-bubble size. With ultrafine particles, extremely fine bubbles must be generated to improve attachment. Such bubbles can be generated by *in situ* electrolysis in a modified flotation cell. Electroflotation has been used for some time in waste-treatment applications to float solids from suspensions. Direct current is passed through the pulp within the cell by two electrodes, generating a stream of hydrogen and oxygen bubbles. Considerable work has been done on factors affecting the bubble size on detachment from the electrodes, such as electrode potential, pH, surface tension and contact angle of the bubble on the electrode. On detachment, the majority of bubbles are in the 10–60 μm range, and bubble density can be controlled by current density to yield optimum distribution of ultrafine bubbles as well as adequate froth control. Conventional flotation processes produce bubbles ranging from 0.6 to 1 mm in diameter and there is considerable variation in bubble size.

Some other factors have also been noted in addition to the fine bubbles. For example, the flotation of cassiterite is improved when electrolytic hydrogen is used for flotation. This may be due to nascent hydrogen reducing the surface of the cassiterite to tin, allowing the bubbles to attach themselves.

Although the main applications of electroflotation are in sewage treatment, this technique is capable of selectively floating solids and has been used in the food industry. It may have a future role in the treatment of fine mineral particles.[96]

Agglomeration-Skin Flotation

In agglomeration flotation, the hydrophobic mineral particles are loosely bonded with relatively smaller air-bubbles to form agglomerates, which are denser than water but less dense than the particles wetted by the water. Separation of the agglomerated and non-agglomerated particles is achieved by flowing film gravity concentration. When the agglomerates reach a free water surface, they are replaced by skin-floating individual particles. In skin-flotation, surface tension forces result in the flotation of the hydrophobic particles, while the hydrophilic particles sink, effecting a separation.

In *table flotation*, the reagentised particles are fed onto a wet shaking table. The pulp is diluted to 30–35% solids and is aerated by air jets from a series of drilled pipes arranged above the deck, at right-angles to the riffles, in such a way that the holes are immediately above the material carried by the riffles. The hydrophobic particles form aggregates with the air-bubbles and float to the water surface, from where they skin-float to the normal "tailings" side of the table. The wetted particles become caught in the riffles and discharge at the end of the table where the concentrate normally reports in most shaking table gravity-separation processes.

With table flotation, and other agglomeration processes, it is desirable to film and float the most abundant mineral, if possible, as the capacity of the table is limited to the amount of material that can be carried along the riffles. This is the reverse of the ideal conditions for froth flotation, where it is desirable to film and float the mineral that is least abundant, so as to reduce entrainment of unwanted material to the minimum. This difference renders table flotation most suitable for removing sulphide minerals from pyritic tin ore concentrate, or for the concentration of non-metallic minerals, such as fluorite, baryte, and phosphate rock. Such minerals are often liberated at sizes too coarse for conventional flotation. Agglomeration separations are possible over a wide range of sizes, usually from a maximum of about 1.5 mm in diameter to 150 μm. Minerals with low specific gravities, such as fluorite, have been separated at sizes of up to 3 mm.

Table flotation was used until relatively recently in the treatment of coarse phosphate rock, but has been replaced by similar methods utilising pinched sluices and spirals as the flowing film devices.[97]

Flotation Plant Practice

The Ore and Pulp Preparation

It is inevitable that there will be changes in the character of the ore being fed to a flotation circuit. There should therefore be means available for both observing and adjusting for such changes. Variation in the crystal structure and intergrowth may have an important effect on liberation and optimum mesh of grind. Change in the proportion of associated minerals is a very common occurrence and one which may be largely overcome by blending the ore both before and after crushing has been completed. When the feed is high grade it is relatively easy to produce a highly mineralised froth and high-grade concentrate; when feed is low grade it is harder to maintain a stable froth and it may be necessary to switch one of the final cleaning cells to a lower-grade section if cells and launders have suitable flexibility.

Fluctuation in the nature and proportion of minerals in the run-of-mine ore invariably occurs when ores are withdrawn from more than one location and the variation observed may be further accentuated by partial oxidation of the ore. This may occur from geological changes or from delayed transportation of broken ore from the stope in the mine to the processing plant. Oxidation also commonly occurs as a result of overlong storage in stockpiles or ore-bins and therefore it is necessary to determine how prone a particular ore is to oxidation and to ensure that the holding time is kept well below a critical level. Oxidised ores are softened by lattice decomposition and become more inert to collector reagents and more prone to overgrinding.

Wet grinding is the most important factor contributing to the performance of the flotation circuit. It is therefore of vital importance that the grinding circuit shall provide a reliable means of control as a guarantee that the milled product will allow maximum liberation of the values. In the comminution section of the plant, poor operation in the crushing stage can be offset in grinding. There is, however, no way of offsetting poor grinding practice˙and it is wise to use experienced operators on this section.

The degree of grinding required is determined by testwork and removal of free mineral at the coarsest possible size is always desirable. Modern flotation takes this into account as is evidenced by the general trend towards floating the mineral in stages; first coarse, then fine. The advantages of floating a mineral as coarse as possible include:

(i)   lower grinding costs;
(ii) increased recovery due to decreased slime losses;
(iii) fewer overground particles;
(iv) increased metallurgical efficiency;
(v)   less flotation equipment;
(vi) increased efficiency in thickening and filtration stages.

Laboratory control of grinding must be carried out on a routine basis, by screening and assaying the tailings in order to determine the losses in each size fraction and the reason for their occurrence. It is often found that the largest losses occur in the coarsest particles, due to inadequate liberation, and if grinding all the ore to a finer size improves recovery economically then it should be done. Often the losses occur in the very fine "sub-sieve" fractions, due to overgrinding of heavy mineral values. In this case it may be necessary to "scalp" the grinding circuit and remove heavy minerals which are returned to the mill by the classifier at sizes below optimum grind. This can be done by adding flotation reagents to the mill discharge and removing fine liberated minerals by a unit flotation cell between this and the classifier (Fig. 12.45). Apart from the advantages of reducing overgrinding, the density of the mill discharge can be controlled so as to give optimum

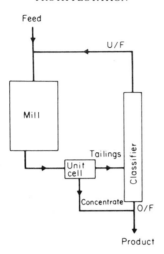

FIG. 12.45 Removal of fine heavy minerals from the grinding circuit by unit flotation cell.

flotation efficiency. In conventional circuits the mill discharge density is controlled according to the cyclone requirements, the cyclone overflows often requiring dewatering before feeding flotation.

Flotation within the grinding circuit, particularly of heavy, coarse lead minerals, is performed at several concentrators,[98] and the aim of Outokumpu's *flash flotation* method is to recover such coarse valuable minerals which would normally be recycled to the grinding circuit via classification. The concentrate produced is final concentrate, needing no further cleaning, and is produced in a specially designed flotation machine — *Skim-air* — which removes the coarse floatable particles while allowing the others to return to the mill for further grinding, thus reducing the amount of valuable mineral lost to fines and increasing the average particle size of the final concentrate. A number of these cells have been installed in Finnish concentrators, with considerable benefits.[99]

As was shown earlier, if the mineral is readily floatable, and is associated with a relatively non-floatable gangue, it may be more economical to produce a coarse final tailings and regrind the resulting low-grade rougher concentrate, which may then be considered almost as a middlings product (Fig. 12.25). The secondary, or regrind operation, treating only a small percentage of the original fine ore feed, can therefore be carried to a size fine enough to liberate. Subsequent flotation then produces the maximum recovery/grade results in the greatest economic return per tonne of ore milled. There is, of course, an upper limit on size at which flotation may be practised effectively, due to the physical limitations of the bubble in lifting coarse particles. Whilst it may be argued that factors such as shape, density and aerophilic properties may be influential, the practical upper limit rarely exceeds 0.5 mm and is usually below 0.3 mm. For a wide variety of minerals, reagents and flotation machines, recovery is greatest for particles in the size range 100–10 μm.[75,77] Below about 10 μm recovery falls steadily. There is no evidence of a critical size below which particles become unfloatable. The reason for the difficulties experienced in selective flotation of fine particles is not fully understood and varies from ore to ore. Very fine particles have relatively high surface area in relation to mass and tend to oxidise readily, or be coated with slimes before reaching the conditioning section, which makes collector adsorption difficult. Particles of low mass tend to be repelled by the slip-stream which surrounds a fast-moving air bubble, and should therefore be offered small, slow-moving bubbles. On the other hand, if small particles are in suspension near the froth column, they tend to overflow with the froth column regardless of their composition, as the downward pull of gravity is offset by the upward force due to the drift of the bubbles.

When the ore value is low, the *slimes* (the ultra-fine fraction which may be detrimental to flotation) are often removed from the granular fraction by passing the feed through de-sliming cyclones, and discarding the overflow. Alternatively, de-sliming can be carried out between flotation stages; for instance, rougher flotation may be followed by a de-sliming operation, which improves recovery in the scavenging stage. If the slimes contain substantial values, they are sometimes treated separately, thus increasing overall recovery.

The White Pine Copper Co. of Michigan, USA treats an ore consisting of chalcocite and native copper finely disseminated in a shale gangue.[100] Rapid flotation of the fine chalcocite and native copper is followed by de-sliming of the primary flotation tailings.

Elimination of these gangue slimes accelerates recovery of the middlings in the scavenging stage (Fig. 12.46).

The operating density of the pulp is determined by testwork, and is influenced by the mean size of particles within the feed. Coarse particles will settle in a flotation cell at a relatively rapid rate, which may be substantially reduced by increasing the volume of particles in the pulp. As a general rule higher-density pulps are applied to coarser sizes. In treating heavy sulphide ores, low-grade rougher concentrates are obtained from pulps of between 30 and 50% solids, whilst reground cleaner concentrates are obtained from pulps of between 10 and 30% solids.

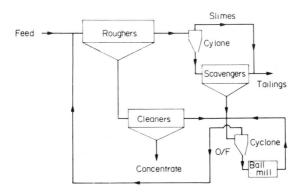

FIG. 12.46. White Pine Copper Co. flotation circuit.

Reagents and Conditioning

Each ore is a unique problem and reagent requirements must be carefully determined by testwork, although it may be possible to obtain guidelines for reagent selection from examples of similar operations. An enormous amount of experience and information is freely available from reagent manufacturers. One vital requirement of a collector or frother is that it becomes totally emulsified prior to usage. Suitable emulsifiers must be used if this condition is not apparent.

Selection of reagents must be followed by careful consideration of the points of administration in the circuit. It is essential that reagents are fed smoothly and uniformly to the pulp, which requires close control of reagent feeding and on the pulp flow-rate. Frothers are always added last where possible; since they do not react chemically they only require dispersion in the pulp, and long conditioning times are unnecessary. Adding frothers early tends to produce a mineralised froth floating on the surface of the pulp during the conditioning stage. This is due to entrained air, which can cause uneven distribution of the collector.

In flotation, the amount of agitation and consequent dispersion are closely associated with the time required for physical and chemical reactions to take place. Conditioning prior to flotation is now considered standard practice and is an important factor in decreasing flotation time. This is perhaps the most economical way of increasing the capacity of a flotation circuit. The minerals are converted to a readily floatable form as a result of ideal conditioning, and therefore a greater volume can be treated. Although it is possible to condition in a flotation machine, it is generally not economical to do so, although it is currently common practice for stage addition to include booster dosage of collector into cell banks, particularly at the transition from rougher to scavenger collection. Machines in the flow-line are often used as conditioners. Agitators are often interposed between the grinding mills and the flotation circuit to smooth out surges in grade and flow rate from the mills. Reagents are often added to these storage reservoirs for conditioning. Alternatively, reagents may be added to the grinding circuit in order to ensure optimum dispersion. The ball mill is a good conditioner and is often used when the collectors are oily and need emulsifying and long conditioning times. The advantage of conditioning in the mill is that the collector is present at the time that new surface is being formed, before oxidation can take place. The disadvantage is that reagent rate control is difficult, as the feed to the mill may have continual minor grade fluctuations, and the mill may have a high circulating load, which can become overconditioned. Where very close control of conditioning time is essential, such as in the selective flotation of complex ores, special conditioning tanks are incorporated into the flow-line (Fig. 12.47). The pulp and reagent are

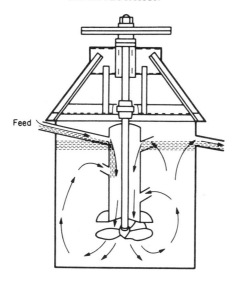

FIG. 12.47. Denver conditioning tank.

fed down the open standpipe and fall on to the propeller, which forces
the mixture downwards and outwards. The outlet at the side of the
tank can be adjusted to give a height sufficient to give the pulp its
desired residence time within the tank.

Stage addition of reagents often yields higher recoveries at
substantially lower cost than if all reagents are added at the same point
in the circuit prior to flotation. The first 75% of the values is normally
readily floatable, providing optimum mesh of grind is achieved. The
remaining values may well be largely composite in nature and will
therefore require more careful reagent conditioning, but perhaps 15%
are sufficiently large or sufficiently rich in value to be recovered
relatively easily. The remaining 10% can potentially affect the whole
economic balance of the process, being both fine in size and low in
values. Because this fraction is such a critical one it must be examined
extremely carefully and regularly and reagent addition must be
carefully and quickly controlled.

When feasible to do so it is usually more desirable to float in an alkaline or neutral circuit. Acid circuits usually require specially constructed equipment to withstand corrosion. It is a common finding that the effectiveness of a separation may occur within very narrow pH limits, in which case the key to success for the whole process lies in the pH control system. In selective flotation where more than one mineral is concentrated, the separation pH may well vary from one stage to the next. This, of course, makes it vitally important to regulate reagents to bring about these conditions and control them accurately.

The first stage of pH control is often undertaken by adding dry lime to the fine ore-bins, which tends to reduce oxidation of sulphide mineral surfaces. Final close pH control may be carried out on the classifier overflow, by the addition of lime as a slurry. The slurry is usually taken from a ring-main, as lime settles out quickly if not kept moving, and forms a hard cement within the pipelines.

Solid flotation reagents can be fed by rotating disc, vibro, and belt feeders, but more commonly reagents are added in liquid form. Insoluble liquids such as pine oil are often fed at full strength, whereas water-soluble reagents are made up to fixed solution strengths, normally about 10%, before addition. Reagent mixing is performed on day shift in most mills, under close supervision, to produce a 24-h supply. The aqueous reagents are usually pumped through ring-mains, from where they are drawn off to feeders as required.

A typical reagent feeder is the *Clarkson bucket feeder*. This has small buckets mounted on a vertical rotating disc, which dip into the liquid in the tank and spill out their contents on to a tipping plate on each revolution. The feed rate can be varied by adjusting the speed, the size of buckets, or more commonly, in operation, the angle of tilt of the plate, which governs the amount of liquid returned to the tank. A number of discs can be driven from one shaft, each dispensing a different reagent from a separate compartment in the tank. A very reliable and accurate feeder for small quantities is the *Peristaltic pump*. Variable-speed rollers squeeze a carrier tube seated in a curved track, thus displacing the reagent along the tube. Flow rates of between 0.1 and 850 ml min^{-1} can be obtained with these feeders.

Small amounts of frother are often injected directly into the pipeline carrying the flotation feed, by positive-displacement piston metering pumps.

Control of Flotation Plants

Automatic control is increasingly being used, the control strategies being almost as numerous as the number of plants involved. The key to effective control is on-line chemical analysis (Chapter 3), which produces real-time analysis of the metal composition of process streams.

Although there are many reports of successful applications, in reality few, if any, plants can claim to be *fully* automatic in the sense of operating unattended over extended periods. The main problems have been in developing process models which will define set-points and limits to accommodate changes in ore-type, mineralogy, texture, chemical composition of the mine water, and contamination of the feed. Control systems have been unsuccessful in some cases due to inadequate maintenance of instrumentation. It is essential, for instance, that pH probes are kept clean, and that all on-line instrumentation is regularly serviced and calibrated. Implementation of control strategies at the plant design stage have rarely been successful as the most significant control variables are often not identified until experience of the plant has been gained. Only then can control strategies based on these variables, and with specific objectives, be successfully attempted. The most successful systems have been those which allow the control room operator to interact with the computer when necessary to adjust set-points and limits. In this respect it is doubtful whether automatic control can achieve better metallurgical efficiency than experienced, conscientious operators. Its great advantage, however, is that the computer is constantly vigilant, not being affected by shift change-overs, tea-breaks and other interruptions which affect the human operator.

A flotation control system consists of various sub-systems, some of which may be manually controlled, while others may have computer controlled loops, but all contributing to the overall control objective.[101-103] The aim should be to improve the metallurgical efficiency, i.e. to produce the best possible grade-recovery curve, and to stabilise the process at the concentrate grade which will produce the most economic return from the throughput (Fig. 12.48), despite disturbances entering the circuit. This has not, as yet, been achieved by computer control alone.

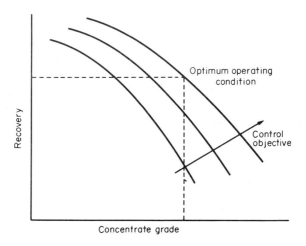

FIG. 12.48. Flotation control objective.

Disturbances caused by variations in feed rate, pulp density and particle size distribution should be minimal if grinding circuit control is effective, such that the prime function of flotation control is to compensate for variations in mineralogy and floatability. The variables which are manipulated either manually or automatically to effect this are reagent and air addition rates, pulp and froth levels, and circulating loads by the control of cell-splits on selected banks.

Control of slurry pH is a very important parameter in many selective flotation circuits, the control loop often being independent of the others, although in some cases the set-point is varied according to changes in flotation characteristics. For automatic control of lime or acid it is important that time delays in the control loop are minimised, which requires reagent addition as close as possible to the point of pH measurement. Since lime is often added to the grinding mills in order to minimise corrosion and to precipitate heavy metal ions from solution, there is a considerable and unacceptable distance–velocity lag if the pH probe is situated in the rougher cells. In such cases, excess lime may be added to the mills and final pH control made by acid addition at a point close to the roughers, such as into the conditioning tank. To prevent a considerable excess of lime being added to the mills, with a subsequent high acid requirement at the roughers, supervisory control can be used to minimise the acid consumption. In the circuit shown in Fig. 12.49, the computer monitors the acid consumption, and if this amount is excessive, changes the lime addition set-point to reduce the input to the grinding mills. This action can be adjusted so as to be very slow, thus minimising instability due to the long distance–velocity lag in the feed-back loop.[104]

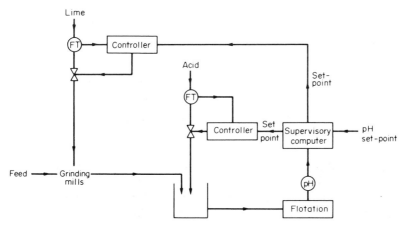

FIG. 12.49. Supervisory control of pH.

Control of collector addition rate is often performed by feed-forward control based on a linear response to assays or tonnage of valuable metal in the flotation feed. Typically, increase in collector dosage increases mineral recovery until a plateau is reached, beyond which further addition may either have no practical effect, or a slight reduction in recovery may occur. The gangue recovery also increases with collector addition such that beyond the plateau region, selectivity is reduced (Fig. 12.50).

The aim of collector control is to maintain the addition rate at the edge of the plateau, the main difficulty being in identifying the optimum point, especially when the response changes due to changes in ore-type, or the interaction with other reagents. For this reason,

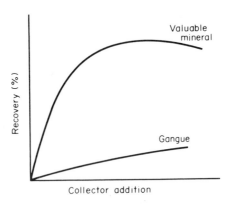

FIG. 12.50. Effect of collector addition.

automatic control, using feed-forward loops, has rarely been successful in the long term. There are many cases of successful semi-automatic control, however, where the operator adjusts the set-point to accommodate changes in ore-type, and the computer controls the reagent addition over fairly narrow limits of feed grade. For example, feed-forward control of copper sulphate activator and xanthate to the zinc roughers has been implemented into the control strategy at Mattagami Lake Mines, Canada.[105] The reagents are varied in proportion to changes in feed grade according to a simple algorithm:

$$\text{Feed rate} = A + (B \times \%\text{Zn in feed})$$

where A and B vary for the different reagents. The operator may change the base amount A as different ore types are encountered. The control system is shown in Fig. 12.51.

Although feed-forward control can provide a degree of stability, stabilisation may be more effective using feed-back data. The distance–velocity lag experienced with feed-back loops utilising tailings assays can be overcome to some extent by making use of the fact that the circuit begins to respond to changes in flotation characteristics immediately the ore enters the banks, and this can be detected by measurements in the first few cells. For instance, control of rougher concentrate grade is a useful strategy, as this strongly influences the final cleaner concentrate grade.

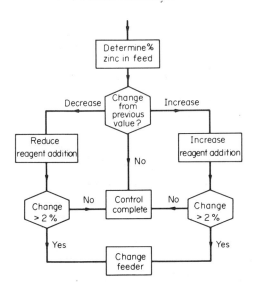

FIG. 12.51. Feed-forward control strategy at Mattagami Lake.

At Mount Isa in Australia, feed-forward control of xanthate addition to the copper roughers was unsatisfactory, as the optimum addition rate was not simply related to the mass of copper in the flotation feed.[106] The assay of concentrate produced in the first four cells of the bank was combined with the four-cell tailings and feed assays to compute the four-cell recovery. It was found that there was a linear response between this recovery and the collector dosage required to maintain the overall recovery at the edge of the plateau. The control strategy, although fairly successful in the short-term, eventually failed when changes in ore-type occurred. Computation of unit process recovery in this way is also subject to appreciable error due to the inherent inaccuracies of on-stream analysis data (see Example 3.14).

The amount of frother added to the flotation system is an important variable, but automatic control has been unsuccessful in many cases, as the nature of the froth is dependent on only very minor changes in frother addition and is much affected by intangible factors such as contamination of the feed, mine water content etc. At low addition rates, the froth is unstable and recovery of minerals is low, whereas increasing frother addition rate has a marked effect on the flotation rate, increasing the weight, and reducing the grade, of concentrate produced. Flowrate of concentrate has been controlled in some systems by regulating the frother addition. The grade is not as sensitive to changes in frother addition, but there may be a good relationship between grade and flowrate. Cascade control can be used, where the concentrate grade controls the concentrate flowrate set-point, which in turn controls the frother addition set-point (Fig. 12.52).

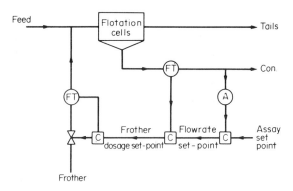

FIG. 12.52.   Cascade control of frother addition.

Air input to the flotation process, and froth depth, are parameters which, like frother addition, affect the recovery of minerals into the concentrate, and can be used to control concentrate grade, tailings grade or mass flowrate of concentrate. Aeration and froth depth do not, however, affect subsequent cleaning operations, as will residual frother carried over from the roughers, and they are often used as primary control variables. Flotation generally responds faster to changes in aeration than to changes in froth depth, and because of this aeration is often a more effective control variable, especially where circulating loads have to be controlled. There is obviously interaction between frother addition, aeration and froth depth, and where computer-controlled loops are used it is necessary to control these variables such that only minor changes are made. This can be done by manipulating only one of these variables, maintaining the others constant at predetermined optimum levels unless the conditions deviate outside acceptable limits, which may vary with ore-type. At Vihanti in Finland (Fig. 12.61), the copper grade of the bulk copper–lead rougher concentrate controls the rate of aeration and frother addition to the roughers and scavengers. Aeration has priority, being the cheaper "reagent" and leaving no residual concentration if used in excess. However, if the addition rate reaches a certain upper limit, then the frother rate is increased.[92]

The importance of froth depth is mainly due to the effect that it has on the gangue content of the concentrate. Free gangue can be carried into the concentrate mainly by mechanical entrainment, and the deeper the froth layer the more drainage of gangue into the cell occurs. Froth depth is very commonly used to control the concentrate grade, an increase in froth depth increasing the grade, but often at the expense of a slight reduction in recovery. Froth depth is often regarded as the difference between the pulp level and the level of the flotation cell overflow lip, and as such is controlled by changing the pulp level. The pulp level is most often measured by bubble tubes, as is the case for detecting slurry sump levels in grinding circuits, and control of the level is effected by either movable weirs, dart valves or pinch valves. Froth level set-points can be cascaded to aeration or frother set-point controllers in order to maintain the required depth. Specification of the actual froth depth requires a knowledge of the level of the froth column surface, which may not coincide with the height of the cell overflow lip. Figure 12.53 shows a device developed at Mattagami Lakes for sensing the level of the froth column, this level controlling the frother dosage set-point.[107] The sensor consists of a set of stainless steel electrodes connected to an electronic circuit which senses the number in contact with the froth. The seven electrodes, one of which is always immersed in the pulp, are of gradually decreasing length, so that the number in contact with the froth is directly proportional to the depth of the froth column.

Similar devices are used at Pyhasalmi in Finland (Fig. 12.60). The addition of copper sulphate activator to the zinc circuit is controlled mainly by the on-stream analysis data but an excess tends to depress

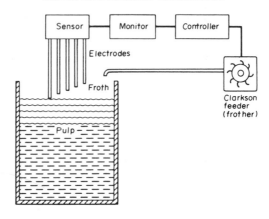

FIG. 12.53. Froth measuring device.

the froth level. The circuit contains several froth level measuring devices which indicate the improper addition of copper sulphate early enough to adjust the frother and sulphate addition to prevent a disturbance.

The ultimate aim of control is to increase the economic efficiency of the process by seeking to optimise performance. Evolutionary optimisation methods (Chapter 3) have been introduced into some systems to control variables which have very marked effects on process efficiency. The control method involves periodically adjusting the set-points of the controlled variables according to a defined procedure, the effect on economic efficiency being calculated and fed back to the operating system. Such methods cannot, however, be fully effective unless satisfactory stabilisation of plant performance can be achieved over long periods.

Herbst et al.[108] have discussed the use of advanced model-based control strategies in flotation, highlighting the advantages of these modern methods over classical control schemes.

The Black Mountain concentrator in South Africa has developed adaptive optimisation to control lead flotation.[109] Optimising control calculates the combination of metal recovery and concentrate grade which will achieve the highest economic return per unit of ore treated under the prevailing conditions. The criterion used to evaluate plant performance is the concept of economic efficiency (Chapter 1), in this case defined as:

$$E.E. = \frac{\text{Revenue derived per tonne of ore at achieved concentrate grade and recovery}}{\text{Revenue derived per tonne of ore at target concentrate grade and recovery}}$$

Target concentrate grade and recovery are calculated from the operating recovery-grade curve, which is continuously updated based on a 24-hour data bank, to allow for changes in the nature of the ore, quality of grinding, etc. Many factors influence the optimum combination of recovery and grade, such as commodity prices, reagent and treatment costs, transport costs, etc. The fundamental principle of adaptive optimisation is that concentrate grade and recovery can be predicted by on-line multivariable linear regression models, the coefficients of the models being continuously updated from the 24-hour data bank. Independent variables that determine grade and recovery can be reagent additions, grades of rougher concentrate, final concentrate and cleaner tailing, feed grade and throughputs. Some independent variables are controllable whilst others are not. The optimising routine is as follows:

— the relevant process variables are collected and filtered and the fit of the model is checked with residuals,
— the concentrate grade and recovery are predicted from the models on a grid of set-points that can be controlled,
— economic efficiencies are calculated for each combination of metal recovery and grade,
— the set-points of the lower level stabilising control loops are changed to those giving the highest predicted economic efficiency.

The optimising routine is performed every half hour at Black Mountain, and its introduction has reportedly brought about considerable improvements.

The Pyhasalmi concentrator has developed optimisation control based on a multi-linear response model, to optimise copper and zinc recoveries, and the balance of these metal values in each concentrate, to provide the highest economic efficiency (smelter value of metal in concentrate/value of metal in feed).[110] This takes into account factors such as the penalties caused by the presence of zinc in the copper concentrate and the increasing costs of transportation due to a low copper content in the concentrate. Cyanide addition is the most influential variable in the copper circuit, while copper sulphate dosage to the zinc rougher bank is adjusted to maximise the economic recovery of the total zinc flotation circuit. The effect of copper sulphate on the rougher concentrate assays and the scavenger tailing assay is determined, and the approach used is to apply multiple linear regression to a three hour history of data stored in the process control computer. With this procedure, the effect of copper sulphate changes on economic recovery can be determined, and therefore the requirement to either increase or decrease copper sulphate to improve the economic recovery is known. Copper sulphate changes, determined by the optimising control system, are usually made every 6–30 min.

A comprehensive control system for a flotation plant requires extensive instrumentation and involves a considerable capital outlay. Figure 12.54 shows the instrumentation requirement for a simple feed-

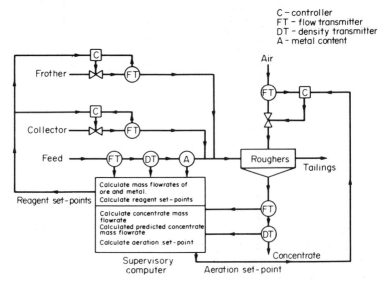

FIG. 12.54. Instrumentation for rougher circuit control.

forward system which could assist in control of a sulphide rougher bank, and Fig. 12.55 shows the instrumentation used in the control of the Mount Isa copper flotation circuit in Queensland, Australia.[106] Although various cascade control loops have been attempted in this circuit, they have been unsuccessful in the long term due to changes in feed conditions, and set-points within the loops are mainly controlled by the operators.

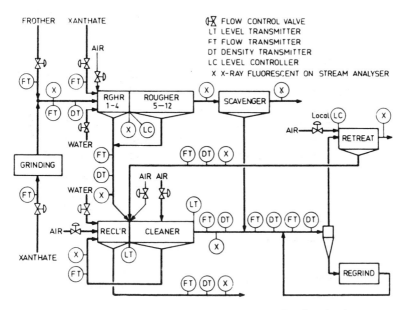

FIG. 12.55. Instrumentation in Mount Isa copper flotation circuit.

Lynch *et al.*[101] have analysed the cost of such installations, which provide potentially significant economic and metallurgical benefits. The majority of plants which have installed instrumentation for manual or automatic control purposes have reported improved metal recoveries varying from 0.5 to 3.0%, sometimes with increased concentrate grades. Reduction in reagent consumptions of between 10 and 20% have also been reported.

Typical Flotation Separations

The expansion of flotation as a method of mineral concentration can be observed from the following data. According to a survey carried out by the US Bureau of Mines, the ore treated by flotation in the US expressed as million tonnes was 180 in 1960, 253 in 1965, 367 in 1970, and 383 in 1975. Of the total material treated in 1975, 66% was sulphides, 7% metal oxides and carbonates, 24% non-metallic minerals, and 3% coal.[111]

Although flotation is increasingly used for the non-metallic and oxidised minerals, the bulk of the tonnage currently processed is sulphide from the ores of copper, lead, and zinc, often associated in complex ore deposits. The treatment of such ores serves as an introduction to the flowsheets encountered in plant practice. Comprehensive reviews of the complete range of sulphide, oxide, and non-metallic flotation separations can be found elsewhere,[2,100,112,115] and good reviews of the flotation of specific materials such as coal,[116] phosphates,[117-120] iron ore,[121-123] cassiterite,[124-127] scheelite,[128] chromium and manganese minerals[129] are also available.

Flotation of Copper Ores

Over 6 million tonnes of copper are produced annually in the Western World, and in 1985 about 17% and 20% of this total production was from the United States and Chile respectively. Significant tonnages were also produced in Canada (12%), Zaire (8%), Zambia (7%) and Peru (6%).[130] In recent years, due to low copper prices, a considerable quantity of mine capacity has been unused, particularly in the United States, where many mines have been forced to shut down or cut back on production. Ore grades in U.S. mines average only 0.6% Cu, compared with 2.2% Cu in Africa and 1.2% Cu in South America, and to convert such low-grade ores to saleable concentrates, especially with current low metal prices and high production costs, requires a high level of technology and control, and a careful balance between concentrate grades, recovery and milling costs.

Copper is characterised by having a number of economic ore minerals (Appendix 1), many of which may occur in the same deposit, and in various proportions according to depth. Copper sulphides in the upper part of an ore-body often oxidise and dissolve in water percolating down the outcrops of the deposit. A typical reaction with chalcopyrite is:

$$2CuFeS_2 + 17O + 6H_2O + CO_2 \rightarrow 2Fe(OH)_3 + 2CuSO_4 + 2H_2SO_4 + H_2CO_3.$$

The residual ferric hydroxide left in this leached zone is called *gossan* and its presence has often been used to identify a copper ore-body. As the water percolates through the zone of oxidation it may precipitate secondary minerals such as malachite and azurite to form an oxidised cap on the deeper primary ore.

The bulk of the dissolved copper, however, usually stays in solution until it passes below the water table into reducing conditions, where the dissolved metals may be precipitated from solution as secondary sulphides, e.g.:

$$CuFeS_2 + CuSO_4 \rightarrow FeSO_4 + 2CuS \text{ (covellite)}$$

$$5FeS_2 + 14CuSO_4 + 12H_2O \rightarrow 5FeSO_4 + 12H_2SO_4 + 7Cu_2S \text{ (chalcocite)}$$

As these secondary sulphide minerals contain relatively high amounts of copper, the grade of the ore in this *zone of supergene enrichment* is increased above that of the underlying primary mineralisation, and where supergene enrichment has been extensive, spectacularly rich copper "bonanzas" are formed.

The earliest copper miners worked the relatively small amounts of metallic copper contained in the oxidised zone of the ore-bodies. The discovery of smelting allowed high-grade oxidised copper minerals to be worked and processed. With improved developments in copper metallurgy, such as matte smelting and conversion, the secondary sulphide supergene zones were mined and processed, these deposits often being shallow and containing 5% or more copper.

The development of froth flotation had an enormous impact on copper mining, enabling the most abundant primary mineral, chalcopyrite, and other sulphides to be efficiently separated from ores of relatively low grade and fine grain size. Another major development was the introduction of vast tonnage open-pit mining methods to the copper industry, allowing the excavation of tens of thousands of tonnes of ore per day. This made economic the processing of huge low-grade bulk copper deposits known as *porphyries*, the most important being found in the United States and South America.

The importance of froth flotation and high-tonnage mining can be seen by considering that until 1907 practically all the copper mined in the United States was from underground vein deposits, averaging 2.5% Cu, whereas at present ore grades in the United States average only 0.6% Cu, and about 50% of the world's copper is produced from porphyry deposits, the rest mainly from vein-type and bedded deposits.

The exact definition of copper porphyry has been the subject of debate amongst geologists for a long time.[131] They are essentially very large oval or pipe-shaped deposits containing on average 140 million tonnes of ore, averaging about 0.8% Cu and 0.015% Mo, and a variable amount of pyrite.[132] All porphyry copper deposits contain at least traces of molybdenite (MoS_2), and in many cases molybdenum is an important by-product. Porphyry copper mineralisation is often referred to as disseminated, and although on a large scale immense volumes of ore may contain disseminated values, on a small scale the occurrence of sulphides is controlled by fractures. Even apparently disseminated sulphide minerals are often aligned with quartz microveinlets, or lie in a chain-like fashion (see Fig. 1.2b). The chains mark early fractures, which have been sealed and camouflaged by quartz and feldspar.[133]

The first deposits of this type to be mined on a large scale were in the south-western states of the United States. It was apparent that the deposits could be economically mined in bulk by large-scale low-cost methods such as block-caving and open-pit methods. This is because the copper minerals are distributed uniformly through large blocks of the deposit so that the expensive selective mining methods which must be used with vein or bedded deposits are not needed. The extent of the ore-body is usually determined by its copper content rather than by geological structure, the copper content tending to decrease away from the core of the mass. The cut-off grade which determines the boundary between ore and waste varies from mine to mine and according to the prevailing economic climate.

Porphyry copper operations are very much influenced by the geology of the ore deposition. Mining necessarily starts in the upper zones of the orebody where secondary alteration has enriched the ore grade, and where the mineralogy allows the production of concentrates often grading more than 40% Cu, at high recovery. High levels of output can be achieved with fairly compact mills and smelters. As the operation matures, however, lower grade primary (*hypogene*) ore is encountered, in which the mineralogy limits concentrate grades to only around 25–30% Cu, and more ore needs to be produced to realise the same net copper output, the alternative being to maintain the current plant throughput while metal output declines. Reagent use and flowsheets often have to be adapted to accommodate these changes in mineralogy. A classic case is the El Teniente mine in Chile, the world's largest underground copper mine, which was developed in one of the largest known copper porphyry deposits on earth (estimated to contain 44 million tonnes of copper in ore grading 0.99% Cu or more). In 1979, the ore, of grade 1.54% Cu, was being mined and processed at the rate of 57,500 tpd to produce a concentrate containing 40% Cu.[134] By 1984, with the secondary supergene zone approaching exhaustion, the ore grade had fallen to 1.4% Cu, and the mining rate had increased to 68,500 t d^{-1}, with a further expansion to 90,000 t d^{-1} being undertaken. It was predicted that the mined grade would fall to 1.2% by the end of the 1980s and to 1.0% by the end of the century.[135]

Although mining and processing of copper porphyries is on a vast scale, concentration of the ore is fairly straightforward, due to the high efficiency of froth flotation, and to the fact that breakage of the ore occurs preferentially at the fracture zones containing the copper sulphides. This means that relatively coarse grinding produces composite particles with much of the valuable mineral exposed, facilitating rougher flotation.

Copper sulphide minerals are readily floatable and respond well to anionic collectors such as xanthates, notably amyl, iso-propyl and butyl. Alkaline circuits of pH 8.5–12 are generally used, with lime controlling the pH and depressing any pyrite present. Frother usage has changed significantly in recent years, away from the natural reagents such as pine oil and cresylic acids, to the synthetic frothers such as the higher alcohols (e.g. MIBC) and polyglycol esters. Cleaning of the rougher concentrates is usually necessary to reach an economic smelter grade (25–50% Cu depending on mineralogy), and rougher concentrates as well as middlings must often be reground for maximum recovery, which is usually between 80 and 90%. Primary grinding is normally to about 50–60% -75 microns, rougher concentrates being reground to 90–100% -75 microns to promote optimum liberation of values. Reagent consumption is generally in the range 1–5 kg of lime per tonne of ore, 0.002–0.3 kg t^{-1} of xanthate and 0.02–0.15 kg t^{-1} of frother.

The largest copper concentrator in the world is operated at Bougainville in Papua New Guinea. The plant has been progressively expanded since initial start-up in 1972 from 80,000 t d^{-1} to 130,000 t d^{-1} to compensate for the lower grade ore encountered as the open-pit deepened. The principal copper mineral in the porphyry deposit is chalcopyrite (Fig. 12.56) with minor amounts of bornite. Gold and silver are also present in the primary ore, which in 1983 graded 0.46% Cu, 0.56 g t^{-1} Au and 1.44 g t^{-1} Ag.[136] Gold occurs in solid solution with sulphide mineralisation and as free gold. As with most porphyry deposits, molybdenum is also present, but the Mo:Cu ratio is lower than that generally found in the North and South American porphyry copper mines.

The flotation circuit is large, but fairly simple. After primary grinding to 80% -250 microns, the ore is conditioned with 7 g t^{-1} of sodium isopropyl xanthate collector, 18 g t^{-1} of methyl isobutyl carbinol (MIBC) and 6 g t^{-1} of Aerofroth 65 frothers, and 500 g t^{-1} of lime, before being fed to the circuit which is shown in Fig. 12.57. A further 4 g t^{-1} of potassium amyl xanthate collector is added to the rougher banks.

Rougher and scavenger concentrates are reground separately before further cleaning, the rougher concentrates being cleaned in three stages, while the reground scavenger concentrates are cleaned four times, the last three stages being in the same cleaning circuit as the rougher concentrate. The concentrate produced contains 28% copper, 27 g t^{-1} of gold and 70 g t^{-1} of silver, at a recovery of 91% copper, 72% gold and 70% silver.

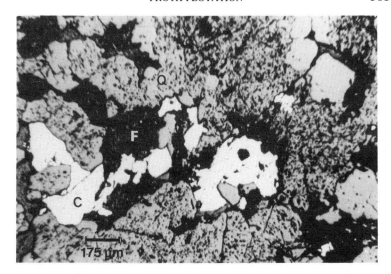

FIG. 12.56. Bougainville porphyry copper ore. Chalcopyrite (C) disseminated in quartz (Q) and feldspar (F).

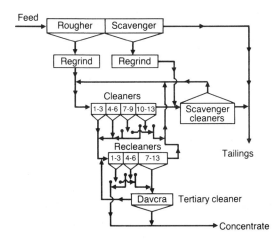

FIG. 12.57. Flotation circuit–Bougainville Copper Ltd.[83]

By-products are important to the economics of copper porphyry operations, and the most important by-product of the North and South American porphyries is molybdenum. Molybdenum occurs as the highly floatable mineral molybdenite, which is separated from the copper minerals after regrinding and cleaning of the copper rougher concentrates. Regrinding to promote optimum liberation needs careful control, as molybdenite is a soft mineral which slimes easily and whose floatability decreases as particles become finer. Rougher concentrates are therefore classified, only coarse cyclone underflows being reground in closed circuit. Cleaned copper concentrates are thickened, after which the copper minerals are depressed allowing molybdenite to be floated into a concentrate which is further cleaned, sometimes in up to twelve stages. Cleaning is important as molybdenite concentrates are heavily penalised by the smelter if they contain copper and other impurities, and the final copper content is often adjusted by leaching in sodium cyanide which easily dissolves chalcocite and covellite and some other secondary copper minerals. Chalcopyrite, however, does not dissolve in cyanide, and in some cases is leached with hot ferric chloride.

Copper depression is achieved by the use of a variety of reagents, sometimes in conjunction with prior heat treatment. Heat treatment is used to destroy residual flotation reagents, and is most commonly achieved by the use of steam injected into the slurry. Depression of chalcopyrite may be effectively accomplished by the use of sodium cyanide, but this reagent is not so effective when chalcocite and bornite are present, in which case depression can be completed by the use of ferro- and ferri-cyanides, or by using "Nokes Reagent", a product of the reaction of sodium hydroxide and phosphorus pentasulphide. This reagent has an instantaneous depressing action on copper minerals and is rapidly consumed, so is added to the circuit in stages. It can be an expensive depressant because of its high (2–5 kg t^{-1} of concentrate) consumption, and is sometimes used in combination with cyanide. Other copper depressants are arsenic Nokes (As_2O_3 dissolved in Na_2S), sodium sulphide, sodium hydrosulphide and thioglycolic acid. The molybdenite is floated using a light fuel oil as collector.

Figure 12.58 shows the molybdenum recovery flowsheet at the Chuquicamata Mine in Chile, the world's biggest copper producer. [137]

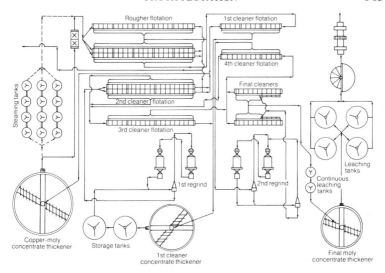

FIG. 12.58. Molybdenum flotation at Chuquicamata.[137]

The copper concentrate, containing 0.8–3% MoS_2 is floated in the rougher circuit after depressing the copper minerals with sodium hydrosulphide.[138] The first cleaner concentrate is recleaned in four to seven stages using 2.5 kg t^{-1} of arsenic Nokes reagent, and regrinding of first and fourth cleaner concentrates, to produce a concentrate containing 55% Mo and 1–2% Cu. This product is then leached with sodium cyanide to reduce the copper content, which is predominantly as chalcopyrite, to below 0.3%. Sodium cyanide is also added to the last two cleaner stages. All flotation cells in the molybdenum plant operate with nitrogen from the smelter oxygen plant, rather than air, the reducing potential considerably lowering the consumption of depressant.[139]

By-products play an important role in the economics of the Palabora Mining Co. in South Africa, which treats a complex carbonatite ore to recover copper, magnetite, uranium and zirconium values. The ore assays about 0.5% Cu, the principal copper minerals being chalcopyrite and bornite, although chalcocite, cubanite ($CuFe_2S_3$) and other copper minerals are present in minor amounts. The flotation feed is coarse—80% −300 microns—due to the high grinding resistance of the magnetite in the ore which would increase grinding costs if ground to a finer size, and due to the fact that the flotation tailings are treated by low-intensity magnetic separation to recover magnetite, and Reichert cone gravity concentration to recover uranothorite and baddeleyite.

The flotation circuit consists of eight separate sections, the last two sections being fed from an autogenous grinding circuit. The first five parallel sections, the original Palabora flowsheet, are fed from conventional mills, each at the rate of 385 t h^{-1} (Fig. 12.59). Flotation feed is conditioned with sodium isobutyl xanthate and frother before being fed to the rougher flotation banks. The more readily floatable minerals, mainly liberated chalcopyrite and bornite, float off in the first few cells, and more collector is added before the final scavenger cells, in order to float off the less floatable particles, such as cubanite,

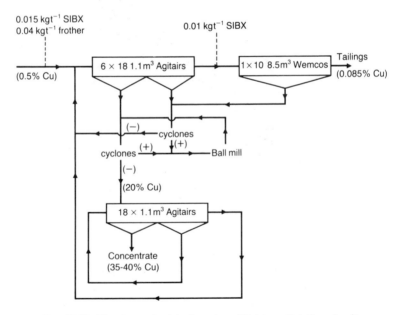

FIG. 12.59. Flowsheet of original section of Palabora flotation circuit.

and in order to attempt to float the less responsive copper minerals, such as valleriite, a copper–iron sulphide containing Mg and Al groups in the crystal lattice. Valleriite occurs intergrown with other sulphide minerals (Fig. 12.60), and due to the fact that it is a very soft mineral, it can lead to poor flotation recoveries. During comminution, breakage occurs along the soft and friable valleriite, leaving grains of other copper sulphides with a valleriite coating, preventing these grains from floating (Fig. 12.61).

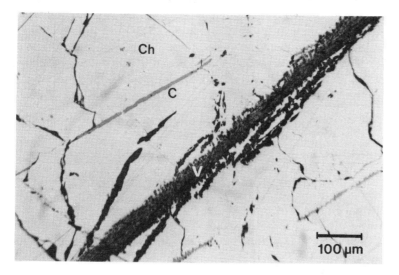

FIG. 12.60. Palabora copper ore. Valleriite (V) and cubanite (C), intergrown with chalcopyrite (Ch).

Rougher and scavenger concentrates are reground to 90% −45 microns, before being fed to the cleaner circuit at a pulp density of 14% solids, this dilution being possible due to the removal of magnetite and other heavy minerals into the tailings, and the fine particle size produced after grinding.

FIG. 12.61. Palabora flotation tailings particle, showing valleriite (dark) forming a coating around chalcopyrite (light).

Oxidised Copper Ores

Due to the nature of copper deposits and mineralisation, it is sometimes possible to selectively mine and process the oxidised cap on the primary zone. Minerals such as malachite and azurite are soluble in dilute sulphuric acid and can be processed economically by acid leaching as a prelude to precipitation of the copper by electrolysis (electrowinning). Processing of such oxidised ores has become more attractive during recent years due to the availability of cheap sulphuric acid produced at smelters as a means of reducing sulphur dioxide emissions into the atmosphere.

In Central Africa significant tonnages of oxidised ore are concentrated by flotation before being leached, ores containing a mixture of sulphide and oxidised minerals being treated by first floating off the sulphides to produce a concentrate for the smelter. Much of the published work on oxide copper minerals is concerned with malachite, and chrysocolla, a copper silicate. The latter is one of the most widely distributed and least understood of all the major copper minerals, being a very difficult mineral to characterise and float.[140] Malachite responds well to flotation techniques and in Central Africa flotation of malachite ores after sulphidisation is successfully practised.[38] Xanthate collector coatings are loosely bound to oxide copper minerals and sulphidisation enhances the flotation process.

A classic example of the treatment of such a mixed ore is provided by the Chingola Division of Nchanga Consolidated Copper Mines in Zambia, which is second only to Chuquicamata in terms of world production. This great mine, situated on the Zambian Copperbelt, treats a high-grade ore (3.4% Cu), from underground and open-pit operations, and containing a variety of copper minerals, mainly chalcocite and malachite, but with minor amounts of chalcopyrite, bornite, covellite, azurite, cuprite, and chrysocolla. The mine started production in 1938, treating underground ore, containing mainly sulphide copper minerals, with a small amount of oxide minerals, by flotation methods.

The method adopted at Nchanga was to first float the sulphide minerals by conventional methods, using xanthate, and then to condition the sulphide tailings in order to activate the oxide copper minerals.

The ore is crushed and ground in two separate mills, one treating underground ore, the other open-pit ore, the concentrator producing flotation concentrates from a composite of the ground products. The flotation feed is conditioned with sodium isopropyl xanthate collector and triethoxybutane (TEB) frother, before being fed to sulphide flotation (Fig. 12.62). Sulphide roughing produces high- and low-grade concentrates which are cleaned separately to produce a high-grade (45–55% Cu) concentrate, which is shipped to the smelter, and a low-grade (12–15%) concentrate which is delivered to the Chingola roaster.

Sulphide rougher tailings are conditioned with a 15% solution of sodium hydrosulphide, and a 1:1 mixture of palm-kernel oil and dieseline, which acts as a combined collector–frother. Oxide rougher

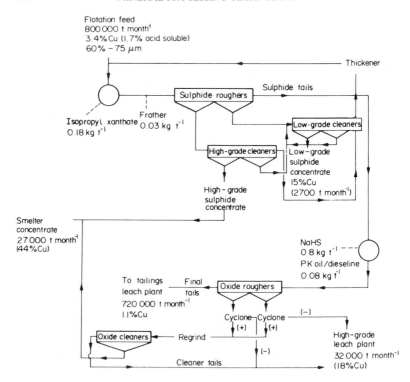

FIG. 12.62. Nchanga flotation flowsheet.

concentrate is classified by cyclones, and the underflows are reground and cleaned to remove residual sulphide copper. Cyclone overflows and cleaner tails are thickened and filtered. This oxide concentrate, containing about 15–20% copper, mainly as malachite and azurite, was originally added to the sulphide concentrate and sent to the smelter. However, with increasing amounts of oxide minerals, particularly with the introduction of large quantities of open pit ore in the 1950s, the reduction in sulphur content in the concentrates caused problems in smelting, as sulphur is required to sustain some of the exothermic reactions in the process. It therefore became necessary to treat the oxide concentrates by a separate route, and in 1952 an acid leach plant was established, incorporating facilities for leaching the concentrates with sulphuric acid and precipitating the copper from solution by electrowinning. By 1969, the total mining rate had increased to over 800,000 t per month, of which some 550,000 t per month came from open-pit sources, and the leach plant was producing over 100,000 tonnes per year of electrolytic copper.

Unfortunately the flotation of "sulphidised" oxide minerals is relatively inefficient, recoveries rarely being above 50%, so that the flotation tailings contain between 0.7 and 0.9% copper, of which 0.5–0.6% is as oxide minerals, particularly the relatively unfloatable minerals cuprite, chrysocolla and pseudo-malachite (a basic copper phosphate), as well as residual malachite and azurite.

In 1968 means of recovering this residual copper were investigated, as 4500 tonnes of copper were being lost per month, and in addition there existed large tailings dams from past operations containing appreciable quantities of oxide minerals. In the United States at that time a well-known technique known as solvent extraction, or liquid ion exchange, had been successfully introduced to copper processing, and this method was adopted by Nchanga, who commenced commissioning of the tailings recovery plant in 1971. By 1974 the world's largest solvent extraction plant commenced treatment of over 10 million tonnes per year of low grade tailings to produce an extra 100,000 tonnes per year of finished copper.[141]

Although relatively expensive, upgrading by chemical methods such as solvent extraction is likely to be developed and used to a progressively greater extent in the future, because of the lower grade and oxidised ores that will have to be processed. The use of such processes alone cannot, however, be economically justified at present. They may find their greatest use, as at Nchanga, in the treatment of tailings hitherto regarded as waste. Figure 12.63 summarises the mineral extraction flowsheet at Nchanga. The tailings from oxide flotation, due to the introduction of solvent extraction, can be regarded as an ore. A deposit of such grade and mineralogy would probably be uneconomic to mine and process but due to the fact that the tailings "ore" incurs no mining cost and has already been comminuted, its exploitation is viable.

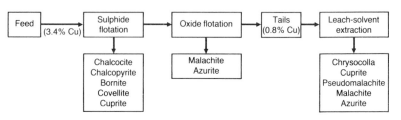

FIG. 12.63. Mineral extraction flowsheet at Nchanga.

Flotation of Lead-Zinc Ores

The bulk of the world's lead and zinc is supplied from deposits which often occur as finely disseminated bands of galena and sphalerite, with varying amounts of pyrite, as replacements in various rocks, typically limestone or dolomite. This banding sometimes allows heavy medium pre-concentration prior to grinding (Fig. 11.13).

Although galena and sphalerite usually occur together in economic quantities, there are exceptions, such as the lead ore-body in S.E. Missouri, U.S.A., where the galena is associated with relatively minor amounts of zinc, and the zinc-rich Appalachian Mountain region, mined in Tennessee and Pennsylvania, where lead production is very small.

Feed grades are typically 1–5% Pb and 1–10% Zn, and although relatively fine grinding is usually needed (often to well below 75 μm), fairly high flotation concentrate grades and recoveries can be achieved. Typically, lead concentrates of 55–70% lead are produced containing 2–7% Zn, and zinc concentrates of 50–60% Zn containing 1–6% Pb. Although galena and sphalerite are the major ore minerals, cerussite ($PbCO_3$), anglesite ($PbSO_4$), marmatite ((Zn, Fe)S) and smithsonite ($ZnCO_3$) can also be significant. In some deposits the value of associated metals, such as silver, cadmium, gold and bismuth, is almost as much as that of the lead and zinc, and lead–zinc ores are the largest sources of silver and cadmium.

Several processes have been developed for the separation of galena from zinc sulphides, but by far the most widely used method is that of two-stage selective flotation, where the zinc and iron minerals are depressed, allowing the galena to float, followed by the activation of the zinc minerals in the lead tailings to allow a zinc float.

Sphalerite (and to a lesser extent pyrite) can become activated by heavy metal ions in solution, which replace metallic zinc on the mineral surfaces by a process of ion-exchange (e.g. equation 12.11). This activated surface can adsorb xanthate and produce a very insoluble heavy metal xanthate which provides the surface with a water-repellant "envelope". Clean sphalerite is not strongly hydrophobic in xanthate solutions, as zinc xanthate has a relatively high solubility, and hence a stable envelope is not formed.

Heavy metal ions are often present in the slurry water, especially if the ore is slightly oxidised. The addition of lime or soda ash to the slurry can precipitate them as relatively insoluble basic salts, thus "de-activating" the sphalerite to some extent. The alkali is usually added to the grinding mills as well as to the lead float conditioner, as it is in the grinding process that many heavy metal ions are released into solution.

Lead flotation is usually performed at a pH of between 9 and 11, lime, being cheap, often being used to control alkalinity. Lime also acts as a strong depressant for pyrite, but it can also depress galena to some extent. Soda ash is sometimes preferred because of this, especially when the pyrite content is relatively low. The Cyprus Anvil concentrator in Canada reported that the single largest advance in metallurgical performance came with the replacement of lime by soda ash as the pH modifier on the lead circuit feed.[142]

The effectiveness of alkalis as deactivators is dependent on the concentration of heavy metal ions in solution, as the basic salts which are precipitated, although of extremely limited solubility, can provide a source of heavy metal ions sufficient to cause sphalerite activation. In most cases, therefore, other depressants are required, the most widely used being sodium cyanide (up to 0.15 kg t^{-1}) and zinc sulphate (up to 0.2 kg t^{-1}), either alone or in combination. These reagents are commonly added to the grinding circuit, as well as to the lead float, and their effectiveness depends very much on pulp alkalinity.

Apart from the reactions with metal ions in solution, cyanide has long been used to dissolve surface copper from activated sphalerite, and can react with iron and zinc xanthates to form soluble complexes, eliminating xanthate from the surfaces of the minerals of these metals. Pyrite is thus depressed with the sphalerite, and cyanide is generally the preferred depressant where soda ash regulates alkalinity and pyrite presence is significant.

The effectiveness of depressants also depends on the concentration and selectivity of the collector. Xanthates are most widely used in lead–zinc flotation, and the longer the hydrocarbon chain, the greater the stability of the metal xanthate in cyanide solutions and the higher the concentration of cyanide required to depress the mineral. If the galena is readily floatable, potassium or sodium ethyl xanthate may be used, together with a "brittle" frother such as M.I.B.C. Sodium isopropyl xanthate may be needed if the galena is tarnished, or if considerable amounts of lime are used to promote pyrite depression. Powerful collectors such as amyl xanthate can be used if the sphalerite is clean and hydrophilic, and are needed where the galena is highly oxidised and floats poorly.

Although cyanides are widely used due to their high degree of selectivity, they do have certain disadvantages. They are toxic and expensive, and they depress and dissolve some of the gold and silver which are often present in economic amounts. For these reasons, zinc sulphate is used in many plants to supplement cyanide. This reduces cyanide consumption (usually to well below 0.1 kg t^{-1}), and a number of mines in the U.S.A. achieve depression by the use of zinc sulphate alone.

After flotation of the galena, the tailings are usually treated with between $0.3–1 \text{ kg t}^{-1}$ of copper sulphate, which reactivates the surface of the zinc minerals (equation 12.11), allowing them to be floated. Lime ($0.5–2 \text{ kg t}^{-1}$) is used to depress pyrite, as it has no depressing effect on the activated zinc minerals, and a high pH (10–12) is used in the circuit. Isopropyl xanthate is perhaps the most commonly used collector, although ethyl, isobutyl, and amyl are also used, sometimes in conjunction with dithiophosphate (aerofloats), dependent on conditions. As activated sphalerite behaves in a similar way to chalcopyrite, thionocarbamates such as Z-200 are also common collectors, selectively floating the zinc minerals from the pyrite.

Careful control of reagent feeding must be observed when copper sulphate is used in conjunction with xanthates, as they react readily with copper ions. Ideally, the minerals should be conditioned with the activator separately, so that when the conditioned slurry enters the collector conditioner there is little residual copper sulphate in solution. Although the activation process is fairly rapid in acidic or neutral conditions, in practice it is usually carried out in an alkaline circuit in order to prevent pyrite activation, and a conditioning time of some 10–15 minutes is required to make full use of the reagent. This is because the alkali precipitates the copper sulphate as basic compounds which are sufficiently soluble to provide a reservoir of copper ions for the activation reaction.

The Sullivan concentrator of Cominco Ltd., British Columbia, operates an interesting flowsheet which includes de-zincing of the lead concentrates and de-leading of the zinc concentrates. The ore is essentially a replacement deposit in argillaceous quartzite, the orebodies being massive fine-grained mixtures of sulphides sometimes interbanded with the country rock. The principal economic minerals are galena and marmatite (7ZnS:FeS), iron being present mainly as pyrrhotite, and to a lesser extent pyrite. Silver is closely associated with the galena and is an important by-product.

The flowsheet is shown in Fig. 12.64. After primary grinding to 55% -74 μm with cyanide, xanthate and lime, the ore is fed to a unit flotation cell, where a mixture of M.I.B.C. and pine oil frothers is added. The pH is maintained at 8.5, and a coarse lead concentrate is floated, and cleaned once. This concentrate, assaying about 65% lead is used as medium in the H.M.S. circuit preceding grinding. The tailing from the coarse lead flotation is ground to 87%–74 μm, and is conditioned with sodium isopropyl xanthate, cyanide, lime and M.I.B.C., before being fed to the lead roughers at a pH of 9.5. Further addition of cyanide and xanthate to the head of the scavenger cells produces a concentrate which is returned to secondary grinding. The lead rougher concentrate is cleaned, the tailings being reground and returned to the lead roughers. The pH in the cleaners is 10.0, and the cleaner concentrate is further cleaned at pH 10.5 to produce a concentrate which contains 10–14% Zn. The final stage of lead flotation is the de-zincing of the second lead cleaner concentrate. After activating the zinc minerals with copper sulphate, the galena is depressed by raising the pH to 11.0 with the addition of lime, and by

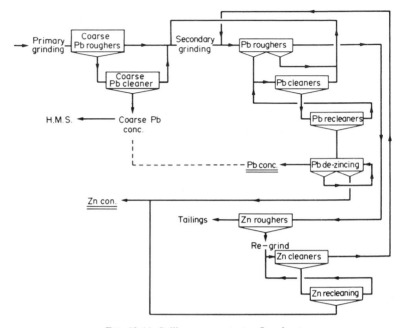

FIG. 12.64. Sullivan concentrator flowsheet.

steam heating the slurry to 30–40°C. A rougher dezincer concentrate is cleaned once in the first few cells of the bank, and the dezincer tailing is the final lead concentrate, assaying about 62% Pb and 4.5% Zn. The lead scavenger tailings are conditioned with about 0.7 kg t^{-1} of copper sulphate, prior to feeding to zinc rougher flotation where xanthate, lime and frother are added to the cells. A rougher concentrate is floated at pH 10.6, and is reground before being fed to the first stage of cleaning. The tailings from this stage, containing 2.5–4% Pb, are pumped back to the head of the lead circuit to allow a better recovery of lead in that concentrate. The cleaner concentrate is recleaned twice, the final concentrate being combined with the dezincer concentrate to produce the final zinc concentrate containing 50% Zn and 4% Pb.

Extremely fine intergrowth between galena and sphalerite inhibits selective flotation separation, and in some cases sphalerite is activated by copper ions in the ore to such an extent that depression of sphalerite fails, even when the most powerful combinations of reagents, such as zinc sulphate and cyanide, are used. Bulk flotation of the lead and zinc minerals may in such cases have a number of economic advantages. Coarse primary grinding is often sufficient with bulk flotation, as the valuable minerals need only be liberated from the gangue, not from each other. The flotation circuit design is normally relatively simple. In contrast, selective flotation calls for finer primary grinding, in order to free the valuable minerals not only from the gangue, but also from each other. This increases mill size and energy requirements; the flotation volume will increase proportionally to the number of selective concentrates.

However, the production of bulk lead–zinc concentrates is only reasonable if there are smelters which are adequately equipped for such concentrates. The only smelting process available is the *Imperial Smelting Process*, which was developed at a time when most of the lead and zinc was recovered from low-pyrite ore deposits. In recent years, however, lead and zinc are increasingly being recovered from complex and highly pyritic ores. Bulk concentrates for smelting in the ISP should be low in iron, as iron is recovered in the smelter slag. An increase in iron content increases slag production, and correspondingly increases zinc losses, as the slag carries about 5% zinc. Furthermore, a high iron content increases smelter energy consumption. When smelter revenues are compared, the highest revenues are achieved when selective concentrates are produced. Even mixing selective concentrates into a bulk concentrate will yield higher revenues than bulk concentrates produced by direct flotation. This is because better selectivity between non-ferrous minerals and pyrite is achieved by the optimal conditions adapted to the separation of galena and pyrite in the first, and sphalerite and pyrite in the second step. The chemical conditions in a bulk flotation cannot be adjusted to meet both conditions simultaneously if a high amount of pyrite is present. It has been shown that, although selective and more expensive than bulk flotation, the increase in revenues gained is often much higher than the additional operating costs.[143]

Bulk flotation followed by separation can sometimes be used, although in most cases the activated sphalerite and pyrite in the bulk concentrate are covered with a layer of collector, and are difficult to depress unless extremely large amounts of reagent are used. This is especially the case if copper sulphate has been used to activate the sphalerite; cyanide will react with residual copper ions in solution. Every attempt is made at plants using bulk flotation to use the minimum collector feed for the bulk flotation step, which can lead to low recoveries. Bulk flotation is performed at Zinkgruvan, Sweden's largest zinc mine.[144] Grinding is autogenous and the lead ions released during grinding activate the sphalerite to such an extent that deactivation by alkali is not practical at this stage. The flotation plant consists of bulk flotation and lead flotation stages, each circuit consisting of rougher, scavenger and cleaner steps. The galena and sphalerite are floated with 0.12 kg t^{-1} of potassium ethyl xanthate, no activator being required. After five stages of cleaning, the concentrate is conditioned with 0.6 kg t^{-1} of $ZnSO_4$ to depress the sphalerite, and the galena is floated at pH 10 with potassium ethyl xanthate. After six stages of cleaning, and further additions of $ZnSO_4$, a lead concentrate of 65% Pb and a zinc concentrate of 55% Zn are produced.

An interesting bulk-selective flowsheet is operated at the Tochibora mine, in Japan,[145] which has an annual output of 960,000 t of ore, grading 4.3% Zn. 0.3% Pb and 22 g t^{-1} Ag. Pyrite is not present to any extent in the ore, the principal gangue minerals being hedenbergite ($CaFeSi_2O_6$), quartz, calcite and epidote. Crushed ore is ground to 80% passing 75 μm, and is conditioned with Na_2CO_3 and $CuSO_4$ before bulk flotation at pH 9.4. Sodium ethyl xanthate is used as collector and pine oil as frother. After cleaning the bulk flotation concentrates, the slurry is conditioned with NaCN and activated carbon, after which galena is floated, the tailings being the zinc concentrate. The lead concentrate is fed into trommels, the oversize forming a graphite by-product concentrate, while the undersize is fed to shaking tables. The table middlings and tailings are recycled to the differential flotation circuit, the cleaned concentrate being the final lead concentrate. Concentrates grading 60.7% Zn and 65.3% Pb are obtained at recoveries of 93.3% Zn and 80.2% Pb.

Flotation of Copper–Zinc and Copper–Lead–Zinc Ores

The production of separate concentrates from ores containing economic amounts of copper, lead and zinc is complicated by the similar metallurgy of chalcopyrite and activated zinc minerals. The mineralogy of many of these ores is a complex assembly of finely disseminated and intimately associated chalcopyrite, galena and sphalerite in a gangue consisting predominantly of pyrite or pyrrhotite (often 80–90%), quartz and carbonates. Such massive sulphide ores, of volcanosedimentary origin, are also a valuable source of silver and gold.

The complex Cu–Pb–Zn ores represent 15% of total world production and 7.5% of the world copper reserves, these percentages being higher for zinc.[146] Grades of ore mined average 0.3–3% Cu, 0.3–3% Pb, 0.2–10% Zn, 3–100 g t^{-1} silver, and 0–10 g t^{-1} gold.

The major processing problems encountered are related specifically to the mineralogy of the assemblies. Due to the extremely fine dissemination and interlocking of the minerals, extensive fine grinding is often needed, usually to well below 75 μm. There are notable exceptions to this such as at Bleikvassli in Norway where a primary grind of 80%–240 μm is adequate, with no regrinding.[147] In the New Brunswick deposits in Canada, however, grinding to 80%–40 μm is required in certain areas, optimum mineral recoveries being in the range 10–25 μm. Such extensive fine grinding is extremely energy intensive (in the order of 50 kWh t^{-1}), and the large surface area produced leads to high reagent consumptions, the release of metal ions into solution which reduces flotation selectivity, and a greater tendency for surface oxidation. Oxidation is particularly serious with galena, which is often overground in closed circuit grinding, being the heaviest mineral in the complex ores.

In most cases, concentrates are produced at relatively poor grades and recoveries, typical grades being:

	%Cu	%Pb	%Zn
Copper concentrates	20–30	1–10	2–10
Lead concentrates	0.8–5	35–65	2–20
Zinc concentrates	0.3–2	0.4–4	45–55

Recoveries of 40–60% for copper, 50–60% for lead, and 70–80% for zinc are reported for New Brunswick deposits.[148] Smelting charges become excessive with contaminated concentrates, as very rarely is a metal paid for when it is not in its proper concentrate and penalties are often imposed for the presence of zinc and lead in copper concentrates. Silver and gold are well paid for in copper and lead concentrates, whereas payment in zinc concentrates is often zero. Direct sale of the concentrates to custom smelters is necessary where the size of the orebody precludes the development of a specialised smelter complex, such as that at the Ronnskar works of Boliden, Sweden, where a collection of metallurgical plants facilitates the transfer or recycling of residues and by-products from one process stage to another for the recovery of all metal values.[149]

The overall revenue for a mine exploiting such deposits can be very low compared to the relatively high contained value of the ore. Gray[150] has shown the economic limitations of processing complex ores by a standard route by comparing the concentrator performance at two Australian mines: North Broken Hill and Woodlawn. The former mine realises about 56% of the potential ore value in payments received, whereas Woodlawn realises only about 27% of the ore value in payments. The disparity in the two balances is almost solely due to the differences in recovery resulting from the much greater mineralogical complexity of the Woodlawn deposits. Deposits with such complex mineralogy are to be found in many parts of the world, whereas deposits with mineralogy comparable with North Broken Hill are now rare. The metallurgist's task is to characterise each deposit quantitatively and systematically and then to select the economically optimum combination of process steps to suit the characteristics. Imre and Castle[151] have also comprehensively reviewed the exploitation strategies for complex Cu–Pb–Zn orebodies, discussing the interaction and optimisation of the beneficiation and extractive metallurgical flowsheets and the options for extractive metallurgy in processing complex sulphides including pyrite. Barbery[81] has also discussed the many potential processing options available for treating complex sulphides, concluding that it is likely, for some years, that combined processes will be developed, linking physical separation processes with hydrometallurgy for maximum efficiency in recovering values into concentrates that are well paid by conventional existing smelters. In turning such integrated treatment concepts into reality, the fundamental question will be: is one flowsheet, involving one set of processes, capital and operating costs superior to another treatment approach with a different set of costs and metallurgical performance? Further, it is necessary to assess the impact of different product grades from the integrated process on subsequent downstream processes. The answer to this question, although critical, is likely to be very complex, and McKee[152] analyses the role of computer analysis in answering such questions.

Flotation is, at present, the only method that can be used to beneficiate the complex sulphide ores, and a wide variety of flowsheets are in use, some involving sequential flotation, others bulk flotation of copper and lead minerals followed by separation. Bulk flotation of all the economic sulphides from pyrite has also been studied. Although bulk flotation has certain advantages, it has been shown that the requirements for adequate galena flotation as well as those for selective flotation of sphalerite from pyrite are difficult to meet in a single bulk circuit, and better metallurgical efficiency can be obtained by floating, and then mixing, separate copper–lead and zinc concentrates. However, the main disadvantage is that a concentrate having no market is produced for which new metallurgical processes have to be developed.[81]

In the flotation of *copper–zinc ores*, where lead is absent, or is not present in economic quantities, lime is almost universally used to control alkalinity at pH 8–12, and to deactivate the zinc minerals by precipitation of heavy metal ions. In a few cases, the addition of lime to the mills and flotation circuit is sufficient to prevent the flotation of zinc minerals, but in most cases supplementary depressants are required. Sodium cyanide is often added in small quantities $(0.01–0.05 \text{ kg t}^{-1})$ to the mills and cleaners; if present in large amounts, chalcopyrite is also depressed. Zinc sulphate is also used in conjunction with cyanide, and in some cases sodium sulphite (or bisulphite) or sulphur dioxide depressants are used. The surfaces of pyrite and sphalerite can adsorb sulphite ions, which prevent collector adsorption, and the reducing action of the sulphite ions can prevent oxidation and dissolution of copper, hence preventing activation of zinc and iron minerals.

After conditioning, the copper minerals are floated using either xanthates, or if the mineralogy allows, a selective copper collector such as isopropyl thionocarbamate. Typically, copper concentrates contain 20–30% Cu, and up to 5% Zn. Copper flotation tailings are activated with copper sulphate, and zinc minerals are floated as described in the previous section.

Due to the very close control of reagent additions required in copper–zinc separations, on-stream X-ray analysis of plant flow-streams is being increasingly used together with some form of automatic control. A good example is the Pyhasalmi concentrator in Finland (Fig. 12.65), which is highly automated, and involves sequential flotation of copper, zinc and pyrite.[92] The copper circuit consists of conventional roughing and scavenging, followed by three cleaning stages, the tailings passing to the zinc flotation circuit. Despite the use of cyanide $(0.025 \text{ kg t}^{-1})$ and zinc sulphate (1.45 kg t^{-1}), a problem in the copper circuit is the natural activation of sphalerite by copper-bearing water; because of this a flotation time of about 20 minutes is required for satisfactory copper recovery (about 90%) and the copper concentrate contains about 25% Cu and 3.5% Zn. Reagent additions are controlled automatically according to set-points regulated by on-stream analysis of copper, zinc and iron contents in various flowstreams. Due to the varying quality of the ore caused by fluctuating quantities of activated zinc minerals, cyanide addition is the most important variable affecting the economic recovery and is controlled from the set-points to keep the zinc content of the copper concentrate at a minimum while maintaining optimum copper recovery.

The method most widely used to treat ores containing economic amounts of lead, copper and zinc is to initially float a bulk lead–copper concentrate, while depressing the zinc and iron minerals. The zinc minerals are then activated and floated, while the bulk concentrate is treated by depression of either the copper or lead to produce separate concentrates.

The bulk float is performed in an alkaline circuit, usually pH 7.5–9.5, lime, in conjunction with depressants such as cyanide and zinc sulphate, being added to the mills and bulk circuit. Depression of zinc and iron sulphides is sometimes supplemented by the addition of small

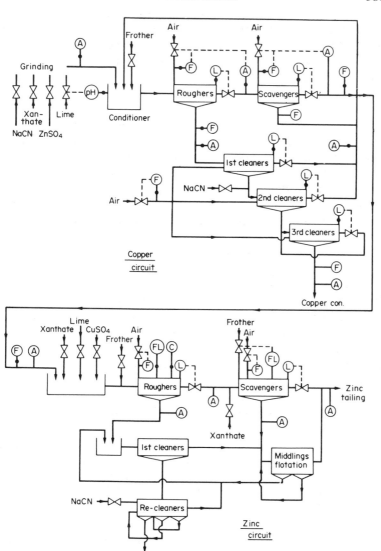

FIG. 12.65. Pyhasalmi flotation circuit. F = flowrate; L = level; A = assay;
FL = froth level; C = conductivity.

amounts of sodium bisulphite or sulphur dioxide to the cleaning stages, although these reagents should be used sparingly as they can also depress galena.

The choice and dosage of collector used for bulk flotation are critical not only for the bulk flotation stage, but also for the subsequent separation. Xanthates are commonly used, and while a short-chain collector such as ethyl xanthate gives high selectivity in floating galena and chalcopyrite, and permits efficient copper–lead separation, it does not allow high recoveries into the bulk concentrate, particularly of the galena. Much of the lost galena subsequently floats in the zinc circuit, contaminating the concentrate, as well as representing an economic loss. Because of this, a powerful collector such as amyl or isobutyl xanthate is commonly used, and very close control of the dosage is required. Usually, fairly small amounts, of between 0.02–0.06 kg t^{-1}, are used, as an excess makes copper–lead separation difficult and large amounts of depressant are required, which may depress the floating mineral, contaminating lead and copper concentrates.

Although the long chain collectors improve bulk recovery, they are not as selective in rejecting zinc, and sometimes a compromise between selectivity and recovery is needed, and a collector such as sodium isopropyl xanthate is chosen, Dithiophosphates, either alone or in conjunction with xanthates, are also used as bulk float collectors, and small amounts of thionocarbamate may be used to increase copper recovery.

The choice of method for separating the copper from the lead minerals depends on the response of the minerals and the relative abundance of the copper and lead minerals. It is preferable to float the mineral present in least abundance, and galena depression is usually performed when the ratio of lead to copper in the bulk concentrate is greater than unity.

Lead depression is also undertaken if economic amounts of chalcocite or covellite are present, as these minerals do not respond to depression by cyanide, or if the galena is oxidised or tarnished and does not float readily. It may also be necessary to depress the lead minerals if the concentration of copper ions in solution is high, due to the presence of secondary copper minerals in the bulk concentrate. The standard copper depressant, sodium cyanide, combines with these ions to form complex cuprocyanides (equation 12.24), thus reducing free cyanide ions available for copper depression. Increase in cyanide addition only serves to accelerate the dissolution of secondary copper minerals.

Depression of galena is achieved using sodium dichromate, sulphur dioxide, and starch in various combinations, whereas copper minerals are depressed using cyanide, or cyanide–zinc complexes. Table 12.4 shows methods of depression used at various concentrators.[153]

Depression of galena by the addition of sodium dichromate at high pH is still used in many plants. The hydrophobic character of the xanthate layer on the galena surface is inhibited by the formation of

TABLE 12.4. METHODS OF SEPARATING COPPER–LEAD CONCENTRATES

Mine	Approx Ratio Pb/Cu	Depres- sed	Method	Copper Cu	Copper Pb	Lead Pb	Lead Cu	Reference
St. Joe Minerals Corp. Missouri, U.S.A.	45:1	Pb	SO$_2$–Starch– Dichromate	26	4	74	0.5	154
Brunswick Mining, Canada	20:1	Pb	SO$_2$–Starch– Heat	23	6	31	0.6	155
Garpenburg, Sweden	10:1	Pb	Dichromate	13	14	57	0.8	144
Asarco Inc., Buchans Unit, Canada	6:1	Pb	SO$_2$– Dichromate	27	7	57	2.6	156
Minera Madrigal, Peru	2:1	Pb	Dichromate	30	4	68	3.5	157
Mattabi Mines, Canada	1:1	Pb	SO$_2$–Starch– Heat	24	4	51	1.6	158
Vihanti, Finland	1:1	Pb	Dichromate	25	2	42	1.6	92
Western Mines, Myra Falls, Canada	1:1	Cu	Cyanide	28	3	43	3.8	159
Willroy Mines, Canada	0.3:1	Cu	Cyanide	24	3	34	5.3	160
Geco Div., Noranda, Canada	Very low	Cu	Cyanide	27	1	48	7.3	161

hydrated lead chromate.[(162)] At Vihanti (Fig. 12.66), the galena is depressed by the addition of 0.01 kg t^{-1} of sodium dichromate to the bulk concentrate. After copper flotation, the separation tailings are further floated to remove residual copper, the cleaner tailings producing the final lead concentrate. Although there is no automatic control of the separation circuit, the rate of addition of dichromate is critical, as an excess is returned to the rougher feed with the cleaner tailing which depresses lead into the zinc circuit.

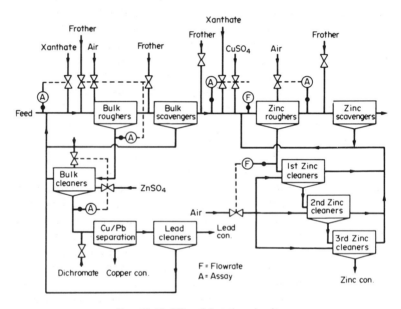

FIG. 12.66. Vihanti flotation circuit.

Although the amount of dichromate used is only small (0.01–0.2 kg t^{-1}), chromate ions can cause environmental pollution, and other methods of depression are sometimes preferred. Depression of galena by sulphite adsorption is the most widely used method, sulphur dioxide, either as liquid or gas, being added to the bulk concentrate; sodium sulphite is less commonly used. In many cases, causticised starch is added in small amounts as an auxiliary depressant, but tends to depress the copper if insufficient sulphur dioxide is used. The sulphur dioxide reduces the pH to between 4 and 5.5, the slightly acidic conditions cleaning the surfaces of the copper minerals, thus aiding their floatability. Small amounts of dichromate may be added to the circuit to supplement lead depression. At St. Joe Minerals Corp.,[154] this is added to the final stages of the five-stage cleaning circuit.

In some plants, galena depression is aided by heating the slurry to about 40°C by steam injection. Kubota *et al.*[163] showed that galena can be completely depressed, with no reagent additions, by raising the slurry temperature above 60°C, and this method is being used by the Dowa Mining Company in Japan.[164-5] The xanthate adsorbed on the galena is removed, but that on the chalcopyrite surface remains. It is thought that preferential oxidation of the galena surface at high temperature is the mechanism for depression. At Woodlawns in Australia, the lead concentrate originally assayed 30% Pb, 12% Zn, 4% Cu, 300 ppm Ag, and 20% Fe, and received very unfavourable smelter terms.[166] Heat treatment of the concentrate at 85°C for 5 minutes, followed by reverse flotation, gave a product containing 35% Pb, 15% Zn, 2.5% Cu, 350 ppm Ag and 15% Fe, with improved sales terms.

At the Brunswick Mining concentrator in Canada[155] (Fig. 12.67), the bulk copper–lead concentrate is conditioned for 20 minutes with 0.03 kg t^{-1} of a wheat dextrine–tannin extract mixture to depress the galena, and 0.03 kg t^{-1} of activated carbon to absorb excess reagents and contaminants, and then the pH is lowered to 4.8 with liquid SO_2. The slurry is further conditioned for 20 minutes at this low pH, then 0.005 kg t^{-1} of thionocarbamate is added to float the copper minerals. The rougher concentrate is heated by steam injection to 40°C, and is then cleaned three times to produce a copper concentrate containing 23% Cu, 6% Pb and 2% Zn. The lead concentrate produced is further upgraded by regrinding the copper separation tails, and then heating the slurry with steam to 85°C, and conditioning for 40 minutes. Xanthate and dithiophosphate collectors are then added to float pyrite. The rougher concentrate produced is reheated to 70°C and is cleaned once. The hot slurry from the lead upgrading tailing contains about 32.5% Pb, 13% Zn and 0.6% Cu, and after cooling is further treated to float a lead–zinc concentrate, leaving a final lead concentrate of 36% Pb and 8% Zn.

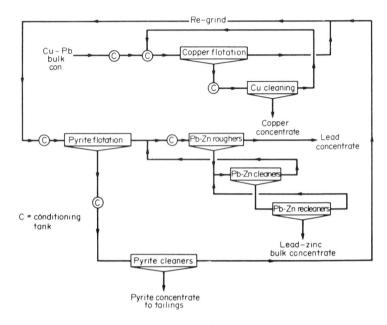

FIG. 12.67. Brunswick mining flotation circuit.

In general, where the ratio of lead to copper in the bulk concentrate is less than unity, depression of the copper minerals by sodium cyanide may be preferred. Where standard cyanide solution may cause unacceptable dissolution of precious metals and small amounts of secondary copper minerals, a cyanide–zinc complex can sometimes be used to reduce these losses. At Morococha in Peru,[167] a mixture of sodium cyanide, zinc oxide and zinc sulphate is used, which allows a recovery of 75% of the 120 g t^{-1} of silver in the ore.

Close alkalinity control is necessary when using cyanides, a pH of between 7.5–9.5 commonly being used, although the optimum value may be higher, dependent on the ore. Cyanide depression is not used if economic quantities of chalcocite or covellite are present in the bulk concentrate, since it has little depressing action on these minerals. As cyanide is a very effective sphalerite depressant, most of the zinc reporting to the bulk concentrate is depressed into the copper concentrate, which may incur smelter penalties. Cyanide, however, has little action on galena, allowing effective flotation of the galena from the chalcopyrite, and hence a low-lead–copper concentrate. Lead is never paid for in a copper concentrate, and is often penalised.

In a few cases, adequate metallurgical performance cannot be achieved by semi-bulk flotation, and sequential selective flotation must be performed. This necessarily increases capital and operating costs, as the bulk of the ore—the gangue minerals—are present at each stage in separation, but it allows selective use of reagents to suit the mineralogy at each stage. Sequential separation is required where there is a marked difference in floatability between the copper and lead minerals, which makes bulk rougher flotation and subsequent separation of the minerals in the bulk concentrate difficult, as at the Black Mountain Broken Hill concentrator in South Africa.[168] In Australia, sequential separation is performed at Cobar Mines Ltd.[169] Metallurgical development at Woodlawns in Australia has been an ongoing process. The original circuit, designed to depress lead with dichromate, was never effective for various reasons, and a combination of bulk and sequential flotation is being used.[166,170] The feed, containing roughly 1.3% Cu, 5.5% Pb, and 13% Zn, is conditioned with SO_2, starch, sodium metabisulphite and a dithiophosphate collector, after which a copper concentrate is produced, which is cleaned twice. The copper tailings are conditioned with lime, NaCN, starch, and sodium secondary butyl xanthate, prior to flotation of a lead concentrate which contains the less floatable copper minerals. This concentrate is reverse cleaned by steam heating to 85°C prior to flotation of the copper minerals with no further reagent addition. The floated copper minerals are pumped to the initial copper cleaning circuit. Lead rougher tailings feed the zinc roughing circuit.

The general flowsheet for sequential flotation involves conditioning the slurry with SO_2 at low pH (5–7), and using a selective collector such as ethyl xanthate, dithiophosphate, or thionocarbamate, which allows a copper concentrate which is relatively low in lead to be floated. The copper tailings are conditioned with lime or soda ash, xanthate, sodium cyanide and/or zinc sulphate, after which a lead concentrate is produced, the tailings being treated with copper sulphate prior to zinc flotation.

References

1. Jones, M. H., and Woodcock, J. T. (eds.), *Principles of Mineral Flotation*, Australas. Inst. Min. Metall., Victoria, 1984.
2. Glembotskii, V. A., Klassen, V. I., and Plaksin, I. N., *Flotation, Primary Sources*, New York, 1972.
3. Leja, J., *Surface Chemistry of Froth Flotation*, Plenum Press, New York (1982).
4. Sutherland, K. L., and Wark, I. W., *Principles of Flotation*, Australian I.M.M. (1955).
5. King, R. P. (ed.), *The Principles of Flotation*, S. Afr. I.M.M. (1982).
6. Fuerstenau, M. C., et al, *Chemistry of Flotation*, AIMME, New York, 1985.
7. Ives, K. J. (ed.), *The Scientific Basis of Flotation*, Martinus Nijhoff Publishers, The Hague, 1984.
8. Schulze, H. J., *Physico-chemical Elementary Processes in Flotation*, Elsevier Science Publishing Co., Amsterdam, 1984.
9. Ranney, M. W. (ed.), *Flotation Agents and Processes—Technology and Applications*, Noyes Data Corp., New Jersey, 1980.
10. Crozier, R. D., Plant reagents. Part 1: Changing pattern in the supply of flotation reagents. *Mining Mag.*, 202 (Sept. 1984).
11. Wrobel, S. A., Economic flotation of minerals, *Min. Mag.* **122**, 281 (Apr. 1970).
12. Houot, R., *et al.*, Selective flotation of phosphatic ores having a siliceous and/or a carbonated gangue, *Int. J. Min. Proc.*, **14**, 245 (June 1984).
13. Broekaert, E., *et al.*, New processes for cassiterite ore flotation, in *Mineral Processing and Extractive Metallurgy*, eds. M. J. Jones and P. Gill, p. 453, IMM, London, 1984.
14. Houot, R., *et al.*, Industrial sulphonates and barite flotation, *Trans. Inst. Min. Metall.*, **94**, C195 (Dec. 1985).
15. de Cuyper, J., and Broekaert, E., Flotation of a complex ore containing barite and fluorspar, using alkyl-sulphates and sulphonates as respective collectors, in *Proc. 1st Int. Min. Proc. Symp.*, **1**, 222, Izmir (1986).
16. Baldauf, H., *et al*, Alkane dicarboxylic acids and aminoaphthol sulphonic acids—a new reagent regime for cassiterite flotation, *Int. J. Min. Proc.*, **15**, 117 (Aug. 1985).
17. Collins, D. N., *et al.*, Use of alkyl imino-bis-methylene phosphonic acids as collectors for oxide and salt-type minerals, in *Reagents in the Minerals Industry*, eds. M. J. Jones and R. Oblatt, p. 1, IMM, London, 1984.
18. Shaw, D. R., Dodecyl mercaptan: a superior collector for sulphide ores, *Mining Engng.*, **33**, 686 (June 1981).
19. Ackerman, P. K., *et al*, Importance of reagent purity in evaluation of flotation collectors, *Trans. Instn. Min. Metall.*, **95**, C165 (Sept. 1986).
20. Woods, R., Electrochemistry of sulphide flotation, in *Flotation: A.M. Gaudin Memorial Volume*, ed. Fuerstenau, M. C., Vol. 1, p. 298, AIMME, New York, 1976.
21. Shergold, H. L., Flotation in mineral processing, in *The Scientific Basis of Flotation*, ed. K. J. Ives, p. 229, Martinus Nijhoff Publishers, The Hague, 1984.
22. Jones, M. H., and Woodcock, J. T., Decomposition of alkyl dixanthogens in aqueous solutions, *Int. J. Min. Proc.* **10**, 1 (1983).
23. Fuerstenau, D. W., and Mishra, R. K., On the mechanism of pyrite flotation with xanthate collectors, in *Complex Sulphide Ores* (ed. M. J. Jones), I.M.M., London (1980).

24. Finkelstein, N. P. and Poling, G. W., The role of dithiolates in the flotation of sulphide minerals, *Miner. Sci. Eng.* **9**, 177 (1977).
25. Poling, G. W. Reactions between thiol reagents and sulphide minerals, in *Flotation: A. M. Gaudin Memorial Volume*, vol. 1, p. 334, A.I.M.M.E. (1976).
26. Woods, R., and Richardson, P. E., The flotation of sulphide minerals–electrochemical aspects. in *Advances in Mineral Processing*, ed. P. Somasundaran, chap. 9, SME, Colorado, 1986.
27. Pritzker, M. D., *et al.*, Solution and flotation chemistry of sulphide minerals, *Can. Met. Quarterly*, **24**(1), 27 (1985).
28. Rao, S. R., *Xanthates and Related Compounds*, Marcel Dekker, New York (1971).
29. Granville, A., *et al.*, Review of reactions in the flotation system galena-xanthate-oxygen, *Trans. I.M.M. Sec. C*, **81**, 1 (March 1972).
30. Harris, G. H., Xanthates, in *Encyclopaedia of Chemical Technology* **22**, 419, Wiley, New York (1970).
31. Mingione, P. A., Use of dialkyl and diaryl dithiophosphate promoters as mineral flotation reagents, in *Reagents in the Minerals Industry*, ed. M. J. Jones and R. Oblatt, p. 19, IMM, London, 1984.
32. Chander, S., and Fuerstenau, D. W., On the floatability of sulphide minerals with thiol collectors: the chalcocite/diethyl dithiophosphate system, *Proc. 11th Int. Min. Proc. Cong., Cagliari* 583 (1975).
33. Goold, L. A., The reaction of sulphide minerals with thiol collectors, *NIM Report No.* 1439, Johannesburg (1972).
34. Yarar, B., *et al.*, Electrochemistry of the galena-diethyl dithiocarbamate-oxygen flotation system, *Trans. I.M.M. Sec. C*, **78**. 181, 1969.
35. Jiwu, M., *et al.*, Novel frother-collector for flotation of sulphide minerals- CEED, in *Reagents in the Minerals Industry*, ed. M. J. Jones and R. Oblatt, p. 287, IMM, London, 1984.
36. Jones, M. H., and Woodstock, J. T., Spectrophotometric determination of Z-200 in flotation liquors, *Proc. Aust. I.M.M.* **231**, 11, Sept. 1969.
37. Ackerman, P. K., et al., Effect of alkyl substituents on performance of thionocarbamates as copper sulphide and pyrite collectors, in *Reagents in the Minerals Industry*, ed. M. J. Jones and R. Oblatt, p. 69, IMM, London, 1984.
38. Fuerstenau, D. W., and Raghavan, S., Surface chemical properties of oxide copper minerals. in *Advances in Mineral Processing*, ed. P. Somasundaran, chap. 23, p. 395, SME Inc., Littleton, 1986.
39. Somasundaran, P., Interfacial chemistry of particulate flotation, *A.I.Ch.E. Symposium Series no.* 150, **71**, 1 (1975).
40. Soto, H., and Iwasaki, I., Flotation of apatite from calcareous ores with primary amines, *Minerals and Metallurgical Processing*, **2**, 160 (Aug. 1985).
41. Crozier, R. D., Frother function in sulphide flotation, *Min. Mag.* 26 (Jan. 1980).
42. Lekki, J., and Laskowski, J., A new concept of frothing in flotation systems and general classification of flotation frothers, *Proc. 11th Int. Min. Proc. Cong., Cagliari*, 427 (1975).
43. Girczys, J., and Laskowski, J., Selectivity in sphalerite-marcasite flotation during processing of Pb–Zn–Fe sulphide ores under alkaline conditions with copper sulphate solution, in *Proc. XVth Int. Min. Proc. Cong.*, **2**, 167, Cannes (1985).
44. Fuerstenau, D. W., *et al.*, Sulphidization and flotation behaviour of anglesite, cerussite and galena, in *Proc. XVth Int. Min. Proc. Cong.*, **2**, 74, Cannes (1985).
45. Malghan, S. G., Role of sodium sulphide in the flotation of oxidized copper, lead, and zinc ores, *Minerals & Metallurgical Processing*, **3**, 158 (Aug. 1986).

46. Parsonage, P. G., Effects of slime and colloidal particles on the flotation of galena, in *Flotation of Sulphide Minerals*, ed. K.S.E. Forssberg, p. 111. Elsevier, Amsterdam, 1985.

47. Shin, B. S., and Choi, K. S. Adsorption of sodium metasilicate on calcium minerals, *Minerals and Metallurgical Processing*, **2**, 223 (Nov. 1985).

48. Pavlica, J., *et al.*, Industrial application of ferro-sulphate and sodium cyanide in depressing zinc minerals, in *Proc. 1st Int. Min. Proc. Symp.*, **1**, 183, Izmir (1986).

49. Ser, F., and Nieto, J. M., Sphalerite separation from Cerro Colorado copper–zinc bulk concentrate, in *Proc. XVth Int. Min. Proc. Cong.*, **3**, 247, Cannes (1985).

50. Konigsman, K. V., Flotation techniques for complex ores, in *Complex Sulfides*, ed. A. D. Zunkel *et al.*, p. 5, TMS-AIME, Pennsylvania, 1985.

51. Broman, P. G., *et al.*, Experience from the use of SO₂ to increase the selectivity in complex sulphide ore flotation, in *Flotation of Sulphide Minerals*, ed. K.S.E. Forssberg, p. 277, Elsevier, Amsterdam, 1985.

52. Nagaraj, D. R., *et al.*, Structure-activity relationships for copper depressants, *Trans. Instn. Min. Metall.*, **95**, C17 (March 1986).

53. Agar, G. E., Copper sulphide depression with thioglycollate and trithiocarbonate, *CIM Bulletin*, **77**, 43 (Dec. 1984).

54. Hayes, R. A., *et al.*, Collectorless flotation of sulphide minerals, *Miner. Process and Ext. Met. Rev.*, **2**, 1 (1987).

55. Buckley, A. N., *et al.*, Investigation of the surface oxidation of sulphide minerals by linear potential sweep voltammetry and X-ray photoelectron spectroscopy, *Flotation of Sulphide Minerals*, ed. K. S. E. Forssberg, p. 41, Elsevier, Amsterdam, 1985.

56. Luttrell, G. H., and Yoon, R. H., The collectorless flotation of chalcopyrite ores using sodium sulphide. *Int. J. Min. Proc.*, **13**, 271 (Nov. 1984).

57. Guy, P. J., and Trahar, W. J., The effects of oxidation and mineral interaction on sulphide flotation, in *Flotation of Sulphide Minerals*, ed. K. S. E. Forssberg, p. 91, Elsevier, Amsterdam, 1985.

58. Nakazawa, H., and Iwasaki, I., Effect of pyrite-pyrrhotite contact on their floatabilities, *Minerals and Metallurgical Processing*, **2**, 206 (Nov. 1985).

59. Kocabag, D., and Smith, M. R., The effect of grinding media and galvanic interactions upon the flotation of polymetallic ores, in *Complex Sulfides*, eds. A. D. Zunkel *et al.*, p. 55, TMS-AIME, Pennsylvania, 1985.

60. Learmont, M. E., and Iwasaki, I., Effect of grinding media on galena flotation, *Minerals and Metallurgical Processing*, **1**, 136 (Aug. 1984).

61. Adam, K., and Iwasaki, I., Effects of polarisation on the surface properties of pyrrhotite, *Minerals and Metallurgical Processing*, **1**, 246 (Nov. 1984).

62. Woods, R., Electrochemistry of sulphide flotation, in *Principles of Mineral Flotation*, eds. M. H. Jones and J. T. Woodcock, p. 91, Australas. I.M.M., Victoria, 1984.

63. Heimala, S., *et al.*, New potential controlled flotation process developed by Outokumpu, *Proc. XVth Int. Min. Proc. Cong.*, **3**, 88, Cannes (1985).

64. Konigsman, K. V., Flotation techniques for complex ores, in *Complex Sulfides*, eds. A. D. Zunkel *et al.*, p. 5, TMS-AIME, Pennsylvania, 1985.

65. Williams, S. R., and Phelan, J. M., Process development at Woodlawn Mines, in *Complex Sulfides*, eds. A. D. Zunkel *et al.*, p. 293, TMS-AIME, Pennsylvania, 1985.

66. Chander, S., Oxidation/reduction effects in depression of sulphides—a review, *Minerals and Metallurgical Processing*, **2**, 26 (Feb. 1985).
67. MacDonald, R. D., and Brison, R. J., Applied research in flotation, *Froth Flotation 50th Anniversary Volume*, 298, AIMME, New York, 1962.
68. Finch, J. A., Kitching, R., and Robertson, K. S., Laboratory simulation of a closed-circuit grind for a heterogeneous ore, *CIM Bulletin* **72**, 198 (Mar. 1979).
69. Agar, G. E., and Kipkie, W. B., Predicting locked cycle flotation test results from batch data, *CIM Bulletin* **71**, 119 (Nov. 1978).
70. Dorenfeld, A. C., Flotation circuit design, *Froth Flotation 50th Anniversary Volume*, 365, AIMME, New York, 1962.
71. Steane, H. A., Coarser grind may mean lower metal recovery but higher profits, *Can. Min. J.* **97**, 44 (May 1976).
72. Mori, S., *et al.*, Kinetic studies of fluorite flotation, in *Proc. XVth Int. Min. Proc. Cong.*, **3**, 154, Cannes (1985).
73. Dowling, E. C., *et al.*, Model discrimination in the flotation of a porphyry copper ore, *Minerals & Metallurgical Processing*, **2**, 87 (May 1985).
74. Agar, G. E., The optimization of flotation circuit design from laboratory rate data, in *Proc. XVth Int. Min. Proc. Cong.*, **2**, 100, Cannes (1985).
75. Trahar, W. J., and Warren, L. J., The floatability of very fine particles—a review, *Int. J. Min. Proc.*, **3**, 103 (1976).
76. Hemmings, C. E., An alternative viewpoint on flotation behaviour of ultrafine particles, *Trans. Inst. Min. Metall.*, **89**, C113 (Sept. 1980).
77. Trahar, W. J., A rational interpretation of the role of particle size in flotation, *Int. J. Min. Proc.* **8**, 289 (1981).
78. Agar, G. E., *et al.*, Optimising the design of flotation circuits, *CIM Bulletin*, **73**, 173 (1980).
79. Lindgren, E., and Broman, P., Aspects of flotation circuit design, *Concentrates* **1**, 6 (1976).
80. Hardwicke, G. B., *et al.*, Granby Mining Corporation, in *Milling Practice in Canada*, CIM Special vol. 16 (1978).
81. Barbery, G., Complex sulphide ores: processing options, in *Mineral Processing at a Crossroads—problems and prospects*, eds. B. A. Wills and R. W. Barley, p. 157, Martinus Nijhoff Publishers, Dordrecht, 1986.
82. Young, P., Flotation machines, *Min. Mag.* **146**, 35 (Jan. 1982).
83. Tilyard, P. A., Process developments at Bougainville Copper Ltd, *Bull. Proc. Australas. Inst. Min. Metall.*, **291**, 33 (March 1986).
84. Cienski, T., and Coffin, V., Column flotation operation at Mines Gaspe molybdenum circuit, *Can. Min. J.* **102**, 28 (March 1981).
85. Dobby, G. S., *et al.*, Column flotation: some plant experience and model development, in *Automation for Mineral Resource Development*, eds. A. W. Norrie and D. R. Turner, p. 259, Pergamon Press, Oxford, 1986.
86. Dobby, G. S., and Finch, J. A., Flotation column scale-up and modelling, *CIM Bulletin*, **79**, 89 (May 1986).
87. McKay, J. D., *et al.*, Column flotation of Montana chromite ore, *Minerals and Metallurgical Processing*, **3**, 170 (Aug. 1986).
88. Chironis, N. P., Cell creates microbubbles to latch on to finer coal, *Coal Age*, **91**, 62 (Aug. 1986).
89. Malinovskii, V. A., *et al.*, Technology of froth separation and its industrial applications, *Trans. 10th Int. Min. Proc. Cong.*, paper 43 (London 1973).

90. Harris, C. C., Flotation machines, in *Flotation: A. M. Gaudin Memorial Volume*, p. 753, AIME, New York (1976).
91. Sorensen, T. C., Large Agitair flotation machines design and operation, *XIV Int. Min. Proc. Cong.*, Paper No. VI-10, CIM, Toronto, Canada (Oct. 1982).
92. Wills, B. A., Pyhasalmi and Vihanti concentrators, *Min. Mag.* 176 (Sept. 1983).
93. Niitti, T., and Tarvainen, M., Experiences with large Outokumpu flotation machines, *Proc. XIV Int. Min. Proc. Cong.*, Paper No. VI-7, CIM, Toronto, Canada (Oct. 1982).
94. Arbiter, N., and Harris, C. C., Flotation machines, *Froth Flotation 50th Anniversary Volume*, p. 347, AIMME, New York, 1962.
95. Gaudin, A. M., *Flotation*, McGraw-Hill, New York, 1957.
96. Matis, K. A., and Gallios, G. P., Dissolved-air and electrolytic flotation, in *Mineral Processing at a Crossroads—problems and prospects*, eds. B. A. Wills and R. W. Barley, p. 37, Martinus Nijhoff Publishers, Dordrecht, 1986.
97. Moudgil, B. M., and Barnett, D. H., Agglomeration-skin flotation of coarse phosphate rock, *Mining Engng.* 283 (Mar. 1979).
98. Scales, M., Coarse flotation reduces overgrinding, *Can. Min. J.*, **105**, 19 (March 1984).
99. Anon, Flash flotation, *Int. Mining*, **3**, 14 (May 1986).
100. Tveter, E. C., and McQuiston, F. W., Plant practice in sulphide mineral flotation. *Froth Flotation 50th Anniversary Volume*, p. 382, AIMME, New York, 1962.
101. Lynch, A. J., Johnson, N. W., Manlapig, E. V., and Thorne, C. G., *Mineral and Coal Flotation Circuits*, Elsevier Scientific Publishing Co., Amsterdam (1981).
102. Wills, B. A., Automatic control of flotation, *Engng. & Min. J.*, **185**, 62 (June 1984).
103. Paakkinen, V. E., and Cooper, H. R., Flotation process control, in *Computer Methods for the 80's in the Minerals Industry* (ed. A. Weiss), A.I.M.M.E., New York, 1979.
104. Le Guen, F., The control of a flotation process, *CIM Bulletin*, 113 (April 1975).
105. Konigsmann, K. V., Hendriks, D. W., and Daoust, C., Computer control of flotation at Mattagami Lake Mines, *CIM Bulletin*, 117 (March 1976).
106. Fewings, J. H., Slaughter, P. J., Manlapig, E. V., and Lynch, A. J., The dynamic behaviour and automatic control of the chalcopyrite flotation circuit at Mount Isa Mines Ltd, *Proc. XIII Int. Min. Proc. Cong.*, 405, Warsaw (1979).
107. Kitzinger, F., Rosenblum, F., and Spira, P., Continuous monitoring and control of froth level and pulp density, *Mining Engng.* **31**, 310 (April 1979).
108. Herbst, J. A., *et al.*, Strategies for the control of flotation plants, in *Design and Installation of Concentration and Dewatering Circuits*, eds. A. L. Mular and M. A. Anderson, ch. 36, p. 548, SME Inc., Littleton, 1986.
109. Twidle, T. R., *et al.*, Optimising control of lead flotation at Black Mountain, *Proc. XVth Int. Min. Proc. Cong.*, **3**, 189, Cannes (1985).
110. Miettunen, J., The Pyhasalmi Concentrator- 13 years of computer control, in *Proc. 4th IFAC Symp. on Automation in Mining, Mineral and Metal Processing*, p. 391, Finnish Soc. Aut. Control, Helsinki, 1983.
111. Anon, US flotation: cost of reagents averaged 21 c per tonne in 1975, *Eng. Min. J.* 36 (Dec. 1976).
112. Thom, C., Standard flotation separations. *Froth Flotation 50th Anniversary Volume*, p. 328, AIMME, New York, 1962.
113. Baarson, R. E., Ray, C. L., and Treweek, H. B., Plant practice in non-metallic

mineral flotation, *Froth Flotation 50th Anniversary Volume*, p. 427, AIMME, New York, 1962.

114. Fuerstenau, M. C. (ed.), *Flotation: A. M. Gaudin Memorial Volume*, Vol. 2, AIMME, New York, 1976.

115. Jordan, T. S., *et al.*, Non-sulphide flotation: principles and practice, in *Design and Installation of Concentration and Dewatering Circuits*, eds. A. L. Mular and M. A. Anderson, chap. 2, p. 16, SME Inc., Littleton, 1986.

116. Ozbayoglu, G., Coal flotation, in *Mineral Processing Design*, eds. B. Yarar and Z. M. Dogan, p. 76, Martinus Nijhoff, Dordrecht, 1987.

117. Hsieh, S. S., and Lehr, J. R., Beneficiation of dolomitic Idaho phosphate rock by the TVA diphosphonic acid depressant process, *Minerals and Metallurgical Processing*, **2**, 10 (Feb. 1985).

118. Lawver, J. E., *et al.*, New techniques in beneficiation of the Florida phosphates of the future, *Minerals and Metallurgical Processing*, **1**, 89 (Aug. 1984).

119. Anon, Phosphates—a review of processing techniques, *World Mining Equip.*, **10**, 40 (April 1986).

120. Moudgil, B. M., Advances in phosphate flotation, in *Advances in Mineral Processing*, ed. P. Somasundaran, chap. 25, p. 426, SME Inc., Littleton, 1986.

121. Iwasaki, I., Iron ore flotation, theory and practice, *Mining Engng.* **35**, 622 (June 1983).

122. Hout, R., Beneficiation of iron ore by flotation, *Int. J. Min. Proc.*, **10**, 183 (1983).

123. Nummela, W., and Iwasaki, I., Iron ore flotation, in *Advances in Mineral Processing*, ed. P. Somasundaran, chap. 18, p. 308, SME Inc., Littleton, 1986.

124. Rabelink, T. B. M., The physical nature of tin flotation—the Billiton Minerals story, *Mine & Quarry* 45 (Sept. 1982).

125. Anon, Advances in modern mineral processing: tin flotation, *World Mining Equip.*, **9**, 26 (March 1985).

126. Lepetic, V. M., Cassiterite flotation: a review, in *Advances in Mineral Processing*, ed. P. Somasundaran, chap. 19, p. 343, SME Inc., Littleton, 1986.

127. Senior, G. D., and Poling, G. W., The chemistry of cassiterite flotation, in *Advances in Mineral Processing*, ed. P. Somasundaran, chap. 13, p. 229, SME Inc., Littleton, 1986.

128. Beyzavi, A. N., A contribution to scheelite flotation, taking particularly into account calcite-bearing scheelite ores, *Erzmetall.* **38**, 543 (Nov. 1985).

129. Fuerstenau, M. C., *et al.*, Flotation behaviour of chromium and manganese minerals, in *Advances in Mineral Processing*, ed. P. Somasundaran, chap. 17, p. 289, SME Inc., Littleton, 1986.

130. Crowson, P., and Thompson, M., Copper, *Mining Annual Review*, 27 (1986).

131. Lacy, W. C., *Porphyry Copper Deposits*, Australian Mineral Foundation Inc., 1974.

132. Sutolov, A., *Copper Porphyries*, University of Utah Printing·Services, 1974.

133. Edwards, R., and Atkinson, K., *Ore Deposit Geology*, Chapman & Hall, London, 1986.

134. Dayton, S., Chile: where major new copper output can materialise faster than any place else, *Engng. & Min. J.*, **180**, 68 (Nov. 1979).

135. Burger, J. R., Chile: World's largest copper producer is expanding, *Engng. & Min. J.*, **185**, 33 (Nov. 1984).

136. Sassos, M. P., Bougainville Copper, *Engng. & Min. J.*, **184**, 56 (Oct. 1983).

137. Sisselman, R., Chile's Chuquicamata: looking to stay no. 1 in copper output, *Engng. & Min, J.*, **8**, 59 (Aug. 1978)
138. Shirley, J., and Sutolov, A., Byproduct molybdenite, in *SME Mineral Processing Handbook*, ed. N. L. Weiss, sec. 16–17 (2), AIMME, 1985.
139. Crozier, R. D., Codelco's development plans for Chuquicamata and El Teniente, *Mining Mag.*, **155**, 460 (Nov. 1986).
140. Laskowski, J., *et al.*, Studies on the flotation of chrysocolla, *Mineral Processing and Tech. Review*, **2**, 135 (1985).
141. Holmes, J., Solvent extraction process is major advance in base metal recovery, *Optima*, **23**, 47 (March 1973).
142. Wallinger, W. N., Current operating practice at the Cyprus Anvil Concentrator, *CIM Bulletin* **71**, 134 (Jan. 1978).
143. Bergmann, A., and Haidlen, U., Economical aspects of bulk and selective flotation, in *Flotation of Sulphide Minerals*, ed. K. S. E. Forssberg, Elsevier, Amsterdam, 1985.
144. Anon., Swedish mills—flowsheets, operating data, *World Mining* 30, 137 (Oct. 1977).
145. Anon., Kamioka Mine, *Mining Mag.*, 387 (Nov. 1984).
146. Cases, J. M., Finely disseminated complex sulphide ores, in *Complex Sulphide Ores* (ed. M. J. Jones), p. 234, IMM (1980).
147. Anon., Bleikvassli and Mofjell, *Mining Mag.* 427 (Nov. 1980).
148. Stemerowicz, A. I., and Leigh, G. W., Flotation techniques for producing high recovery bulk Zn–Pb–Cu–Ag concentrates from a New Brunswick massive sulphide ore, *CANMET Rep.* 79–8, Aug. 1978.
149. Barbery, G. *et al.*, Exploitation of complex sulphide deposits: a review of processing options from ore to metals, in *Complex Sulphide Ores* (ed. M. J. Jones) p. 135, IMM (1980).
150. Gray, P. M. J., Metallurgy of the complex sulphide ores, *Mining Mag.*, 315 (Oct. 1984).
151. Imre, U., and Castle, J. F., Exploitation strategies for complex Cu–Pb–Zn orebodies, in *Mineral Processing and Extractive Metallurgy*, eds. M. J. Jones and P. Gill, p. 473, IMM, London, 1984.
152. McKee, D. J., Future applications of computers in the design and control of mineral beneficiation circuits, in *Automation for Mineral Resource Development*, p. 175, Pergamon Press, Oxford, 1986.
153. Wills, B. A., The separation by flotation of copper–lead–zinc sulphides, *Min. Mag.* 36 (Jan. 1984).
154. Clifford, K. L., *et al.*, Galena–Sphalerite–Chalcopyrite flotation at St. Joe Minerals Corporation, *Mining. Engng.* **31**, 180 (Feb. 1979).
155. McTavish, S., Flotation practice at Brunswick Mining, *CIM Bulletin* 115 (Feb. 1980).
156. Powell, C. R., Asarco Inc., Buchans Unit, in *Milling Practice in Canada* (ed. D. E. Pickett), p. 157, CIM (1978).
157. Wyllie, R. J. M., Minera Madrigal goes deeper for more tonnage, *World Mining*, **33**, 60 (Oct. 1980).
158. Allan, W., and Bourke, R. D., Mattabi Mines Ltd., in *Milling Practice in Canada* (ed. D. E. Pickett), p. 175, CIM (1978).
159. Eccles, A. G., Western Mines Ltd., in *Milling Practice in Canada* (ed. D. E. Pickett), p. 200, CIM (1978).

160. Bradley, F., *et al.*, Willroy Mines Ltd., in *Milling Practice in Canada* (ed. D. E. Pickett), p. 203, CIM (1978).
161. Brooks, L. S., and Barnett, C., Noranda Mines Ltd—Geco Division, in *Milling Practice in Canada* (ed. D. E. Pickett), p. 182, CIM (1978).
162. Cecile, J. L., *et al.*., Galena depression with chromate ions after flotation with xanthates: a kinetic and spectrometry study, in *Complex Sulphide Ores* (ed. M. J. Jones), p. 159, IMM (1980).
163. Kubota, T., *et al.*, A new method for copper–lead separation by raising pulp temperature of the bulk float, *Proc. XI Int. Min. Proc. Cong.*, Cagliari, Istituto di Arte Mineraria, Cagliari, 1975.
164. Anon, Kosaka Mine and Smelter, *Mining Mag.*, 403 (Nov. 1984).
165. Anon, Hanaoka Mine, *Mining Mag.*, 414 (Nov. 1984).
166. Burns, C. J., *et al.*, Process development and control at Woodlawn Mines, *14th Int. Min. Proc. Cong.*, Paper IV–18, CIM, Toronto, Canada (Oct. 1982).
167. Pazour, D. A. Morococha—five product mine shows no signs of dying, *World Mining* **32,** 56 (Nov. 1979).
168. Beck, R. D., and Chamart, J. J., The Broken Hill concentrator of Black Mountain Mineral Development Co. (Pty) Ltd, South Africa, in *Complex Sulphide Ores* (ed. M. J. Jones), p. 88, IMM (1980).
169. Seaton, N. R., Copper–lead–zinc ore concentration at Cobar Mines Pty Ltd., Cobar, N. S. W., in *Mining and Metallurgical Practices in Australasia* (ed. J. T. Woodcock), Aust. I.M.M., (1980).
170. Roberts, A. N., *et al.*, Metallurgical development at Woodlawn Mines, Australia, in *Complex Sulphide Ores* (ed. M. J. Jones), p. 128, IMM (1980).

CHAPTER 13

MAGNETIC AND HIGH-TENSION SEPARATION

Introduction

Magnetic and high-tension separators are being considered in the same chapter, as there is often a possibility of an overlap in the application of the two processes. For example, as can be seen later, there is often great debate as to which form of separation is best suited at various stages to the treatment of heavy mineral sand deposits.

Magnetic Separation

Magnetic separators exploit the difference in magnetic properties between the ore minerals and are used to separate either valuable minerals from non-magnetic gangue, e.g. magnetite from quartz, or magnetic contaminants or other valuable minerals from the non-magnetic values. An example of this is the tin-bearing mineral cassiterite, which is often associated with traces of magnetite or wolframite which can be removed by magnetic separators.

All materials are affected in some way when placed in a magnetic field, athough with most substances the effect is too slight to be detected. Materials can be classified into two broad groups, according to whether they are attracted or repelled by a magnet:

Diamagnetics are *repelled* along the lines of magnetic force to a point where the field intensity is smaller. The forces involved here are very small and diamagnetic substances cannot be concentrated magnetically.

Paramagnetics are *attracted* along the lines of magnetic force to points of greater field intensity. Paramagnetic materials can be concentrated in high-intensity magnetic separators. Examples of paramagnetics which are separated in commercial magnetic separators are ilmenite ($FeTiO_3$), rutile (TiO_2), wolframite (((Fe, Mn)WO_4), monazite (rare earth phosphate), siderite ($FeCO_3$), pyrrhotite (FeS), chromite ($FeCr_2O_4$), hematite (Fe_2O_3), and manganese minerals.

Some elements are themselves paramagnetic, such as Ni, Co, Mn, Cr, Ce, Ti, O, and the Pt group metals, but in most cases the paramagnetic properties of minerals are due to the presence of iron in some ferromagnetic form.

Ferromagnetism can be regarded as a special case of paramagnetism, involving very high forces. Ferromagnetic materials have very high susceptibility to magnetic forces and retain some magnetism when removed from the field (*remanence*). They can be concentrated in low-intensity magnetic separators and the principal ferromagnetic mineral separated is magnetite (Fe_3O_4), although hematite (Fe_2O_3) and siderite ($FeCO_3$) can be roasted to produce magnetite and hence give good separation. The removal of "tramp" iron from ores can also be regarded as a form of low-intensity magnetic separation.

It is not intended to review the theory of magnetism in any depth, as this is amply covered in many of the textbooks on elementary physics.

The unit of measurement of *magnetic flux density* or *magnetic induction* (B) (the number of lines of force passing through a unit area of material) is the *tesla* (T).

The magnetising force which induces the lines of force through a material is called the *field intensity* (H), and by convention has the units ampere metre^{-1} (1 ampere metre^{-1} = $4\pi \times 10^{-7}$ tesla).

The *intensity of magnetisation* or the *magnetisation* (M ampere/m) of a material relates to the magnetisation induced in the material, and:

$$B = \mu_0(H + M) \qquad (13.1)$$

the constant of proportionality, μ_0 being the *permeability of free space*, and having the value of $4\pi \times 10^{-7}$ T.m/A. In vacuum, $M = 0$, and it is extremely low in air, such that equation 13.1 becomes:

$$B = \mu_0 H \qquad (13.2)$$

so that the value of field intensity is virtually the same as that of flux density, and the term magnetic field intensity is then often loosely used. However, when dealing with the magnetic field inside materials, particularly ferromagnetics that concentrate the lines of force, the value of the induced flux density will be much higher than the field intensity, and it must be clearly specified which term is being referred to.

Magnetic susceptibility (*S*) is the ratio of the intensity of magnetisation produced in the material to the magnetic field which produces the magnetisation, i.e.:

$$S = M/H. \qquad (13.3)$$

Combining equations 13.1 and 13.3:

$$B = \mu_0 H (1 + S)$$

or $\qquad\qquad B = \mu\mu_0 H \qquad (13.4)$

where $\mu = 1 + S$, and is a dimensionless number known as the *relative permeability*.

For paramagnetic materials, *S* is a small positive constant, and is a negative constant for diamagnetic materials. Figure 13.1 shows plots of induced magnetisation (*M*) versus the strength of the external field (*H*), for paramagnetic (hematite) and diamagnetic (quartz) materials. Both plots show straight line relationships between *M* and *H*, in each case the slope representing the magnetic susceptibility (*S*) of the material, i.e. about 0.01 for hematite and around −0.001 for quartz.

The magnetic susceptibility of a ferromagnetic material is dependent on the magnetic field, decreasing with field strength as the material becomes *saturated*. Figure 13.2 shows a plot of *M* versus *H* for magnetite, showing that at an applied field of 1 T the magnetic susceptibility is about 0.35, and saturation occurs at about 1.5T. Many high-intensity magnetic separators use iron cores and frames to produce the desired magnetic flux concentrations and field strengths. Iron saturates magnetically at about 2–2.5 T, and the non-linear ferromagnetic relationship between inducing field strength and

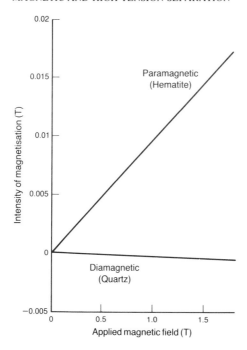

FIG. 13.1. Magnetisation curves for paramagnetic and diamagnetic materials.

magnetisation intensity necessitates the use of very large currents in the energising coils, sometimes up to hundreds of amperes.

The capacity of a magnet to lift a particular mineral is dependent not only on the value of the field intensity, but also on the *field gradient*, i.e. the rate at which the field intensity increases towards the magnet surface. Because paramagnetic minerals have higher magnetic permeabilities than the surrounding media, usually air or water, they concentrate the lines of force of an external magnetic field. The higher the magnetic susceptibility, the higher is the field density in the particle and the greater is the attraction up the field gradient towards increasing field strength. Diamagnetic minerals have lower magnetic susceptibility than their surrounding medium and hence expel the lines of force of the external field. This causes their expulsion in the

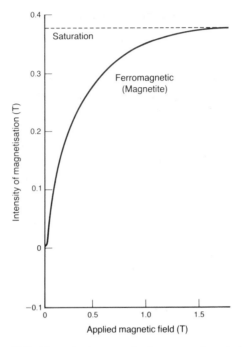

FIG. 13.2. Magnetisation curve for ferromagnetic material.

direction down the gradient of the field towards decreasing field strength. This negative diamagnetic effect is usually orders of magnitude smaller than the positive paramagnetic attraction.[1]

It can be shown that

$$F \propto H\frac{\mathrm{d}H}{\mathrm{d}l},$$ (13.5)

where F is the force on the particle, H is the field intensity, and $\mathrm{d}H/\mathrm{d}l$ is the field gradient.

Thus in order to generate a given lifting force, there are an infinite number of combinations of field and gradient which will give the same effect. Production of a high field gradient as well as high intensity is therefore an important aspect of separator design.

Magnetic Separators

Design

Certain elements of design are incorporated in all magnetic separators, whether they are low or high intensity, wet or dry. The prime requirement, as has already been mentioned, is the provision of a high-intensity field in which there is a steep field strength gradient. In a field of uniform flux, such as in Fig. 13.3(a), magnetic particles will orient themselves, but will not move along the lines of flux. The most straightforward method for producing a converging field is by providing a V-shaped pole above a flat pole, as in Fig. 13.3(b). The tapering of the upper pole concentrates the magnetic flux into a very small area giving high intensity. The lower flat pole has the same total magnetic flux distributed over a larger area. Thus there is a steep field gradient across the gap by virtue of the different intensity levels.

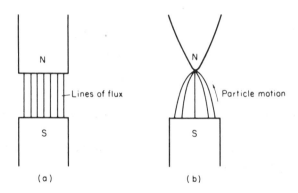

FIG. 13.3 (a) Field of uniform flux, (b) converging field.

Another method of producing a high field gradient is by using a pole which is constructed of alternate magnetic and non-magnetic laminations (Fig. 13.4).

Provision must be incorporated in the separator for regulating the intensity of the magnetic field so as to deal with various types of material. This is easily achieved in electromagnetic separators by

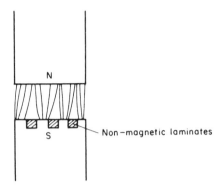

FIG. 13.4. Production of field gradient by laminated pole.

varying the current, while with permanent magnets the interpole distance can be varied.

Commercial magnetic separators are continuous-process machines and separation is carried out on a moving stream of particles passing into and through the magnetic field. Close control of the speed of passage of the particles through the field is essential, which rules out free fall as a means of feeding. Belts or drums are very often used to transport the feed through the field.

The introduction into a magnetic field of particles which are highly susceptible concentrates the lines of force so that they pass through them (Fig. 13.5).

Since the lines of force converge to the particles, a high field gradient is produced which causes the particles themselves to behave as magnets, thus attracting each other. Flocculation, or agglomeration, of the particles can occur if they are small and highly susceptible and if the field is intense. This has great importance as these magnetic "flocs" can entrain gangue and can bridge the gaps between magnet poles, reducing the efficiency of separation. Flocculation is especially serious with dry separating machines operating on fine material. If the ore can be fed through the field in a monolayer, this effect is much less serious, but, of course, the capacity of the machine is drastically reduced. Flocculation is often minimised by passing the material through consecutive magnetic fields, which are usually arranged with

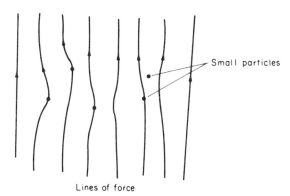

Small particles

Lines of force

FIG. 13.5. Concentration of flux on mineral particles.

successive reversal of the polarity. This causes the particle to turn through 180°, each reversal tending to free the entrained gangue particles. The main disadvantage of this method is that flux tends to leak from pole to pole, reducing the effective field intensity. Recent work has shown that the generation of low-frequency vibrations in the zone of separation in the pulp may improve the selectivity of the process.[2]

Provision for collection of the magnetic and non-magnetic fractions must be incorporated into the design of the separator. Rather than allow the magnetics to contact the pole-pieces, which would cause problems of detachment, most separators are designed so that the magnetics are attracted to the pole-pieces, but come into contact with some form of conveying device, which carries them out of the influence of the field, into a bin or belt. Non-magnetic disposal presents no problems, free fall from a conveyor into a bin often being used. Middlings are readily produced by using a more intense field after the removal of the highly magnetic fraction.

Types of Magnetic Separator

Magnetic separators can be classified into low- and high-intensity machines, which may be further classified into dry-feed and wet-feed separators.

Low-intensity separators are used to treat ferromagnetic materials and some highly paramagnetic minerals.

Low-intensity Magnetic Separation

Dry low-intensity magnetic separation is confined mainly to the concentration of coarse sands which are strongly magnetic, the process being known as *cobbing*, and often being carried out in drum separators. Below the 0.5-cm size range, dry separation tends to be replaced by wet methods, which produce much less dust loss and usually a cleaner product. Low-intensity wet separation is now widely used for purifying the magnetic medium in the heavy medium separation process (see Chapter 11), as well as for the concentration of ferromagnetic sands.

Drum separators are the most common machines in current use for cleaning the medium in HMS circuits and are widely used for concentrating finely ground iron ore. They consist essentially of a rotating non-magnetic drum (Fig. 13.6) containing three to six stationary magnets of alternating polarity. These were initially electromagnets, but are now virtually always permanent magnets, utilising modern ceramic magnetic alloys, which retain their intensity for an indefinite period. Separation is again by the "pick-up" principle. Magnetic particles are lifted by the magnets and pinned to the drum and are conveyed out of the field, leaving the gangue in the tailings compartment. Water is introduced into the machine to provide a current which keeps the pulp in suspension. Field intensities of up to 0.7 T at the pole surfaces can be obtained in this type of separator.

The drum separator shown in Fig. 13.6 is of the *concurrent* type, whereby the concentrate is carried forward by the drum and passes through a gap where it is compressed and dewatered before leaving the separator. This design is most effective for producing an extremely clean magnetic concentrate from relatively coarse materials and is widely used in heavy medium recovery systems.

The separator shown in Fig. 13.7 is of the *counter-rotation* type, where the feed flows in the opposite direction to the rotation. This type is used in roughing operations, where occasional surges in feed must be

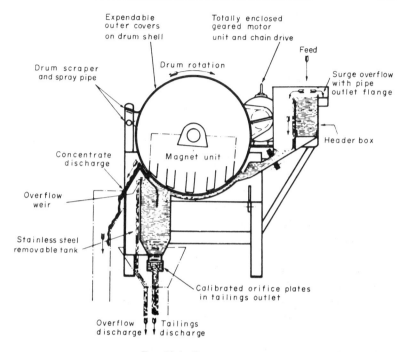

FIG. 13.6. Drum separator.

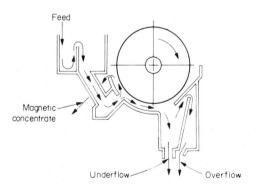

FIG. 13.7 Counter-rotation drum separator.

handled, where magnetic material losses are to be held to a minimum, while an extremely clean concentrate is not required, and when high solids loading is encountered.

Figure 13.8 shows a *counter-current* separator, where the tailings are forced to travel in the opposite direction to the drum rotation and are discharged into the tailings chute. This type of separator is designed for finishing operations on relatively fine material, or particle size less than about 250 μm.

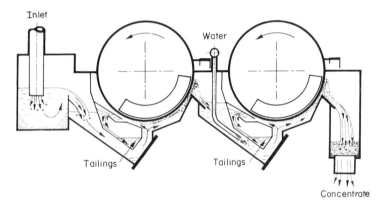

FIG. 13.8. Counter-current separator.

Drum separators are widely used to treat low-grade taconite ores, which contain 40–50% Fe, mainly as magnetite, but in some areas with hematite, finely disseminated in bands in hard siliceous rocks. Very fine grinding is necessary to free the iron minerals that produce a concentrate that requires pelletising before being fed to the blast furnaces.

In a typical flowsheet the ore is ground progressively finer, the primary grind usually being undertaken autogenously, or by rod mill, followed by magnetic separation in drum separators. The magnetic concentrate is reground and again treated in drum separators. This concentrate may be further reground, followed by a third stage of magnetic separation. The tailings from each stage of magnetic separation are either rejected, or in some cases treated by spiral or Reichert cone concentrators to recover hematite.

At Palabora, the tailings from copper flotation (Fig. 12.59) are deslimed, after which the $+105$ μm material is treated by Sala drum separators to recover 95% of the magnetite at a grade of 62% Fe.

One of the oldest types of separator used to concentrate moderately magnetic ores is the *cross-belt separator* (Fig. 13.9).

FEED

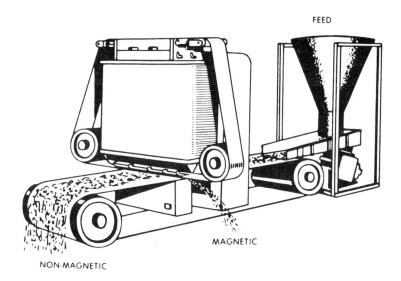

MAGNETIC

NON-MAGNETIC

FIG. 13.9. Cross-belt separator.

Dry material is fed in a uniform layer on to the conveyor belt and is carried between the poles of the magnetic system, this consisting of two or more horseshoe electromagnets, the poles of which are arranged one above the other. The poles of the upper magnets are wedge-shaped while the lower poles are flat. This concentrates the field and attracts the paramagnetic minerals toward the wedge-shaped poles. The cross-belts serve to prevent the magnetic particles from adhering to the poles and carry them out of the field.

A modification of the cross-belt separator is the *disc separator*, manufactured by Boxmag-Rapid Ltd., which permits a much smaller air-gap than the belt separator and a greater degree of selectivity. The operating principle lies in the provision of a series of discs, incorporating concentrating grooves, revolving above a conveyor belt and magnetised by induction from powerful stationary electromagnets situated below the belt (Fig. 13.10). Each disc is situated such that the leading edge provides a greater air-gap than the trailing edge, permitting each disc to extract and separate two magnetic products of different permeability. Progressive intensification of the magnetic field is obtained in the direction of the feed-belt travel by vertical adjustment of the discs and current control on each electromagnet. The sized ore is spread evenly on to the moving belt and passes through the graduated magnetic zones, the magnetics being extracted and separately discharged in the order of their magnetic susceptibility. The tails pass forward and are discharged at the end of the conveyor belt. Models having one, two, or three discs are manufactured and it is possible to obtain very precise selectivity between minerals differing only slightly from one another in magnetic response, since current, disc height and tilt are all adjustable when the machine is in operation. With normal settings and currents the flux on the discs is usually in the range 0.8–1.5 T, which is strong enough to pick up many paramagnetic

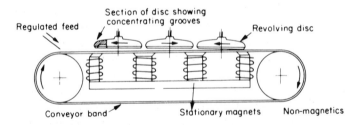

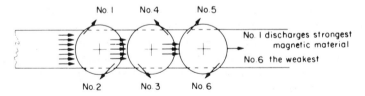

FIG. 13.10. Disc separator—diagrammatic.

minerals. Such separators are used on beach sand deposits in India, Sri Lanka, and Australia, and on the alluvial Malaysian tin ores, where paramagnetic ilmenite and monazite are separated from the non-magnetic cassiterite.

High-intensity Separators

Very weakly paramagnetic minerals can only be effectively removed from an ore feed if high-intensity fields of 2 T and more can be produced.

Until the 1960s high-intensity separation was confined solely to dry ore, having been used commercially since about 1908.

Induced roll separators (Fig. 13.11) are widely used to treat beach sands, wolframite, tin ores, glass sands, and phosphate rock. They have also been used to treat feebly magnetic iron ores, principally in Europe. The roll, on to which the ore is fed, is composed of phosphated steel laminates compressed together on a non-magnetic stainless steel shaft. By using two sizes of lamination, differing slightly in outer diameter, the roll is given a serrated profile which promotes the high field intensity and gradient required. Field strengths of up to 2.2 T are attainable in the gap between feed pole and roll. Non-magnetic particles are thrown off the roll into the tailings compartment, whereas magnetics are gripped, carried out of the influence of the field and deposited into the magnetics compartment. The gap between the feed pole and rotor is adjustable and is usually decreased from pole to pole to take off successively more weakly magnetic products. The setting of the splitter plates cutting into the trajectory of the discharged material is obviously of great importance.

The *Permroll*[3] is a roll separator which uses powerful permanent magnets capable of generating field densities and gradients similar to those occurring in the narrow air-gaps of the induced roll separator. The magnetic field is generated without an air-gap, and the separator is able to handle highly magnetic as well as weakly paramagnetic particles, since a belt passes over the magnetic roll, preventing direct roll–particle contact.

Dry high-intensity separation is largely limited to ores containing

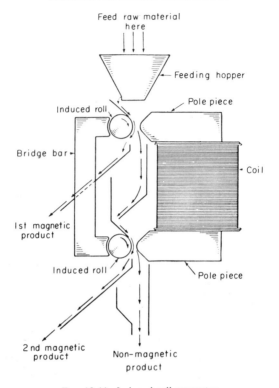

FIG. 13.11. Induced roll separator.

little if any material finer than about 75 μm. The effectiveness of separation on such fine material is severely reduced by the effects of air currents, particle–particle adhesion and particle–rotor adhesion.

Without doubt the greatest advance in the field of magnetic separation was the development of continuous high-intensity wet separators.[4] This reduces the minimum particle size for efficient separation allowing ores to be concentrated magnetically that cannot be concentrated effectively by dry high-intensity methods, because of the fine grinding necessary to ensure complete liberation of the magnetic fraction. In some flowsheets, expensive drying operations can be eliminated by using a wet concentration system.

The first models of high-intensity wet magnetic separators were designed on the principle of the induced roll machines. A typical model is the Gill separator, which was introduced in 1964[5] (Fig. 13.12). The laminated grooved rotor rotates about a vertical axis at the centre of a set of electromagnetic pole pieces. The feed slurry flows down along the grooves between a pole piece and the rotor, the tailings passing through into the non-magnetics box and the magnetics being carried forward out of the influence of the field, to be deposited in the magnetics box. Although not a true high-intensity machine, generating a maximum field of less than 1.4 T, the Gill separator has been used in Australia and Malaysia for separating highly magnetic ilmenite from heavy mineral concentrates, but it will not treat weakly magnetic hematite ores effectively.

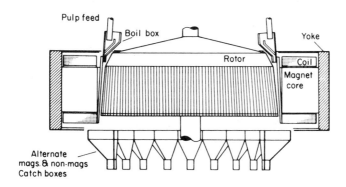

FIG. 13.12. Side view of four-pole Gill wet magnetic separator.

Perhaps the most well-known separator now being used to treat fine hematite ores is the Jones wet high-intensity magnetic separator, the design principle of which is utilised in many other types of wet separator used today.

The machine consists of a strong main frame (Fig. 13.13), made of structural steel. The magnet yokes are welded to this frame, with the electromagnetic coils enclosed in air-cooled cases. The actual separation takes place in the plate boxes which are on the periphery of the one or two rotors attached to the central roller shaft. The feed,

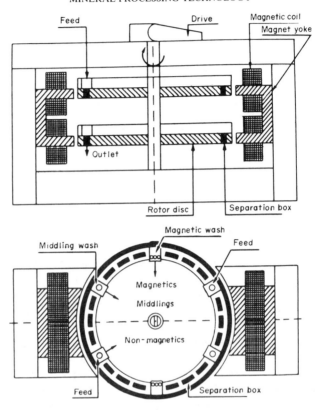

FIG. 13.13. Operating principle of the Jones high-intensity wet magnetic separator.

which is a thoroughly mixed slurry, flows through the separator via
fitted pipes and launders into the plate boxes (Fig. 13.14), which are
grooved to concentrate the magnetic field at the tip of the ridges.
Feeding is continuous due to the rotation of the plate boxes on the
rotors and the feed points are at the leading edges of the magnetic
fields. Each rotor has two symmetrically disposed feed points.

The feebly magnetic particles are held by the plates, whereas the
remaining non-magnetic pulp passes straight through the plate boxes
and is collected in a launder. Before leaving the field any entrained

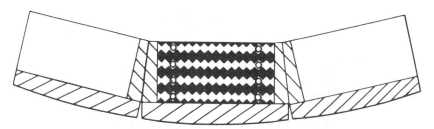

FIG. 13.14. Plan of Jones plate box showing grooved plates and spacer bars.

non-magnetics are washed out by low-pressure water and are collected as a middlings product.

When the plate boxes reach a point midway between the two magnetic poles, where the magnetic field is essentially zero, the magnetic particles are washed out with high pressure scour water sprays operating at up to 5 bar (Fig. 13.15). Field intensities of over 2 T can be produced in these machines. The production of a 1.5 T field requires an electric power consumption in the coils of 16 kW per pole. Of the 4 t of water used with every tonne of solids, approximately 90% is recycled.

Wet high-intensity magnetic separation has its greatest use in the concentration of low-grade iron ores containing hematite, where they frequently replace flotation methods, although the trend towards magnetic separation has been slow in North America, mainly due to the very high capital cost of such separators. It has been shown[6] that the capital cost of flotation equipment for concentrating weakly magnetic ore is about 20% that of a Jones separator installation, although flotation operating costs are about three times higher. Total cost depends on terms for capital depreciation; over 10 years or longer the high-intensity magnetic separator may be the most attractive process. Additional costs for water treatment may also boost the total for a flotation plant. One of the largest applications of high-intensity wet magnetic separation is at the Companhia Vale de Rio Doce plant in Itabira, Brazil, where Jones separators are used to treat 120 t h^{-1} of -150-μm specular hematite ore[7] (Fig. 13.16)

Various other designs of wet high-intensity separator have been produced, a four-pole machine being manufactured by Boxmag-Rapid

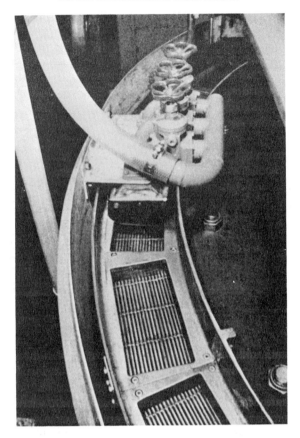

Fig. 13.15. Jones separator—magnetic wash.

Ltd. The plate boxes in this design are an array of magnetic stainless steel "wedge-bars" similar to those used in fine screening (Fig. 13.17).

In addition to their large-scale application for the recovery of hematite, wet high-intensity separators are now in operation for a wide range of duties, including removal of magnetic impurities from cassiterite concentrates, removal of fine magnetic from asbestos, removal of magnetic impurities from scheelite concentrates, purification of talc, the recovery of wolframite and non-sulphide

FIG. 13.16. Jones separator treating Brazilian hematite ore.

molybdenum-bearing minerals from flotation tailings, and the treatment of heavy mineral beach sands. They have also been successfully used for the recovery of gold and uranium from cyanidation residues in South Africa.[8] These residues contain some free gold, while some of the fine gold is locked in sulphides, mainly pyrite, and in various silicate minerals. The free gold can be recovered by further cyanidation, while flotation can recover the pyritic gold. Magnetic separation can be used to recover some of the free gold, and much of the silicate-locked gold, due to the presence of iron impurities and coatings.

The paramagnetic properties of some sulphide minerals, such as chalcopyrite and marmatite, have been exploited by applying wet high-intensity magnetic separation to augment differential flotation processes commonly used to separate these minerals from less magnetic or non-magnetic sulphides.[9] Testwork showed that a Chilean copper concentrate could be upgraded from 23.8% to 30.2%

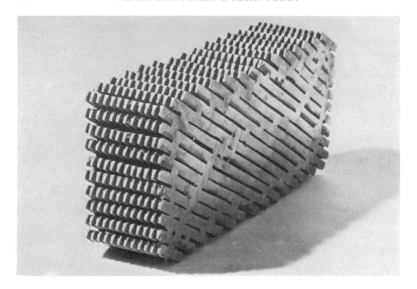

FIG. 13.17. Section through Boxmag-Rapid grid assembly.

Cu, at 87% recovery. This was done by separating the chalcopyrite from pyrite in a field of 2 T. In Cu – Pb separation operations, it was found that chalcopyrite and galena could be effectively separated with field strengths as low as 0.8 T. When the process was applied to the de-coppering of a molybdenite concentrate, it was possible to reduce the copper content from 0.8% to 0.5% with over 97% Mo recovery.

High-gradient Magnetic Separators

In order to separate paramagnetic minerals of extremely low magnetic susceptibility, high magnetic forces must be generated. These forces can be produced by increasing the magnetic field strength, and in conventional high-intensity magnetic separators use is made of the ferromagnetic properties of iron to generate a high B-field (induced field) many hundreds of times greater than the applied H-field, with a minimum consumption of electrical energy. The working field occurs in air-gaps in the magnetic circuit, the disadvantage being that the volume of iron required is many times greater than the gap volume where separation takes place. The steel plates in a Jones separator, for example, occupy up to 60% of the process volume. Thus high-intensity magnetic separators using conventional iron circuits tend to be very massive and heavy in relation to their capacity. A large separator may contain over 200 t of iron to carry the flux, hence capital and installation costs are extremely high.

As iron saturates at around 2–2.5 T, conventional iron circuits are of little value for generating fields above about 2 T. Such fields can only be generated by the use of high H-fields produced in solenoids, but the energy consumption is extremely high and there are problems in cooling the solenoid.

An alternative is to increase the magnetic force by increasing the value of the magnetic field gradient. Instead of using one large convergent field in the gap of a magnetic circuit, the uniform field of a solenoid is used (Fig. 13.18). The core, or working volume, is filled with a matrix of secondary poles, such as ball bearings, or wire wool, the latter filling only about 10% of the working volume. Each secondary pole, due to its high permeability, can produce a maximum field strength of the order of 2 T, but more importantly, each pole produces, in its immediate vicinity, high field gradients of up to 1 T mm^{-1}. Thus a multitude of high gradients across numerous small gaps, centred around each of the secondary poles, is achieved.

The solenoid can be clad externally with an iron frame to form a continuous return path for the magnetic flux, thus reducing the energy consumption for driving the coil by a factor of about 2.

The matrix is held in a canister into which the slurry is fed. As particles are captured, the ability of the matrix to extract particles is reduced. Periodically the magnetic field can be removed and the matrix flushed with water to remove the captured material.

An inherent disadvantage of high gradient separators is that an increase in field gradient necessarily reduces the working gap between secondary poles, the magnetic force having only a short reach, usually not more than 1 mm. It is therefore necessary to use gaps of only about 2 mm between poles, such that the matrix separators are best suited to the treatment of very fine particles. They are used mainly in the kaolin

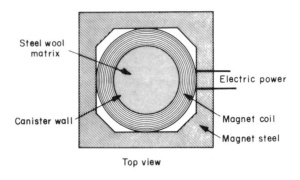

Top view

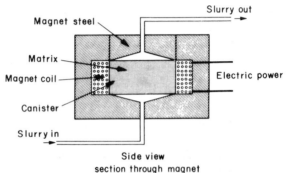

Side view
section through magnet

FIG. 13.18. High-gradient magnetic separator.

industry, for removing micron-sized particles which contain iron. Several large separators, with the ferromagnetic matrix contained in baskets approximately 2 m in diameter are in commercial use in the United States and in Cornwall, England. They operate with fields of 2 T, and have capacities ranging between 10 and 80 t h^{-1} depending on the final clay quality desired.

One of the most important factors which will effect coal preparation policy in the future is the environmental issue associated with acid rain and its link with sulphur emissions from fossil fuels. Sulphur occurs in coal in three forms. It is part of the coal substance (organic sulphur), or occurs as the minerals pyrite and marcasite, or as sulphates. The most important factor for the engineer is the pyritic sulphur content, as technology is not yet sufficiently developed to consider the removal of organic sulphur. If pyrite can be liberated by fine crushing to around 1 mm, then froth flotation or gravity methods can be used to remove it from the coal. However, if very fine crushing is necessary to liberate the pyrite, then high-gradient magnetic separation is a possibility. Increased international interest is at present being shown by coal preparation engineers in coal – liquid mixtures as a replacement for conventional hydrocarbon fuels such as diesel oil and natural gas. A typical coal – water mixture consists of pulverised coal of less than 50 microns particle size, and low ash content (2–6%) dispersed in an aqueous slurry, with a pulp density of between 50 and 80% solids.[10] In order to produce these mixtures it is necessary to treat good quality coal by fine grinding and deep cleaning to remove ash and sulphur. High-gradient magnetic separation is capable of removing pyrite from pulverised coal, and much work is currently being performed on a variety of coal types.

Superconducting Separators

Undoubtedly the future developments and applications of magnetic separation in the mineral industry will lie in the use of high magnetic forces. Matrix separators with very high field gradients and multiple small working gaps can draw little advantage from field strengths above the saturation levels of the secondary poles. However, "open-gradient" separators, with large working volumes to accommodate coarser particles at high capacity, need to use the highest possible field strengths in order to generate the high magnetic forces required to treat feebly paramagnetic particles. Field strengths in excess of 2 T can only be generated economically by the use of *superconducting magnets*.[1]

Certain alloys have the property of presenting no resistance to electric currents at extremely low temperatures. An example is niobium–tantalum at 4.2 K, the temperature of liquid helium. Once a current commences to flow through a coil made from such a superconducting material, it will continue to flow without being connected to a power source, and the coil will become, in effect, a permanent magnet. Superconducting magnets can produce extremely intense and uniform magnetic fields, of up to 15 T.

In 1986 a superconducting high-gradient magnetic separator was designed and built by Eriez Magnetics to process kaolinite clay in the United States.[11] This machine will use only about 0.007 kW in producing 5 T of flux, the ancillary equipment needed requiring another 20 kW. In comparison, a conventional 2 T high-gradient separator of similar throughput would need about 250 kW to produce the flux, and at least another 30 kW to cool the magnet windings.

The 5 T machine is an assembly of concentric components (Fig. 13.19). A removable processing canister is installed in a processing chamber located at the centre of the assembly. This is surrounded by a double-walled, vacuum-insulated container that accommodates the superconductive magnet's niobium/titanium–tantalum winding, and the liquid helium coolant. A thermal shield, cooled with liquid

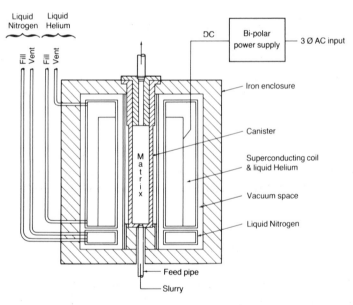

FIG. 13.19. 5 T superconducting magnetic separator.

nitrogen to 77K, limits radiation into the cryostat. In operation, the supply of slurry is periodically cut off, the magnetic field is shut down, and the canister backwashed with water to clear out accumulated magnetic contaminants.

An open-gradient drum magnetic separator with a superconducting magnet system has also been designed and built[12] (Fig. 13.20).

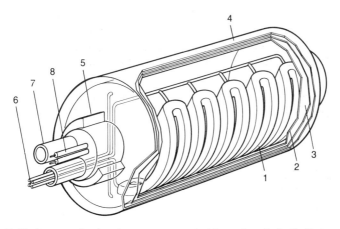

FIG. 13.20. Superconducting drum separator. 1. Magnetic coils 2. Radiation shield 3. Vacuum tank 4. Drum 5. Plain bearing 6. Helium supply 7. Vacuum line 8. Current supply.

Although separation is identical to that in conventional drum separators, the magnetic flux density at the drum surface can reach 3.2 T, generated by the superconductive magnet assembly within the drum.

High-tension Separation

High-tension separation utilises the difference in electrical conductivity between the various minerals in the ore feed. Since almost all minerals show some difference in conductivity it would appear to represent the universal concentrating method. In practice, however, the method has fairly limited application, and its greatest use is in separating some of the minerals found in heavy sands from beach or stream placers. The fact that the feed must be perfectly dry imposes limitations on the process, but it also suffers from the same great disadvantage as dry magnetic separators—the capacity is very small for finely divided material. For most efficient operation, the feed should be in a layer, one particle deep, which severely restricts the throughput if the particles are as small as, say, 75 μm.

The first mineral separation processes utilising high voltage were virtually true electrostatic processes employing charged fields with little or no current flow. High-tension separation, however, makes use of a comparatively high rate of electrical discharge, with electron flow and gaseous ionisation having major importance.

The attraction of particles carrying one kind of charge toward an electrode of the opposite charge is known as the "lifting effect", as such particles are lifted from the separating surface toward the electrode. Materials which have a tendency to become charged with a definite polarity may be separated from each other by the use of the lifting effect even though their conductivities may be very similar. As an example, quartz assumes a negative charge very readily and may be separated from other poor conductors by an electrode which carries a positive charge. Pure electrostatic separation is relatively inefficient, even with very clean mineral, and is sensitive to changes of humidity and temperature.

A large percentage of the commercial applications of high-tension separation has been made using the "pinning effect", in which non-conducting mineral particles, having received a surface charge from the electrode, retain this charge and are pinned to the oppositely charged separator surface by positive–negative attraction. Figure 13.21 shows a Carpco laboratory high-tension separator, which makes use of the pinning effect to a high degree in combination with some lifting effect. Figure 13.22 shows the principle of separation diagrammatically.

The mixture of ore minerals, of varying susceptibilities to surface charge, is fed on to a rotating drum made from mild steel, or some other conducting material, which is earthed through its support

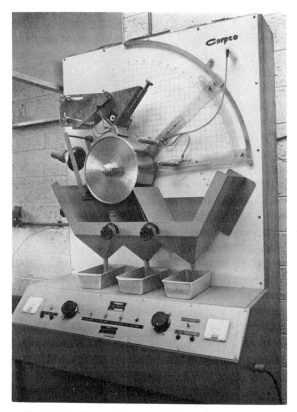

FIG. 13.21. Carpco laboratory high-tension separator.

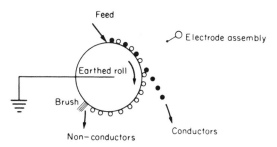

FIG. 13.22. Principle of high-tension separation.

bearings. An electrode assembly, comprising a brass tube in front of which is supported a length of fine wire, spans the complete length of the roll, and is supplied with a fully rectified DC supply of up to 50 kV, usually of negative polarity. The voltage supplied to the assembly should be such that ionisation of the air takes place. This can often be seen as a visible *corona discharge*. Arcing between the electrode and the roll must be avoided, as this destroys the ionisation. When ionisation occurs, the minerals receive a spray discharge of electricity which gives the poor conductors a high surface charge, causing them to be attracted to and pinned to the rotor surface. The particles of relatively high conductivity do not become charged as readily, as the charge rapidly dissipates through the particles to the earthed rotor. These particles of higher conductivity follow a path, when leaving the rotor, approximating to the one which they would assume if there were no charging effect at all.

The electrode assembly is designed to create a very dense high-voltage discharge. The fine wire of the assembly is placed adjacent to and parallel to the large diameter electrode and is mechanically and electrically in contact with it. This fine wire tends to discharge readily, whereas the large tube tends to have a short, dense, non-discharging field. This combination creates a very strong discharge pattern which may be "beamed" in a definite direction and concentrated to a very narrow arc. The effect on the minerals passing through the beam is very strong and is due largely to gaseous ions which are created due to the high-voltage gradient in the field of the corona.

A combination of the effects of pinning and lifting can be created by using a static electrode large enough to preclude corona discharge, following the electrode. The conducting particles, which are flung from the rotor, are attracted to this electrostatic electrode, and the compound process produces a very wide and distinct separation between the conducting and non-conducting particles.

Table 13.1 shows typical minerals which are either pinned to or thrown from the rotor during high-tension separation.

To cater for such an extensive range of minerals, all the parameters influencing separation must be readily adjusted while the separator is performing. These variables include the roll speed, the position of the

TABLE 13.1. TYPICAL BEHAVIOUR OF MINERALS IN
HIGH-TENSION SEPARATORS

Minerals pinned to rotor	Minerals thrown from rotor
Apatite	Cassiterite
Barite	Chromite
Calcite	Diamond
Corundum	Fluorspar
Garnet	Galena
Gypsum	Gold
Kyanite	Hematite
Monazite	Ilmenite
Quartz	Limonite
Scheelite	Magnetite
Sillimonite	Pyrite
Spinel	Rutile
Tourmaline	Sphalerite
Zircon	Stibnite
	Tantalite
	Wolframite

electrode wire with respect to the electrode tube, the position of the electrode assembly with respect to the roll, variation of the DC voltage and polarity, the splitter plate position, the feed rate and heating of the feed. Heating the feed is important, since best results are generally obtained only with very dry material. This is particularly difficult in high humidity regions. It is not often that a single pass will sufficiently enrich an ore and Fig. 13.23 shows a typical flowsheet, where the falling particles are deflected to lower sets of rollers and electrodes until the required separation has taken place.

High-tension separators operate on feeds containing particles of between 60 and 500 μm in diameter. Particle size influences separation behaviour, as the surface charges on a coarse grain are lower in relation to its mass than on a fine grain. Thus a coarse grain is more readily thrown from the roll surface, and the conducting fraction often contains a small proportion of coarse non-conductors. Similarly, the finer particles are most influenced by the surface charge, and the non-conducting fraction often contains some fine conducting particles.

Final cleaning of these products is often carried out in purely electrostatic separators, which employ the "lifting effect" only. Modern electrostatic separators are of the *plate*, or *screen* type, the

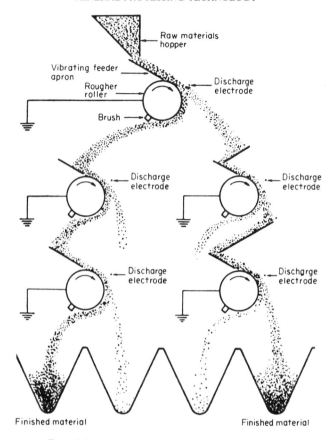

FIG. 13.23. Arrangement of separators in practice.

former being used to clean small amounts of non-conductors from a predominantly conducting feed, while the screen separators remove small amounts of conductors from a mainly non-conducting feed. The principle of operation is the same for both types of separator. The feed particles gravitate down a sloping, grounded plate into an electrostatic field induced by a large, oval, high-voltage electrode (Fig. 13.24). The electrostatic field is effectively shorted through the conducting particles, which are lifted towards the charged electrode in order to

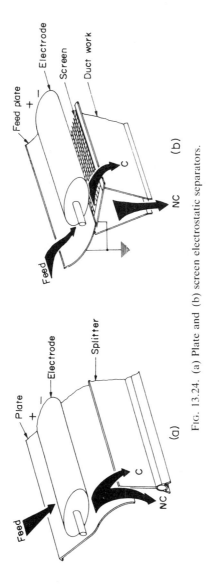

Fig. 13.24. (a) Plate and (b) screen electrostatic separators.

decrease the energy of the system. Non-conductor grains are poorly affected by the field. The fine grains are most affected by the lifting force, and so fine conductors are preferentially lifted to the electrode, whereas coarse non-conductors are most efficiently rejected. This is the converse of the separation which takes place in the high-tension separators, where most effective separation of fine non-conductors from coarse conductors takes place; a combination of high-tension separators as primary roughers, followed by final cleaning in electrostatic separators, is therefore used in many flowsheets. Since the magnitude of the forces involved in electrostatic separation is very low, the separators are designed for multiple passes of the non-conductors (Fig. 13.25).

It was mentioned earlier that there is some possibility of an overlap in the application of magnetic and high-tension separators, particularly in the processing of heavy mineral sand deposits. Table 13.2 shows some of the common minerals present in such alluvial deposits, along with their properties related to magnetic and high-tension separation. Mineral sands are commonly mined by floating dredges, feeding floating concentrators at up to 2000 t h^{-1} or more. Figure 13.26 shows a typical dredge and floating concentrator operating at Richards Bay in South Africa. Such concentrators, consisting of a complex circuit of sluices, spirals, or Reichert cones, upgrade the heavy mineral content to around 90%, the feed grades varying from less than 2%, up to 20% heavy mineral in some cases. The gravity concentrate is then transferred to the separation plant for recovery of the minerals by a combination of gravity, magnetic and high tension methods. Flowsheets vary according to the properties of valuable minerals present, wet magnetic separation often preceding high-tension separation where magnetic ilmenite is the dominant mineral. A generalised flowsheet for such a separation is shown in Fig. 13.27. Low-intensity drum separators remove any magnetite from the feed, after which high-intensity wet magnetic separators separate the monazite and ilmenite from the zircon and rutile. Drying of these two fractions is followed by high-tension separation to produce final separation, athough further cleaning is sometimes carried out by electrostatic separators. For example, screen electrostatic separators may be used to clean the zircon and monazite concentrates, removing fine

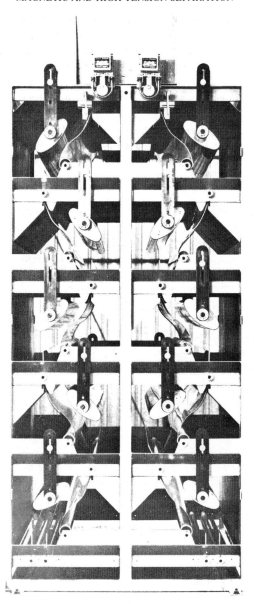

FIG. 13.25. Plate electrostatic separator with two-start, ten electrodes.

TABLE 13.2. TYPICAL BEACH SAND
MINERALS

Magnetics	Non-magnetics
Magnetite—T	Rutile—T
Ilmenite—T	Zircon—P
Garnet—P	Quartz—P
Monazite—P	

T = thrown from high-tension separator surface.
P = pinned to high-tension separator surface.

conducting particles from these fractions. Similarly, plate electrostatic separators could be used to reject coarse non-conducting particles from the rutile and ilmenite concentrates.

FIG. 13.26. Heavy mineral sand mining and pre-concentration plant.

Figure 13.28 shows a simplified circuit used by Associated Minerals Consolidated Ltd, on the east coast of Australia.[13] After cleaning the heavy mineral concentrate by the use of spirals and tables, the concentrate is dried and passed onto high-tension separators. After

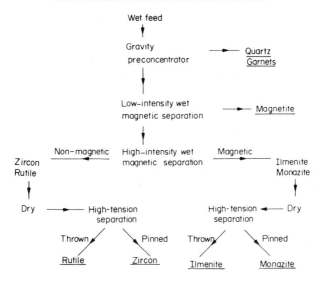

FIG. 13.27. Typical beach sand treatment flowsheet.

several stages of cleaning, high-intensity induced roll separators are used to treat the conducting and non-conducting products. The conducting rutile concentrate is cleaned by the use of electrostatic plate separators, the non-conducting zircon being cleaned by electrostatic screen separators. Since any remaining silica, being non-magnetic and non-conducting, reports to the zircon concentrate, this is removed by a combination of pneumatic and air tabling (see Chapter 10). Similar flowsheets are used in South-East Asia for the treatment of alluvial cassiterite deposits, which are also sources of minerals such as ilmenite, monazite and zircon.

Magnetic separators are commonly used for up-grading low-grade iron ores, wet high-intensity separation often replacing the flotation of hematite. A combination of magnetic and high-tension separation is used at the Scully Mine of Wabush Mines, in Canada.[14] The ore, grading about 35% Fe, is a quartz–specular hematite–magnetite schist, and after crushing and autogenous grinding to -1 mm, is fed to banks of rougher and cleaner spiral concentrators (Fig. 13.29). The

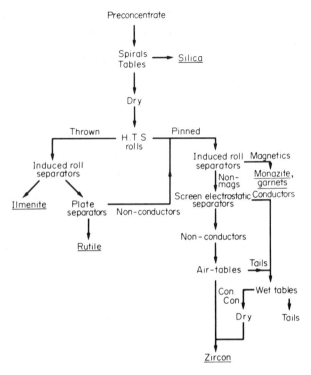

FIG. 13.28. Simplified flowsheet of separation plant of Associated Minerals
Consolidated Ltd.

spiral concentrate is filtered and dried, and cleaned in Carpco
high-tension roll separators. The spiral tailings are thickened, and
further treated by magnetic drum separators to remove residual
magnetite, followed by Jones wet high-intensity separators, which
remove any remaining hematite. The magnetic concentrates are
classified and dried, and blended with the high-tension product, to give
a final concentrate of about 66% Fe. Cleaning of only the gravity
tailings by magnetic separation is preferred, as relatively small
amounts of magnetic concentrate have to be dealt with, the bulk of the
material being unaffected by the magnetic field. Similarly, relatively

little material is pinned to the rotor in the high-tension treatment of the gravity concentrate, the iron minerals being unaffected by the ionic field.

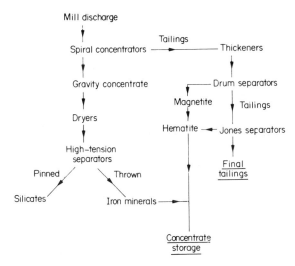

Fig. 13.29. Flowsheet of Scully concentrator.

References

1. Cohen, H. E., Magnetic separation, in *Mineral Processing at a Crossroads*, eds. B. A. Wills and R. W. Barley. p. 287 Martinus Nijhoff Publishers, Dordrecht, 1986.
2. Kuzev, L., and Stoev, S., Vibromagnetic separation of iron ores, *World Mining Equip.*, **11,** 84 (Dec. 1986).
3. Arvidson, B. R., and Barnea, E., Recent advances in dry high-intensity permanent magnet separator technology, *Proc. XIVth Int. Min. Proc. Cong.*, paper IX-7, CIM, Toronto (Oct. 1982).
4. Lawver, J. E., and Hopstock, D. M., Wet magnetic separation of weakly magnetic minerals, *Minerals Sci. Engng.*, **6,** 154 (1974).
5. Hudson, S. B., The Gill high intensity wet magnetic separator, *Proc. VIIIth Int. Min. Proc. Cong.*, paper B-6, Leningrad, 1968.
6. White, L., Swedish symposium offers iron ore industry an overview of ore dressing developments, *Engng. Min. J.* **179,** 71 (Apr. 1978).

7. Bartnick, J. A., Stone, W. J. D., and Zabel, W. H., Superconcentrate production by the Jones separator: capital and operating costs, *Int. Symposium of Iron and Steel Industry*, Brazilia, 1973.
8. Corrans, I. J., *et al*, The performance of an industrial wet high-intensity magnetic separator for the recovery of gold and uranium, *J. S. Afr. Inst. Min. Metall.*, **84,** 57 (March 1984).
9. Tawil, M. M. E., and Morales, M. M., Application of wet high intensity magnetic separation to sulphide mineral beneficiation, in *Complex Sulfides*, ed. A. D. Zunkel, p. 507, TMS-AIME, Pennsylvania, 1985.
10. Wills, B. A., Coal preparation, *Mining Annual Rev.*, 266 (1985).
11. Stefanides, E. J., Superconducting magnets upgrade paramagnetic particle removal, *Design News* (May 1986).
12. Unkelbach, K. H., and Kellerwessel, H., A superconductive drum type magnetic separator for the beneficiation of ores and minerals, in *Proc. XVth Int. Min. Proc. Cong.*, **1,** 371, Cannes (1985).
13. Canning, A. E., Dry mineral sand separation plants of Associated Minerals Consolidated Ltd, New South Wales and Queensland, in *Mining and Metallurgical Practices in Australasia* (ed. J. T. Woodcock), Aust. I. M. M., 1980.
14. Anon., Canadian iron mines contending with changing politics, restrictive taxes, *Engng. Min. J.* 72 (Dec 1974).

CHAPTER 14

ORE SORTING

Introduction

Ore sorting is the original concentration process, having probably been used by the earliest metal workers several thousand years ago. It involves the appraisal of *individual* ore particles and the rejection of those particles that do not warrant further treatment.

Hand sorting has declined in importance due to the need to treat large quantities of low-grade ore which requires extremely fine grinding. Hand sorting of some kind, however, is still practised at many mines, even though it may only be the removal of large pieces of timber, tramp iron, or unexploded dynamite from the run-of-mine ore.

Electronic ore-sorting equipment was first produced in the late 1940s, and although its application is fairly limited, it is an important technique for the processing of certain minerals.[1]

Electronic Sorting

Sorting is feasible when the ore is economically liberated at a fairly coarse size, usually greater than about 10 mm. Sorters assess the difference in a specific physical property between the particles and send a signal to a mechanical or electronic device which removes the valuable particles from the stream. It is essential, therefore, that a distinct difference in the required physical property is apparent between the valuable minerals and gangue.

The particle surfaces must be thoroughly washed before sorting, so that blurring of the signal does not occur and, as it is not practical to attempt to feed very wide rock size ranges to a single machine, the feed must undergo preliminary sizing. The ore must be fed in a monolayer, as display of individual particles to the sorting device must be effected.

There are many particle characteristics which are presently used to activate the sensing device. *Photometric* sorting is the mechanised form of hand-sorting, in which the ore is divided into components of differing value by visual examination.

The basis of the photometric sorter (Fig. 14.1) is a laser light source and sensitive photomultiplier, used in a scanning system to detect light reflected from the surfaces of rocks passing through the sorting zone. Electronic circuitry analyses the photomultiplier signal, which changes with the intensity of the reflected light and produces control signals to actuate the appropriate valves of an air-blast rejection device to remove certain particles selected by means of the analysing process. The sorter is fully automatic and can be attended by one operator on a part-time basis. Typical throughput per machine ranges from 50 t h^{-1} for a $-65 + 30$ mm feed to 200 t h^{-1} for a $-150 + 70$ mm material.

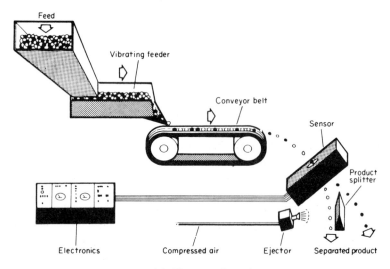

Fig. 14.1. Photometric sorting.

The Gunson's Sortex MP80 machine (Fig. 14.2) was probably the first sorter to employ microprocessor technology.[2] The sorter can handle minerals in the size range 10–150 mm at feed rates of up to 150 t h^{-1}.

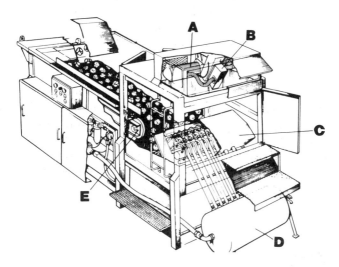

FIG. 14.2. Sortex MP80 machine showing A, quartz-halogen lamps; B, solid-state camera; C, ejector bank; D, air receiver; E, feed belt drum motor.

The feed is vibrated from the surge hopper onto a continuous 0.8-m wide belt travelling at 2.1 m s^{-1} to form a monolayer of randomly spaced particles moving at the same velocity. At the end of the belt, the feed is thrown into free flight and passes through the viewing zone which is brightly lit by quartz-halogen lamps.

The lamps are housed in a cabinet, together with a solid-state camera focused on the full width of the feed path. The sensing element of the camera consists of a horizontal array of 1024 photosensors which give a very high resolution over the 0.8 m width. As the mineral particle moves across the viewing area, it is scanned one thousand times per second, and a complete picture of it is built up and stored. The information is modified by an electronic pre-filter and then fed to the microprocessor for the decision-making process, which is based on a virtually complete spectral picture of the exposed surface. The microprocessor evaluates the information to decide whether or not to eject the particle from the stream. The size of the particle and its position in the stream is also evaluated. If the particle is to be ejected, the microprocessor will first automatically select the correct ejector valve according to the particle's position in the flight path. It then controls the time delay necessary for the particle to reach the ejection zone after leaving the viewing area, and then computes the duration of air jet so that it is proportional to the particle size. One or more valves may be selected depending on particle size.

The RTZ Ore Sorters Model 16 photometric sorter has been used successfully in industry since 1976 on a wide range of ore types.[3] The sensing system is positioned at the point where the rocks leave the main feed belt, and consists of a rotating mirror and a photomultiplier (Fig. 14.3). High-intensity light from the laser is reflected off the mirror drum and onto the ore stream. The light reflected from the particles passes back to the mirror drum and onto the photomultiplier from where electrical impulses are relayed to the processor. The number of facets on the mirror drum and the speed of rotation are such

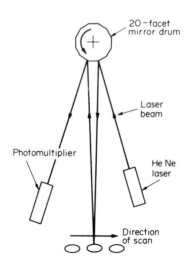

FIG. 14.3. Model 16 laser beam scanning.

that the laser beam scans the ore stream 2000 times per second, equivalent to one scan every 2 mm of rock length

Photometric sorting has many applications, although it is more often used for the separation of industrial minerals than for metalliferous ores. Present uses include the separation of magnesite, barite, talc, limestone, marble, gypsum, flint, and, in the metalliferous field, the recovery of wolframite and scheelite from quartz.[4]

An RTZ ore-sorting machine at the Doornfontein mine in South Africa is used for treating gold ores.[5] Rocks having white or grey quartz pebbles in a darker matrix are accepted, while quartzite, ranging from light green, through olive green to black, is rejected. Most of the gold occurs in rocks in the "accept" category. Uniform distribution of the ore entering the sorter is achieved by the use of tandem vibrating feeders and the ore is washed on the second feeder to remove slimes which may affect the light-reflecting qualities.

Electronic sorting has a very important use in the recovery of diamonds. Diamond ores are pre-concentrated by HMS prior to recovery of the diamonds by either electronic sorting, grease tabling (Chapter 12), or a combination of the two. In some cases, grease tabling cannot be used, as due to a surface coating on the diamonds they do not adhere to the grease. Use is made in electronic sorting of the fact that diamonds fluoresce under a beam of X-rays. The fluorescence is detected by a photomultiplier which initiates an air-blast electronically to divert the diamond from the main stream.[6] Typically, sorting produces an enrichment of the diamond ore by several thousand to one, and, being automatic, can be operated with very little supervision, so increasing security. It is not entirely universally used, however, as some diamonds do not readily fluoresce, so traditional grease-tabling must also be incorporated in such cases.

Electronic sorters are being used to treat uranium ores. Uranium is a strong emitter of gamma-rays, which can be readily detected by a commercial NaI(Th) scintillation detector placed at up to 25 mm from the rock surface. The sorting belt (Fig. 14.4) travels at 1 m s^{-1} and any radioactive material is deflected by a blast of compressed air. If the gamma count is obtained on a rock and the mass of the rock and the geometry of the counting system compensated for, then the gamma count can be interpreted in terms of the uranium grade of the rock.

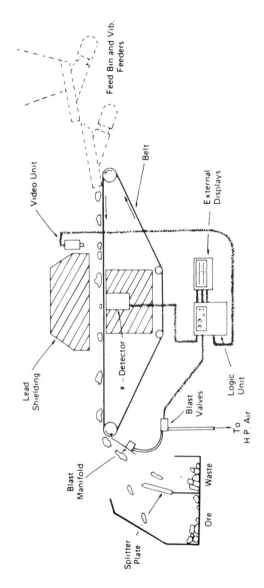

FIG. 14.4. Radiometric sorting system based on gamma-ray emission.

Sorting is achieved by rejecting any piece of ore appearing to have a uranium grade less then the cut-off grade. Machines are available for 50–150 mm and for 25–50 mm feed sizes. They are being used to sort uranium in South Africa,[7] Australia and Canada.

Neutron absorption separation has been used for the sorting of boron minerals.[8] The ore is delivered by a conveyor belt between a slow neutron source and a scintillation neutron detector. The neutron flux attenuation by the ore particles is detected and used as the means of sorting. The method is most applicable in the size range 25–150 mm. Boron minerals are easy to sort by neutron absorption since the neutron capture cross-section of the boron atom is very large compared with those of common associated elements and thus the neutron absorption is almost proportional to the boron content of the particles.

Photoneutron separation is recommended for the sorting of beryllium ores, since when a beryllium isotope in the mineral is exposed to gamma radiation of a certain energy, a photoneutron is released and this may be detected by scintillation or by a gas counter.

The RTZ Ore Sorters Model 19 conductivity/magnetic response sorter remotely measures the electrical conductivity and magnetic susceptibility of individual particles.[9] The detectors respond to only slight variations in such properties, making the sorter suitable for pre-concentration of a wide variety of ores, including sulphides, oxides and native metals. Sorters are available to suit feed sizes from 25 mm to 150 mm, with capacities of up to 120 t h^{-1} at the upper size limit. A vibrating feeder deposits the feed onto the feed belt assembly (Fig. 14.5), the rock travelling on parallel feed belts to the centrifugal accelerator, which spaces the particles longitudinally, after which they are deposited on the main sorting belt. The detectors assess the conductivity and magnetic properties of the particles and relay this information to the microprocessor. As the particles leave the main belt, an optical system measures their individual size and shape. The processor then calculates the grade of the particle from its size and electrical and magnetic properties. This grade is compared with a pre-selected cut-off value, after which the processor activates the appropriate valves in the air blast manifold to deflect selected particles from their normal free fall trajectory.

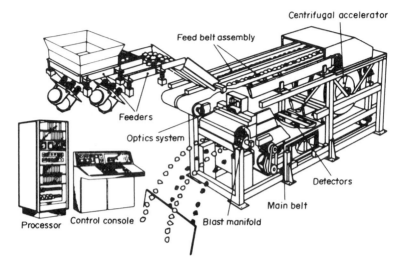

FIG. 14.5. Conductivity/magnetic response sorter.

"Precon" is a preconcentrating device developed by Outokumpu in Finland.[10] It can be used to treat heterogeneous ores where the average content of valuable metals is at least 0.4–0.5%, and where there is at least a 5% difference in the sum metal content between individual ore and waste lumps. The most suitable metals are chromium, iron, cobalt, nickel, copper and zinc, or any combination of these. "Precon" uses a gamma scattering analysis to evaluate the sum metal content of the lump to be measured, the measuring time for each lump being as short as 20–50 msec, dependent on the size of the lump. After measuring the quantity of radiation scattered by the lump, the measurement is compared with a preset accept/reject threshold, after which it is either accepted or blasted aside by an air battery. Throughput is dependent on lump size, the unit being able to handle lumps from 35 mm to 150 mm in size. With 35-mm lumps the capacity is about 7 t h^{-1}, increasing to 40 t h^{-1} with 150-mm lumps.

"Precon" has been installed at Outokumpu's Hammaslahti mine, and treats primary crushed ore. The ore breccia and veins are composed of pyrrhotite and chalcopyrite, with some sphalerite in places, and the ore grade averages about 1.2% copper, although the variation in the grade is wide—from 0.2 to 4% copper. In some stopes there is heavy dilution of the ore with black schists which occur in the hanging wall. After primary crushing the ore is screened, and the 80–150-mm fraction is fed to "Precon", which rejects about 25% as waste. The accepted fraction, grading about 1.24% copper, joins the bypassed material and moves on to secondary crushing. The rejected material grades about 0.2% copper.

Equipment to sort asbestos ore has been developed.[11] The detection technique is based on the low thermal conductivity of asbestos fibres and uses sequential heating and infra-red scanning to detect the asbestos seams. A similar machine has been installed at King Island Scheelite in Tasmania, where the scheelite is sensed by its fluorescence under ultra-violet radiation.

References

1. Sassos, M. P., Mineral sorters, *Engng. & Min. J.*, **185,** 68 (June 1985).
2. Anon., Micro-processor speeds optical sorting of industrial minerals, *Mine & Quarry* **9,** 48 (Mar. 1980).
3. Anon, Photometric ore sorting, *World Mining* **34,** 42 (April 1981).
4. Anon, Photometric sorting at Mount Carbine wolframite mine, Queensland, *Min. Mag.* 28 (Jan. 1979).
5. Keys, N. J., *et al.*, Photometric sorting of ore on a South African gold mine. *J. S. Afr. IMM J'burg* **75,** 13 (Sept. 1974).
6. Anon., New generation of diamond recovery machines developed in South Africa, *S. A. Min. Eng. J.* 17 (May 1971).
7. Anon., Radiometric sorters for Western Deep Levels gold mine, South Africa, *Min. J.* 132 (Aug. 1981).
8. Mokrousov, V. A., *et al.*, Neutron-radiometric processes for ore beneficiation, *Proc. XIth Int. Min. Proc. Cong., Cagliari*, 1249 (1975).
9. Anon., New ore sorting system, *Min. J.* 446 (Dec. 11, 1981).
10. Kennedy, A., Mineral processing developments at Hammaslahti, Finland, *Mining Mag.*, 122 (Feb. 1985).
11. Collier, D., *et al.*, Ore sorters for asbestos and scheelite, *Proc. 10th Int. Min. Proc. Cong., London*, 1973.

CHAPTER 15

DEWATERING

Introduction

With few exceptions, most mineral-separation processes involve the use of substantial quantities of water and the final concentrate has to be separated from a pulp in which the water/solids ratio may be high.

Dewatering, or solid-liquid separation, produces a relatively dry concentrate for shipment. Partial dewatering is also performed at various stages in the treatment, so as to prepare the feed for subsequent processes.

Dewatering methods can be broadly classified into three groups:

(a) sedimentation;
(b) filtration;
(c) thermal drying.

Sedimentation is most efficient when there is a large density difference between liquid and solid. This is always the case in mineral processing where the carrier liquid is water. Sedimentation cannot always be applied in hydrometallurgical processes, however, because in some cases the carrier liquid may be a high-grade leach liquor having a density approaching that of the solids. In such cases, filtration may be necessary.

Dewatering in mineral processing is normally a combination of the above methods. The bulk of the water is first removed by sedimentation, or thickening, which produces a thickened pulp of perhaps 55–65% solids by weight. Up to 80% of the water can be separated at this stage. Filtration of the thick pulp then produces a moist filter cake of between 80 and 90% solids, which may require thermal drying to produce a final product of about 95% solids by weight.

Sedimentation

Rapid settling of solid particles in a liquid produces a clarified liquid which can be decanted, leaving a thickened slurry, which may require further dewatering by filtration.

The settling rates of particles in a fluid are governed by Stokes' or Newton's laws, depending on the particle size (Chapter 9). Very fine particles, of only a few microns diameter, settle extremely slowly by gravity alone, and centrifugal sedimentation may have to be performed. Alternatively, the particles may be agglomerated, or *flocculated*, into relatively large lumps, called *flocs*, that settle out more rapidly.

Coagulation and Flocculation

Little more than 30 years ago the word flocculant meant very little to the mineral processor. It was more likely to be used with reference to either an inorganic salt, in which case it was really a coagulant, or alternatively a naturally occurring polymeric substance such as starch or guar gum. In both cases the substance was applied as a settling aid. Nowadays the term flocculant is more likely to be used in describing high molecular weight, water-soluble polymers of the type generically known as polyacrylamides.[1,2,3]

Coagulation causes extremely fine colloidal particles to adhere directly to each other. All particles exert mutual attraction forces, known as *London–Van der Waals' forces*, which are effective only at very close range. Normally, the adhesion due to these forces is prevented by the presence around each particle of an electrically charged atmosphere, which generates repulsion forces between particles approaching each other. There is, therefore, in any given system a balance between the attractive forces and the electrical repulsion forces present at the solid–liquid interface (Fig. 15.1).

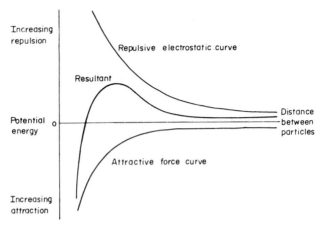

FIG. 15.1. Potential energy curves for two particles approaching each other.

In any given system the electrical charges on the particle surfaces will be of the same sign, aqueous suspensions of pH 4 and above generally being negative. Positively charged surfaces occur mainly in strong acid solutions.

The repulsion forces not only prevent coagulation of the particles, but also retard their settlement by keeping them in constant motion, this effect being more pronounced the smaller the particle. Coagulants are electrolytes having an opposite charge to the particles, thus causing charge neutralisation when dispersed in the system, allowing the particles to come into contact and adhere as a result of molecular forces. Inorganic salts have long been used for this purpose, and as counter-ions in aqueous systems are most frequently positively charged, salts containing highly charged cations, such as Al^{+++}, Fe^{+++}, and Ca^{++}, are mainly used. Lime, or sulphuric acid, depending on the surface charge of the particles, can also be used to cause coagulation. Most pronounced coagulation occurs when the particles have zero charge in relation to the suspending medium, this occurring when the *zeta potential* is zero. The nature of the zeta potential can be seen from Fig. 15.2, which shows a model of the *electrical double layer* at the surface of a particle.[4] The surface shown has a negative charge,

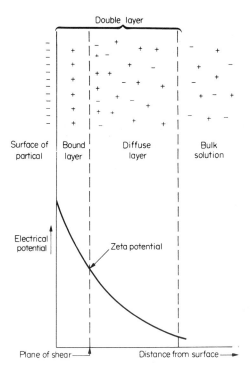

FIG. 15.2. The electrical double layer.

such that positive ions from solution will be attracted to it, forming a bound layer of positive ions—known as the *Stern layer*, and a *diffuse layer* of counter ions decaying in concentration with increasing distance until the solution equilibrium concentration is attained. These layers of ions close to the surface constitute the electrical double layer. When a particle moves in the liquid, shear takes place between the bound layer, which moves with the particle, and the diffuse layer, the potential at the plane of shear being known as the zeta potential. The magnitude of the zeta potential depends on the surface potential and the concentration and charge of the counter-ions. In general, the greater the counter-ion charge and counter-ion concentration, the lower is the zeta potential, although ions of high charge may cause complete charge reversal; therefore optimum doses of electrolyte are critical.

Flocculation involves the formation of much more open agglomerates than those resulting from coagulation and relies upon molecules of reagent acting as bridges between separate suspended particles. The reagents used to form the "bridges" are long-chain organic polymers, which were formerly natural materials, such as starch, glue, gelatine, and guar gum, but which are now increasingly synthetic materials, loosely termed polyelectrolytes. The majority of these are anionic in character but some of them are non-ionic, and some cationic, but these form a minor proportion of the commercially available products of today's flocculant market. Inorganic salts are not able to perform this bridging function, but they are sometimes used in conjunction with an organic reagent as a cheaper means of charge neutralisation, although an ionic polyelectrolyte can and often does perform both functions.

The polyacrylamides, which vary widely in molecular weight and charge density, are extensively used as flocculants. The charge density refers to the percentage of the acrylic monomer segments which carry a charge. For instance, if the polymer is uncharged it comprises n similar segments of the acrylic monomer. The polymer is thus a homopolymer–polyacrylamide:

$$\left[\begin{array}{c} NH_2 \\ | \\ C{=}O \\ | \\ -CH_2-CH \end{array} \right]_n$$

If the acrylic monomer is completely hydrolysed with NaOH, the product comprises n segments of sodium acrylate—an anionic polyelectrolyte, having a charge density of 100%:

$$\left[\begin{array}{c} O-Na^+ \\ | \\ C{=}O \\ | \\ -CH_2-CH \end{array} \right]_n$$

Charge density may be controlled in manufacture between the limits 0–100%, to produce a polyacrylamide of anionic character, weak or strong, depending on the degree of hydrolysis:

$$\left[\left(\begin{array}{c} CH_2-CH \\ | \\ C=O \\ | \\ NH_2 \end{array}\right)_x - \begin{array}{c} CH_2-CH \\ | \\ C=O \\ | \\ O-Na^+ \end{array}\right]_y$$

By similar chemical reactions, polymers of cationic character can be produced. It would be expected that, since most suspensions encountered in the minerals industry contain negatively charged particles, cationic polyelectrolytes, where the cation adsorbs to the particles, would be most suitable. Although this is true for charge neutralisation purposes, and attraction of the polymer to the particle surface, it is not necessarily true for the "bridging" role of the flocculant. For bridging, the polymer must be strongly adsorbed, and this is promoted by chemical groups having good adsorption characteristics, such as amide groups. The majority of commercially available polyelectrolytes are anionic, since these tend to be of higher molecular weight than the cationics, and are less expensive.

The mode of action of the anionic polyacrylamide depends on a segment of the very long molecule being adsorbed on the surface of a particle, leaving a large proportion of the molecule free to be adsorbed on another particle, so forming an actual molecular linkage, or bridge, between particles (Fig. 15.3).

While only one linkage is shown in Fig. 15.3, in practice many such interparticle bridges are formed, linking a number of particles together. The factors influencing the degree of flocculation are the efficiency or strength of adsorption of the polymer on the surface, the degree of agitation during flocculation and the subsequent agitation, which can result in breakdown of flocs.[5]

The maximum effect of a flocculant is achieved at an optimum dosage rate; excess polymer can cause dispersion of the particles due to floc breakdown. Due to the fragile nature of the flocs, flocculating agents are not successful with hydrocyclones, while success with

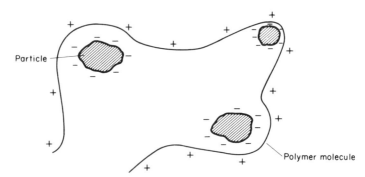

FIG. 15.3. Action of an anionic polyelectrolyte.

centrifuges can only be achieved with special techniques for a limited range of applications. Even pumping of the flocculated slurry may destroy the flocs due to rupture of the long-chain molecules.

Polyelectrolytes are normally made up to stock solutions of about 0.5–1%, which are diluted to about 0.01% before adding to the slurry. The diluted solution must be added at enough points in the stream to ensure its contact with every portion of the system. A shower pipe is frequently used for this purpose (Fig. 15.4).

Mild agitation is essential at the addition points, and shortly thereafter, to assist in flocculant dispersion in the process stream. Care should be taken to avoid severe agitation after the flocs have been formed.

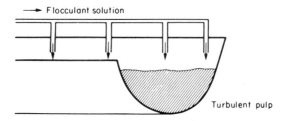

FIG. 15.4. Typical method of flocculant addition.

Selective Flocculation

The treatment of finely disseminated ores often results in the production of ultra-fine particles, or slimes, which respond poorly to conventional separation techniques, and are often lost in the process tailings. *Selective* flocculation of the desired minerals in the pulp, followed by separation of the aggregates from the dispersed material, is a potentially important technique, although plant applications are at present rare.[6] Although attempts have been made to apply selective flocculation to a wide range of ore types, the bulk of the work has been concerned with its application to the treatment of clays, iron, phosphate, and potash ores. A prerequisite for the process is that the mineral mixture must be stably dispersed prior to the addition of a high molecular weight polymer, which selectively adsorbs on only one of the constituents of the mixture. Selective flocculation is then followed by removal of the flocs of one component from the dispersion.

The greatest amount of work on selective flocculation has been concerned with the treatment of fine-grained non-magnetic oxidised taconites, which has led to the development of Cleveland Cliffs Iron Company's 10 million tonne per year operation in the United States. The finely intergrown ore is autogenously ground to 85% -25 μm with caustic soda and sodium silicate, which act as dispersants for the fine silica. The ground pulp is then conditioned with a corn-starch flocculant which selectively flocculates the hematite. About one-third of the fine silica is removed in a de-slime thickener, together with a loss of about 10% of the iron values. Most of the remaining coarse silica is removed from the flocculated underflow by reverse flotation, using an amine collector.[7]

Gravity Sedimentation

Gravity sedimentation, or *thickening*, is the most widely applied dewatering technique in mineral processing, and it is a relatively cheap, high-capacity process, which involves very low shear forces, thus providing good conditions for flocculation of fine particles.

The *thickener* is used to increase the concentration of the suspension by sedimentation, accompanied by the formation of a clear liquid. In most cases the concentration of the suspension is high and hindered settling takes place. Thickeners may be batch or continuous units, and consist of relatively shallow tanks from which the clear liquid is taken off at the top, and the thickened suspension at the bottom. The *clarifier* is similar in design, but is less robust, handling suspensions of much lower solid content than the thickener.

The continuous thickener consists of a cylindrical tank, the diameter ranging from about 2 m to 200 m in diameter, and of depth 1–7 m. Pulp is fed into the centre via a *feed-well* placed up to 1 m below the surface, in order to cause as little disturbance as possible (Fig. 15.5). The clarified liquid overflows a peripheral launder, while the solids which settle over the entire bottom of the tank are withdrawn as a thickened pulp from an outlet at the centre. Within the tank are one or more rotating radial arms, from each of which are suspended a series of blades shaped so as to rake the settled solids towards the central outlet. On most modern thickeners these arms rise automatically if the torque exceeds a certain value, thus preventing damage due to overloading. The blades also assist the compaction of the settled particles and produce a thicker underflow than can be achieved by simple settling. The solids in the thickener move continuously downwards, and then inwards towards the thickened underflow outlet, while the liquid

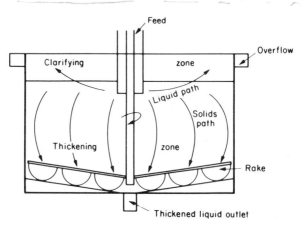

FIG. 15.5. Flow in a continuous thickener.

moves upwards and radially outwards. In general, there is no region of constant composition in the thickener.[8]

Thickener tanks are constructed of steel, concrete, or a combination of both, steel being most economical in sizes of less than 25 m in diameter. The tank bottom is often flat, while the mechanism arms are sloped towards the central discharge. With this design, settled solids must "bed-in" to form a false sloping floor. Steel floors are rarely sloped to conform with the arms because of expense. Concrete bases and sides become more common in the larger-sized tanks. In many cases the settled solids, because of particle size, tend to slump and will not form a false bottom. In these cases the floor should be concrete and poured to match the slope of the arms. Tanks may also be constructed with sloping concrete floors and steel sides, and earth bottom thickeners are in use, which are generally considered to be the lowest cost solution for thickener bottom construction.[9]

The method of supporting the mechanism depends primarily on the tank diameter. In relatively small thickeners, of diameter less than about 45 m, the drive head is usually supported on a superstructure spanning the tank, with the arms being attached to the drive shaft. Such machines are referred to as *bridge* or *beam* thickeners (Fig. 15.6). The underflow is usually drawn from the apex of a cone located at the centre of the sloping bottom.

A common arrangement for larger thickeners, of up to about 180 m in diameter, is to support the drive mechanism on a stationary steel or concrete centre column. In most cases, the rake arms are attached to a drive cage, surrounding the central column, which is connected to the drive mechanism. The thickened solids are discharged through an annular trench encircling the centre column (Fig. 15.7).

Figure 15.8 shows an 80-m-diameter thickener of this type.

In the *traction thickener*, a single long arm is mounted with one end on the central support column while to the other are fixed traction wheels that run on a rail on top of the tank wall. The wheels are driven by motors which are mounted on the end of the arm and which therefore travel around with it. This is an efficient and economical design since the torque is transmitted through a long lever arm by a simple drive. They are manufactured in sizes ranging from 60 m to approximately 120 m in diameter.

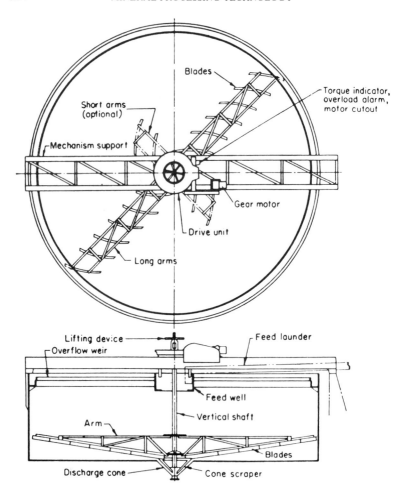

FIG. 15.6. Thickener with mechanism supported by superstructure.

Cable thickeners have a hinged rake arm fastened to the bottom of the drive cage or centre shaft. The hinge is designed to give simultaneous vertical and horizontal movement of the rake arm. The rake arm is *pulled* by cables connected to a torque or drive arm structure, which is rigidly connected to the centre shaft at a point just

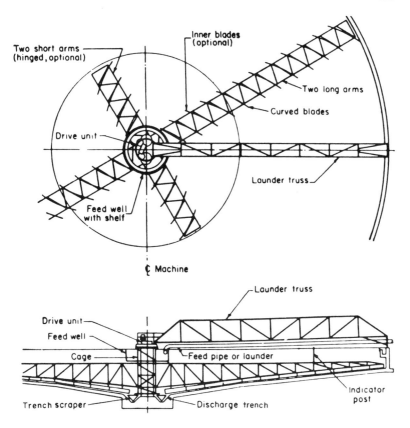

FIG. 15.7. Thickener with mechanism supported by centre column.

below the liquid level. The rake is designed to automatically lift when the torque developed due to its motion through the sludge rises. This design allows the rake arm to find its own efficient working level in the sludge, where the torque balances the rake weight.

In all thickeners the speed of the raking mechanism is normally about 8 m min^{-1} at the perimeter, which corresponds to about 10 rev h^{-1} for a 15-m-diameter thickener. Power consumption is thus extremely low, such that even a 60-m unit may require only a 10-kW motor. Wear and maintenance costs are correspondingly low.

FIG. 15.8. 80-m diameter centre-column-supported thickener.

The underflow is usually withdrawn from the central discharge by pumping, although in clarifiers the material may be discharged under the hydrostatic head in the tank. The underflow is usually collected in a sludge-well in the centre of the tank bottom, from where it is removed via piping through an underflow tunnel. The underflow lines should be as short and as straight as possible to reduce the risk of choking, and this can be achieved, with large tanks, by taking them up from the sludge-well through the centre column to pumps placed on top, or by placing the pumps in the base of the column and pumping up from the bottom. This has the advantage of dispensing with the expensive underflow tunnel. A development of this is the *caisson thickener* in which the centre column is enlarged sufficiently to house a central control room; the pumps are located in the bottom of the column, which also contains the mechanism drive heads, motors, control panel, underflow suction, and discharge lines. The interior of the caisson can be a large heated room. The caisson concept has lifted the possible ceiling on thickener sizes; at present they are manufactured in sizes up to 180 m in diameter.

Underflow pumps are often of the *diaphragm* type;[10] these are positive action pumps for medium heads and volumes, and are suited to the handling of thick viscous fluids. They can be driven by electric motor through a crank mechanism, or by directly acting compressed air. A flexible diaphragm is oscillated to provide suction and discharge through non-return valves, and variable speed can be achieved by changing either the oscillating frequency or the stroke. In some plants, variable-speed pumps are connected to nucleonic density gauges on the thickener underflow lines, which control the rate of pumping to maintain a constant underflow density. The thickened underflow is pumped to filters for further dewatering.

Thickeners often incorporate substantial storage capacity so that, for instance, if the filtration section is shut down for maintenance, the concentrator can continue to feed material to the dewatering section. During such periods the thickened underflow should be recirculated into the thickener feed-well. At no time should the underflow cease to be pumped, as chokage of the discharge cone rapidly occurs.

Since capital is the major cost of thickening, selection of the correct size of thickener for a particular application is important.

The two primary functions of the thickener are the production of a clarified overflow and a thickened underflow of the required concentration.

For a given throughput the clarifying capacity is determined by the thickener diameter, since the surface area must be large enough so that the upward velocity of liquid is at all times lower than the settling velocity of the slowest-settling particle which is to be recovered. The degree of thickening produced is controlled by the residence time of the particles and hence by the thickener depth.

The solids concentration in a thickener varies from that of the clear overflow to that of the thickened underflow being discharged. Although the variation in concentration is continuous, the concentrations at various depths may be grouped into four zones, as shown in Fig. 15.9.

When materials settle with a definite interface between the suspension and the clear liquid, as is the case with most flocculated

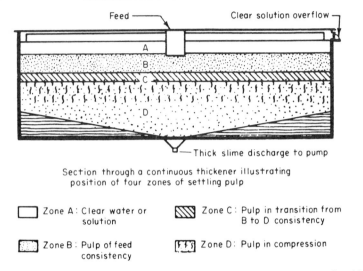

FIG. 15.9. Concentration zones in a thickener. (From *Chemical Engineers' Handbook* by J. H. Perry, McGraw-Hill, 1963.)

mineral pulps, the solids-handling capacity determines the surface area. Solids-handling capacity is defined as the capacity of a material of given dilution to reach a condition such that the mass rate of solids leaving a region is equal to or greater than the mass rate of solids entering the region. The attainment of this condition with a specific dilution depends on the mass subsidence rate being equal to or greater than the corresponding rise rate of displaced liquid. A properly sized thickener containing material of many different dilutions, ranging from the feed to the underflow solids contents, has adequate area such that the rise rate of displaced liquid at any region never exceeds the subsidence rate.[11]

The satisfactory operation of the thickener as a clarifier depends upon the existence of a clear-liquid overflow at the top. If the clarification zone is too shallow, some of the smaller particles may escape in the overflow. The volumetric rate of flow of liquid upwards is equal to the difference between the rate of feed of liquid and the rate of removal in the underflow. Hence the required concentration of solids in the underflow, as well as the throughput, determines the conditions in the clarification zone.[8]

The method developed by Coe and Clevenger[12] is commonly employed to determine surface area when the material settles with a definite interface.

If F is the liquid-to-solids ratio by weight at any region with the thickener, D is the liquid-to-solids ratio of the thickener discharge, and W t h^{-1} of dry solids are fed to the thickener, then $(F - D)W$ t h^{-1} of liquid moves upwards to the region from the discharge.

The velocity of this liquid current is thus

$$\frac{(F - D)W}{AS} \tag{15.1}$$

where A is the thickener area (m^2) and S is the specific gravity of the liquid (kg l^{-1}).

Because this upward velocity must not exceed the settling rate of the solids in this region, at equilibrium

$$\frac{(F - D)W}{AS} = R, \tag{15.2}$$

where R is the settling rate (m h^{-1}).
The required thickener area is therefore

$$A = \frac{(F - D)W}{RS}. \tag{15.3}$$

From a complete set of R and F values the area required for various dilutions may be found by recording the initial settling rate of materials with dilutions ranging from that of the feed to the discharge. The dilution corresponding to the maximum value of A represents the minimum solids-handling capacity and is the critical dilution.

In using this method the initial constant sedimentation rate is found through tests in graduated cylinders using dilutions ranging from the feed dilution to the underflow dilution, the rate of fall of the interface between the thickened pulp and clarified solution being timed.

Once the required surface area is established, it is necessary to apply a safety factor to the calculated area. This should be at least two.

The Coe and Clevenger method requires multiple batch tests at different arbitrary pulp densities before an acceptable unit area can be selected. The Kynch model[13] offers a way of obtaining the required area from a single batch-settling curve, and is the basis of several thickening theories, which have been comprehensively reviewed by Pearse.[14]

The Talmage and Fitch method[15] applies Kynch's mathematical model to the problem of thickener design. The results of a batch-settling test are plotted linearly as mudline (interface between settled pulp and clear water) height against time (Fig. 15.10).

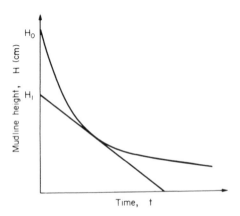

FIG. 15.10. Batch-settling curve.

Talmage and Fitch showed that by constructing a tangent to the curve at any point, then if H is the intercept of the tangent on the ordinate,

$$CH = C_0H_0,$$ (15.4)

where C_0 kg l^{-1} is the original feed *solids concentration*, H_0 cm is the original mudline height, and H is the mudline height corresponding to a uniform slurry of concentration C kg l^{-1}, at the point where the tangent was taken. Therefore, for any selected point on the settling curve, the local concentration can be obtained from equation 15.4, and the settling rate from the gradient of the tangent at that point. Thus a set of data of concentration against settling rate can be obtained from the single batch-settling curve.

For a pulp of solids concentration C kg l^{-1}, the volume occupied by the solids in 1 l of pulp is C/d, where d kg l^{-1} is the specific gravity of dry solids.

Therefore the weight of water in 1 l of pulp

$$= 1 - C/d = \frac{d - C}{d} \cdot$$

Therefore the water/solids ratio by weight

$$= \frac{d - C}{dC} \cdot$$

For pulps of concentrations C kg l^{-1} of solids, and C_u kg l^{-1} of solids, the difference in water/solids ratio

$$= \frac{d - C}{dC} - \left(\frac{d - C_u}{dC_u} \right)$$

$$= \frac{1}{C} - \frac{1}{C_u} \cdot$$

Therefore the values of concentration obtained, C, and the settling rates, R, can be substituted in the Coe and Clevenger equation (15.3), i.e.

$$A = \left(\frac{1}{C} - \frac{1}{C_u} \right) \frac{W}{RS}, \tag{15.5}$$

where C_u is the underflow solids concentration.

A simplified version of the Talmage and Fitch method is offered by determining the point on the settling curve where the solids go into compression. This point corresponds to the limiting settling conditions and controls the area of thickener required. In Fig. 15.11, C is the compression point, and a tangent is drawn to the curve at this point, intersecting the ordinate at H. A line is drawn parallel to the abscissa, cutting the ordinate at H_u; H_u corresponds to the intersection on the ordinate of a tangent from a point C_u on the curve, where C_u is the

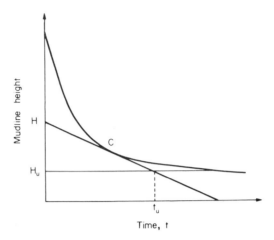

FIG. 15.11. Modified Talmage and Fitch construction.

solids concentration of the thickener underflow. The tangent from C intersects this line at a time corresponding to t_u. H_u can be calculated from equation 15.4.

The required thickener area (equation 15.5)

$$= \frac{W(1/C - 1/C_u)}{(H - H_u)/t_u},$$

where $(H - H_u)/t_u$ is the gradient of the tangent at point C, i.e. the settling rate of the particles at the compression point concentration. Since $CH = C_0 H_0$,

$$A = \frac{W[(H/C_0 H_0) - (H_u/C_0 H_0)]}{(H - H_u)/t_u} = W \frac{t_u}{C_0 H_0}. \qquad (15.6)$$

In most cases, the compression point concentration will be less than that of the underflow concentration. In cases where this is not so, then the tangent construction is not necessary, and t_u is the point where the underflow line crosses the settling curve. In many cases, the point of compression on the curve is clear, but when this is not so, a variety of methods have been suggested for its determination.[16,17]

The Coe and Clevenger and modified Talmage and Fitch methods are the most widely used in the metallurgical industry to predict thickener area requirements. Both methods have limitations, the Talmage and Fitch technique relying critically on identifying a compression point, and both must be used in conjunction with empirical safety factors. Generally, the Coe and Clevenger method tends to underestimate the thickener area requirement, whilst the Talmage and Fitch method tends to overestimate. It is usually better to overestimate in design to allow for feed fluctuations and increase in production, and because of this, and its relative experimental simplicity, the Talmage and Fitch method is often preferred, providing that a compression point is readily identifiable.

The mechanism of thickening has been far less well expressed in mathematical terms than the corresponding clarifying mechanisms. The depth of the thickener is therefore usually determined by experience. The diameter is usually large compared with the depth, and therefore a large ground area is required. *Tray thickeners* (Fig. 15.12) are sometimes installed to save space. In essence, a tray thickener is a series of unit thickeners mounted vertically above one another. They operate as separate units, but a common central shaft is utilised to drive the sets of rakes.

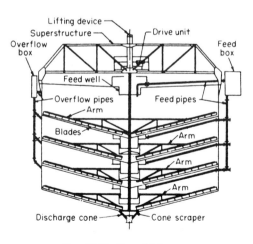

FIG. 15.12. Tray thickener.

High-capacity Thickeners

Conventional thickeners suffer from the disadvantage that large floor areas are required, since the throughput depends above all on the area, while depth is of minor importance.

In recent years, machines known as "high-capacity" thickeners have been introduced by various manufacturers. Many varieties exist, and the machines are typified by a reduction in unit area requirement from conventional installations.[18,19]

The "Enviro-Clear" thickener developed by Envirotech Corporation is typical[20] (Fig. 15.13).

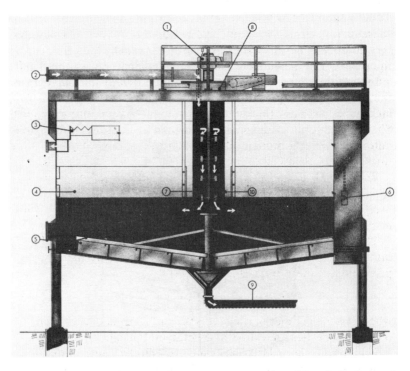

FIG. 15.13. Enviro-Clear high-capacity thickener. 1, mixer drive; 2, feed pipe; 3, overflow launder; 4, inclined settling plates; 5, rake arm; 6, level sensor; 7, flocculant feed pipe; 8, drive unit with overload control; 9, sludge discharge; 10, mixing chamber.

The feed enters via a hollow drive shaft where flocculant is added and is rapidly dispersed by staged mechanical mixing. This staged mixing action is said to improve thickening since it makes most effective use of the flocculant. The flocculated feed leaves the mixing chambers and is injected into a blanket of slurry where the feed solids are further flocculated by contacting previously flocculated material. Direct contact between rising fluid and settling solids, which is common to most thickeners, is averted with slurry blanket injection. Radially mounted inclined plates are partially submerged in the slurry blanket; the settling solids in the slurry blanket slide downwards along the inclined plates, producing faster and more effective thickening than vertical descent. The height of the slurry blanket is automated through the use of a level sensor.

Tailings water reclamation at Bougainville Copper Ltd in Papua New Guinea is performed using four 24.5-m diameter Enviro-Clear thickeners.[21] It is estimated that the high-capacity thickeners occupy an area only 5% of that which would be required for conventional thickeners operating without flocculant. Thickener control is totally automatic, all four thickeners operating on rake torque control. Sludge bed level is detected by a level sensor and is controlled automatically at a predetermined set-point by a pinch valve on the thickener underflow line.

The lamella thickener[22] utilises a nest of inclined parallel plates which reduce settling distance and at the same time increase effective area. The floor-space requirement of the lamella thickener is only about 20% of that of the conventional thickener (Fig. 15.14).

The inclined parallel trays allow the settled solids to slide by gravity into a hopper. The effective settling area is therefore the horizontal projection of these trays (Fig. 15.15), i.e.

$$A_{\text{eff}} = nA \cos \alpha, \tag{15.7}$$

where n is the number of trays, A is the surface area of each plate, and α is the angle between trays and horizontal plane.

The whole lamella pack can be vibrated intermittently or continuously when treating sticky sludges. The feed enters the thickener through a bottomless feed-box at a point which determines the relationship between clarification and thickening areas. The area

FIG. 15.14. Lamella thickener equivalent in capacity to a 10-m-diameter conventional unit.

below the feed-box outlet is for thickening while the area above is for clarification. The discharge liquid flows upward and is removed through specially designed boxes, thus providing a uniform flow distribution between the lamella plates.

Compression in a conventional thickener is provided by surface load, depth of solids, and retention time. As the retention time in a lamella thickener is short, low-amplitude vibrations are applied to the hopper, which play an important part in providing solids compression and flow.

One of the plants which have installed lamella thickeners is Eldorado Nuclear's Beaverlodge operation in northern Saskatchewan. The thickener installed has an effective area of 100 m^2 and has

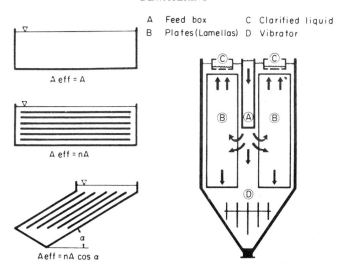

A Feed box   C Clarified liquid
B Plates(Lamellas) D Vibrator

FIG. 15.15. Principle of lamella thickener.

replaced the former 250-m² lime slurry thickener. The underflow densities have been equivalent to those obtained previously in the conventional thickener and overflow clarities have been superior under steady operating conditions. The unit may be shut down, drained, and restarted without requiring the usual period for forming the bed.[23]

Centrifugal Sedimentation

Centrifugal separation can be regarded as an extension of gravity separation, as the settling rates of particles are increased under the influence of centrifugal force. It can, however, be used to separate emulsions which are normally stable in a gravity field.

Centrifugal separation can be performed either by hydrocyclones or centrifuges.

The simplicity and cheapness of the hydrocyclone (Chapter 9) make it very attractive, although it suffers from restrictions with respect to the solids concentration which can be achieved and the relative proportions of overflow and underflow into which the feed may be split. Generally the efficiency of even a small-diameter cyclone falls off rapidly at very fine particle sizes and particles smaller than about 10 μm in diameter will invariably appear in the overflow, unless they are very heavy. Flocculation of such particles is not possible, since the high shear forces within a cyclone rapidly break up any agglomerates. The cyclone is therefore inherently better suited to classification rather than thickening.

By comparison, centrifuges are much more costly and complex, but have a much greater clarifying power and are generally more flexible. Much greater solids concentrations can be obtained than with the cyclone.

Various types of centrifuge are used industrially,[8,11,24] the *solid bowl scroll centrifuge* having widest use in the minerals industry due to its ability to discharge the solids continuously.

The basic principles of a typical machine are shown in Fig. 15.16. It consists essentially of a horizontal revolving shell or bowl, cylindroconical in form, inside which a screw conveyor of similar section rotates in the same direction at a slightly higher or lower speed. The feed pulp is admitted to the bowl through the centre tube of the revolving-screw conveyor. On leaving the feed pipe the slurry is immediately subjected to a high centrifugal force causing the solids to settle on the inner surface of the bowl at a rate which depends on the rotational speed

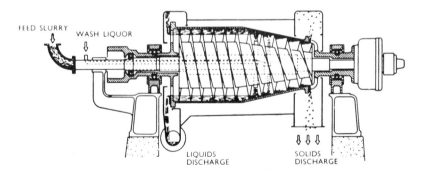

FIG. 15.16. Continuous solid bowl scroll centrifuge.

employed, this normally being between 1600 and 8500 rev min^{-1}. The separated solids are conveyed by the scroll out of the liquid and discharged through outlets at the smaller end of the bowl. The solids are continuously dewatered by centrifugal force as they proceed from the liquid zone to the discharge. Excess entrained liquor drains away to the pond circumferentially through the particle bed.

When the liquid reaches a predetermined level it overflows through the discharge ports at the larger end of the bowl.

The actual size and geometry of these centrifuges varies according to the throughput required and the application. The length of the cylindrical section largely determines the clarifying power and is thus made a maximum where overflow clarity is of prime importance. The length of the conical section, or "beach", decides the residual moisture content of the solids, so that a long shallow cone is used where maximum dryness is required.

Centrifuges are manufactured with bowl diameters ranging from 15 cm to 150 cm, the length generally being about twice the diameter. Throughputs vary from about 0.5 to 50 m^3 h^{-1} of liquid and from about 0.25 to 100 t h^{-1} of solids depending on the feed concentration, which may vary widely from 0.5 to 70% solids, and on the particle size, which may range from about 12 mm to as fine as 2 μm, or even less when flocculation is used. The wide application of flocculation is limited by the tendency of the scroll action to damage the flocs and thus redisperse the fine particles. The moisture content in the product varies widely, typically being in the range 5–20%.

Filtration

Filtration is the process of separating solids from liquid by means of a porous medium which retains the solid but allows the liquid to pass. The theory of filtration has been comprehensively reviewed mathematically elsewhere[8,11] and will not be covered here.

The conditions under which filtration are carried out are many and varied and the choice of the most suitable type of equipment will depend on a large number of factors. Whatever type of equipment is used, a *filter cake* gradually builds up on the medium and the resistance to flow progressively increases throughout the operation. Factors affecting the rate of filtration include:[8]

(a) The pressure drop from the feed to the far side of the filter medium. This is achieved in *pressure filters* by applying a positive pressure at the feed end and in *vacuum filters* by applying a vacuum to the far side of the medium, the feed side being at atmospheric pressure.

(b) The area of the filtering surface.

(c) The viscosity of the filtrate.

(d) The resistance of the filter cake.

(e) The resistance of the filter medium and initial layers of cake.

Filtration in mineral processing applications normally follows thickening. The thickened pulp may be fed to storage agitators from where it is drawn off at a uniform rate to the filters. Flocculants are sometimes added to the agitators in order to aid filtration. Slimes have an adverse effect on filtration, as they tend to "blind" the filter medium; flocculation reduces this and increases the voidage between particles, making filtrate flow easier. The lower molecular weight flocculants tend to be used in filtration, as the flocs formed by high molecular weight products are relatively large, and entrain water within the structure, increasing the moisture content of the cake, even though pick-up may be improved. Smaller flocs are formed with the lower molecular weight flocculants, which have a higher shear-resistance, and the resultant filter cake is a uniform porous structure which allows rapid dewatering, yet prevents migration of the finer particles through the cake to the medium.[4] Other filter aids are used to reduce the liquid surface tension, thus assisting flow through the medium.

The Filter Medium

The choice of the filter medium is often the most important consideration in assuring efficient operation of a filter. Its function is generally to act as a support for the filter cake, while the initial layers of cake provide the true filter. The filter medium should be selected primarily for its ability to retain solids without blinding. It should be mechanically strong, corrosion resistant, and offer as little resistance to flow of filtrate as possible. Relatively coarse materials are normally used and clear filtrate is not obtained until the initial layers of cake are formed, the initial cloudy filtrate being recycled.

Filter media are manufactured from cotton, wool, linen, jute, nylon, silk, glass fibre, porous carbon, metals, rayon and other synthetics, and miscellaneous materials such as porous rubber. Cotton fabrics are by far the most common type of medium, primarily because of their low initial cost and availability in a wide variety of weaves. They can be used to filter solids as fine as 10 μm.

Filtration Tests

It is not normally possible to forecast what may be accomplished in the filtration of an untested product, therefore preliminary tests have to be made on representative samples of the pulp before the large-scale plant is designed. Tests are also commonly carried out on pulps from existing plants, to assess the effect of changing operating conditions, filter aids, etc. A simple vacuum *filter leaf* test circuit is shown in Fig. 15.17.

The filter leaf, consisting of a section of the industrial filter medium, is connected to a filtrate receiver equipped with a vacuum gauge. The receiver is connected to a vacuum pump. If the industrial filter is to be a continuous vacuum filter, this operation must be simulated in the test. The cycle is divided into three sections—cake formation (or "pick-up"), drying, and discharge. Sometimes pick-up is followed by a period of washing and the cake may also be subjected to compression during drying. While under vacuum, the test leaf is submerged for the pick-up period in the agitated pulp to be tested. The leaf is then removed and held with the drainpipe down for the allotted drying time. The cake can then be removed, weighed, and dried. The daily filter capacity can then be determined by the dry weight of cake

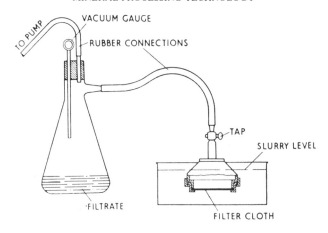

FIG. 15.17. Laboratory test filter.

per unit area of test leaf multiplied by the daily number of cycles and the filter area.

Types of Filter

Cake filters are the type most frequently used in mineral processing, where the recovery of large amounts of solids from fairly concentrated slurries is the main requirement. Those where the main requirement is the removal of small amounts of solid from relatively dilute suspensions are known as screening or clarification filters.

Cake filters may be pressure, vacuum, batch, or continuous types.

Pressure filters. Because of the virtual incompressibility of solids, filtration under pressure may have advantages over vacuum. Higher flow rates and better washing and drying may result from the higher pressures that can be used. However, the continuous removal of solids from the pressure-filter chamber can be extremely difficult and consequently, although continuous pressure filters do exist, the vast majority operate as batch units.

Filter presses are the most frequently used type of pressure filter. They are made in two forms—the plate and frame press and the recessed plate or chamber press.

The *plate and frame press* (Fig. 15.18) consists of plates and frames arranged alternately. The hollow frame is separated from the plate by the filter cloth. The filter press is closed by means of a screw or hydraulic piston device and compression of the filter cloth between plates and frames helps to prevent leakages. A tight chamber is therefore formed between each pair of plates. The slurry is introduced to the empty frames of the press through a continuous channel formed by the holes in the corners of the plates and frames. The filtrate passes through the cloth and runs down the grooved surfaces of the plates and is removed through a continuous channel. The cake remains in the frame and, when the frame is full, the filter cake can be washed, after which the pressure is released and the plates and frames separated one by one. The filter cake in the frames can then be discharged, the filter press closed again and the cycle repeated.

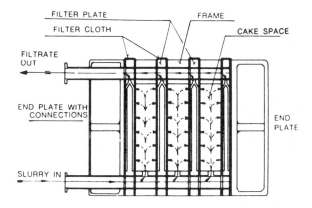

FIG. 15.18. Plate and frame filter press.

The *chamber press* (Fig. 15.19) is similar to the plate and frame type except for the fact that the filter elements consist solely of the recessed filter plates. The individual filter chambers are therefore formed between successive plates. All the chambers are connected by means

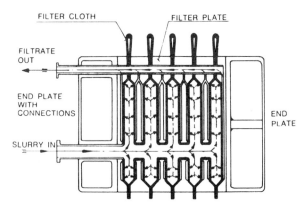

FIG. 15.19. Chamber or recessed plate filter press.

of a comparatively large hole in the centre of each plate. The filter cloth with a central hole covers the plate and slurry is led through the inlet channel. The clear filtrate passing through the cloth is removed by means of smaller holes in the plate, the cake gradually depositing in the chambers.

Recessed plate filters are most widely used for treating slurries with high solids contents. They afford easier cake discharge than plate and frame presses, and are better suited to mechanisation.[25,26] They are the preferred system for dewatering fine tailings in coal preparation plants, as they are the only device which can produce perfectly clean water and a solid cake which can easily be handled.[27] Automatic chamber press filtration has gained acceptance in Europe, and is also being used elsewhere for the dewatering of flotation concentrates. Pressure filtration results in significantly lower filter cake moisture contents, which decreases or eliminates the necessity of filter cake drying using thermal methods.

At Vihanti in Finland, copper and lead concentrates are alternately filtered in batches on a fully automatic Larox chamber filter, which was introduced in 1982. This has replaced the copper and lead disc filters and thermal dryers, and has reduced dewatering energy from 2.3 kWh t^{-1} in 1978 to 1.1 kWh t^{-1} in 1983. The filter cake contains 6–7% moisture.[28] Similar filters have been installed in a number of

North American base metal concentrators, for dewatering coarse-grained concentrates.[29-33]

Vacuum filters. There are many different types of vacuum filter, but they all incorporate filter media suitably supported on a drainage system, beneath which the pressure is reduced by connection to a vacuum system. Vacuum filters may be batch or continuous.[34-36]

Batch vacuum filters. The *leaf filter* has a number of leaves, each consisting of a metal framework or a grooved plate over which the filter cloth is fixed (Fig. 15.20).

Numerous holes are drilled in the pipe framework, so that when a vacuum is applied, a filter cake builds up on both sides of the leaf. A number of leaves are generally connected and are first immersed in slurry held in a filter feed tank and are then mechanically transferred, if required, to a wash tank and then to a cake-receiving vessel where the cake is removed by replacing the vacuum by air pressure (Fig. 15.21).

Although simple to operate, these filters require considerable floor space and suffer from the possibility of sections of cake dropping from the leaves during transport from tank to tank. They are now used only for clarification, i.e. the removal of small amounts of suspended solids from liquors.

Horizontal leaf, or *tray filters*, work in much the same manner as a laboratory Buchner filter and consist of rectangular pans having a false bottom of filter medium. They are filled with pulp, the vacuum is applied until the cake is dry, when the pan is inverted, being supported on pivots, the vacuum is disconnected and low-pressure air introduced under the filter medium to remove the cake.

Continuous vacuum filters. These are the most widely used filters in mineral processing applications and fall into three classes—drums, discs, and horizontal filters.

The *rotary-drum filter* (Fig. 15.22) is the most widely used type in industry, finding application both where cake washing is required and where it is unnecessary.

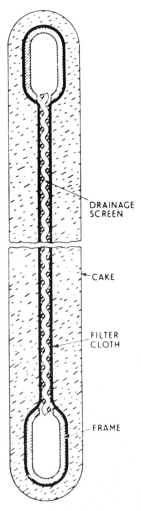

FIG. 15.20. Cross-section of typical filter leaf.

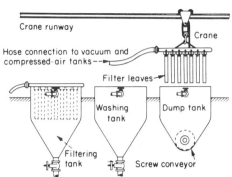

FIG. 15.21. Typical leaf filter circuit. (From *Chemical Engineers' Handbook* by J. H. Perry, McGraw-Hill, 1963.)

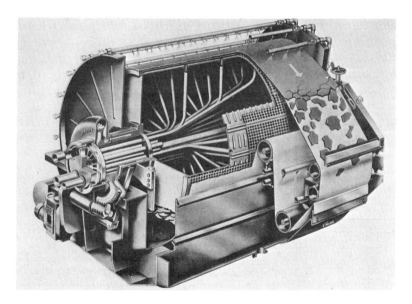

FIG. 15.22. Rotary-drum filter with belt discharge.

The drum is mounted horizontally and is partially submerged in the filter trough, or "boot", into which the slurry is fed and maintained in suspension by agitators. The periphery of the drum is divided into compartments, each of which is provided with a number of drain lines, which pass through the inside of the drum, terminating at one end as a ring of ports which are covered by a rotary valve to which vacuum is applied. The filter medium is wrapped tightly round the drum surface which is rotated at low speed, usually in the range 0.1–0.3 rev min^{-1}, but up to 3 rev min^{-1} for very free-filtering materials.

As the drum rotates, each compartment goes through the same cycle of operations, the duration of each being determined by the drum speed, the depth of submergence of the drum, and the arrangement.of the valve. The normal cycle of operations consists of filtration, drying and discharge, but it is possible to introduce other operations into the basic cycle, such as cake washing and cloth cleaning.

Various methods are used for discharging the solids from the drum, depending on the material being filtered. The most common form makes use of a reversed blast of air, which lifts the cake so that it can be removed by a knife, without the latter actually contacting the medium. Another method is string discharge, where the filter cake is formed on an open conveyor—the strings—which are in contact with the filter cloth in the filtration, washing, drying zones. A further advance on this method is belt discharge, as shown in Fig. 15.22, where the filter medium itself leaves the filter and passes over the external roller, before returning to the drum. This has a number of advantages in that very much thinner cakes can be handled, with consequently increased filtration and draining rates and hence better washing and dryer products. At the same time, the cloth can be washed on both sides by means of sprays before it returns to the drum (Fig. 15.23), thus minimising the extent of blinding. Cake washing is usually carried out by means of sprays or weirs, which cover a fairly limited area at the top of the drum.

The capacity of the vacuum pump will be determined mainly by the amount of air sucked through the cake during the washing and drying periods when, in most cases, there will be a simultaneous flow of both liquid and air. A typical layout is shown in Fig. 15.24, from which it is seen that the air and liquid are removed separately.

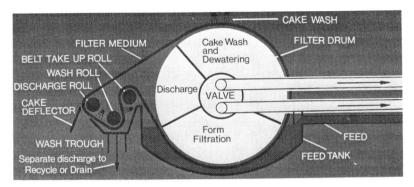

FIG. 15.23. Belt discharge filter with cloth washing.

The barometric leg should be at least 10 m high to prevent liquid being sucked into the vacuum pump.

Variations on standard drum filters to enable them to handle coarse, free-draining, quick-settling materials, include top feed units where the material is distributed across the drum at about top dead centre and discharged at between 90 and 180 degrees from the feed point.

The principle of operation of *disc filters* (Fig. 15.25) is similar to that of rotary drum filters. The solids cake is formed on both sides of the circular discs, which are connected to the horizontal shaft of the

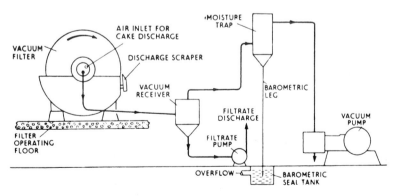

FIG. 15.24. Typical rotary-drum filter system.

FIG. 15.25. Rotary-disc filters.

machine. The discs rotate and lift the cake above the level of the slurry in the trough, whereupon the cake is suction-dried and is then removed by a pulsating air blow with the assistance of a scraper. The discs can be located along the shaft at about 30-cm centres and consequently a large filtration area can be accommodated in a small floor space. Cost per unit area is thus lower than for drum filters, but cake washing is virtually impossible and the disc filter is not as adaptable as a drum filter.

The *horizontal belt filter* (Fig. 15.26) consists of an endless perforated rubber drainage deck supporting a separate belt made from a suitable filter cloth. At the start of the horizontal travel, slurry flows by gravity on to the belt. Filtration immediately commences, due partly to gravity and partly to the vacuum applied to the suction boxes which are in contact with the underside of the drainage deck during the course of its travel.

The cake which forms is dewatered, dried by drawing air through it,

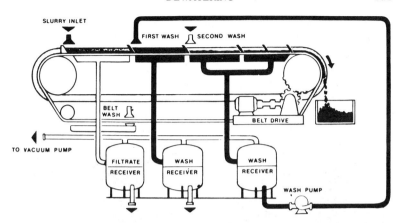

Fig. 15.26. Horizontal belt filter.

and then discharged as the belt reverses over a small-diameter roller. If required, one or more washes can be incorporated.

The main disadvantage of this machine is the heavy wear and tear of the flexible drainage belt which results from it being dragged across the vacuum boxes whilst the latter are in suction. This results in a loss of vacuum and hence poor drainage.

The *Pannevis belt filter* is a development in which the medium is in the form of an endless horizontal belt, the upper part of which acts as the filter; beneath this section are several mobile vacuum trays. Suction is applied intermittently and, as soon as suction commences, each tray attaches itself to the belt and moves along with it, during which time filtration occurs. Upon interruption of the suction, the trays are released and return to their starting-points; thus their motion is reciprocating.

The applications for horizontal belt filters are increasing. They are particularly suited to hydrometallurgical circuits where metal values are dissolved in alkali or acid. These values can be recovered from waste solids by filtration of the leached slurry and countercurrent washing.[24] Large belt filters are in operation on cyanide-leached gold ore and acid-leached uranium ore, and twenty-six 80 m^2 filters are being used to treat large tonnages of leached copper tailings at

Nchanga in Zambia, in the world's largest filter plant.[37] Belt filters are also suited for concentrated slurries of fast settling products, where efficient washing is required. In addition to their low installed capital cost when compared with disc, drum- and press-type filters, relatively low operating costs mean that these filters offer a particularly cost-effective and reliable solution to filtration problems, especially with low-value materials such as mine tailings. Work on coal slurries has shown that horizontal belt vacuum filtration should produce lower cake moistures than those from rotary vacuum filtration and at a reduced cost per tonne.[38]

Drying

The drying of concentrates prior to shipping is the last operation performed in the mineral-processing plant. It reduces the cost of transport and is usually aimed at reducing the moisture content to about 5% by weight. Dust losses are often a problem if the moisture content is lower.

Rotary thermal dryers are often used. These consist of a relatively long cylindrical shell mounted on rollers and driven at a speed of up to 25 rev min^{-1}. The shell is at a slight slope, so that material moves from the feed to discharge end under gravity. Hot gases, or air, are fed in either at the feed end to give parallel flow or at the discharge to give counter-current flow.

The method of heating may be either direct, in which case the hot gases pass through the material in the dryer, or indirect, where the material is in an inner shell, heated externally by hot gases. The direct-fired is the dryer most commonly used in the minerals industry, the indirect-fired type being used when the material must not contact the hot combustion gases. Parallel flow dryers (Fig. 15.27) are used in the majority of current operations because they are more fuel efficient and have greater capacity than counterflow types.[39] Since heat is applied at the feed end, build-up of wet feed is avoided, and in general these units are designed to dry material to not less than 1% moisture. Since counter-flow dryers apply heat at the discharge end, a completely dry product can be achieved, but its use with heat-sensitive

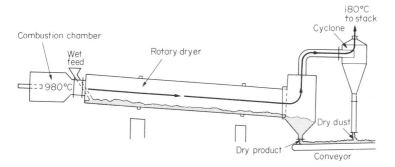

FIG. 15.27. Direct fired, parallel flow rotary dryer (after Kram[39]).

materials is limited because the dried material comes into direct contact with the heating medium at its highest temperature.

The product from the dryers is often stockpiled, before being loaded on to trucks or rail-cars as required for shipment. The containers may be closed, or the surface of the contents sprayed with a skin-forming solution, in order to eliminate dust losses.[40]

References

1. Pearse, M. J., Synthetic flocculants in the mineral industry—types available, their uses and advantages, in *Reagents in the Minerals Industry*, ed. M. J. Jones and R. Oblatt, p. 101, IMM, London, 1984.
2. Adamson, G. F. S., Some recent papers on flocculation, *Mine & Quarry*, **10**, 42 (March 1981).
3. Hunter, T. K., and Pearse, M. J., The use of flocculants and surfactants for dewatering in the mineral processing industry, *Proc. IVth Int. Min. Proc. Cong.*, Paper IX–11, CIM, Toronto (Oct. 1982).
4. Moss, N., Theory of flocculation, *Mine & Quarry* **7**, 57 (May 1978).
5. Lightfoot, J., Practical aspects of flocculation, *Mine & Quarry*, **10**, 51 (Jan./Feb. 1981).
6. Read, A. D., and Hollick, G. T., Selective flocculation, *Mine & Quarry*, **9**, 55 (Apr. 1980).
7. Paananen, A. D., and Turcotte, W. A., Factors influencing selective flocculation—desliming practice at the Tilden Mine, *Mining Engng.* **32**, 1244 (Aug. 1980).
8. Coulson, J. M., and Richardson, J. F., *Chemical Engineering*, vol. 2, Pergamon Press, Oxford, 1968.
9. Hsia, E. S., and Reinmiller, F. W., How to design and construct earth bottom thickeners, *Trans. Soc. Min. Engrs.* 36 (Aug. 1977).

10. Anon., Pumps for the mining industry, *Min. Mag.* 569 (June 1978).

11. Perry, J. H., *Chemical Engineers' Handbook*, 4th ed., McGraw-Hill, New York, 1963.

12. Coe, H. S., and Clevenger, G. H., Methods for determining the capacities of slime-settling tanks, *Trans. AIMME* **55**, 356 (1916).

13. Kynch, C. J., A theory of sedimentation, *Trans. Faraday Soc.* **48**, 166 (1952).

14. Pearse, M. J., *Gravity Thickening Theories: A Review*, Warren Spring Lab. Report LR 261 (MP), 1977.

15. Talmage, W. P., and Fitch, E. B., Determining thickener unit areas, *Ind. Engng. Chem.* **47**, 38 (Jan. 1955).

16. Fitch, E. B., Gravity separation equipment—clarification and thickening, in *Solid-Liquid Separation Equipment Scale-up* (ed. D. B. Purchas), Uplands Press, Croydon, 1977.

17. Pearse, M. J., *Laboratory Procedures for the Choice and Sizing of Dewatering Equipment in the Mineral Processing Industry*, Warren Springs Lab. Report LR 281 (MP), 1978.

18. Keane, J. M., Sedimentation: theory, equipment, and methods, *World Mining* **32**, 44 (Nov. 1979).

19. Keane, J. M., Recent developments in solids/liquid separation, *World Mining* 110 (Oct. 1982).

20. Emmett, R. C., and Klepper, R. P., Technology and performance of the hi-capacity thickener, *Mining Engng.* **32**, 1264 (Aug. 1980).

21. Deans, B. L., and Glatthaar, J. W., Small thickeners yield big gains at Bougainville, *Engng, & Min. J.*, **187**, 36 (Oct. 1986).

22. Anon., Liquid–solid separation device, *Min. Mag.*, 59 (July 1974).

23. Pickett, D. E., and Joe, E. G., Canadian advances in 1973—milling and process metallurgy, *Can. Min. J.* 21 (Mar. 1974).

24. Bragg, R., Filters & centrifuges, *Min. Mag.* 90 (Aug. 1983).

25. Kurita, T., and Suwa, S., How the fully automatic filter press has developed, *Filtration and Separation* 109 (Mar./Apr. 1978).

26. Anon., Automatic chamber filter dewaters large tonnages, *World Mining* 9 (July 1982).

27. Jones, G. S., Pressure filtration of coal tailings, *Mixing & Dewatering in the Minerals Industry*, MIRO, London, Session 2, 1985.

28. Wills, B. A., Pyhasalmi and Vihanti concentrators, *Min. Mag.* 176 (Sept. 1983).

29. Pazour, D. A., Saving energy in beneficiation and refining, *World Mining* **35**, 44 (April 1982).

30. Anon., Pressure filters reduce operating and capital costs, *World Mining* **35**, 50 (Nov. 1982).

31. Garon, M., and Nesset, J. E., Dewatering of base metal concentrates by pressure filtration, *Proc. XIVth Int. Min. Proc. Cong.*, Paper IX-5, CIM, Toronto, Oct. 1982.

32. Nesset, J. E., Dewatering Brunswick concentrates by pressure filtration, *CIM Bulletin* 103 (July 1982).

33. Konigsmann, K. V., Trends in concentrate dewatering in mills of the Noranda group, *CIM Bulletin*, **77**, 47 (Dec. 1984).

34. Moos, S. M., and Dugger, R. E., Vacuum filtration: available equipment and recent innovations, *Min. Engng.* **31**, 1473 (Oct. 1979).

35. Keleghan, W., Vacuum filtration: Part 1, *Mine & Quarry*, **15**, 51 (Jan./Feb. 1986).

36. Keleghan, W. T. H., The practice of vacuum filtration, *Mine & Quarry*, **15,** 38 (March 1986).

37. Anon., Large horizontal belt filters for Zambian tailings project, *Min. J.* 268 (April 22, 1983).

38. Vickers, F., et al, An alternative to rotary vacuum filtration for fine coal dewatering, *Mine & Quarry*, **14,** 25 (Oct. 1985).

39. Kram, D. J., Drying, calcining, and agglomeration, *Eng. & Min. J.* **181,** 134 (June 1980).

40. Kolthammer, K. W., Concentrate drying, handling and storage, in *Mineral Processing Plant Design* (ed. A. L. Mular and R. B. Bhappu), p. 601, A.I.M.M.E., New York, 1978.

CHAPTER 16

TAILINGS DISPOSAL

Introduction

The disposal of mill tailings is a major environmental problem, which is becoming more serious with the increasing exploration for metals and the working of lower-grade deposits. Apart from the visual effect on the landscape of tailings disposal, the major ecological effect is usually water pollution, arising from the discharge of water contaminated with solids, heavy metals, mill reagents, sulphur compounds, etc.

The nature of tailings varies widely; they are usually transported and disposed of as a slurry of high water content, but they may be composed of very coarse dry material, such as the float fraction from heavy medium plants. Due to the lower costs of mining from open pits, ore from such locations is often of very low grade, resulting in the production of large amounts of very fine tailings. The physical characteristics of the ore in relation to milling control the final particle size of the tailings and there has been no research aimed at reducing tailings-disposal problems by increasing the mean size.

Methods of Disposal of Tailings

The methods used to dispose of tailings have developed due to environmental pressures, changing milling practice, and realisation of profitabie applications. Early methods included discharge of tailings into rivers and streams, which is still practised at some mines, and the dumping of coarse dewatered tailings on to land. The many nineteenth-century tips seen in Cornwall and other parts of Britain are evidence of this method. Due to the damage caused by such methods, and the much finer grinding necessary on most modern ores, other techniques have been developed. The most satisfactory way of dealing with tailings is to make positive use of them, such as reprocessing in order to recover additional values (see Chapter 1), or to use them as a useful product in their own right, e.g. the use of coarse (20–30 mm) HMS float as railway ballast and aggregate.

It is common practice in underground mines, in which the method of working requires the filling of mined-out areas, to return the coarser fraction of the mill tailings underground. This method has been used since the beginning of the century in South Africa's gold mines. Back-filling worked-out stopes reduces the volume of tailings which must be impounded on the surface, but not all tailings are suited as back-fill material. It is invariably necessary to de-slime the tailings, the resultant slimes, which may account for up to 50% of the total weight, requiring surface disposal.[1] Some tailings swell or shrink after the fill has been placed, and some have the useful property of being self-cementing, which removes the necessity of adding cement to the back-fill, which is common practice prior to placement underground. The use of back-fill can cause surface disposal problems in that borrowed fill may have to be used to construct the tailings impoundment, as the coarse fraction of the tailings, which is often used for construction, has been removed.

Back-fill methods have not been applied to the large amounts of tailings produced by open-pit mining methods, as this would entail temporary storage during the life of the mine prior to disposal in the worked-out pit and the most widely used method is to contain the tailings within a purpose-built dam. The impoundment must provide safe and economical storage for the required volume of tailings and permit the construction and operation of pollution control facilities.

Tailings Dams

The design, construction, and operation of tailings dams is rapidly becoming a major consideration for most new mining developments, as well as for many existing operations.[2,3]

It is economically advantageous to site the impoundment close to the mine, but this imposes limits on site selection. The ground underlying the dam must be structurally sound and able to bear the weight of the impoundment. If such a site cannot be found close to the mine, it may be necessary to pump the tailings, at a high pulp density, to a suitable location. The largest tailings line in current use is 70 km long, at a Japanese mine.[4]

Tailings dams may be built across river valleys, or as curved or multisided dam walls on valley sides, this latter design facilitating drainage. On flat, or gently sloping ground, lagoons are built with walls on all sides of the impoundment.

The disposal of tailings adds to the production costs, so it is essential to make disposal as cheap as possible. This requirement led initially to the development of the once commonly used *upstream method* of tailings-dam construction, so named because the centre line of the dam moves upstream into the pond.

In this method, a small starter dam is placed at the extreme downstream point (Fig. 16.1) and the dam wall is progressively raised on the upstream side. The tailings are discharged by spigoting off the top of the starter dyke and, when the initial pond is nearly filled, the dyke is raised and the cycle repeated. Various methods are used to raise the dam; material may be taken from the dried surface of the previously deposited tailings and the cycle repeated, or more commonly the wall may be built from the coarse fraction of the tailings,

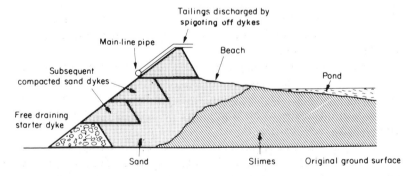

FIG. 16.1. Upstream tailings dam.

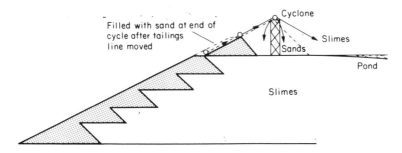

FIG. 16.2. Construction of upstream tailings dam using cyclones.

separated out by cyclones, or spigots, the fines being directed into the pond (Figs. 16.2 and 16.3).

The main advantages of the upstream construction are the low cost and the speed with which the dam can be raised by each successive dyke increment.

The method suffers from the disadvantage that the dam wall is built on the top of previously deposited unconsolidated slimes retained behind the wall. There is a limiting height to which this type of dam can be built before failure occurs and the tailings flow out and, because of this, the upstream method of construction is now less commonly used.

The *downstream method* is a relatively new development which has evolved as a result of efforts to devise methods for constructing larger and safer tailings dams. It is essentially the reverse of the upstream method, in that as the dam wall is raised, the centreline shifts downstream and the dam remains founded on coarse tailings (Fig. 16.4). Most procedures involve the use of cyclones to produce sand for the dam construction.

Downstream dam building is the only method that permits design and construction of tailings dams to acceptable engineering standards. All tailings dams in seismic areas, and all major dams, regardless of their location, should be constructed using some form of the downstream method. The major disadvantage of the technique is the large amount of sand required to raise the dam wall. It may not be possible, especially in the early stages of operation, to produce sufficient sand volumes to maintain the crest of the tailings dam above

FIG. 16.3. Construction of tailings dam wall utilising cyclone underflows.

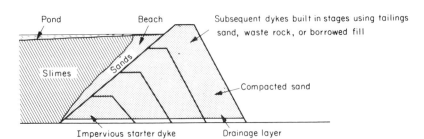

FIG. 16.4. Downstream tailings dam.

the rising pond levels. In such cases, either a higher starter dam is required or the sand supply must be augmented with borrowed fill, such procedures increasing the cost of tailings disposal.

The *centre-line method* (Fig. 16.5) is a variation of that used to construct the downstream dam and the crest remains in the same horizontal position as the dam wall is raised. It has the advantage of requiring smaller volumes of sand-fill to raise the crest to any given height. The dam can thus be raised more quickly and there is less trouble keeping it ahead of the tailings pond during the early stages of construction. Care, however, must be exercised in raising the upstream face of the dam to ensure that unstable slopes do not develop temporarily.

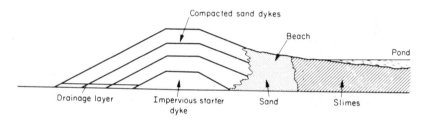

FIG. 16.5. Centre-line tailings dam.

Very stable tailings dams can be constructed from open-pit over-burden, or waste rock, according to the local circumstances. An example is shown in Fig. 16.6. Since the tailings are not required for the dam construction, they may be fed into the pool without separation of the sands from the slimes. In some cases the output of overburden may not be sufficient to keep the dam crest above the tailings pond, and it may be necessary to combine waste rock and tailings sand-fills to produce a safe economical dam.

An interesting method of disposal is in use at the Ecstall (Kidd Creek) operation of Texasgulf Canada Ltd.[5] The tailings disposal area consists of 3000 acres enclosed by a gravel dyke. Mill tailings are thickened and pumped to a central spigoting location inside the dam. The system is designed to build a mountain of tailings in the central area and thus keep the height of the perimeter dyke to a minimum.

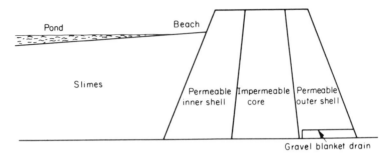

FIG. 16.6. Dam constructed from overburden.

Erosion of dams due to wind and rain can effect the stability and produce environmental problems. Many methods are used to combat this, such as vegetation of the dam banks[6] and chemical stabilisation to form an air- and water-resistant crust.

There is little doubt that tailings dams have a visual impact on the environment due to their regular geometric shape. Perhaps the most conspicuous is the downstream type, whose outer wall is continually being extended, and cannot be re-vegetated until closure. There are, however, few reasons why dam walls should not be landscaped at some stage in their life, and many dams are now being designed to permit early visual integration with the environment.[7] An example is the impoundments at Flambeau, North Wisconsin, USA,[8] where a rock-fill dam wall 18 m high, 24 m wide at the crest, and 111 m wide at the base, has been designed to minimise both visual and pollution effects (Fig. 16.7). The wall consists of a clay core, with the downstream side faced with non-pyritic rock and covered with top-soil, permitting re-vegetation and consequently reduced visual impact.

The most serious problem associated with the disposal of tailings is the release of polluted water, and this has been extensively investigated.[9] The main effects of pollution are due to the effluent pH, which may cause ecological changes; dissolved heavy metals, such as copper, lead, zinc, etc., which can be lethal to fish-life if allowed to enter local water-courses; mill reagents, which are usually present in only very small quantities, but, nevertheless, may be harmful; and

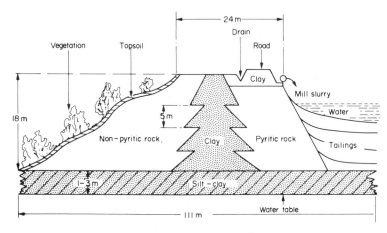

FIG. 16.7. Flambeau impoundment.

suspended solids, which should be minimal if the tailings have spent long residence times in the dam, thus allowing the solids to settle and produce a clear decant.

Figure 16.8 shows a generalised representation of water gain and loss at a tailings impoundment.[7]

With the exception of precipitation and evaporation, the rates and volumes of the water can be controlled to a large extent. It is more

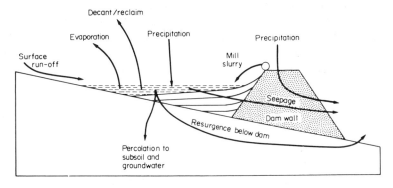

FIG. 16.8. Water gain and loss in a typical tailings dam.

satisfactory to attempt to prevent the contamination of natural waters rather than to purify them afterwards, and if surface run-off to the dam is substantial, then interception ditches should be installed. It is difficult to quantify the amount of water lost to groundwater, but this can be minimised by selecting a site with impervious foundations, or by sealing with an artificial layer of clay. Seepage through the dam wall is often minimised by an impervious slimes layer on the upstream face of the dam, but this is expensive, and many mines prefer to encourage free-drainage of the dam through a pervious, chemically barren material. In the case of upstream dams, this can be a barren starter dyke, while with downstream and centre-line constructions, a free-draining gravel blanket can be used. A small seepage pond with impervious walls and floors situated below the main dam can collect this water, from where it can be pumped back into the tailings pond. If the dam wall is composed of metal-bearing rock, or sulphide tailings, the seepage is often highly contaminated due to its contact with the solid tailings, and may have to be treated separately.

The tailings are often treated with lime in order to neutralise acids and precipitate heavy metals as insoluble hydroxides before pumping to the dam. Such treated tailings may be thickened and the overflow, free of heavy metals, returned to the mill (Fig. 16.9), thus reducing the water and pollutant input to the tailings dam.

Assuming good control of the above inputs and outputs of dam water, the most important factor in achieving pollution control is the method used to remove surplus water from the dam. Decant facilities are required on all dams, to allow excess free water to be removed.

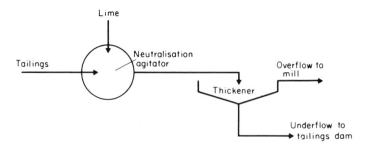

FIG. 16.9. Treatment of tailings with lime.

Inadequate decant design has caused many major dam failures.[10] Many older dams used decant towers with discharge lines running through the base of the dam to a downstream pump-house. Failures of such structures were common due to the high pressures exerted on the pipelines, leading to uncontrolled losses of fluids and tailings downstream. Floating, or movable, pumphouses situated close to the tailings pond are now in common use.

Recycling of water from the decant is becoming more important due to pressures from governments and environmentalists.[11] As much water as possible must be reclaimed from the tailings pond for re-use in the mill and the volume of fresh make-up water used must be kept to a minimum. The difference between the total volume of water entering the tailings pond and the volume of water reclaimed plus evaporation losses must be stored with the tailings in the dam. If that difference exceeds the volume of the voids in the stored tailings, there becomes a surplus of free water that can build up to tremendous quantities over the life of a mine. A typical dam-reclaim system is shown in Fig. 16.10.

The main disadvantage of water reclamation is the recirculation of pollutants to the mill, which can interfere with processes such as flotation. It has been shown,[12] however, that milling problems with recycled water are confined mainly to complex selective flotation circuits, due to the use of various activators and depressants, which

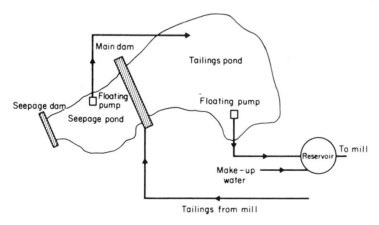

FIG. 16.10. Water-reclamation system.

must be used in controlled amounts in sequence and which may be present in the recycled water. Water treatment can usually overcome this, at little or no extra cost, because similar treatment would be required for the effluent discharge in any case.

Complexes of metals with cyanide and ammonia are especially prone to stabilisation and solubilisation in caustic solution and may require special treatment other than straightforward neutralisation by lime. Although natural degradation occurs to some extent, this is of little value in many cases during the winter months, when the tailings ponds may be ice-covered, and several processes have been developed to treat cyanide-bearing effluent.[13] Alkaline chlorination, whereby cyanide is oxidised to cyanate, has perhaps received the greatest attention,[14] but cyanides can also be effectively destroyed by oxidation with ozone[15] or hydrogen peroxide, by reactions with sulphur dioxide and air,[16] and by electrochemical treatment, ion-exchange, and volatilisation of hydrogen cyanide. In the latter method, which has been proven full-scale in the mining industry, the tailings are acidified to produce hydrogen cyanide. This is volatalised by intensive air-sparging, while simultaneously recovering the evolved gas in a lime solution for recycling. The aerated, acidified barren solution is then reneutralised to precipitate the metal ions.

The mineralogical nature of the tailings often provides natural pollution control. For instance, the presence of alkaline gangue minerals such as limestone can render metals less soluble and neutralise oxidation products. Such ores thus present less problems than sulphide ores associated with neutral-acid gangues, which oxidise to produce sulphuric acid, and apart from acidifying the water, consume dissolved oxygen as well.[17] Chemical treatment of the effluent is essential, neutralisation by lime usually being performed, which precipitates the heavy metals, and promotes flocculation as well as reducing acidity.

References

1. Down, C. G., and Stocks, J., Methods of tailings disposal, *Min. Mag.* 345 (May 1977).

2. Klohn, E. J., Current tailings dam design and construction methods, *Min. Engng.* **33,** 798 (July 1981).
3. Vick, S. G., Siting and design of tailings impoundments, *Min. Engng.* **33,** 653 (June 1981).
4. McElvain, R. E., and Cave, I., Transportation of tailings, in *Tailings Disposal Today* (ed. C. L. Aplin and G. O. Argall), Miller Freeman, San Francisco, 1973.
5. Amsden, M. P., The Ecstall Concentrator, *CIM Bulletin* **67,** 105 (May 1974).
6. Hill, J. R. C., and Nothard, W. F., The Rhodesian approach to the vegetating of slimes dams, *Jl S Afr. IMM* **74,** 197 (1973).
7. Down, C. G., and Stocks, J., Environmental problems of tailings disposal, *Min. Mag.* 25 (July 1977).
8. Shilling, R. W., and May, E. R., Case study of environmental impact—Flambeau project, *Min. Cong. J.* **63,** 39 (1977).
9. Anon., Air and water pollution controls, *Eng. & Min. J.* **181,** 156 (June 1980).
10. Kealy, C. D., Safe design for metal tailings dams, *Min. Cong. J.* **59,** 51 (1973).
11. Scott, J. S., The development of national wastewater regulations and guidelines for the mining industry, *CIM Bull.* 15 (Feb. 1976).
12. Pickett, D. E. and Joe, E. G., *Waste Recycling Experience in Canadian Mills*, Report MPI(P) 73–14, Mines Branch, Ottawa (1973).
13. Scott, J. S., and Ingles, J. C., Removal of cyanide from gold mill effluents, *Can. Min. J.* **102,** 57 (March 1981).
14. Eccles, A. G., Pollution control at Western Mines Myra Falls operations, *CIM Bulletin* 141 (Sept. 1977).
15. Jeffries, L. F., and Tczap, A., Homestake's Grizzly Gulch tailings disposal project, *Min. Cong. J.* 23 (Nov. 1978).
16. Lewis, A., New Inco Tech process attacks toxic cyanides, *Engng. & Min. J.*, **185,** 52 (July 1984).
17. Down, C. G. and Stocks, J., *Environmental Impact of Mining*, Applied Science Publishers, London 1977.

APPENDIX I

METALLIC ORE MINERALS

Metal	Main applications	Ore minerals	Formula	% metal	Sp. gr.	Occurrence/associations
ALUMINIUM	Where requirements are lightness, high electrical and thermal conductivity, corrosion resistance, ease of fabrication. Forms high tensile strength alloys	BAUXITE Diaspore Gibbsite Boehmite	AlO(OH) Al(OH)$_3$ AlO(OH)	—	3.2–3.5 2.3–2.4 3.0–3.1	Bauxite, which occurs massive, is a mixture of minerals such as diaspore, gibbsite, and boehmite with iron oxides and silica. Occurs as residual earth from weathering and leaching of rocks in tropical climates
ANTIMONY	Flame resistant properties of oxide used in textiles, fibres, and other materials. Alloyed with lead to increase strength for accumulator plates, sheet and pipe. Important alloying element for bearing and type metals	STIBNITE	Sb$_2$S$_3$	71.8	4.5–4.6	Main ore mineral. Commonly in quartz grains and in limestone replacements. Associates with galena, pyrite, realgar, orpiment, and cinnabar

Element	Uses	Mineral	Formula		Density	Occurrence
ARSENIC	Limited use in industry. Small amounts alloyed with copper and lead to toughen the metals. In oxide form, used as insecticide	Arsenopyrite	FeAsS	46.0	5.9-6.2	Widely distributed in mineral veins, with tin ores, tungsten, gold, and silver, sphalerite and pyrite. Since production of metal is in excess of demand, is commonly regarded as gangue
		Realgar	AsS	70.1	3.5	Often associate in mineral veins in minor amounts
		Orpiment	As_2S_3	61	3.4-3.5	
BERYLLIUM	Up to 4% Be alloyed with copper to produce high tensile alloys with high fatigue, wear, and corrosion resistance, which are used to make springs, bearings, and valves, and spark-proof tools. Used for neutron absorption in nuclear industry. Used in electronics for speakers and styli	Beryl	$Be_3Al_2Si_6O_{18}$	5	2.6-2.8	Only source of the metal. Often mined as gemstone—emerald, aquamarine. Commonly occurs as accessory mineral in coarse-grained granites (pegmatites) and other similar rocks. Also in calcite veins and mica schists. As similar density to gangue minerals; difficult to separate other than by hand-sorting
BISMUTH	Pharmaceuticals; low melting point alloys for automatic safety devices, such as fire-sprinklers. Improves	Native	Bi	100	9.7-9.8	Minor amounts in veins associated with silver, lead, zinc, and tin ores
		Bismuthinite	Bi_2S_3	81.2	6.8	Occurs in association with magnetite, pyrite,

Metal	Main applications	Ore minerals	Formula	% metal	Sp. gr.	Occurrence/associations
	casting properties when alloyed with tin and lead					chalcopyrite, galena, and sphalerite, and with tin and tungsten ores. Majority of bismuth produced as by-product from smelting and refining of lead and copper
CADMIUM	Rust-proofing of steel, copper, and brass by electroplating and spraying; production of pigments; negative plate in alkali accumulators; plastic stabilisers	Greenockite	CdS	77.7	4.9–5.0	Found in association with lead and zinc ores, and in very small quantities with many other minerals. Due to volatility of the metal, mainly produced during smelting and refining of zinc, as a by-product
CAESIUM	Low ionization potential utilised in photoelectric cells, photomultiplier tubes, spectrophotometers, infra-red detectors. Minor pharmaceutical use	POLLUCITE	$Cs_4Al_4Si_9$ $O_{26}.H_2O$	10.0	2.9	Occurs in pegmatites of complex mineralogical character. Rare mineral.
		Lepidolite (Lithium mica)	$K(Li,Al)_3(Si,Al)_4$ $O_{10}(OH,F)_2$	—	2.8–2.9	Occurs in pegmatites, often in association with tourmaline and spodumene. Often carries traces of rubidium and caesium
CHROMIUM	Used mainly as alloying element in steels to give resistance to wear,	CHROMITE	$FeCr_2O_4$	46.2	4.1–5.1	Occurs in olivine and serpentine rocks, often concentrated

Element	Minerals	Formula	Metal %	Sp. gr.	Uses	Occurrence
CHROMIUM					corrosion, heat, and to increase hardness and toughness. Used for electroplating iron and steel. Chromite used as refractory with neutral characteristics. Used in production of bichromates and other salts in tanning, dyeing, and pigments	sufficiently into layers or lenses to be worked. Due to its durability, it is sometimes found in alluvial sands and gravels
COBALT	Smaltite Cobaltite	$CoAs_2$ $CoAsS$	28.2 35.5	5.7–6.8 6.0–6.3	Used as alloying element for production of high-temperature steels and magnetic alloys. Used as catalyst in chemical industry. Cobalt powder used as cement in sintered carbide cutting tools	Smaltite and cobaltite occur in veins, often together, with arsenopyrite, silver, calcite, and nickel minerals
	Carrolite Linnaeite	$CuCo_2S_4$ Co_3S_4	20.5 58	4.8–5.0 4.8–5.0		Carrolite and linnaeite sometimes occur in small amounts in copper ores. Cobalt is usually only minor constituent in ores such as lead, copper, and nickel and extracted as by-product
COPPER	CHALCOPYRITE	$CuFeS_2$	34.6	4.1–4.3	Used where high electrical or thermal conductivity important. High corrosion resistance and easy to fabricate.	Main ore mineral. Most often in veins with other sulphides, such as galena, sphalerite, pyrrhotite, pyrite, and also cassiterite.

Metal	Main applications	Ore minerals	Formula	% metal	Sp. gr.	Occurrence/associations
	Used in variety of alloys—brasses, bronzes, aluminium bronzes, etc.					Common gangue minerals quartz, calcite, dolomite. Disseminated with bornite and pyrite in porphyry copper deposits
		CHALCOCITE	Cu_2S	79.8	5.5–5.8	Often associated with cuprite and native copper
		BORNITE	Cu_5FeS_4	63.3	4.9–5.4	Associates with chalcopyrite and chalcocite in veins
		COVELLITE	CuS	66.5	4.6	Sometimes as primary sulphide in veins, but more commonly as secondary sulphide with chalcopyrite, chalcocite, and bornite
		CUPRITE	Cu_2O	88.8	5.9–6.2	Found in oxidised zone of deposits, with malachite, azurite, and chalcocite
		MALACHITE	$CuCO_3.Cu(OH)_2$	57.5	4.0	Frequently associated with azurite, native copper, and cuprite in oxidised zone
		Native	Cu	100	8.9	Occurs in small amounts with other copper minerals
		Tennantite	$Cu_8As_2S_7$ (variable)	57.5 (variable)	4.4–4.5	Tennantite and tetrahedrite found in

Element	Mineral	Formula	% metal	Density	Uses	Occurrence
COPPER	Tetrahedrite	$4Cu_2S.Sb_2S_3$	52.1	4.4–5.1		veins with silver, copper, lead, and zinc minerals. Tetrahedrite more widespread and common in lead–silver veins
	Azurite	$2CuCO_3.Cu(OH)_2$	55	3.8–3.9		Occurs in oxidised zone. Not as widespread as malachite
	Enargite	$Cu_3A_5S_4$	48.4	4.4		Associates with chalcocite, bornite, covellite, pyrite, sphalerite, tetrahedrite, baryte, and quartz in near-surface deposits
GALLIUM	Occurs in some zinc ores, but no important ore minerals	—	—	—	Electronics industry for production of light-emitting diodes. Used in electronic memories for computers	About 90% of production is a direct by-product of alumina output. Also found in coal ash and flue dusts
GERMANIUM	Argyrodite	$3Ag_2S.GeS_2$	8.3	6.1	Electronics industry	Occurs with sphalerite, siderite, and marcasite. No important ore minerals. Chief source is cadmium fume from sintering zinc concentrates
GOLD	NATIVE	Au	85–100 (invariably alloyed with silver and copper, and	12–20	Jewellery, monetary use, electronics, dentistry, decorative plating	Disseminated in quartz grains, often with pyrite, chalcopyrite, galena, stibnite, and arsenopyrite. Also

Metal	Main applications	Ore minerals	Formula	% metal	Sp. gr.	Occurrence/associations
				other metals)		found alluvially in stream or other sediments. South African "banket" is consolidated alluvial deposit
		Sylvanite	$(AuAg)Te_2$	24.5	7.9–8.3	Tellurides occurring in Kalgoorlie gold ores of Western Australia
		Calaverite	$AuTe_2$	43.6	9.0	
HAFNIUM	Naval nuclear reactors, flashbulbs, ceramics, refractory alloys, and enamels	No ore minerals		—	—	Produced as co-product of zirconium sponge
INDIUM	Electronics, component of low-melting-point alloys and solders, protective coating on silverware and jewellery	Occurs as trace element in many ores		—	—	Recovered from residues and flue dusts from some zinc smelters
IRON	Iron and steel industry	HEMATITE	Fe_2O_3	70	5–6	Most important iron ore. Occurs in igneous rocks and veins. Also as ooliths or cementing material in sedimentary rocks
		MAGNETITE	Fe_3O_4	72.4	5.5–6.5	The only ferromagnetic mineral. Widely distributed in several environments,

IRON

Mineral	Formula	%	S.G.	Occurrence
				including igneous and metamorphic rocks; and beach-sand deposits
Goethite	$Fe_2O_3.H_2O$	62.9	4.0–4.4	Widespread occurrence, associated with hematite and limonite
Limonite	Hydrous ferric oxides	Variable 48–63	3.6–4.0	Natural rust, chief constituent being goethite. Often associates with hematite in weathered deposits
Siderite	$FeCO_3$	48.3	3.7–3.9	Occurs massive in sedimentary rocks and as gangue mineral in veins carrying pyrite, chalcopyrite, galena
Pyrrhotite	FeS	61.5 (variable)	4.6	The only magnetic sulphide mineral. Occurs disseminated in igneous rocks, commonly with pyrite, chalcopyrite, and pentlandite
Pyrite	FeS_2	46.7	4.9–5.2	One of most widely distributed sulphide minerals. Used for production of sulphuric acid, but often regarded as gangue

Metal	Main applications	Ore minerals	Formula	% metal	Sp. gr.	Occurrence/associations
LEAD	Batteries, corrosion resistant pipes and linings, alloys, pigments, radiation shielding	GALENA	PbS	86.6	7.4–7.6	Very widely distributed, and most important lead ore. Occurs in veins, often with sphalerite, pyrite, chalcopyrite, tetrahedrite, and gangue minerals such as quartz, calcite, dolomite, baryte, and fluoride. Also in pegmatites, and as replacement bodies in limestone and dolomite rocks, with garnets, feldspar, diopside, rhodonite, and biotite. Often contains up to 0.5% Ag, and is important source of this metal
		Cerussite	$PbCO_3$	77.5	6.5–6.6	In oxidised zone of lead veins, associated with galena, anglesite, smithsonite, and sphalerite
		Anglesite	$PbSO_4$	68.3	6.1–6.4	Occurs in oxidation zone of lead veins
		Jamesonite	$Pb_4FeSb_6S_{14}$	50.8	5.5–6.0	Occurs in veins with

galena, sphalerite, pyrite, stibnite

Element	Uses	Mineral	Composition	% metal	Sp. gr.	Remarks
LITHIUM	Lightest metal. Lithium carbide used in production of aluminium. Used as base in multipurpose greases; used in manufacture of lithium batteries. Large application in ceramics industry. Very little use in metallic form	SPODUMENE	$LiAlSi_2O_6$	3.7	3.0–3.2	Occurs in pegmatites with lepidolite, tourmaline, and beryl
		Amblygonite	$2LiF.Al_2O_3.P_2O_5$	4.7	3.0–3.1	Rare mineral occurring in pegmatites with other lithium minerals
		Lepidolite	$LiF.KF.Al_2O_3.3SiO_2$	1.9	2.8–3.3	Mica occurring in pegmatites with other lithium minerals
		Tourmaline	Complex borosilicate of Al, Na, Mg, Fe, Li, Mn	—	3.0–3.2	Not a commercial source of metal. Some crystals used as gems. Occurs in granite pegmatites, schists, and gneisses
MAGNESIUM	Small amounts used in aluminium alloys to increase strength and corrosion resistance. Used to desulphur blast-furnace iron. Added to cast-iron to produce nodular iron. Used in cathodic protection, as a reagent in petrol processing and as reducing agent in titanium, and zirconium production. Structural uses where	Dolomite	$MgCa(CO_3)_2$	13	2.8–2.9	Most magnesium extracted from brine, rather than ore minerals. Mineral used in manufacture of refractories. Occurs as gangue mineral in veins with galena and sphalerite. Also occurs widely as rock-forming mineral
		Magnesite	$MgCO_3$	29	3.0–3.2	Used mainly for cement and refractory bricks. Often associates with serpentine

Metal	Main applications	Ore minerals	Formula	% metal	Sp. gr.	Occurrence/associations
	lightness required—magnesium die-castings	Carnallite	$KMgCl_3.6H_2O$	9	1.6	Occurs with halite and sylvine
		Brucite	$Mg(OH)_2$	42	2.4	Occurs in dolomitic limestones and veins with talc, calcite, and in serpentine
MANGANESE	Very important ferro-alloy. About 95% of output used in steel and foundry industry. Balance mainly in manufacture of dry cells and chemicals	PYROLUSITE	MnO_2	63.2	4.5–5.0	Often found in oxidised zone of ore deposits containing manganese. Also in quartz veins and manganese nodules
		Manganite	Mn_2O_3	62.5	4.2–4.4	Occurs in association with baryte, pyrolusite, and goethite and in veins in granite
		Braunite	$3Mn_2O_3.MnSiO_3$	78.3	4.7–4.8	Occurs in veins with other manganese minerals
		Psilomelane	Mixture of Mn oxides	—	3.3–4.7	Found with pyrolusite and limonite in sediments or quartz veins
MERCURY	Electrical apparatus, scientific instruments, manufacture of paint, electrolytic cells, solvent for gold, manufacture of drugs and chemicals	CINNABAR	HgS	86.2	8.0–8.2	Only important mercury mineral. Occurs in fractures in sedimentary rocks with pyrite, stibnite, and realgar. Common gangue minerals are

Metal / Uses	Mineral	Formula	%	Density	Occurrence
MOLYBDENUM Main use as ferro-alloy. Metal also used in manufacture of electrodes and furnace parts. Also used as catalyst corrosion inhibitor, additive to lubricants	MOLYBDENITE	MoS_2	60	4.7–4.8	quartz, calcite, baryte, and chalcedony. Widely distributed in small quantities. Occurs in granites and pegmatites with wolfram and cassiterite
	Wulfenite	$PbMoO_4$	26.2	6.5–7.0	Found in oxidised zone of lead and molybdenum ores. Commonly with anglesite, cerrusite, and vanadinite
NICKEL Very important ferro-alloy due to its high corrosion resistance (stainless steels). Also alloyed with many non-ferrous metals—chromium, aluminium, manganese. Used for electroplating steels, as base for chromium plate. Pure metal corrosion resistant, and resists alkali attack. Is non-toxic and used for food handling and pharmaceutical equipment	PENTLANDITE	$(FeNi)S$	22.0	4.6–5.0	Occurs invariably with chalcopyrite, and often intergrown with pyrrhotite, millerite, cobalt, selenium, silver, and platinum metals
	GARNIERITE	Hydrated Ni-Mg silicate	25–30	2.4	Often occur massive or earthy, in decomposed serpentines, often with chromium ores, deposits being known as "lateritic"
	Niccolite	$NiAs$	44.1	7.3–7.7	Occurs in igneous rocks with chalcopyrite, pyrrhotite, and nickel sulphides. Also in veins with silver, silver–arsenic, and cobalt minerals

Metal	Main applications	Ore minerals	Formula	% metal	Sp. gr.	Occurrence/associations
		Millerite	NiS	64.8	5.3–5.7	Occurs as needle-like radiating crystals in cavities and as replacement of other nickel minerals. Also in veins with other nickel minerals and sulphides
NIOBIUM (Columbium)	Important ferro-alloy. Added to austenitic stainless steels to inhibit intergranular corrosion at high temperatures	PYROCHLORE (Microlite)	$(Ca,Na)_2(Nb,Ta)_2$ $O_6(O,OH,F)$	—	4.2–6.4	Occurs in pegmatites associated with zircon and apatite. Pyrochlore is name given to niobium-rich minerals, and microlite to tantalum-rich minerals
		COLUMBITE (Tantalite)	$(Fe,Mn)(Nb,Ta)_2O_6$	—	5.0–8.0	In granite pegmatites with cassiterite wolframite, spodumene tourmaline, feldspar, and quartz. Columbite is name given to niobium-rich, and tantalite to tantalum rich-minerals in series
PLATINUM GROUP (Platinum Palladium	Platinum and palladium have wide use in jewellery and dentistry. Platinum, due to its					Platinum group metals occur together in nature as native metals or alloys

NATIVE PLATINUM	Pt	45–86	21.5 (pure)	Platinum alloyed with other platinum group metals, iron, and copper. Occurs disseminated in igneous rocks, associates with chromite, and copper ores. Found in lode and alluvial deposits	high melting point and corrosion resistance, is widely used for electrical contact material and in manufacture of chemical crucibles, etc. Also wide use as a catalyst. Iridium is also used in jewellery and in dental alloys and in electrical industry. Long life platinum–iridium electrodes used in helicopter spark-plugs. Rhodium used in thermocouples, and platinum–palladium–rhodium catalysts are used in control of automobile emissions. Osmium, the heaviest metal known, with a melting point of 2200°C, and ruthenium, have little commercial importance
PLATINUM GROUP (Osmium, Iridium, Rhodium, Ruthenium)					
SPERRYLITE	$PtAs_2$	56.6	10.6	Occurs in pyrrhotite deposits and in gold-quartz veins. Also with covellite and limonite	
Osmiridium	Alloy of Os–Ir	—	19.3–21.1	Found in small amounts in some gold and platinum ores, where it is recovered as by-product	
RADIUM	See URANIUM MINERALS	—	—	Constituent of uranium minerals	Industrial radiography, treatment of cancer, and production of luminous paint

Metal	Main applications	Ore minerals	Formula	% metal	Sp. gr.	Occurrence/associations
RARE EARTHS	The cerium subgroup is the most important industrially. Rare earths used as catalysts in petroleum refining, iron–cerium alloys used as cigarette-lighter flints. Used in ceramics and glass industry and in production of colour televisions	MONAZITE BASTNAESITE	Rare earth and thorium phosphate (Ce,La)(CO$_3$)F	—	4.9–5.2	*See* Thorium Often found in pegmatites, veins and carbonatite plutons
		Xenotime	YPO$_4$	48.4 (Y)	4.4–5.1	Source of yttrium. Wide occurrence as accessory mineral, often in pegmatites, and alluvial deposits, associated with monazite, zircon, rutile, ilmenite, and feldspars
RHENIUM	Used as catalyst in production of low-lead petrol. Used as catalyst with platinum. Used extensively in thermocouples, temperature controls, and heating elements. Also used as filaments in electronic apparatus	Molybdenite	MoS$_2$	—	4.7–4.8	Rhenium occurs associated with molybdenite in porphyry copper deposits, and is recovered as by-product
RUBIDIUM	Rubidium and caesium largely interchangeable in properties and uses, although latter usually preferred to meet present small industrial demand	*See* CAESIUM				Rubidium widely dispersed as minor constituent in major caesium minerals

Element	Uses	Variety	Formula	%	S.G.	Remarks
SILICON	Used in steel industry and as heavy medium alloy as ferro-silicon. Also used to de-oxidise steels. Metal used as semi-conductor	QUARTZ	SiO_2	46.9	2.65	Commonest mineral, forming 12% of earth's crust. Essential constituent of many rocks, such as granite and sandstone, and virtually sole constituent of quartzite rock
SELENIUM	Used in manufacture of fade-resistant pigments, photo-electric apparatus, in glass production, and various chemical applications. Alloyed with copper and steel to improve machineability	Naumanite Clausthalite Eucairite Berzelianite	Ag_2Se $PbSe$ $(AgCu)_2Se$ Cu_2Se	26.8 27.6 18.7 38.3	8.0 8.0 7.5 6.7	Selenides occur associated with sulphides, and bulk of selenium recovered as by-product from copper sulphide ores
SILVER	Sterling ware, jewellery, coinage, photographic and electronic products, mirrors, electroplate, and batteries	ARGENTITE	Ag_2S	87.1	7.2–7.4	Closely associated with lead, zinc, and copper ores, and bulk of silver produced as by-product from smelting such ores
		Native	Ag	100 (max.)	10.1–11.1	Usually alloyed with copper, gold, etc., and occurs in upper part of silver sulphide deposits
		Cerargyrite	$AgCl$	75.3	5.8	Occurs in upper parts of silver veins together with native silver and cerussite

Metal	Main applications	Ore minerals	Formula	% metal	Sp. gr.	Occurrence/associations
TANTALUM	Used in certain chemical and electrical processes due to extremely high corrosion resistance. Used in production of special steels used for medical instruments. Used for electrodes, and tantalum carbide used for cutting tools. Used in manufacture of capacitors	PYROCHLORE TANTALITE	See NIOBIUM			As well as ore minerals, certain tin slags becoming important source of tantalum
TELLURIUM	Used in production of free machining steels, in copper alloys, rubber production, and as catalyst in synthetic fibre production	Sylvanite Calaverite	} See GOLD			Produced with selenium as by-product of copper refining. These metal tellurides, which are important gold ores, and other tellurides of bismuth and lead, are most important sources of tellurium
THALLIUM	Very poisonous, and finds limited outlet as fungicide and rat poison. Thallium salts used in Clerici solution, an important heavy liquid	Occurs in some zinc ores, but no ore minerals	—	—	—	By-product of zinc refining

Element	Mineral	Formula	%	S.G.	Occurrence	Uses
THORIUM	MONAZITE	$(Ce,La,Th)PO_4$	—	4.9–5.4	Although occurring in lode deposits in igneous rocks such as granites, the main deposits are alluvial, beach-sand deposits being most prolific source. Occurs associated with ilmenite, rutile, zircon, garnets, etc.	Radioactive metal. Used in electrical apparatus, and in magnesium–thorium and other thorium alloys. Oxide of importance in manufacture of gas-mantles, and used in medicine
	Thorianite	$ThO_2.U_3O_8$	21	9.3	Occurs in some beach-sand deposits	
TIN	CASSITERITE	SnO_2	78.6	6.8–7.1	Found in lode and alluvial deposits. Lode deposits in association with wolfram, arsenopyrite, copper, and iron minerals. Alluvially, often associated with ilmenite, monazite, zircon, etc.	Main use in manufacture of tin-plate, for production of cans, etc. Important alloy in production of solders, bearing-metals, bronze, type-metal, pewter, etc.
TITANIUM	ILMENITE	$FeTiO_3$	31.6	4.5–5.0	Accessory mineral in igneous rocks especially gabbros and norites. Economically concentrated into alluvial sands, together with rutile, monazite zircon	Due to its high strength and corrosion resistance, about 80% of titanium produced used in aircraft and aerospace industries. Also used in power-station heat-exchanger

Metal	Main applications	Ore minerals	Formula	% metal	Sp. gr.	Occurrence/associations
	tubing and in chemical and desalination plants	RUTILE	TiO_2	60	4.2	Accessory mineral in igneous rocks, but economic deposits found in alluvial beach-sand deposits
TUNGSTEN	Production of tungsten carbide for cutting, drilling, and wear-resistant applications. Used in lamp filaments, electronic parts, electrical contacts, etc. Important ferro-alloy, producing tool and high-speed steels	WOLFRAM	$(Fe,Mn)WO_4$	50	7.1–7.9	Occurs in veins in granite rocks, with minerals such as cassiterite, arsenopyrite, tourmaline, galena, sphalerite, scheelite, and quartz. Also found in some alluvial deposits
		SCHEELITE	$CaWO_4$	63.9	5.9–6.1	Occurs under same conditions as wolfram. Also occurs in contact metamorphic deposits
URANIUM	Nuclear fuel	PITCHBLENDE (URANINITE)	UO_2 (variable-partly oxidised to U_3O_8)	80–90	8–10	Most important uranium and radium ore. Occurs in veins with tin, copper, lead, and arsenic sulphides, and radium
		Carnotite	$K_2(UO_2)_2(VO_4)_2.3H_2O$ (approx.)	Variable	4–5	Secondary mineral found in sedimentary rocks, also in pitchblende deposits. Source of radium

Metal	Mineral	Formula	% metal	S.G.	Remarks	Uses
URANIUM	Autunite	$Ca(UO_2)_2(PO_4)_2.10–12H_2O$	49	3.1	Occur together in oxidised zones as	
	Torbernite	$Cu(UO_2)_2(PO_4)_2.12H_2O$	48	3.5	secondary products from other uranium minerals	
VANADIUM	PATRONITE	VS_4 (approx.)	28.5	—	Ocurs with nickel and molybdenum sulphides and asphaltic material	Important ferro-alloy. Vanadium used in manufacture of special steels, such as high-speed tool steels. Increases strength of structural steels—used for oil and gas pipelines. Vanadium–aluminium master alloys used in preparation of some titanium-based alloys. Vanadium compounds used in chemical and oil industries as catalysts. Also used as glass-colouring agent and in ceramics
	CARNOTITE	*See* URANIUM	Variable	4–5	*See* URANIUM	
	Roscoelite (Vanadium mica)	$H_8K(MgFe)(AlV)_4(SiO_3)_{12}$	Variable	2.9	Frequently with carnotite	
	Vanadinite	$(PbCl)Pb_4(PO_4)_3$	Variable	6.6–7.1	Occurs in oxidation zone of lead, and lead–zinc deposits. Also with other vanadium minerals in sediments	
ZINC	SPHALERITE	ZnS	67.1	3.9–4.1	Most common zinc ore mineral, frequently associated with galena, and copper sulphides in vein deposits. Also occurs in limestone replacements, with	Corrosion protective coatings on iron and steel ("galvanising"). Important alloying metal in brasses and zinc die-castings. Used to manufacture

Metal	Main applications	Ore minerals	Formula	% metal	Sp. gr.	Occurrence/associations
	corrosion-resistance paints, pigments, fillers, etc.					pyrite, pyrrhotite, and magnetite
		Smithsonite (Calamine)	$ZnCO_3$	52	4.3–4.5	Mainly occurs in oxidised zone of ore deposits carrying zinc minerals. Commonly associated with sphalerite, galena, and calcite
		Hemimorphite (Calamine)	$Zn_4Si_2O_7(OH)_2.H_2O$	54.3	3.4–3.5	Found associated with smithsonite accompanying the sulphides of zinc, iron, and lead
		Marmatite	$(Zn,Fe)S$	46.5–56.9	3.9–4.2	Found in close association with galena
		Franklinite Zincite Willemite	Oxide of Fe, Zn, Mn ZnO Zn_2SiO_4	Variable 80.3 58.5	5.0–5.2 5.4–5.7 4.0–4.1	Franklinite, zincite, and willemite occur together in a contact metamorphic deposit at Franklin, New Jersey, in a crystallite limestone, where the deposit is worked for zinc and manganese
ZIRCONIUM	Used, alloyed with iron, silicon, and tungsten, in nuclear reactors, and for removing	ZIRCON	$ZrSiO_4$	49.8	4.6–4.7	Widely distributed in igneous rocks, such as granites. Common constituent of residues

of various sedimentary
rocks, and occurs in
beach sands associated
with ilmenite, rutile,
and monazite

oxides and nitrides
from steel. Used in
corrosion-resistant
equipment in chemical
plants

MPT—X

APPENDIX II

COMMON NON-METALLIC ORES

Material	Uses	Main ore minerals	Formula	Sp. gr.	Occurrence
ANHYDRITE	Increasing importance as a fertilizer, and in manufacture of plasters, cements, sulphates, and sulphuric acid	ANHYDRITE	$CaSO_4$	2.95	Occurs with gypsum and halite as a saline residue. Occurs also in "cap rock" above salt domes, and as minor gangue mineral in hydrothermal metallic ore veins
APATITE	See PHOSPHATES				
ASBESTOS	Heat resistance materials, such as fire-proof fabrics and brake-linings. Also asbestos cement products, sheets for roofing and cladding, fire-proof paints, etc.	CHRYSOTILE (Serpentised asbestos)	$Mg_3Si_2O_5(OH)_4$	2.5–2.6	Fibrous serpentine occurring as small veins in massive serpentine
		CROCIDOLITE	$Na_2(Mg,Fe,Al)_5Si_8O_{22}(OH)_2$	3.4	Fibrous riebeckite, or blue asbestos, occurring as veins in bedded ironstones
		AMOSITE	$(Mg,Fe)_7Si_8O_{22}(OH)_2$	~3.2	Fibrous anthophyllite, occurring as long fibres in certain metamorphic rocks
		ACTINOLITE	$Ca_2(Mg,Fe)_5Si_8O_{22}(OH)_2$	3.0–3.4	True asbestos, occurring in schists and in some igneous rocks as alteration product of pyroxene

Group	Mineral	Uses	Formula	S.G.	Occurrence
BADDELEYITE	BADDELEYITE	Ceramics, abrasives, refractories, polishing powders, and manufacture of zirconium chemicals	ZrO_2	5.4–6.0	Mainly found in gravels with zircon, tourmaline, corundum, ilmenite, and rare-earth minerals
BARYTES	BARYTE	Main use in oil- and gas-well drilling industry in finely ground state as weighting agent in drilling muds. Also in manufacture of barium chemicals, and as filler and extender in paint and rubber industries	$BaSO_4$	4.5	Most common barium mineral, occurring in vein deposits as gangue mineral with ores of lead, copper, zinc, together with fluorite, calcite, and quartz. Also as replacement deposit of limestone and in sedimentary deposits
BORATES	BORAX	Used in manufacture of insulating fibreglass, as fluxes for manufacture of glasses and enamels. Borax used in soap and glue industries, in cloth manufacture and tanning. Also used as preservatives, antiseptics, and in paint driers	$Na_2B_4O_7.10H_2O$	1.7	An evaporite mineral, precipitated by the evaporation of water in saline lakes, together with halite, sulphates, carbonates, and other borates in arid regions
	KERNITE		$Na_2B_4O_7.4H_2O$	1.95	Very important source of borates. Occurrence as borax
	COLEMANITE		$Ca_2B_6O_{11}.5H_2O$	2.4	In association with borax, but principally as a lining to cavities in sedimentary rocks
	ULEXITE		$NaCaB_5O_9.8H_2O$	1.9	Occurs with borax in lake deposits. Also with gypsum and rock salt
	SASSOLINE		H_3BO_3	1.48	Occurs with sulphur in volcanoes and in hot lakes and lagoons

Material	Uses	Main ore minerals	Formula	Sp. gr.	Occurrence
		BORACITE	$Mg_3B_7O_{13}Cl$	2.95	Occurs in saline deposits with rock-salt, gypsum, and anhydrite
CALCIUM CARBONATE	Many uses according to purity and character. Clayey variety used for cement, purer variety for lime. Marble for building and ornamental stones. Used as smelting flux, and in printing processes. Chalk and lime applied to soil as dressing. Transparent calcite (Iceland spar), used in construction of optical apparatus	CALCITE	$CaCO_3$	2.7	Calcite is common and widely distributed mineral, often occuring in veins, either as main constituent, or as gangue mineral with metallic ores. It is a rock-forming mineral, which is mainly quarried as the sedimentary rocks limestone and chalk, and metamorphic rock marble
CHINA CLAY	Manufacture of porcelain and china. Used as filler in manufacture of paper, rubber, and paint	KAOLINITE	$Al_2Si_2O_5(OH)_4$	2.6	A secondary mineral produced by the alteration of aluminous silicates, and particularly of alkali feldspars
CHROMITE	Used as refractory in steel-making furnaces	CHROMITE	*See* CHROMIUM MINERALS (Appendix I)		
CORUNDUM	Abrasive. Next to diamond, is hardest known mineral. Coloured variety used as gemstones	CORUNDUM (Emery)	Al_2O_3	3.9–4.1	Occurs in several ways. Original constituent of various igneous rocks, such as syenite. Also in metamorphic rocks such

				Occurrence	Uses
				as marble, gneiss, and schist. Occurs also in pegmatites and in alluvial deposits. Impure form is emery, containing much magnetic and hematite	
CRYOLITE	CRYOLITE	Na_3AlF_6	3.0	Occurs in pegmatite veins in granite with siderite, quartz, galena, sphalerite, chalcopyrite, fluorite, cassiterite, and other minerals. Only real deposit in Greenland	Used as flux in manufacture of aluminium by electrolysis
DIAMOND	DIAMOND (Bort)	C	3.5	Distributed sporadically in kimberlite pipes. Also in alluvial beach and river deposits. Bort is grey to black and opaque, and is used industrially	Gemstone. Used extensively in industry for abrasive and cutting purposes—hardest known mineral. Used for tipping drills in mining and oil industry
DOLOMITE	DOLOMITE	$CaMg(CO_3)_2$	2.8–2.9	Rock-forming mineral. Occurs as gangue mineral in veins containing galena and sphalerite	Important building material. Also used for furnace linings and as flux in steel-making
EMERY					See CORUNDUM
EPSOM SALTS	EPSOMITE	$MgSO_4{\cdot}7H_2O$	1.7	Usually as encrusting masses on walls of caves or mine workings. Also in oxidised zone of pyrite deposits in arid regions	Medicine and tanning
FELDSPAR	ORTHOCLASE (Isomorphous forms—Microline,	$KAlSi_3O_8$	2.6	Most abundant of all minerals, and most important rock-forming	Used in manufacture of porcelain, pottery, and glass. Used in production

Material	Uses	Main ore minerals	Formula	Sp. gr.	Occurrence
FELDSPAR	of glazes on earthenware, etc., and as mild abrasive	Sanidine, and Adularia—the *potassic* feldspars)			mineral. Widely distributed, mainly in igneous, but also in metamorphic and sedimentary rocks
		ALBITE	$NaAlSi_3O_8$	2.6	
		ANORTHITE	$CaAl_2Si_2O_8$	2.74	
		(*Plagioclase* feldspars form series having formula ranging from $NaAlSi_3O_8$ to $CaAl_2Si_2O_8$, changing progressively from Albite, through oligoclase, andesine, labradorite, and bytownite to anorthite)			
FLUORSPAR	Mainly as flux in steelmaking. Also for manufacture of specialised optical equipment, production of hydrofluoric acid, and fluorocarbons for aerosols. Colour-banded variety known as *Blue-John* used as semi-precious stone	FLUORITE	CaF_2	3.2	Widely distributed, in hydrothermal veins and replacement deposits, either alone, or with galena, sphalerite, barytes, calcite, and other minerals
GARNET	Mainly as abrasive for sandblasting of aircraft components, and for wood polishing. Also certain varieties used as gemstones	PYROPE	$Mg_3Al_2(SiO_4)_3$	3.7	Widely distributed in metamorphic and some igneous rocks. Also commonly found as constituent of beach and river deposits
		ALMANDINE	$Fe_3Al_2(SiO_4)_3$	4.0	
		GROSSULAR	$Ca_3Al_2(SiO_4)_3$	3.5	
		ANDRADITE	$Ca_3Fe_2(SiO_4)_3$	3.8	
		Spessartite	$Mn_3Al_2(SiO_4)_3$	4.2	
		Uvarovite	$Ca_3Cr_2(SiO_4)_3$	3.4	
GRAPHITE (Plumbago)	Manufacture of foundry moulds, crucibles, and paint; used as lubricant	GRAPHITE	C	2.1–2.3	Occurs as disseminated flakes in metamorphic rocks derived from rocks

Mineral	Uses	Composition	S.G.	Occurrence
	and as electric furnace electrodes			with appreciable carbon content. Also as veins in igneous rocks and pegmatites
GYPSUM	Used in cement manufacture, as a fertiliser, and as filler in various materials such as paper, rubber, etc. Used to produce *plaster of Paris*	$CaSO_4.2H_2O$	2.3	Evaporate mineral, occurring with halite and anhydrite in bedded deposits
ILMENITE	About 90% of ilmenite produced is used for manufacture of titanium dioxide, a pigment used in pottery manufacture	*See* TITANIUM MINERALS (Appendix I)		
MAGNESITE	Used as refractory for steel furnace linings, and in production of carbon dioxide and magnesium salts	*See* MAGNESIUM MINERALS (Appendix I)		
MICA	Used for insulating purposes in electrical apparatus. Ground mica used in production of roofing material, and in lubricants, wall-finishes artificial stone, etc. Powdered mica gives "frost" effect on Christmas cards and decorations	MUSCOVITE $KAl_2(AlSi_3O_{10})(OH,F)_2$	2.8–2.9	Widely distributed in igneous rocks, such as granite and pegmatites. Also in metamorphic rocks—gneisses and schists. Also in sedimentary sandstones, clays, etc.
		PHLOGOPITE $KMg_3(AlSi_3O_{10})(OH,F)_2$	2.8–2.85	Most commonly in metamorphosed limestones, also in igneous rocks rich in magnesia

Material	Uses	Main ore minerals	Formula	Sp. gr.	Occurrence
MICA		Biotite	$K(Mg,Fe)_3(AlSi_3O_{10})(OH,F)_2$	2.7–3.3	Widely distributed in granite, syenite and diorite. Common constituent of schists and gneisses and of contact metamorphic rocks
PHOSPHATES	Main use as fertilizers. Small amounts used in production of phosphorous chemicals	APATITE	$Ca_5(PO_4)_3(F,Cl,OH)$	3.1–3.3	Occurs as accessory mineral in wide range of igneous rocks, such as pegmatites. Also in metamorphic rocks, especially metamorphosed limestones and skarns. Principal constituent of fossil bones in sedimentary rocks
		PHOSPHATE ROCK	Complex phosphates of Ca, Fe, Al		Most extensive phosphate rock deposits associated with marine sediments, typically glauconite-bearing sandstones, limestones, and shales. *Guano* is an accumulation of excrement of sea-birds, found mainly on oceanic islands
POTASH	Used as fertilisers, and source of potassium salts. Nitre also used in explosives manufacture (*saltpetre*)	SYLVINE	KCl	2.0	Occurs in bedded evaporite deposits with halite and carnallite
		CARNALLITE	$KMgCl_3.6H_2O$	1.6	In evaporite deposits with sylvine and halite

Mineral	Formula / Reference	Density	Occurrence	Uses
ALUNITE	$KAl_3(SO_4)_2(OH)_6$	2.6	Secondary mineral found in areas where volcanic rocks containing potassic feldspars have been altered by acid solutions	
NITRE	KNO_3	2.1	Occurs in soils in arid regions, associated with gypsum, halite, and nitratine	
QUARTZ	*See* SILICON MINERALS (Appendix I)			Building materials, glass making, pottery, silica bricks, ferro-silicon, etc. Used as abrasive in scouring soaps, sandpaper, toothpaste, etc. Due to its piezo-electric properties, quartz crystals widely used in electronics
ROCK SALT (HALITE)	$NaCl$	2.2	Occurs in extensive stratified evaporite deposits, formed by evaporation of land-locked seas in geological past. Associates with other water soluble minerals, such as sylvine, gypsum, and anhydrite	Culinary and preserving uses. Wide use in chemical manufacturing processes
RUTILE	*See* TITANIUM MINERALS (Appendix I)			Production of welding rod coatings, and titanium dioxide, a pigment used in pottery manufacture
SERPENTINE	$Mg_3Si_2O_5(OH)_4$	2.5–2.6	Secondary mineral formed from minerals such as	Used as building stone and other ornamental work.

Material	Uses	Main ore minerals	Formula	Sp. gr.	Occurrence
SERPENTINE	Fibrous varieties source of asbestos (see ASBESTOS)				olivine and orthopyroxene. Occurs in igneous rocks containing these minerals, but typically in serpentites, formed by alteration of olivine-bearing rocks
SILLIMANITE MINERALS (Aluminium silicates)	Raw material for high-alumina refractories, for iron and steel industry, and other metal smelters. Also used in glass industry, and as insulating porcelains for spark-plugs, etc.	KYANITE (Disthene)	Al_2SiO_5	3.5–3.7	Typically in regionally metamorphosed schists and gneisses, together with garnet, mica, and quartz. Also in pegmatites and quartz veins associated with schists and gneisses
		ANDALUSITE	Al_2SiO_5	3.1–3.2	In metamorphosed rocks of clayey composition. Also as accessory mineral in some pegmatites, with corundum, tourmaline, and topaz
		SILLIMANITE	Al_2SiO_5	3.2–3.3	Typically in schists and gneisses produced by high-grade regional metamorphism
		Mullite	$Al_6Si_2O_{13}$	3.2	Rarely found in nature, but synthetic mullite produced in many countries
SULPHUR	Production of fertilizers, sulphuric acid, insecticides, gunpowder,	NATIVE SULPHUR	S	2.0–2.1	In craters and crevices of extinct volcanoes. In sedimentary rocks, mainly

Ore	Uses	Mineral	Composition	G	Occurrence
	sulphur dioxide, etc.	Pyrite	See Iron Minerals (Appendix I)		limestone in association with gypsum. Also in cap rock of salt domes, with anhydrite, gypsum, and calcite
Talc	As filler for paints, paper, rubber, etc. Used in plasters, lubricants, toilet powder, French chalk. Massive varieties used for sinks, laboratory table tops, acid tanks, etc.	Talc	$Mg_3Si_4O_{10}(OH)_2$	2.6–2.8	Secondary mineral formed by alteration of olivine, pyroxene, and amphibole, and occurs along faults in magnesium rich rocks. Also occurs in schists, in association with actinolite. Massive talc known as *soapstone* or *steatite*
Vermiculite	Outstanding thermal and sound insulating properties, light, fire-resistant, and inert—used principally in building industry	Vermiculite	$Mg_3(Al,Si)_4O_{10}$ $(OH)_2 \cdot 4H_2O$	2.3–2.4	Occurs as an alteration product of magnesian micas, in association with carbonatites
Witherite	Source of barium salts. Small quantities used in pottery industry	Witherite	$BaCO_3$	4.3	Not of wide occurrence. Sometimes accompanies galena in hydrothermal veins, together with anglesite and baryte
Zircon Sand	Used in foundries, refractories, ceramics and abrasives, and in chemical production	Zircon	See Zirconium Minerals (Appendix I)		

BASIC COMPUTER PROGRAMS

The programs in this appendix have been written in standard BASIC, and so should be compatible with most microcomputer systems.

Each program listing is followed by a typical example. The programs have been run on a BBC Model B microcomputer, and certain print format commands, which are specific to BBC BASIC have been underlined in the listings, and explained by a REM statement. These commands can be substituted by keywords specific to any other system.

Program Index

GY   Determination of sample size by Gy's formula

```
10 REM SAMPLE SIZE BY GY FORMULA- B.WILLS 1982
20 REM FROM P.GY-SAMPLING OF PARTICULATE MATERIALS
   ELSEVIER(1979) CHAPTER 2
30PRINT:PRINTTAB(9);"GY'S SAMPLING FORMULA"
40PRINTTAB(9);"*********************"
50PRINT:PRINT:PRINT:PRINT"The program will output the minimum"
60PRINT"practical sampling weights required"
70PRINT"at each stage of sampling. The weight"
80PRINT"given is that obtained from Gy's"
90PRINT"formula multiplied by a safety factor"
100PRINT"of 2. For routine sampling, a confidence"
110PRINT"level of 95% in the results would be"
120PRINT"acceptable, but for research purposes,"
130PRINT"or where great sampling accuracy is"
140PRINT"required, then a 99% level of confidence"
150PRINT"would be required."
160PRINT:PRINT:PRINT:INPUT"  Press RETURN to start program "z$
170PRINT:PRINT"INPUT THE FOLLOWING DATA: ":PRINT
180INPUT"Level of confidence,95% or 99% (95/99)   "B$
190PRINT"Number of sampling stages,including"
200 INPUT"primary sampling   "N
210 INPUT"Approx. assay,%   "A
220INPUT"Metal content of valuable mineral,%   "Q
230INPUT"Max.acceptable relative error,%   "C
240PRINT"Approx.maximum grain size of valuable "
250INPUT"mineral,cms.   "L
260INPUT"Is sample alluvial gold ore? Y/N   "C$
270INPUT"Density of valuable mineral,kg/l   "r
280INPUT"Mean density of gangue minerals,kg/l   "t
290PRINT:PRINT:PRINT"What is the estimated size range of the"
300PRINT"material? (i.e. d/d', where d = 95%"
310PRINT"undersize, and d' = 95% oversize.)"
320PRINT:PRINT"If d/d' is >4, then input 'W'"
330PRINT:PRINT"If d/d' is >2<=4, then input 'X'"
340PRINT:PRINT"If d/d' is >1<=2, then input 'Y'"
350PRINT:PRINT"If d/d' is 1, then input 'Z'"
360PRINT"If d/d' is 1, then input 'Z'"
370PRINT:PRINT:INPUTD$
380DIM d(N):PRINT
390PRINT"Top size of bulk material (cms.)"
400INPUTd(1):PRINT:PRINT
410IF N=1 THEN 460
420FOR B=2 TO N
430PRINT"Top size at sample stage ";B;" (cms.)"
440INPUTd(B):PRINT:PRINT
450NEXT B
460 GOSUB 1350
470IF AC$="WRONG" THEN GOSUB 1550:GOTO460
480IF B$="95" THEN F=2
490a=A/Q
500IF B$="99" THEN F=3
510IFC$="N" THEN f=0.5
520IFC$="Y" THEN f=0.2
530IF D$="W" THEN g1=0.25
540IF D$="X" THEN g1=0.5
550IF D$="Y" THEN g1=0.75
560IF D$="Z" THEN g1=1
570ST=C/(100*F)
580m=(1-a)*((1-a)*r+a*t)/a
590PRINT:PRINT"SAMPLE";TAB(13);"TOP SIZE";TAB(29);"SAMPLE"
```

```
600PRINT"STAGE";TAB(16);"cms.";TAB(29);"WEIGHT"
610PRINT:PRINT
620FOR C=1 TO N
630IF C>1 THEN g=0.25:GOTO650
640g=g1
650REM LARGE SIZE DIST. AFTER CRUSHING
660d=d(C)
670S2=ST*ST/N
680 GOSUB 1320
690M=2*f*g*l*m*d*d*d/S2
700IF M>=1000 THEN M=M/1000:E$="kg":GOTO720
710IF M<1000 THEN E$="g"
720PRINTTAB(2);C;TAB(16);d;
730@%=131594:REM BBC FORMAT TO GIVE 2 DECIMAL PLACES
740PRINTTAB(29);M;E$
750@%=10: REM BBC FORMAT TO RETURN TO NORMAL OUTPUT FORMAT
760NEXT C
770PRINT:PRINT:PRINT"Press return for next part of program"
780INPUTFF$:PRINT
790PRINT"The weights obtained so far assume that"
800PRINT"there has been an equal statistical"
810PRINT"error at each stage of sampling. The"
820PRINT"amount of material required at the"
830PRINT"coarser sizes can, however, be reduced"
840PRINT"by increasing the amount of sample"
850PRINT"taken at the finer sizes. Thus the error";
860PRINT"will be reduced at the finer sizes,"
870PRINT"allowing a greater error at the coarser"
880PRINT"sizes, but maintaining a constant total"
890PRINT"sample error. In order to obtain such"
900PRINT"weights, the next part of the program"
910PRINT"allows the weight taken at any one"
920PRINT"sampling stage to be changed, the effect";
930PRINT"on the other sampling stages then being"
940PRINT"displayed. Press return after each "
950PRINT"change in order to make further changes,";
960PRINT"until satisfactory weights are obtained."
970PRINT"Re-run program if number of sampling"
980PRINT"stages is to be changed. "
990PRINT"Press RETURN to continue."
1000INPUTas$:DIM W(N)
1010DIM S(N)
1020SL=ST*ST
1030FOR Z=N TO 2 STEP -1
1040g=0.25
1050PRINT:PRINT"Suggested weight for stage ";Z;" (gms)"
1060INPUTW(Z):d=d(Z)
1070 GOSUB 1320
1080S(Z)=2*f*g*l*m*d*d*d/W(Z)
1090SL=SL-S(Z)
1100FOR Y=Z-1 TO 1 STEP -1
1110IF Y=1 THEN g=g1
1120d=d(Y)
1130 GOSUB 1320
1140S(Y)=SL/(Z-1)
1150W(Y)=2*f*g*l*m*d*d*d/S(Y)
1160NEXT Y
1170PRINT:PRINT"SAMPLE";TAB(13);"TOP SIZE";TAB(29);"SAMPLE"
1180PRINT"STAGE";TAB(16);"cms.";TAB(29);"WEIGHT"
1190PRINT:PRINT:FOR K=1 TO N
1200PRINT TAB(2);K;TAB(16);d(K);
1210IF W(K)>=1000 THEN W(K)=W(K)/1000:E$="kg":GOTO 1230
```

```
1220IF W(K)<1000 THEN E$="g"
1230@%=131594: REM BBC FORMAT TO GIVE 2 DECIMAL PLACES
1240PRINTTAB(29);W(K);E$
1250@%=10:NEXT K :REM BBC FORMAT TO RETURN TO NORMAL OUTPUT
1260IF Z=2 THEN 1290
1270INPUT"Press return for further weight change"O$
1280NEXT Z
1290INPUT"Press return to repeat procedure"a$
1300GOTO1020
1310REM SUBROUTINE
13201=(L/d)^0.5
1330IF 1>1 THEN 1=1
1340 RETURN
1350REM SUBROUTINE
1360PRINT:PRINT:PRINT"               DATA CHECK"
1370PRINT"          **********"
1380PRINT"ITEM";TAB(20)"INPUT";TAB(30)"CODE"
1390PRINT:PRINT:PRINT"Conf.level";TAB(20)B$;TAB(30)"1"
1400PRINT"No.stages";TAB(20)N;TAB(30)"2"
1410PRINT"Assay";TAB(20)A;TAB(30)"3"
1420PRINT"Metal in mineral";TAB(20)Q;TAB(30)"4"
1430PRINT"Assay error";TAB(20)C;TAB(30)"5"
1440PRINT"Grain size";TAB(20)L;TAB(30)"6"
1450PRINT"Gold ore?";TAB(20)C$;TAB(30)"7"
1460PRINT"Mineral S.G.";TAB(20)r;TAB(30)"8"
1470PRINT"Gangue S.G.";TAB(20)t;TAB(30)"9"
1480PRINT"Size range";TAB(20)D$;TAB(30)"10"
1490FOR AB=1 TO N
1500PRINT"Size at stage ";AB;TAB(20)d(AB);TAB(30)10+AB
1510NEXT AB
1520PRINT:PRINT:PRINT"Press RETURN if data correct."
1530INPUT"Input WRONG if data requires correction   "AC$
1540 RETURN
1550REM SUBROUTINE
1560PRINT:PRINT:PRINT"               DATA CORRECTION"
1570PRINT"          ***************"
1580INPUT"Input code of item to be corrected   "C1
1590IF C1=1 ORC1=7 C1=10 THEN1720
1600INPUT"Input corrected value   "AE
1610IF C1=2 THEN N=AE
1620IF C1=3 THEN A=AE
1630IF C1=4 THENQ=AE
1640IF C1=5 THEN C=AE
1650IF C1=6THENL=AE
1660IFC1=8THENr=AE
1670IFC1=9THENt=AE
1680FORAF=1 TO N
1690IFC1=10+AF THEN d(AF)=AE
1700NEXTAF
1710GOTO1760
1720INPUT"Input corrected value   "AE$
1730IF C1=1 THEN B$=AE$
1740IFC1=7 THEN C$=AE$
1750IFC1=10THEN D$=AE$
1760INPUT"Any more corrections? Y/N   "AG$
1770IF AG$="Y" THEN 1580
1780 RETURN
```

```
>RUN

           GY'S SAMPLING FORMULA
           *********************


The program will output the minimum
practical sampling weights required
at each stage of sampling. The weight
given is that obtained from Gy's
formula multiplied by a safety factor
of 2. For routine sampling, a confidence
level of 95% in the results would be
acceptable, but for research purposes,
or where great sampling accuracy is
required, then a 99% level of confidence
would be required.



           Press RETURN to start program

INPUT THE FOLLOWING DATA:

Level of confidence,95% or 99% (95/99)   95
Number of sampling stages,including
primary sampling   4
Approx. assay,%  5
Metal content of valuable mineral,%  86.6
Max.acceptable relative error,%  2
Approx.maximum grain size of valuable
mineral,cms.  0.015
Is sample alluvial gold ore? Y/N    N
Density of valuable mineral,kg/l  7.5
Mean density of gangue minerals,kg/l  2.65



What is the estimated size range of the
material? (i.e. d/d', where d = 95%
undersize, and d' = 95% oversize.)

If d/d' is >4, then input 'W'

If d/d' is >2<=4, then input 'X'

If d/d' is >1<=2, then input 'Y'

If d/d' is 1, then input 'Z'
If d/d' is 1, then input 'Z'


?W

Top size of bulk material (cms.)
?2.5


Top size at sample stage 2 (cms.)
?0.5


Top size at sample stage 3 (cms.)
```

?0.1

Top size at sample stage 4 (cms.)
?0.004

```
                DATA CHECK
                **********
ITEM                INPUT       CODE


Conf.level           95           1
No.stages             4           2
Assay                 5           3
Metal in mineral     86.6         4
Assay error           2           5
Grain size          1.5E-2        6
Gold ore?             N           7
Mineral S.G.         7.5          8
Gangue S.G.          2.65         9
Size range            W          10
Size at stage 1      2.5         11
Size at stage 2      0.5         12
Size at stage 3      0.1         13
Size at stage 4      4E-3        14
```

Press RETURN if data correct.
Input WRONG if data requires correction

```
SAMPLE          TOP SIZE        SAMPLE
STAGE             cms.          WEIGHT


    1              2.5         1426.11kg
    2              0.5          25.51kg
    3              0.1          456.35g
    4              4E-3          0.08g
```

Press return for next part of program
?

The weights obtained so far assume that
there has been an equal statistical
error at each stage of sampling. The
amount of material required at the
coarser sizes can, however, be reduced
by increasing the amount of sample
taken at the finer sizes. Thus the error
will be reduced at the finer sizes,
allowing a greater error at the coarser
sizes, but maintaining a constant total
sample error. In order to obtain such
weights, the next part of the program
allows the weight taken at any one
sampling stage to be changed, the effect
on the other sampling stages then being
displayed. Press return after each
change in order to make further changes,
until satisfactory weights are obtained.
Re-run program if number of sampling
stages is to be changed.
Press RETURN to continue.

?

Suggested weight for stage 4 (gms)
?1.0

```
SAMPLE          TOP SIZE          SAMPLE
STAGE           cms.              WEIGHT


   1             2.5              1090.13kg
   2             0.5               19.50kg
   3             0.1              348.84g
   4             4E-3               1.00g
```
Press return for further weight change

Suggested weight for stage 3 (gms)
?500

```
SAMPLE          TOP SIZE          SAMPLE
STAGE           cms.              WEIGHT


   1             2.5              946.99kg
   2             0.5               16.94kg
   3             0.1              500.00g
   4             4E-3               1.00g
```
Press return for further weight change

Suggested weight for stage 2 (gms)
?50000

```
SAMPLE          TOP SIZE          SAMPLE
STAGE           cms.              WEIGHT


   1             2.5              570.06kg
   2             0.5               50.00kg
   3             0.1              500.00g
   4             4E-3               1.00g
```

RECVAR   Estimation of errors in recovery calculations

```
10REM CALCULATION OF VARIANCE IN TWO PRODUCT RECOVERY
20REM B.A.WILLS 10 JULY 1984
30INPUT"FEED COMPONENT VALUE? "f
40INPUT"CONCENTRATE COMPONENT VALUE? "c
50INPUT"TAILING COMPONENT VALUE? "t
60INPUT"FEED COMPONENT STANDARD DEVIATION? "SF
70INPUT"CON. COMPONENT STANDARD DEVIATION? "SC
80INPUT"TAILS COMPONENT STANDARD DEVIATION? "ST
 90R=100*(f-t)*c/(f*(c-t))
100A=10000/(f*f*(c-t)*(c-t))
110B=c*c*t*t/(f*f)
120C=(f-t)*(f-t)*t*t/((c-t)*(c-t))
130D=c*c*(c-f)*(c-f)/((c-t)*(c-t))
140S2=(A*B*SF*SF)+(A*C*SC*SC)+(A*D*ST*ST)
150S=SQR(S2)
160PRINT:PRINT:PRINT:PRINT"RECOVERY = ";R;"%"
170PRINT"VARIANCE (R) = ";S2
180PRINT"STANDARD DEVIATION (R) = ";S
```

```
RUN
FEED COMPONENT VALUE? 3.5
CONCENTRATE COMPONENT VALUE? 18.0
TAILING COMPONENT VALUE? 1.0
FEED COMPONENT STANDARD DEVIATION? 0.14
CON. COMPONENT STANDARD DEVIATION? 0.36
TAILS COMPONENT STANDARD DEVIATION? 0.08


RECOVERY = 75.6302521%
VARIANCE (R) = 5.73339371
STANDARD DEVIATION (R) = 2.39445061
```

MASSVAR   Estimation of errors in two-product mass flowrate

```
10REM CALCULATION OF VARIANCE IN TWO PRODUCT MASS FLOWRATE
20REM B.A.WILLS 27 JUNE 1984
30INPUT"FEED COMPONENT VALUE? "f
40INPUT"CONCENTRATE COMPONENT VALUE? "c
50INPUT"TAILING COMPONENT VALUE? "t
60INPUT"FEED COMPONENT STANDARD DEVIATION? "SF
70INPUT"CON. COMPONENT STANDARD DEVIATION? "SC
80INPUT"TAILS COMPONENT STANDARD DEVIATION? "ST
90C=100*(f-t)/(c-t)
100A=100/(c-t)
110B=100*(f-t)/((c-t)*(c-t))
120D=100*(c-f)/((c-t)*(c-t))
130S2=(A*A*SF*SF)+(B*B*SC*SC)+(D*D*ST*ST)
140S=SQR(S2)
150PRINT:PRINT:PRINT:PRINT"CON./FEED = ";C;"%"
160PRINT"VARIANCE (C) = ";S2
170PRINT"STANDARD DEVIATION (C) = ";S
180 RDV=S/C
190PRINT"RELATIVE STANDARD DEVIATION (S.D./C)"
200PRINT"= ";RDV
```

```
RUN
FEED COMPONENT VALUE? 0.92
CONCENTRATE COMPONENT VALUE? 0.99
TAILING COMPONENT VALUE? 0.69
FEED COMPONENT STANDARD DEVIATION? 0.01
CON. COMPONENT STANDARD DEVIATION? 0.01
TAILS COMPONENT STANDARD DEVIATION? 0.01



CON./FEED = 76.6666666%
VARIANCE (C) = 18.2469136
STANDARD DEVIATION (C) = 4.27164062
RELATIVE STANDARD DEVIATION (S.D./C)
= 5.57170515E-2
```

MAXREC Maximising the accuracy of two-product recovery calculations

```
10REM MAXIMISING THE ACCURACY OF RECOVERY CALCS
20REM B.A.WILLS 11 SEPT 1984
30PRINT:PRINT:PRINT"The simple node streams will be termed"
40PRINT"'feed' 'concentrate' and 'tailings'."
50PRINT"The recovery of the component into the"
60PRINT"concentrate stream will be calculated."
70PRINT:PRINT:PRINT:PRINT"Input name of recovery component.    "
80INPUTR$
90PRINT:PRINT:PRINT"Input ";R$;" assay in feed":INPUTf
100PRINT:PRINT"Input estimated standard deviation ":INPUTSf:Vf=Sf*Sf
110PRINT:PRINT"Input ";R$;" assay in concentrate":INPUTc
120PRINT:PRINT"Input estimated standard deviation ":INPUTSc:Vc=Sc*Sc
130PRINT:PRINT"Is ";R$;" assay in tails known?":INPUT"Y/N?    "B$
140 IF B$="Y" THEN 170
150IF B$<>"N" THEN 130
160GOTO 300
170PRINT:PRINT"Input ";R$;" assay in tails":INPUTt
180PRINT:PRINT"Input estimated standard deviation ":INPUTSt:Vt=St*St
190Rr=100*c*(f-t)/(f*(c-t))
200A=100*100/(f*f*(c-t)*(c-t))
210B=c*c*t*t/(f*f)
220C=t*t*(f-t)*(f-t)/((c-t)*(c-t))
230D=c*c*(c-f)*(c-f)/((c-t)*(c-t))
240Vrr=A*(Vf*B+C*Vc+D*Vt)
250 E=100*100/((c-t)*(c-t))
260F=(f-t)*(f-t)/((c-t)*(c-t))
270G=(c-f)*(c-f)/((c-t)*(c-t))
280M=100*(f-t)/(c-t)
290Vm=E*(Vf+F*Vc+G*Vt)
300DIM C$(30)
310PRINT:PRINT:PRINT"Input data for possible mass components"
320IFB$<>"Y"THEN PVm=10000000000 :PM=0:GOTO340
330PVm=Vm:PM=M:PC$=R$:PVr=Vrr
340N=0
350N=N+1
360PRINT:PRINT"Input name of component":INPUTC$(N)
370PRINT:PRINT"Input ";C$(N);" assay in feed":INPUTa
380 PRINT:PRINT"Input estimated standard deviation ":INPUTSa
390Va=Sa*Sa
400PRINT:PRINT"Input ";C$(N);" assay in concentrate":INPUTb
410PRINT:PRINT"Input estimated standard deviation":INPUTSb:Vb=Sb*Sb
420PRINT:PRINT"Input ";C$(N);" assay in tails":INPUTd
430PRINT:PRINT"Input estimated standard deviation":INPUTSd:Vd=Sd*Sd
440M=100*(a-d)/(b-d)
450H=100*100/((b-d)*(b-d))
460J=(a-d)*(a-d)/(b-d)
470K=(b-a)*(b-a)/((b-d)*(b-d))
480Vm=H*(Va+J*Vb+K*Vd)
490Q=100*100*c*c/((b-d)*(b-d)*f*f)
500U=(a-d)*(a-d)/((b-d)*(b-d))
510Y=(b-a)*(b-a)/((b-d)*(b-d))
520Z=(a-d)*(a-d)/(c*c)
530I=(a-d)*(a-d)/(f*f)
540Vr=Q*(Va+U*Vb+Y*Vd+Z*Vc+I*Vf)
550IF Vm<PVm THEN PVm=Vm:PM=M:PVr=Vr:PC$=C$(N)
560PRINT:PRINT"Any more components?":INPUT"Y/N?    "Y$
570 IF Y$="Y" THEN 350
580PRINT:PRINT"Mass fraction component is "PC$
590PRINT:PRINT"Recovery of ";R$;" is:"
```

```
600PRINTPM*c/f
610PRINT"Standard deviation is:"
620PRINTSQR(PVr)
630PRINT:PRINT"% feed to concentrate is:"
640PRINTPM
650PRINT"Standard deviation is:"
660PRINTSQR(PVm)
670 IF B$<>"Y"THEN END
680PRINT:PRINT:PRINT"Recovery from ";R$;" assays is:"
690PRINTRr
700PRINT"Standard deviation is:"
710PRINTSQR(Vrr)
```

RUN

The simple node streams will be termed
'feed' 'concentrate' and 'tailings'.
The recovery of the component into the
concentrate stream will be calculated.

Input name of recovery component.
?TIN

Input TIN assay in feed
?0.92

Input estimated standard deviation
?0.01

Input TIN assay in concentrate
?0.99

Input estimated standard deviation
?0.01

Is TIN assay in tails known?
Y/N? Y

Input TIN assay in tails
?0.69

Input estimated standard deviation
?0.01

Input data for possible mass components

Input name of component

```
?DILUTION RATIO

Input DILUTION RATIO assay in feed
?4.87

Input estimated standard deviation
?0.025

Input DILUTION RATIO assay in concentrate
?1.77

Input estimated standard deviation
?0.025

Input DILUTION RATIO assay in tails
?15.73

Input estimated standard deviation
?0.025

Any more components?
Y/N?  N

Mass fraction component is DILUTION RATIO

Recovery of TIN is:
83.7127819
Standard deviation is:
1.26665371

% feed to concentrate is:
77.7936963
Standard deviation is:
0.230349727


Recovery from TIN assays is:
     82.5
Standard deviation is:
3.40754706
```

CONMAT   Construction of circuit connection matrix

```
10REM CONNECTION MATRIX   B.A. WILLS 11 MAY 1984
20D$="3":INPUT"NUMBER OF STREAMS IN CIRCUIT? "S
30PRINT:INPUT"NUMBER OF NODES? "N
40DIM C(S,N)
50PRINT"ENTER DATA FOR EACH STREAM. LAST STREAM"
60PRINT:PRINT"IS REFERENCE (100 UNITS). IF A STREAM"
70PRINT:PRINT"EITHER DOES NOT LEAVE (i.e. A FEED) OR"
80PRINT:PRINT"DOES NOT ENTER (i.e. A PRODUCT) ANY NODE"
90PRINT:PRINT"IN THE CIRCUIT, THEN ENTER 0 FOR THE"
100PRINT:PRINT"NODE"
110FOR A=1 TO S
120PRINT"STREAM ";A;
130INPUT" ENTERS NODE:- "B:IF B=0 THEN 150
140C(A,B)=1
150INPUT"AND LEAVES NODE:- "D:IF D=0 THEN 170
160C(A,D)=-1
170NEXT
180FOR E=1 TO N
190FOR F=1 TO S
200IF C(F,E)=-1 THEN PRINT" -1";:GOTO220
210PRINT"   ";C(F,E);
220NEXTF
230PRINT" "
240NEXTE
250FD=0
260PRINT:PRINT:PRINT
270FOR G=1 TO S
280J=0
290FOR H=1 TO N
300J=C(G,H)+J
310NEXT H
320IFD$=""THEN 520
330 IF J=1 THEN PRINT"STREAM ";G;" IS A FEED":FD=FD+1:GOTO350
340 IF J=-1 THEN PRINT"STREAM ";G;" IS A PRODUCT":GOTO350
350NEXTG
360TNP=0:TNN=0
370FOR L=1 TO N
380NP=0:NN=0
390FOR M=1 TO S
400IFC(M,L)=1 THEN NP=NP+1
410IFC(M,L)=-1 THEN NN=NN+1
420NEXTM
430IF NP=0 OR NN=0 THENPRINT"MISTAKE AT NODE ";L:END
440TNP=TNP+NP-1
450TNN=TNN+NN-1
460NEXT L
470PRINT"CIRCUIT CONSISTS OF ";TNN;" SIMPLE SEPARATORS AND ";
    TNP;" SIMPLE JUNCTIONS"
480NS=2*(FD+TNN)-1
490IF NS<S THEN PRINT"MINIMUM NUMBER OF STREAMS THAT MUST BE
    SAMPLED, INCLUSIVE OF FEEDS AND PRODUCTS, IS ";NS
500IF NS=S THEN PRINT"ALL STREAMS MUST BE SAMPLED"
510IF NS>S THEN PRINT"ALL STREAMS MUST BE SAMPLED":PRINT"MASS
    FLOWS ON ";NS-S; " STREAMS ARE REQUIRED TO SUPPLEMENT
    REFERENCE MASS"
520PRINT:INPUT"Press RETURN to re-print connection matrix"
    D$:GOTO180
```

```
RUN
NUMBER OF STREAMS IN CIRCUIT? 11

NUMBER OF NODES? 4
ENTER DATA FOR EACH STREAM. LAST STREAM

IS REFERENCE (100 UNITS). IF A STREAM

EITHER DOES NOT LEAVE (i.e. A FEED) OR

DOES NOT ENTER (i.e. A PRODUCT) ANY NODE

IN THE CIRCUIT, THEN ENTER 0 FOR THE

NODE
STREAM 1 ENTERS NODE:- 1
AND LEAVES NODE:- 0
STREAM 2 ENTERS NODE:- 0
AND LEAVES NODE:- 1
STREAM 3 ENTERS NODE:- 2
AND LEAVES NODE:- 1
STREAM 4 ENTERS NODE:- 3
AND LEAVES NODE:- 1
STREAM 5 ENTERS NODE:- 4
AND LEAVES NODE:- 1
STREAM 6 ENTERS NODE:- 0
AND LEAVES NODE:- 2
STREAM 7 ENTERS NODE:- 0
AND LEAVES NODE:- 2
STREAM 8 ENTERS NODE:- 0
AND LEAVES NODE:- 3
STREAM 9 ENTERS NODE:- 0
AND LEAVES NODE:- 3
STREAM 10 ENTERS NODE:- 0
AND LEAVES NODE:- 4
STREAM 11 ENTERS NODE:- 0
AND LEAVES NODE:- 4
   1  -1  -1  -1  -1   0   0   0   0   0   0
   0   0   1   0   0  -1  -1   0   0   0   0
   0   0   0   1   0   0   0  -1  -1   0   0
   0   0   0   0   1   0   0   0   0  -1  -1


STREAM 1 IS A FEED
STREAM 2 IS A PRODUCT
STREAM 6 IS A PRODUCT
STREAM 7 IS A PRODUCT
STREAM 8 IS A PRODUCT
STREAM 9 IS A PRODUCT
STREAM 10 IS A PRODUCT
STREAM 11 IS A PRODUCT
CIRCUIT CONSISTS OF 6 SIMPLE SEPARATORS AND 0 SIMPLE JUNCTIONS
ALL STREAMS MUST BE SAMPLED
MASS FLOWS ON 2 STREAMS ARE REQUIRED TO SUPPLEMENT REFERENCE MASS
```

GAUSSEL   Solution of linear equations

```
10 @%=&20209: REM FORMATS PRINTING AND SETS 2 DECIMAL PLACES
20 INPUT"Enter the number of equations : "N
30 DIM A(N,N+1)
40 E1=1E-60:D=1
50 FOR R=1 TO N
60 PRINT"Input row no. ";R;": ";
70 FOR C=1 TO N+1
80 INPUT A(R,C)
90 NEXT:PRINT:NEXT:PRINT:PRINT
100 PRINT"The augmented matrix is :":PRINT
110 FOR I=1 TO N:FOR J=1 TO N+1:PRINT;A(I,J);"
    ";:PRINT:NEXT J:PRINT:NEXT I:PRINT
120 REM SEARCH FOR PIVOT
130 C=1:R=1
140 P=A(R,C):P1=R
150 FOR I=R TO N
160 IF ABS(A(I,C))<ABS(P) THEN 180
170 P=A(I,C):P1=I
180 NEXT I
190 REM INTERCHANGE ROWS
200 IF ABS(P)<E1 THEN GOTO 360
210 IF P1=R THEN 250
220 D=-D
230 FOR K=C TO N+1
240 B=A(P1,K):A(P1,K)=A(R,K):A(R,K)=B:NEXT K
250 REM NORMALIZE PIVOT
260 D=D*P
270 IFP=OTHEN360
280 FOR K=C TO N+1:A(R,K)=A(R,K)/P:NEXT K
290 IF R=N THEN 370
300 REM COLUMN REDUCTION
310 FOR I1=R+1 TO N:B=A(I1,C)
320 IF ABS(B)<=E1 THEN 340
330 FOR K=C TO N+1:A(I1,K)=A(I1,K)-A(R,K)*B:NEXT K
340 NEXT I1
350 R=R+1:C=C+1:GOTO 140
360 PRINT"Determinant is 0, thus no unique solution":END
370 REM BACK SUBSTITUTION
380 PRINT"The upper diagonal matrix is :":PRINT
390 FOR I=1 TO N:FOR J=1 TO N+1:PRINT;A(I,J);"
    ";:PRINT:NEXT J:PRINT:NEXT I
400 FOR K=1 TO N-1:FOR J=1 TO K
410 A(N-K,N+1)=A(N-K,N+1)-A(N-K,N+1-J)*A(N+1-J,N+1)
420 NEXTJ:NEXT K
430 PRINT"The solution vector is : ":PRINT
440 FOR I=1 TO N:PRINT;A(I,N+1);"    ";:NEXT I:PRINT:PRINT
450 PRINT"and the determinant is : ";D
```

```
RUN
Enter the number of equations : 8
Input row no. 1.00: ?-1
?0
?0
?0
?0
?0
?1
?0
?-1

Input row no. 2.00: ?-1
?-1
?0
?-1
?0
?0
?0
?0
?0

Input row no. 3.00: ?0
?1
?-1
?0
?-1
?0
?0
?0
?0

Input row no. 4.00: ?0
?0
?0
?1
?0
?-1
?0
?-1
?0

Input row no. 5.00: ?0
?0
?0
?0
?1
?0
?-1
?1
?0
```

```
Input row no. 6.00: ?0
?0.51
?-0.12
?0
?-4.2
?0
?0
?0
?0

Input row no. 7.00: ?0
?0
?0
?16.1
?0
?-25
?0
?-2.1
?0

Input row no. 8.00: ?0
?0
?-0.12
?0
?0
?-25
?0
?0
?-1.5
```

The augmented matrix is :

```
-1.00    0.00    0.00    0.00    0.00    0.00    1.00    0.00   -1.00
-1.00   -1.00    0.00   -1.00    0.00    0.00    0.00    0.00    0.00
 0.00    1.00   -1.00    0.00   -1.00    0.00    0.00    0.00    0.00
 0.00    0.00    0.00    1.00    0.00   -1.00    0.00   -1.00    0.00
 0.00    0.00    0.00    0.00    1.00    0.00   -1.00    1.00    0.00
 0.00    0.51   -0.12    0.00   -4.20    0.00    0.00    0.00    0.00
 0.00    0.00    0.00   16.10    0.00   -25.00    0.00   -2.10    0.00
 0.00    0.00   -0.12    0.00    0.00   -25.00    0.00    0.00   -1.50
```

The upper diagonal matrix is :

```
1.00    1.00    0.00    1.00    0.00    0.00    0.00    0.00    0.00
0.00    1.00   -1.00    0.00   -1.00    0.00    0.00    0.00    0.00
0.00    0.00    1.00    1.00    1.00    0.00    1.00    0.00   -1.00
0.00    0.00    0.00    1.00    0.00   -1.55    0.00   -0.13    0.00
0.00    0.00    0.00    0.00    1.00    0.15    0.10    0.01   -0.10
0.00    0.00    0.00    0.00    0.00    1.00   -0.00   -0.00    0.06
0.00    0.00    0.00    0.00    0.00    0.00    1.00   -0.90   -0.10
0.00    0.00    0.00    0.00    0.00    0.00    0.00    1.00    0.04
```

The solution vector is :

```
0.94    -1.05    -0.95    0.11    -0.10    0.06    -0.06    0.04
```

and the determinant is : -1550.42

LAGRAN   Reconciliation of excess data by non-weighted least squares

```
1OREM LAGRAN BY B.A.WILLS 5 OCT.1984
2OREM SIMPLE NODE ADJUSTMENT BY LEAST SQUARES:MODE 7
3OREM FOLLOWED BY LAGRANGIAN MULTIPLIERS
4OPRINT"Nature of input (e.g.feed,rougher con,":PRINT"etc":INPUTF$
5OPRINT"Nature of product 1 (e.g concentrate,":PRINT"cyclone
   o/f,etc":INPUTP$
6OPRINT"Nature of product 2":INPUTQ$
7OPRINT"Number of components (e.g.assays, water/solids,size
   fractions,etc)":INPUTN
80DIMC$(N):DIMF(N):DIMP(N):DIMQ(N):DIMAF(N):DIMAP(N):DIMAQ(N):
   DIMX(N):DIME(N)
90FOR A=1 TO N
100 PRINT"Name of component ";A
110INPUTC$(A)
120PRINTC$(A);" in ";F$:INPUTF(A)
130PRINTC$(A);" in ";P$:INPUTP(A)
140PRINTC$(A);" in ";Q$:INPUTQ(A)
150NEXTA
160REM CALC X AND RESIDUAL E
170REM BEST FIT X =D/G WHERE D=SUM OF (F-Q)(P-Q)/SUM OF SQR(P-Q)
180D=0:G=0
190FOR B=1 TO N
200X(B)=100*(F(B)-Q(B))/(P(B)-Q(B))
210D=D+(((F(B)-Q(B))*(P(B)-Q(B))))
220G=G+((P(B)-Q(B))*(P(B)-Q(B)))
230NEXTB
240XB=D/G:REM XB=BEST FIT X
250K=1+(XB*XB)+((1-XB)*(1-XB))
260REM CALCS OF ADJUSTED COMPONENTS
270FOR J=1 TO N
280E(J)=F(J)-(XB*P(J))-(Q(J)*(1-XB))
290AF(J)=F(J)-(E(J)/K)
300AP(J)=P(J)+(E(J)*XB/K)
310AQ(J)=Q(J)+((E(J)*(1-XB))/K)
320NEXT J
330 REM PRINT RESULTS (B.B.C. FORMAT)
340PRINTTAB(25);"ACTUAL";TAB(63);"ADJUSTED"
350PRINT"COMPONENT";
360PRINT TAB(15);"INPUT";TAB(24);"PROD.1";TAB(34);"PROD.2";
   TAB(48);"X";TAB(55);"INPUT";TAB(64);"PROD.1";TAB(74);"PROD.2"
370 @%=&2020A:REM SETS 2 DECIMAL PLACES AND FIELD FORMAT
380PRINT:PRINT
390FOR Z=1 TO N
400PRINTC$(Z),F(Z),P(Z),Q(Z),X(Z),AF(Z),AP(Z),AQ(Z)
410NEXTZ
420PRINT:PRINT:PRINT"X= ";P$;"/";F$;" AS %"
430PRINT:PRINT:PRINT"BEST FIT VALUE OF ";P$;"/";F$;" IS ";XB*100;"%"
440PRINT:PRINT:PRINT"INPUT=";F$;", PROD.1=";P$;", PROD.2=";Q$
```

```
>RUN
Nature of input (e.g.feed,rougher con,
etc
?FEED
Nature of product 1 (e.g concentrate,
cyclone o/f,etc
?CONCENTRATE
Nature of product 2
?TAILING
Number of components (e.g.assays, water/solids,size fractions,etc)
?6
Name of component 1
?TIN
TIN in FEED
?21.9
TIN in CONCENTRATE
?43.0
TIN in TAILING
?6.77
Name of component 2
?IRON
IRON in FEED
?3.46
IRON in CONCENTRATE
?5.5
IRON in TAILING
?1.76
Name of component 3
?SILICA
SILICA in FEED
?58.0
SILICA in CONCENTRATE
?25.1
SILICA in TAILING
?75.3
Name of component 4
?SULPHUR
SULPHUR in FEED
?0.11
SULPHUR in CONCENTRATE
?0.12
SULPHUR in TAILING
?0.09
Name of component 5
?ARSENIC
ARSENIC in FEED
?0.36
ARSENIC in CONCENTRATE
?0.38
ARSENIC in TAILING
?0.34
Name of component 6
?TIO2
TIO2 in FEED
?4.91
TIO2 in CONCENTRATE
?9.24
TIO2 in TAILING
?2.07
```

> COMPONENT	INPUT	ACTUAL PROD.1	PROD.2	X	INPUT	ADJUSTED PROD.1	PROD.2
TIN	21.90	43.00	6.77	41.76	20.78	43.41	7.47
IRON	3.46	5.50	1.76	45.45	3.25	5.58	1.89
SILICA	58.00	25.10	75.30	34.46	57.16	25.41	75.83
SULPHUR	0.11	0.12	0.09	66.67	0.10	0.12	0.09
ARSENIC	0.36	0.38	0.34	50.00	0.36	0.38	0.34
TIO2	4.91	9.24	2.07	39.61	4.79	9.28	2.15

X= CONCENTRATE/FEED AS %

BEST FIT VALUE OF CONCENTRATE/FEED IS 37.03%

INPUT=FEED, PROD.1=CONCENTRATE, PROD.2=TAILING

WEGHTRE   Reconciliation of excess data by weighted least squares

```
1OREM WEGHTRE BY B.A.WILLS 20 FEBRUARY 1985
20REM ESTIMATION OF BEST FLOW RATE BY WEIGHTED RESIDUALS
3OREM LEAST SQUARES FOLLOWED BY LAGRANGIAN METHOD
40PRINT"Nature of input (e.g.feed,rougher con,":PRINT"etc":INPUTF$
50PRINT"Nature of product 1 (e.g concentrate,":PRINT"cyclone o/f,etc":INPUTP$
60PRINT"Nature of product 2":INPUTQ$
70PRINT"Number of components (e.g.assays, water/solids,size fractions,etc)":I
NPUTN
80PRINT"Do all components have equal relative":INPUT"error Y/N?  "FR$
90IF FR$="Y" THEN E=1
100DIMC$(N):DIMF(N):DIMP(N):DIMQ(N):DIMAF(N):DIMAP(N):DIMAQ(N):DIMX(N):DIME(N)
:DIM VF(N):DIMVP(N):DIMVQ(N)
110FOR A=1 TO N
120 PRINT"Name of component ";A;" (e.g.%Sn,":PRINT"water/solids,%125-250m,etc"
130INPUTC$(A)
140PRINTC$(A);" in ";F$:INPUTF(A)
150IF FR$="Y" THEN 170
160INPUT"Estimated relative standard deviation %"E
170VF(A)=E*E*F(A)*F(A)/10000
180PRINTC$(A);" in ";P$:INPUTP(A)
190IF FR$="Y" THEN 210
200INPUT"Estimated relative standard deviation %"E
210VP(A)=E*E*P(A)*P(A)/10000
220PRINTC$(A);" in ";Q$:INPUTQ(A)
230IF FR$="Y" THEN 250
240INPUT"Estimated relative standard deviation %"E
250VQ(A)=E*E*Q(A)*Q(A)/10000
260NEXTA
270REM CALC X
280REM BEST FIT X =D/G WHERE D=SUM OF (F-Q)(P-Q)/SUM OF SQR(P-Q)
290D=0:G=0
300FOR B=1 TO N
310X(B)=100*(F(B)-Q(B))/(P(B)-Q(B))
320D=D+(((F(B)-Q(B))*(P(B)-Q(B))))
330G=G+((P(B)-Q(B))*(P(B)-Q(B)))
340NEXTB
350XB=D/G:REM XB=BEST FIT X WITH NO WEIGHTING
360 REM CALCULATION OF WEIGHTED BEST FIT
370CB=XB
380DW=0:GW=0:C=CB
390FOR H=1 TO N
400VR=VF(H)+(C*C*VP(H))+((1-C)*(1-C)*VQ(H))
410GW=GW+((P(H)-Q(H))*(P(H)-Q(H)))/VR
420DW=DW+(((F(H)-Q(H))*(P(H)-Q(H))))/VR
430NEXT H
440CB=DW/GW:REM WEIGHTED ESTIMATE
450 IF ABS(CB-C)<0.005 THEN 470
460 GOTO 380
470XB=CB
480REM CALCS OF ADJUSTED COMPONENTS
490FOR J=1TON
500K=VF(J)+(XB*XB*VP(J))+((1-XB)*(1-XB)*VQ(J))
510E(J)=F(J)-(XB*P(J))-(Q(J)*(1-XB))
520AF(J)=F(J)-(E(J)*VF(J)/K)
530AP(J)=P(J)+(E(J)*XB*VP(J)/K)
540AQ(J)=Q(J)+((E(J)*(1-XB)*VQ(J))/K)
550A=E(J)*VF(J)/K:B=E(J)*XB*VP(J)/K:C=E(J)*(1-XB)*VQ(J)/K
560NEXT J
570PRINTTAB(25);"ACTUAL";TAB(63);"ADJUSTED"
580PRINT"COMPONENT";
590PRINTTAB(15);"INPUT";TAB(24);"PROD.1";TAB(34);"PROD.2";TAB(48);"X";TAB(55);
"INPUT";TAB(64);"PROD.1";TAB(74);"PROD.2"
```

```
600 @%=&2020A:REM SETS 2 DECIMAL PLACES AND FORMATS FIELD WIDTH
610PRINT:PRINT
620FOR Z=1 TO N
630PRINTC$(Z),F(Z),P(Z),Q(Z),X(Z),AF(Z),AP(Z),AQ(Z)
640NEXTZ
650PRINT:PRINT:PRINT:PRINT:PRINT"X= ";P$;"/";F$;" AS %"
660PRINT:PRINT:PRINT"BEST FIT VALUE OF ";P$;"/";F$;" IS ";XB*100;"%"
670PRINT:PRINT:PRINT"INPUT=";F$;", PROD.1=";P$;", PROD.2=";Q$
```

```
RUN
Nature of input (e.g.feed,rougher con,
etc
?FEED
Nature of product 1 (e.g concentrate,
cyclone o/f,etc
?CONCENTRATE
Nature of product 2
?TAILING
Number of components (e.g.assays, water/solids,size fractions,etc)
?6
Do all components have equal relative
error Y/N?   Y
Name of component 1 (e.g.%Sn,
water/solids,%125-250m,etc
?TIN
TIN in FEED
?21.9
TIN in CONCENTRATE
?43.0
TIN in TAILING
?6.77
Name of component 2 (e.g.%Sn,
water/solids,%125-250m,etc
?IRON
IRON in FEED
?3.46
IRON in CONCENTRATE
?5.5
IRON in TAILING
?1.76
Name of component 3 (e.g.%Sn,
water/solids,%125-250m,etc
?SILICA
SILICA in FEED
?58.0
SILICA in CONCENTRATE
?25.1
SILICA in TAILING
?75.3
Name of component 4 (e.g.%Sn,
water/solids,%125-250m,etc
```

```
?SULPHUR
SULPHUR in FEED
?0.11
SULPHUR in CONCENTRATE
?0.12
SULPHUR in TAILING
?0.09
Name of component 5 (e.g.%Sn,
water/solids,%125-250m,etc
?ARSENIC
ARSENIC in FEED
?0.36
ARSENIC in CONCENTRATE
?0.38
ARSENIC in TAILING
?0.34
Name of component 6 (e.g.%Sn,
water/solids,%125-250m,etc
?TIO2
TIO2 in FEED
?4.91
TIO2 in CONCENTRATE
?9.24
TIO2 in TAILING
?2.07
```

| | | ACTUAL | | | | ADJUSTED | |
COMPONENT	INPUT	PROD.1	PROD.2	X	INPUT	PROD.1	PROD.2
TIN	21.90	43.00	6.77	41.76	21.80	43.16	6.78
IRON	3.46	5.50	1.76	45.45	3.36	5.61	1.78
SILICA	58.00	25.10	75.30	34.46	55.87	25.26	77.40
SULPHUR	0.11	0.12	0.09	66.67	0.10	0.12	0.09
ARSENIC	0.36	0.38	0.34	50.00	0.36	0.38	0.34
TIO2	4.91	9.24	2.07	39.61	4.98	9.13	2.06

X= CONCENTRATE/FEED AS %

BEST FIT VALUE OF CONCENTRATE/FEED IS 41.30%

INPUT=FEED , PROD.1=CONCENTRATE, PROD.2=TAILING

WILMAN   Reconciliation of excess data by variances in mass equations

```
1OREM WILMAN BY B.A.WILLS 2O FEBRUARY 1985
2OREM ESTIMATION OF BEST FLOW RATE BY VARIANCE IN COMPONENT EQUATIONS
3OREM DATA ADJUSTMENT BY LAGRANGIAN MULTIPLIERS
4OPRINT"Nature of input (e.g.feed,rougher con,":PRINT"etc":INPUTF$
5OPRINT"Nature of product 1 (e.g concentrate,":PRINT"cyclone o/f,etc":INPUTP$
6OPRINT"Nature of product 2":INPUTQ$
7OPRINT"Number of components (e.g.assays, water/solids,size fractions,etc)":I
NPUTN
    8OPRINT"Do all components have equal relative":INPUT"error Y/N?  "FR$
  90 IF FR$="Y" THEN E=1
  100DIMC$(N):DIMF(N):DIMP(N):DIMQ(N):DIMAF(N):DIMAP(N):DIMAQ(N):DIMX(N):DIME(N)
:DIM VF(N):DIMVP(N):DIMVQ(N)
  110FOR A=1 TO N
  120 PRINT"Name of component ";A;" (e.g.%Sn,":PRINT"water/solids,%125-250m,etc"
  130INPUTC$(A)
  140PRINTC$(A);" in ";F$:INPUTF(A)
  150 IF FR$ ="Y" THEN 170
  160INPUT"Estimated relative standard deviation %"E
  170VF(A)=E*E*F(A)*F(A)/10000
  180PRINTC$(A);" in ";P$:INPUTP(A)
  190 IF FR$="Y" THEN 210
  200INPUT"Estimated relative standard deviation %"E
  210VP(A)=E*E*P(A)*P(A)/10000
  220PRINTC$(A);" in ";Q$:INPUTQ(A)
  230 IF FR$="Y" THEN 250
  240INPUT"Estimated relative standard deviation %"E
  250VQ(A)=E*E*Q(A)*Q(A)/10000
  260NEXTA
  270REM CALC X
  280FOR B=1 TO N
  290X(B)=100*(F(B)-Q(B))/(P(B)-Q(B))
  300NEXTB
  310 REM CALCULATION OF WEIGHTED BEST FIT
  320DW=0:GW=0
  330FOR H=1 TO N
  340AA=VF(H)/((P(H)-Q(H))*(P(H)-Q(H)))
  350BB=VP(H)*(F(H)-Q(H))*(F(H)-Q(H))/((P(H)-Q(H))*(P(H)-Q(H))*(P(H)-Q(H))*(P(H)
-Q(H)))
  360CC=VQ(H)*(P(H)-F(H))*(P(H)-F(H))/((P(H)-Q(H))*(P(H)-Q(H))*(P(H)-Q(H))*(P(H)
-Q(H)))
  370VC=AA+BB+CC
  380 DW=DW+((F(H)-Q(H))/((P(H)-Q(H))*SQR(VC)))
  390GW=GW+(1/SQR(VC))
  400NEXT H
  410XB=DW/GW
  420REM CALCS OF ADJUSTED COMPONENTS
  430FOR J=1TON
  440 K=VF(J)+(XB*XB*VP(J))+((1-XB)*(1-XB)*VQ(J))
  450E(J)=F(J)-(XB*P(J))-(Q(J)*(1-XB))
  460AF(J)=F(J)-(E(J)*VF(J)/K)
  470AP(J)=P(J)+(E(J)*XB*VP(J)/K)
  480AQ(J)=Q(J)+((E(J)*(1-XB)*VQ(J))/K)
  490A=E(J)*VF(J)/K:B=E(J)*XB*VP(J)/K:C=E(J)*(1-XB)*VQ(J)/K
  500NEXT J
  510PRINTTAB(25);"ACTUAL";TAB(63);"ADJUSTED"
  520PRINT"COMPONENT";
  530PRINT TAB(15);"INPUT";TAB(24);"PROD.1";TAB(34);"PROD.2";TAB(48);"X";TAB(55)
;"INPUT";TAB(64);"PROD.1";TAB(74);"PROD.2"
```

```
540 @%=&2020A: REM SETS 2 DECIMAL PLACES AND FORMATS FIELD WIDTH
550PRINT:PRINT
560FOR Z=1 TO N
570PRINTC$(Z),F(Z),P(Z),Q(Z),X(Z),AF(Z),AP(Z),AQ(Z)
580NEXTZ
590PRINT:PRINT:PRINT:PRINT:PRINT"X= ";P$;"/";F$;" AS %"
600PRINT:PRINT:PRINT"BEST FIT VALUE OF ";P$;"/";F$;" IS ";XB*100;"%"
610PRINT:PRINT:PRINT"INPUT=";F$;", PROD.1=";P$;", PROD.2=";Q$
```

```
RUN
Nature of input (e.g.feed,rougher con,
etc
?FEED
Nature of product 1 (e.g concentrate,
cyclone o/f,etc
?CONCENTRATE
Nature of product 2
?TAILING
Number of components (e.g.assays, water/solids,size fractions,etc)
?6
Do all components have equal relative
error Y/N?  Y
Name of component 1 (e.g.%Sn,
water/solids,%125-250m,etc
?TIN
TIN in FEED
?21.9
TIN in CONCENTRATE
?43.0
TIN in TAILING
?6.77
Name of component 2 (e.g.%Sn,
water/solids,%125-250m,etc
?IRON
IRON in FEED
?3.46
IRON in CONCENTRATE
?5.5
IRON in TAILING
?1.76
Name of component 3 (e.g.%Sn,
water/solids,%125-250m,etc
?SILICA
SILICA in FEED
?58.0
SILICA in CONCENTRATE
?25.1
SILICA in TAILING
?75.3
Name of component 4 (e.g.%Sn,
water/solids,%125-250m,etc
```

```
?SULPHUR
SULPHUR in FEED
?0.11
SULPHUR in CONCENTRATE
?0.12
SULPHUR in TAILING
?0.09
Name of component 5 (e.g.%Sn,
water/solids,%125-250m,etc
?ARSENIC
ARSENIC in FEED
?0.36
ARSENIC in CONCENTRATE
?0.38
ARSENIC in TAILING
?0.34
Name of component 6 (e.g.%Sn,
water/solids,%125-250m,etc
?TIO2
TIO2 in FEED
?4.91
TIO2 in CONCENTRATE
?9.24
TIO2 in TAILING
?2.07
```

COMPONENT	INPUT	ACTUAL PROD.1	PROD.2	X	INPUT	ADJUSTED PROD.1	PROD.2
TIN	21.90	43.00	6.77	41.76	22.00	42.84	6.76
IRON	3.46	5.50	1.76	45.45	3.38	5.58	1.77
SILICA	58.00	25.10	75.30	34.46	55.55	25.29	77.68
SULPHUR	0.11	0.12	0.09	66.67	0.10	0.12	0.09
ARSENIC	0.36	0.38	0.34	50.00	0.36	0.38	0.34
TIO2	4.91	9.24	2.07	39.61	5.02	9.07	2.06

X= CONCENTRATE/FEED AS %

BEST FIT VALUE OF CONCENTRATE/FEED IS 42.24%

INPUT=FEED, PROD.1=CONCENTRATE, PROD.2=TAILING

CYCLONE   Selection and performance of standard Krebs cyclones

```
10REM HYDROCYCLONE CALCULATIONS BY KREBS-MULAR-JULL FORMULAE
20REM B.A.WILLS 30 NOVEMBER 1986
30PRINT:PRINT:PRINTTAB(7);"HYDROCYCLONE CALCULATIONS"
40PRINTTAB(7);"*************************"
50PRINT:PRINT:PRINT"A. Determination of cut-point (and"
60PRINT"capacity) of standard cyclone of known":PRINT"diameter"
70PRINT:PRINT:PRINT"B. Determination of diameter of cyclone"
80PRINT"needed to give required cut-point"
90PRINT:PRINT:PRINT:INPUT"Input A, or B    "A$
100PRINT:PRINT"INSERT FOLLOWING FEED DATA:"
110PRINT:PRINT:INPUT"S.G. of dry solids, kg/l    "S
120PRINT:PRINT"Feed % solids by weight"
130INPUT"(If only slurry density known, input 0)    "x
140 IF x<>0 THEN D=100*S/((100*S)+x-(x*S)):GOTO 170
150PRINT:PRINT:INPUT"Slurry density, kg/l    "D
160x=100*S*(D-1)/(D*(S-1))
170V=x*D/S
180PRINT:PRINT:PRINT:PRINT"Input the cyclone feed pressure in kPa"
190PRINT"(1 psi= 6.895 kPa). If, in the case of"
200PRINT"an operating cyclone (calculation A),"
210PRINT"the pressure is not known, input 0, and"
220PRINT"then input the volumetric flowrate. If"
230PRINT"this is not known, input 0, then input"
240PRINT"the mass flowrate of dry solids."
250PRINT:PRINT:PRINT:INPUT"Cyclone feed pressure, kPa    "P
260 IF P<>0 THEN 310
270PRINT:PRINT:INPUT"Feed flowrate, cu.m/h    "Q
280 IF Q<>0 THEN M=Q*D*x/100:GOTO 310
290PRINT:PRINT:INPUT"Feed mass flowrate, t/h    "M
300Q=100*M/(x*D)
310IF A$="B" THEN 450
320REM CALCULATIONS A
330PRINT:INPUT"Cyclone diameter, cms    "Dc
340 IF P=0 THEN 370
350Q= 0.0094*(P^0.5)*Dc*Dc
360M=Q*D*x/100
370d50=0.77*(Dc^1.875)*EXP(-.301+(.0945*V)-(.00356*V*V)+(.0000684*V*V*V))/((Q^
0.6)*((S-1)^0.5))
380PRINT:PRINT:GOSUB 520
390 GOSUB 580
400IFP<>0 THEN END
410P=(Q^2)/((.0094^2)*(Dc^4))
420P1=P/6.895
430PRINT"Cyclone pressure is ";P;" kPa"
440PRINT"                    (";P1;" psi)":END
450REM CALCULATIONS B
460INPUT"Required cut-point, microns    "d50
470Dc=(d50^1.481)*((S-1)^.741)*(.0094^.889)*(P^.444)/((.77^1.481)*(EXP(-.301+(
.0945*V)-(.00356*V*V)+(.0000684*V*V*V))^1.481))
480Q=0.0094*(P^.5)*Dc*Dc
490M=Q*D*x/100
500PRINT:PRINT:PRINT:GOSUB 520
510 GOSUB 640:END
520REM SUBROUTINE
530IF Q<1 THEN Q=Q*1000:B$="litres/h":GOTO 550
540B$="cu.m/h"
550IF M<1 THEN M=1000*M:C$="kg/h":GOTO 570
560C$="t/h"
570 RETURN
580 REM SUBROUTINE
590@%=131594: REM SETS 2 DECIMAL PLACES
```

```
600PRINT:PRINT"Cyclone cut-point is ";d50;" microns"
610PRINT:PRINT"Mass flowrate is ";M; C$
620PRINT:PRINT"Volumetric flowrate is ";Q; B$
630 RETURN
640 REM SUBROUTINE
650@%=131594: REM SETS 2 DECIMAL PLACES
660PRINT:PRINT"Required cyclone diameter is ";Dc; "cms"
670PRINT:PRINT:PRINT"Volumetric capacity is ";Q; B$
680PRINT:PRINT:PRINT"Solids capacity is ";M; C$:RETURN
```

RUN

```
          HYDROCYCLONE CALCULATIONS
          *************************
```

A. Determination of cut-point (and
capacity) of standard cyclone of known
diameter

B. Determination of diameter of cyclone
needed to give required cut-point

Input A, or B   B

INSERT FOLLOWING FEED DATA:

S.G. of dry solids, kg/l   3.7

Feed % solids by weight
(If only slurry density known, input 0)   50

Input the cyclone feed pressure in kPa
(1 psi= 6.895 kPa). If, in the case of
an operating cyclone (calculation A),
the pressure is not known, input 0, and
then input the volumetric flowrate. If
this is not known, input 0, then input
the mass flowrate of dry solids.

Cyclone feed pressure, kPa   83
Required cut-point, microns   74

Required cyclone diameter is 65.99cms

Volumetric capacity is 372.89cu.m/h

Solids capacity is 293.55t/h

GRAVITY   Evaluation of heavy liquid data

```
10 REM GRAVITY BY B.A.WILLS DEC. 1986
20PRINT:INPUT"Number of densities in test?    "N
30PRINT:PRINT:PRINT"Input, starting with the lowest density,"
40PRINT"a) the liquid density, b) the weight"
50PRINT"floating (in consistent units), and"
60PRINT"c) the assay of the floats (consistent"
70PRINT"units) "
80DIM D(N),W(N+1),A(N+1),M(N+1),ND(N)
90FOR B= 1 TO N
100PRINT:PRINT:INPUT"Density    "D(B)
110PRINT:PRINT:INPUT"Weight floating    "W(B)
120PRINT:PRINT:INPUT"Floats assay    "A(B)
130NEXT
140PRINT:PRINT:PRINT"Sinks weight at density ";D(N):INPUTW(N+1)
150PRINT:PRINT:PRINT"Assay of sinks at density ";D(N):INPUTA(N+1)
160TW=0:TM=0
170FOR CA=1 TO N+1
180TW=TW+W(CA)
190M(CA)=W(CA)*A(CA)
200TM=TM+M(CA)
210NEXT CA
220GOSUB1000
230IF AX$="WRONG" THEN GOSUB 1110:GOTO160
240IFAX$=""THEN260
250IF AX$<>"" THEN 220
260FA=TM/TW
270CW=0:CM=0
280PRINT:PRINT:PRINTTAB(6);"DENSITY";TAB(19);"WEIGHT %";TAB(34);"ASSAY";TAB(46
);"DISTN %";TAB(58);"WEIGHT %";TAB(71);"DISTN %"
290PRINTTAB(64);"CUMULATIVE":PRINT:PRINT
300FOR J=1 TO N+1
310Wtpc=100*W(J)/TW
320Distpc=100*M(J)/TM
330CW=Wtpc+CW
340CM=Distpc+CM
350@%=&2020D:REM SETS 2 DECIMAL PLACES AND FORMATS FIELD WIDTH
360IF J<(N+1) THEN PRINTD(J),;:GOTO380
370 PRINT TAB(8);"+";D(N),;
380PRINT ,Wtpc,A(J),Distpc,CW,CM
390NEXT J
400PRINT:PRINT:PRINTTAB(20);"100.00";TAB(35);FA;TAB(46);"100.00":@%=10:REM RET
URNS TO NORMAL OUTPUT FORMAT
410FOR G=2 TO N
420ND(G)=(D(G)+D(G-1))/2
430NEXT G
440PRINT:PRINT:PRINT"If partition data for vessel is known, then input 'Y'. If
data is not known,"
450PRINT"input 'N', and then input values for separating density and Ep"
460INPUTA$
470IF A$="Y" THEN 660
480PRINT:PRINT:INPUT"Separating density    "d50
490PRINT:PRINT:INPUT"Ecart probable    "Ep
500IF Ep=0 THEN Ep=.000000000000000000000000001
510WS=W(N+1):MS=M(N+1)
520Z=-Ep/(2.303*LOG(0.5))
530FOR X=2 TO N
540 GOSUB 780
550 GOSUB 820
560NEXT X
570 GOSUB 860:PRINT:PRINT:PRINT
```

```
580 IF Ep=.0000000000000000000000000001 THEN PRINT "Ideal performance at S.G.
";d50:GOTO 610
590PRINT"Predicted performance at S.G. ";d50
600PRINT"and Ep of ";Ep
610 GOSUB 940
620 PRINT:PRINT:PRINT"Do you wish to input other values?  Y/N    "
630INPUT"If N return to heavy liquid balance  "V$
640 IF V$<>"N" GOTO480
650 GOTO 270
660WS=W(N+1):MS=M(N+1)
670FOR X= 2 TO n
680PRINT:PRINT:PRINT"Input FRACTION of feed to sinks at S.G. ";ND(X):INPUTc
690 GOSUB 820
700 NEXT X
710 GOSUB 860
720PRINT:PRINT:PRINT"Predicted performance is:"
730 GOSUB 940
740 INPUT"Press RETURN for heavy liquid balance"D$
750GOTO 270
760END
770 REM SUBROUTINES FOLLOW
780IF ND(X)= d50 THEN c=0.5
790IF ND(X)<d50 THEN c = 0.5*EXP((ND(X)-d50)/Z)
800IF ND(X)>d50 THEN c=1-(0.5*EXP((d50-ND(X))/Z))
810 RETURN
820REM a,b, are weights reporting to sinks
830a=W(X)*c:b=M(X)*c
840WS=WS+a:MS=MS+b
850 RETURN
860WF=TW-WS:MF=TM-MS
870WpcF=WF*100/TW
880MpcF=MF*100/TM
890AF=MF/WF
900WpcS=100-WpcF
910MpcS=100-MpcF
920AS=MS/WS
93ORETURN
940@%=&2020A:PRINT:PRINT:REM SETS 2 DECIMAL PLACES AND FORMATS FIELD WIDTH
950PRINTTAB(16);"WT.%";TAB(26);"ASSAY";TAB(35);"DISTN."
960PRINT:PRINT:PRINT"FLOATS",WpcF,AF,MpcF
970PRINT:PRINT:PRINT"SINKS",WpcS,AS,MpcS
980PRINT:PRINT:@%=10: REM RETURNS TO NORMAL OUTPUT FORMAT
990 RETURN
1000 REM CHECK INPUTS
1010PRINT:PRINT:PRINT"                  DATA CHECK"
1020PRINT"             **********"
1030PRINT:PRINT:PRINT"Density","Weight","Assay","Code":PRINT:PRINT
1040FOR AB=1 TO N
1050PRINT;TAB(O)D(AB);TAB(10)W(AB);TAB(20)A(AB);TAB(30)AB
1060NEXTAB
1070PRINT"+";D(N);TAB(10)W(N+1);TAB(20)A(N+1);TAB(30)N+1
1080PRINT:PRINT:PRINT"If data correct, press RETURN"
1090INPUT"If incorrect, input WRONG   "AX$
1100 RETURN
1110 REM CORRECT INPUTS
1120PRINT:PRINT:PRINT"                DATA CORRECTION"
1130PRINT"             ***************"
1140PRINT:PRINT:PRINT:INPUT"Input code of line to be corrected   "AZ
1150IF AZ=N+1 THEN 1170
1160PRINT:PRINT:PRINT:INPUT"Density   "D(AZ)
1170PRINT:PRINT:INPUT"Weight   "W(AZ)
1180PRINT:PRINT:INPUT"Assay   "A(AZ)
1190 RETURN
```

```
RUN

Number of densities in test?    8


Input, starting with the lowest density,
a) the liquid density, b) the weight
floating (in consistent units), and
c) the assay of the floats (consistent
units)

Density    2.55

Weight floating    1.57

Floats assay    0.003

Density    2.60

Weight floating    9.22

Floats assay    0.04

Density    2.65

Weight floating    26.11

Floats assay    0.04

Density    2.70

Weight floating    19.67

Floats assay    0.04

Density    2.75

Weight floating    11.91

Floats assay    0.17

Density    2.80

Weight floating    10.92
```

Floats assay   0.34

Density   2.85

Weight floating   7.87

Floats assay   0.37

Density   2.90

Weight floating   2.55

Floats assay   1.30

Sinks weight at density 2.9
?10.18

Assay of sinks at density 2.9
?9.60

```
           DATA CHECK
           **********
```

Density	Weight	Assay	Code
2.55	1.57	3E-3	1
2.6	9.22	4E-2	2
2.65	26.11	4E-2	3
2.7	19.67	4E-2	4
2.75	11.91	0.17	5
2.8	10.92	0.34	6
2.85	7.87	0.37	7
2.9	2.55	1.3	8
+2.9	10.18	9.6	9

If data correct, press RETURN
If incorrect, input WRONG

DENSITY	WEIGHT %	ASSAY	DISTN %	WEIGHT % CUMULATIVE	DISTN %
2.55	1.57	0.00	0.00	1.57	0.00
2.60	9.22	0.04	0.33	10.79	0.33
2.65	26.11	0.04	0.93	36.90	1.27
2.70	19.67	0.04	0.70	56.57	1.97
2.75	11.91	0.17	1.81	68.48	3.78
2.80	10.92	0.34	3.32	79.40	7.10
2.85	7.87	0.37	2.60	87.27	9.70
2.90	2.55	1.30	2.96	89.82	12.66
+2.90	10.18	9.60	87.34	100.00	100.00

If partition data for vessel is known, then input 'Y'. If data is not known,
input 'N', and then input values for separating density and Ep
?N

Separating density   2.75

Ecart probable   0.07

Predicted performance at S.G. 2.75
and Ep of 7E-2

              WT.%   ASSAY   DISTN.

FLOATS   61.06   0.09   5.09

SINKS   38.94   2.73   94.91

Do you wish to input other values? Y/N

YIELDOP   Optimisation of coal preparation plant performance

```
10 REM YIELD OPTIMISATION BY B.A.WILLS 23 MAY 1985
20 PRINT"Input NEW if new data being entered."
30PRINT"Press RETURN for demonstration data":INPUT DEM$
40IF DEM$="NEW" THEN 60
50 GOSUB 1920:GOTO 620
60 INPUT"Material sized on how many screens? "S
70INPUT"Material split at how many densities? "D
80DIM S$(S):DIM G(D):DIM W(S+1):DIM Mass(S,D+1):DIM Ash(S,D+1):DIM EP(S):DIM
SD(S):DIM WF(S,D+1):DIM WS(S,D+1)
90 PRINT"INPUT SCREEN SIZES COMMENCING WITH"
100PRINT"COARSEST SCREEN"
110FOR A = 1 TO S
120INPUTS$(A):NEXT
130PRINT"INPUT DENSITIES COMMENCING WITH LOWEST"
140PRINT"DENSITY"
150FOR B=1 TO D
160INPUTG(B):NEXT
170PRINT"SIZE ANALYSIS DATA"
180PRINT"Input weight % in each size fraction"
190PRINT"+ ";S$(1):INPUTW(1):IF S=1THEN 230
200FOR C=2 TO S
210PRINTS$(C);"-";S$(C-1):INPUTW(C)
220NEXT
230PRINT"-";S$(S):INPUTW(S+1)
240PRINT"MASS AND ASH ANALYSIS DATA"
250PRINT"Input % weight and % ash content in each"
260PRINT"density fraction"
270PRINT"Size fraction +";S$(1)
280PRINT"Weight % in density fraction -";G(1)
290 PRINT"(In terms of mass of size fraction)":INPUTMass(1,1)
300PRINT"% ash in density fraction -";G(1)
310INPUTAsh(1,1)
320FOR E=1 TO D-1
330PRINT"Weight % in density fraction ";G(E);"-";G(E+1)
340INPUTMass(1,E+1)
350PRINT"% ash in density fraction ";G(E);"-";G(E+1)
360INPUTAsh(1,E+1)
370NEXT E
380PRINT"Weight % in density fraction +";G(D)
390INPUTMass(1,D+1)
400PRINT"% ash in density fraction +";G(D)
410INPUTAsh(1,D+1)
420IF S=1 THEN 600
430FOR F=2 TO S
440PRINT"SIZE FRACTION ";S$(F);"-";S$(F-1)
450PRINT"Weight % in density fraction -";G(1)
460INPUTMass(F,1)
470PRINT"% ash in density fraction -";G(1)
480INPUTAsh(F,1)
490FOR J=1 TO D-1
500PRINT"Weight % in density fraction ";G(J);"-";G(J+1)
510INPUTMass(F,J+1)
520PRINT"% ash in density fraction ";G(J);"-";G(J+1)
530INPUTAsh(F,J+1)
540NEXT J
550PRINT"Weight % in density fraction +";G(D)
560INPUTMass(F,D+1)
570PRINT"% ash in density fraction +";G(D)
580INPUTAsh(F,D+1)
590NEXT F
```

```
600PRINT"% ash in -";S$(S);" fraction"
610INPUTuAsh
620PRINT"PREDICTED ECART PROBABLE MOYEN"
630PRINT"Input expected Ep for each size fraction"
640 GOSUB 860
650PRINT"SEPARATION DENSITIES"
660PRINT"Input density of separation for each"
670PRINT"size fraction (N.B.consistent densities"
680PRINT"for fractions in same vessel)":PRINT:PRINT
690 GOSUB 940
700 GOSUB 1020
710 GOSUB 1150
720 GOSUB 1390
730 INPUT"Change separating conditions? Y/N "CD$
740IF CD$="Y"THEN INPUT "Input code number "CN
750IFCD$="N" THEN END
760IF CN=S+1 THEN 700
770 PRINT"Change density of separation (D)"
780PRINT"or Ep (E) ?"
790INPUT"Input D or E "DE$
800 IF DE$="D" THEN INPUT"New density?"SD(CN)
810 IF DE$="E" THEN INPUT "New Ep?"EP(CN)
820INPUT"Any more changes? Y/N "CZ$
830IFCZ$="N"THEN 710
840IFCZ$="Y"THEN INPUT "Input code number "CN:GOTO760
850GOTO820
860PRINT"Ep for size fraction +";S$(1)
870INPUTEP(1)
880IFS=1 THEN930
890FOR K=2 TO S
900PRINT"Ep for size fraction ";S$(K);"-";S$(K-1)
910INPUTEP(K)
920NEXT
930 RETURN
940PRINT"Separating density for size fraction ":PRINT"+";S$(1)
950INPUTSD(1)
960IFS=1THEN1010
970FOR L=2 TO S
980PRINT"Separating density for size fraction ":PRINTS$(L);"-";S$(L-1)
990INPUTSD(L)
1000NEXT
1010 RETURN
1020PRINT"FLOTATION PERFORMANCE"
1030PRINT"Input expected flotation yield (%)"
1040PRINT"If fines to be discarded press RETURN"
1050PRINT"If fines to be blended directly"
1060PRINT"then input 100"
1070INPUTY
1080 IF Y=0 THEN fAsh=0:GOTO 1140
1090 IF Y=100 THEN ER=1:GOTO 1130
1100PRINT"Input expected enrichment ratio"
1110PRINT"(Ash% in feed/Ash% in concentrate)"
1120INPUTER
1130fAsh=uAsh/ER
1140 RETURN
1150FOR Q=1 TO S
1160IFEP(Q)=0THEN EP(Q)=0.000000000000001
1170ND=G(1)
1180 ZZ=-EP(Q)/(2.303*LOG(0.5))
1190IF ND=SD(Q)THEN Z=0.5
1200IF ND>SD(Q)THEN Z=0.5*(EXP((SD(Q)-ND)/ZZ))
1210IFND<SD(Q)THEN Z=1-(0.5*(EXP((ND-SD(Q))/ZZ)))
```

```
1220WF(Q,1)=W(Q)*Mass(Q,1)*Z/100
1230WS(Q,1)=(W(Q)*Mass(Q,1)/100)-WF(Q,1)
1240FOR R=2 TO D
1250ND=(G(R)+G(R-1))/2
1260IF ND=SD(Q)THEN Z=0.5
1270IF ND>SD(Q)THEN Z=0.5*(EXP((SD(Q)-ND)/ZZ))
1280IFND<SD(Q)THEN Z=1-(0.5*(EXP((ND-SD(Q))/ZZ)))
1290WF(Q,R)=Mass(Q,R)*W(Q)*Z/100
1300WS(Q,R)=(W(Q)*Mass(Q,R)/100)-WF(Q,R)
1310NEXT R
1320ND=G(D)
1330IF ND=SD(Q)THEN Z=0.5
1340IF ND>SD(Q)THEN Z=0.5*(EXP((SD(Q)-ND)/ZZ))
1350WF(Q,D+1)=Mass(Q,D+1)*W(Q)*Z/100
1360WS(Q,D+1)=(W(Q)*Mass(Q,D+1)/100)-WF(Q,D+1)
1370NEXT Q
1380 RETURN
1390PRINT:PRINTTAB(25);"PREDICTED PERFORMANCE":PRINT
1400TFAU=0:TSAU=0:TFW=0:TSW=0
1410PRINT"Size       Wt      Ash      Sepn     Ep        FLOATS         SINKS
     Code"
1420PRINT"range      %       %        dens              Wt%       Ash%    Wt%
Ash%"
1430PRINT"+";S$(1);TAB(10);" ";
1440FAU=0:SAU=0:FW=0:SW=0
1450FOR AA=1 TO D+1
1460FW=FW+WF(1,AA)
1470FAU=FAU+(WF(1,AA)*Ash(1,AA))
1480SW=SW+WS(1,AA)
1490SAU=SAU+(WS(1,AA)*Ash(1,AA))
1500NEXT AA
1510TFAU=TFAU+FAU:TSAU=TSAU+SAU:TFW=TFW+FW:TSW=TSW+SW
1520@%=&20106:REM SETS 2 DECIMAL PLACES AND FORMATS FIELD WIDTH
1530PRINTW(1);;@%=&20108:PRINT(FAU+SAU)/(FW+SW);;@%=&20209:REM SETS 2 DECIMAL P
LACES AND FORMATS FIELD WIDTH
1540PRINTSD(1);;@%=&20207:PRINTEP(1);;@%=&20107:REM SETS 2 DECIMAL PLACES AND F
ORMATS FIELD WIDTH
1550PRINTFW;;@%=&20109:PRINTFAU/FW;;@%=&20109:REM SETS 2 DECIMAL PLACES AND FOR
MATS FIELD WIDTH
1560PRINTSW;;@%=&20108:PRINTSAU/SW;;@%=&20005:PRINT1:REM SETS 2 DECIMAL PLACES
AND FORMATS FIELD WIDTH
1570IFS=1THEN1740
1580FOR AB=2 TOS
1590PRINTS$(AB);"-";S$(AB-1);TAB(10);" ";
1600FW=0:SW=0:FAU=0:SAU=0
1610FOR CC=1 TO D+1
1620FW=FW+WF(AB,CC)
1630FAU=FAU+(WF(AB,CC)*Ash(AB,CC))
1640SW=SW+WS(AB,CC)
1650SAU=SAU+(WS(AB,CC)*Ash(AB,CC))
1660NEXTCC
1670TFAU=TFAU+FAU:TSAU=TSAU+SAU:TFW=TFW+FW:TSW=TSW+SW
1680@%=&20106:REM SETS 2 DECIMAL PLACES AND FORMATS FIELD WIDTH
1690PRINTW(AB);;@%=&20108:PRINT(FAU+SAU)/(FW+SW);;@%=&20209:REM SETS 2 DECIMAL
PLACES AND FORMATS FIELD WIDTH
1700PRINTSD(AB);;@%=&20207:PRINTEP(AB);;@%=&20107:REM SETS 2 DECIMAL PLACES AND
FORMATS FIELD WIDTH
1710PRINTFW;;@%=&20109:PRINTFAU/FW;;@%=&20109:REM SETS 2 DECIMAL PLACES AND FOR
MATS FIELD WIDTH
1720PRINTSW;;@%=&20108:PRINTSAU/SW;;@%=&20005:PRINTAB:REM SETS 2 DECIMAL PLACES
AND FORMATS FIELD WIDTH
1730NEXT AB
```

```
1740XX=Y*W(S+1)/100
1750YY=W(S+1)-XX
1760XZ=XX*fAsh
1770YZ=(uAsh*W(S+1))-XZ
1780TFAU=TFAU+XZ:TSAU=TSAU+YZ:TFW=TFW+XX:TSW=TSW+YY
1790IFY=OTHEN1810
1800IFER=1THENYY=.0000000000000000000000000001
1810IFYY=OTHENYY=.00000000000000000001
1820PRINT"-";S$(S);TAB(10);" ";:@%=&20106:REM SETS 2 DECIMAL PLACES AND FORMATS
FIELD WIDTH
1830PRINTW(S+1);;@%=&20108:PRINTuAsh;TAB(32);"-";TAB(38);"-";:REM SETS 2 DECIMA
L PLACES AND FORMATS FIELD WIDTH
1840@%=&20109:PRINTXX;:@%=&20109:PRINTfAsh;:REM SETS 2 DECIMAL PLACES AND FORMA
TS FIELD WIDTH
1850@%=&20109:PRINTYY;:@%=&20108:PRINTYZ/YY;:REM SETS 2 DECIMAL PLACES AND FORM
ATS FIELD WIDTH
1860@%=&20005:PRINTS+1:REM SETS 2 DECIMAL PLACES AND FORMATS FIELD WIDTH
1870PRINT:PRINT"TOTAL";TAB(12);"100.0";
1880@%=&20108:PRINT(TFAU+TSAU)/(TFW+TSW);TAB(32);"-";TAB(38);"-";:REM SETS 2 DE
CIMAL PLACES AND FORMATS FIELD WIDTH
1890@%=&20109:PRINTTFW;:@%=&20109:PRINTTFAU/TFW;:REM SETS 2 DECIMAL PLACES AND
FORMATS FIELD WIDTH
1900@%=&20109:PRINTTSW;:@%=&20108:PRINTTSAU/TSW:REM SETS 2 DECIMAL PLACES AND F
ORMATS FIELD WIDTH
1910RETURN
1920S=6:D=9
1930DIM S$(S):DIM G(D):DIM W(S+1):DIM Mass(S,D+1):DIM Ash(S,D+1):DIM EF(S):DIM
SD(S):DIM WF(S,D+1):DIM WS(S,D+1)
1940FOR A=1TO6
1950READ S$(A)
1960NEXT
1970DATA 11.2,8,4,2,1,0.5
1980FOR B=1TO9
1990READ G(B):NEXTB
2000DATA 1.2,1.3,1.4,1.5,1.6,1.7,1.8,1.9,2.0
2010FOR C=1TO7
2020READ W(C):NEXT C
2030DATA 15,20,10,18,8,10,19
2040FOR F=1TO6
2050FOR J=1TO10
2060READ Mass(F,J)
2070READ Ash(F,J)
2080NEXT J
2090NEXT F
2100DATA10,2.3,5,4.6,12,5.8,8,7.3,16,10.2,7,12.9,10,15.6,15,20.2,7,25,10,45.6
2110DATA 15,2.6,6,5.1,3,4.9,11,7.5,15,10.4,10,12.8,8,12,22.2,6,25,14,55
2120DATA 6,2,10,4.1,8,5.2,14,5.8,12,11.2,8,13.6,7,16.5,10,21.9,15,22.9,10,48
2130DATA 18,3.6,9,3.9,10,6.3,8,7.8,8,11.1,8,14,16,16.1,5,22.8,10,26,8,35
2140DATA 15,1.9,10,2.6,6,5.8,9,6.9,14,11,6,13,20,14.2,6,19.8,10,24.6,4,42
2150DATA 8,1.3,8,3.6,15,4.9,8,6.2,12,9.9,10,12.1,12,14.1,5,20,8,26,14,58
2160uAsh=35
2170PRINT"Data for heavy liquid analyses at 9 densities and 6 sizes have been e
ntered"
2180PRINT"                              SIZE RANGE MICRONS"
2190PRINT"              +11.2     8-11.2     4-8       2-4       1-2       0
.5-1"
2200PRINT"Density       Wt%    Ash   Wt%  Ash  Wt%  Ash  Wt%  Ash  Wt%  Ash  Wt
% Ash"
2210FOR AS=1 TO10
2220IFAS=1 THEN PRINT TAB(0);"-1.2";TAB(12);" ";:GOTO2250
2230IFAS=10THENPRINTTAB(0);"+2.0";TAB(12);" ";:GOTO2250
2240PRINTTAB(0);G(AS-1);"-";G(AS);TAB(12);" ";
```

```
 2250FORBS=1 TO6
 2260@%=&20105:PRINTMass(BS,AS),Ash(BS,AS);:REM SETS 2 DECIMAL PLACES AND FORMAT
S FIELD WIDTH
 2270NEXT BS
 2280PRINTTAB(0)
 2290NEXTAS
 2300PRINT"WT.%          15.0      20.0     10.0    18.0     8.0
10.0"
 2310PRINT:PRINT"Undersize fraction (-0.5microns) is 19.0% weight and contains 3
5.0% ash"
 2320PRINT"Flotation yield-enrichment ratios are 56-5 67-4 82-3"
 2330 RETURN
```

```
RUN
Input NEW if new data being entered.
Press RETURN for demonstration data
?
Data for heavy liquid analyses at 9 densities and 6 sizes have been entered
                         SIZE RANGE MICRONS
             +11.2     8-11.2     4-8      2-4      1-2     0.5-1
Density     Wt%   Ash  Wt%  Ash  Wt%  Ash  Wt%  Ash  Wt%  Ash  Wt%  Ash
-1.2        10.0  2.3 15.0  2.6  6.0  2.0 18.0  3.6 15.0  1.9  8.0  1.3

1.2-1.3      5.0  4.6  6.0  5.1 10.0  4.1  9.0  3.9 10.0  2.6  8.0  3.6

1.3-1.4     12.0  5.8  3.0  4.9  8.0  5.2 10.0  6.3  6.0  5.8 15.0  4.9

1.4-1.5      8.0  7.3 11.0  7.5 14.0  5.8  8.0  7.8  9.0  6.9  8.0  6.2

1.5-1.6     16.0 10.2 15.0 10.4 12.0 11.2  8.0 11.1 14.0 11.0 12.0  9.9

1.6-1.7      7.0 12.9 10.0 12.8  8.0 13.6  8.0 14.0  6.0 13.0 10.0 12.1

1.7-1.8     10.0 15.6  8.0 16.0  7.0 16.5 16.0 16.1 20.0 14.2 12.0 14.1

1.8-1.9     15.0 20.2 12.0 22.2 10.0 21.9  5.0 22.8  6.0 19.8  5.0 20.0

1.9-2.0      7.0 25.0  6.0 25.0 15.0 22.9 10.0 26.0 10.0 24.6  8.0 26.0

+2.0        10.0 45.6 14.0 55.0 10.0 48.0  8.0 35.0  4.0 42.0 14.0 58.0

WT.%        15.0       20.0      10.0      18.0      8.0      10.0

Undersize fraction (-0.5microns) is 19.0% weight and contains 35.0% ash
Flotation yield-enrichment ratios are 56-5 67-4 82-3
PREDICTED ECART PROBABLE MOYEN
Input expected Ep for each size fraction
Ep for size fraction +11.2
?0.02
Ep for size fraction 8-11.2
?0.02
Ep for size fraction 4-8
?0.04
Ep for size fraction 2-4
?0.04
Ep for size fraction 1-2
?0.05
Ep for size fraction 0.5-1
```

```
?0.06
SEPARATION DENSITIES
Input density of separation for each
size fraction (N.B.consistent densities
for fractions in same vessel)


Separating density for size fraction
+11.2
?1.5
Separating density for size fraction
8-11.2
?1.5
Separating density for size fraction
4-8
?1.6
Separating density for size fraction
2-4
?1.6
Separating density for size fraction
1-2
?1.6
Separating density for size fraction
0.5-1
?1.6
FLOTATION PERFORMANCE
Input expected flotation yield (%)
If fines to be discarded press RETURN
If fines to be blended directly
then input 100
?
```

PREDICTED PERFORMANCE

Size range	Wt %	Ash %	Sepn dens	Ep	FLOATS Wt%	Ash%	SINKS Wt%	Ash%	Code
+11.2	15.0	15.2	1.50	0.02	5.4	5.1	9.6	20.7	1.
8-11.2	20.0	17.7	1.50	0.02	7.1	4.9	12.9	24.6	2.
4-8	10.0	15.8	1.60	0.04	4.9	6.3	5.1	24.8	3.
2-4	18.0	13.4	1.60	0.04	9.6	6.1	8.4	21.6	4.
1-2	8.0	12.0	1.60	0.05	4.2	5.7	3.8	19.0	5.
0.5-1	10.0	16.9	1.60	0.06	5.1	5.9	4.9	28.2	6.
-0.5	19.0	35.0	−	−	0.0	0.0	19.0	35.0	7.
TOTAL	100.0	19.1	−	−	36.2	5.7	63.8	26.7	

```
Change separating conditions? Y/N N
```

KINETIC   Evaluation of data from batch flotation test

```
10 REM BATCH FLOTATION TEST EVALUATION
20REM B.A.WILLS DEC.1986
30TEST=1:PRINT:PRINT
40INPUT"Number of incremental concentrates?  "N
50INPUT"Metal content of valuable mineral?  "M
60PRINT:PRINT:PRINT"Input data for incremental concentrates"
70 DIM W(N):DIM G(N):DIM T(N):DIM TG(N):DIM RM(N):DIM RG(N):DIMSE(N)
80TW=0:TB=0:WGT=0
90FOR A=1 TO N
100PRINT:INPUT"Time (secs.)  "T(A)
110INPUT"Weight of concentrate (g)  "W(A)
120WGT=WGT+W(A)
130TW=TW+W(A)
140INPUT"Grade of concentrate (%metal)  "G(A)
150B=W(A)*G(A)
160TB=TB+B:TG(A)=TB/TW:NEXT
170PRINT:PRINT:INPUT"Weight of tailings (g)  "WT
180WGT=WGT+WT
190INPUT"Grade of tailings (% metal)  "GT
200B=WT*GT:TB=TB+B
210 FA=TB/WGT
220CMU=0:CWC=0
230FOR C=1 TO N
240CMU=CMU+(W(C)*G(C))
250RM(C)=100*CMU/TB
260CWC=CWC+W(C)
270RG(C)=100*CWC*(M-TG(C))/(WGT*(M-FA))
280SE(C)=RM(C)-RG(C)
290NEXT
300PRINT:PRINT:PRINT"CUMULATIVE GRADE RECOVERY DATA"
310PRINT:PRINT:PRINT"S.E. is separation efficiency (Recovery mineral-Recovery
gangue)"
320PRINT:PRINT:PRINT"        Time (s)       Recovery      Con.Grade      Re
covery        S.E."
330PRINT"                   metal %        metal %       gangue %
 %"
340@%=&02010F: REM SETS 2 DECIMAL PLACES AND FORMATS FIELD WIDTH
350FOR D= 1 TO N
360PRINTT(D),RM(D),TG(D),RG(D),SE(D)
370NEXT
380@%=10:PRINT:PRINT:PRINT:INPUT"Press RETURN for kinetic data"GHJ$:REM RETURN
S TO NORMAL OUTPUT FORMAT
390PRINT:PRINT"Required precision of theoretical"
400PRINT"recovery calculation (A/B/C/D/E)"
410PRINT"A.  1.0%":PRINT"B.  0.5%":PRINT"C.  0.2%":PRINT"D.  0.1%":PRINT"E.  0.05%"
:INPUTPR$
420IF PR$="A" THEN SP=-1
430IF PR$="B" THEN SP=-0.5
440IF PR$="C" THEN SP=-0.2
450IF PR$="D" THEN SP=-0.1
460IF PR$="E" THEN SP=-0.05
470PRINT"Realistic upper limit for search for ":INPUT"recovery of mineral  "RS
M
480PRINT"and lower limit? (this must be greater"
490INPUT"than recovery at 2nd time increment)  "RTM
500PRINT:INPUT"upper limit for gangue?  "RSG
510PRINT"and lower limit? (must be greater than"
520INPUT"recovery at 2nd time increment)  "RTG
530GOSUB540
```

```
540VG$="0":MIN=100
550IFTEST=1 THEN 570
560 GOTO 590
570 FOR RI=RSM TO RTM STEP SP
580GOTO600
590FOR RI=RSG TO RTG STEP SP
600ETA=0:EA=0:ET=0:ET2=0:V=N:EA2=0
610FOR B=1 TO V
620IF TEST=1 THEN R=RM(B):GOTO640
630R=RG(B)
640 IF R=RI THEN V=V-1:GOTO740
650 IF R>RI THEN V=V-1:GOTO740
660C=2.303*LOG((RI-R)/RI)
670Q=C*C
680EA2=EA2+Q
690D=T(B)*C
700ETA=ETA+D
710EA=EA+C
720ET=ET+T(B)
730ET2=ET2+(T(B)*T(B))
740NEXTB
750K=-((V*ETA)-(EA*ET))/((V*ET2)-(ET*ET))
760TH=-((K*ET)+EA)/(V*K)
770IF VG$="T"THEN 820
780ER2=EA2+(K*K*ET2)+(V*K*K*TH*TH)+(2*K*K*TH*ET)+(2*K*ETA)+(2*K*TH*EA)
790IFER2<MIN THEN MIN=ER2:MRI=RI
800NEXTRI
810VG$="T":RI=MRI:GOTO600
820IF TEST =1 THEN KM=K:THM=TH:RIM=RI:GOTO840
830KG=K:THG=TH:RIG=RI:GOTO850
840TEST=2:RETURN
850 PRINT:PRINT:PRINT"KINETIC DATA"
860PRINT:PRINT·PRINT"MINERAL:"
870PRINT"Rate constant= ";KM" per sec"
880@%=131594:PRINT"Maximum theoretical recovery= ";RIM"%":REM SETS 2 DECIMAL P
LACES
890PRINT"Time correction= ";THM" secs"
900PRINT:PRINT:PRINT:PRINT"GANGUE:"
910@%=10:PRINT"Rate constant= ";KG" per sec": REM RETURNS TO NORMAL OUTPUT FOR
MAT
920@%=131594:PRINT"Maximum theoretical recovery= ";RIG"%":REM SETS 2 DECIMAL P
LACES
930PRINT"Time correction= ";THG" secs"
940CVP=2.303*LOG(RIM*KM/(RIG*KG))
950TT=(CVP-(KM*THM)+(KG*THG))/(KM-KG)
960PRINT:PRINT:PRINT"Time at which concentration of valuable":PRINT"mineral en
ds=";TT" secs"
970@%=10:PRINT:PRINT:PRINT:PRINT"CHANGE LIMITS OR PRECISION FOR SEARCH?":REM R
ETURNS TO NORMAL OUTPUT FORMAT
980INPUT"Y/N "CL$
990IF CL$="Y"THEN TEST=1::GOTO 390
```

>RUN

Number of incremental concentrates? 9
Metal content of valuable mineral? 86.6

Input data for incremental concentrates

Time (secs.) 30
Weight of concentrate (g) 32.9
Grade of concentrate (%metal) 72.9

Time (secs.) 60
Weight of concentrate (g) 21.2
Grade of concentrate (%metal) 66.9

Time (secs.) 90
Weight of concentrate (g) 18.8
Grade of concentrate (%metal) 51.7

Time (secs.) 120
Weight of concentrate (g) 14.4
Grade of concentrate (%metal) 37.5

Time (secs.) 240
Weight of concentrate (g) 42.8
Grade of concentrate (%metal) 19.2

Time (secs.) 390
Weight of concentrate (g) 35.3
Grade of concentrate (%metal) 2.41

Time (secs.) 600
Weight of concentrate (g) 48.2
Grade of concentrate (%metal) 0.82

Time (secs.) 840
Weight of concentrate (g) 38.7
Grade of concentrate (%metal) 0.82

Time (secs.) 1200
Weight of concentrate (g) 43.3
Grade of concentrate (%metal) 0.49

Weight of tailings (g) 704.4
Grade of tailings (% metal) 0.16

CUMULATIVE GRADE RECOVERY DATA

S.E. is separation efficiency (Recovery mineral—Recovery gangue)

Time (s)	Recovery metal %	Con.Grade metal %	Recovery gangue %	S.E. %
30.0	37.2	72.9	0.6	36.7
60.0	59.3	70.5	1.1	58.2
90.0	74.4	65.7	1.9	72.4
120.0	82.7	61.0	2.8	80.0
240.0	95.5	47.3	6.4	89.1
390.0	96.8	37.7	10.1	86.7
600.0	97.4	29.4	15.2	82.2
840.0	97.9	25.0	19.4	78.5
1200.0	98.3	21.4	24.0	74.2

```
Press RETURN for kinetic data

Required precision of theoretical
recovery calculation (A/B/C/D/E)
A.  1.0%
B.  0.5%
C.  0.2%
D.  0.1%
E.  0.05%
?D
Realistic upper limit for search for
recovery of mineral   100
and lower limit? (this must be greater
than recovery at 2nd time increment)    80

upper limit for gangue?  50
and lower limit? (must be greater than
recovery at 2nd time increment)  15


KINETIC DATA


MINERAL:
Rate constant= 1.70482609E-2 per sec
Maximum theoretical recovery= 95.40%
Time correction= -1.73 secs


GANGUE:
Rate constant= 9.40018856E-4 per sec
Maximum theoretical recovery= 36.10%
Time correction= -27.99 secs


Time at which concentration of valuable
mineral ends=240.47 secs


CHANGE LIMITS OR PRECISION FOR SEARCH?
Y/N  N
```

INDEX

773